# Advances in Experimental Medicine and Biology

Volume 1158

More information about this series at http://www.springer.com/series/5584

Andrea Urbani • Mohan Babu
Editors

# Mitochondria in Health and in Sickness

Springer

*Editors*
Andrea Urbani
Institute of Biochemistry and Clinical Biochemistry
Università Cattolica del Sacro Cuore
Rome, Italy

Fondazione Policlinico Universitario A. Gemelli – IRCCS
Rome, Italy

Mohan Babu
Department of Biochemistry
University of Regina
Regina, Saskatchewan, Canada

ISSN 0065-2598 ISSN 2214-8019 (electronic)
Advances in Experimental Medicine and Biology
ISBN 978-981-13-8366-3 ISBN 978-981-13-8367-0 (eBook)
https://doi.org/10.1007/978-981-13-8367-0



This Springer imprint is published by the registered company Springer Nature Singapore Pte Ltd.
The registered company address is: 152 Beach Road, #21-01/04 Gateway East, Singapore 189721, Singapore

# Contents

# 1 Mitochondrial Metabolism in Cancer. A Tangled Topic. Which Role for Proteomics?

Patrizia Bottoni and Roberto Scatena

**Abstract**

Given the role of mitochondria in modulating many cellular functions, it is not surprising that they can play a crucial role also in molecular pathophysiology of cancer. In particular, the discovery in recent decades of a link between cancer metabolic processes, alterations of mitochondrial DNA, oncogenes and tumor suppressors has led not only to a renaissance of interest in Warburg's pioneering work, but also to a reexamination of his original observations above all in relation to the current knowledge in cancer cell metabolism. It follows that, although mitochondrial contribution to the pathogenesis of cancer has historically tended to be neglected, it is now evident that reprogrammed mitochondria can contribute to a complex bioenergetic adjustment that sustains not only tumor formation but also its progression. Most importantly, cancer cell metabolism seems to have a role in diversified aspects related to cancer pathophysiology (i.e., aggressiveness, recurrence, metastatic dissemination). Hence, it is imperative to always consider cancer cell metabolism, its adaptability, its influences but, above all, its functional heterogeneity in a single tumor, for a really rational and valid approach towards molecular biology of cancer.

**Keywords**

Cancer cell metabolism · Warburg effect · Oxidative phoshorylation · Complex I · Tumor markers · Cancer stem cells · Cancer diagnosis · Cancer therapy

## 1.1 Introduction

For a lot of time mitochondria are simply considered the oxygen consuming electrochemical generators, where in the process of oxidative phosphorylation (OXPHOS) the reduction of O2 is electrochemically coupled to conservation of energy in the form of ATP. However, in addition to the OXPHOS system, these powerhouses of the cell contain a large array of proteins with various functions. Briefly, functionally different oxidases as monoamine oxidases type A and B, monooygenases (cytochrome P450 monooxygenases), dioxygenase (sulfur dioxygenase and trimethyllysine dioxygenase), several hydoxylases, and more, typically characterize mitochondrial proteome. Moreover, different transporters are present as ion transporters including proton pumps, metabolite transporters and so on. Mitochondrial kinases related to energy transfer pathways, the enzymes of the tricarboxylic acid

P. Bottoni · R. Scatena (✉)
Institute of Biochemistry and Clinical Biochemistry, School of Medicine, Catholic University, Rome, Italy
e-mail: patrizia.bottoni@unicatt.it; roberto.scatena@unicatt.it; roberto.scatena@figliesancamillo.it



A. Urbani, M. Babu (eds.), *Mitochondria in Health and in Sickness*, Advances in Experimental Medicine and Biology 1158, https://doi.org/10.1007/978-981-13-8367-0_1

cycle with several dehydrogenases, the fatty acid oxidation enzymes, the biosynthesis of heme and iron-sulfur clusters, but also the proteins related to regulation and signaling of mitophagy and apoptotic processes, and mitochondrial biogenesis further complicate the proteome of this organelle. Finally, mitochondria maintain/contain their own genetic material, known as mitochondrial DNA (mtDNA) that is also regulated by nuclear-encoded mitochondrially targeted proteins. Importantly, this array of macromolecules is not steady and shows typical fluctuations in concentration/activity related to actual functional status of the cell and of organism in general. Moreover, to further complicate this tangled pathophysiology, it should be stressed that these organelles may differ quite considerably between tissues, as firstly showed by Mootha et al. [1] in mouse, indicating that mitochondria are highly customized to serve local cellular physiology. It is also fundamental to consider the nature of the confounding tissue-specific physiology and pathophysiology to really clarify the role of mitochondrial proteome in different diseases in general and in cancer in particular.

This scientific community's renewed interest in the many roles of mitochondria in cellular function and in the possible contribution of various diseases pathogenesis led in the last decade to an explosion of experimental mitochondrial researches, aimed to carefully define their role and contribution. These studies confirmed the important involvement of mitochondria in many diseases, both as primary site of the pathology and as an origin site of an altered cellular signaling with a strong impact on disease pathogenesis. Moreover, these studies also revealed a much more complex scenario than previously appreciated, showing novel ways by which mitochondrial defects may contribute to diseases [2–7]. In fact, the mitochondrial molecular alterations implicated in these diseases not only depend on a simple energy derangement but may also be related to disturbances of permeability/transport, ROS metabolism, calcium homeostasis, thermogenesis, biosynthetic pathways, intracellular signaling and mitDNA integrity.

At present, we are assisting to a deep reevalution of biochemistry, physiology and pathophysiology of mitochondria with relevant implications in pathogenesis of different acute and chronic diseases (from sepsis to cancer by passing for reperfusion injury).

All that confirms the importance of a deep re-evaluation of the tangled mitochondrial proteome and justifies attempts to obtain an accurate inventory of the organelle's protein components. One of the first recent attempts was realized in 2008, by performing in-depth mass spectrometry of mitochondria from 14 organs, associated to epitope tagging/microscopy and Bayesian integration for data assembling (MitoCarta 1.0) [8]. Later, by using a similar methodological approach a new inventory from human and mouse species was realized (MitoCarta 2.0) which showed 1158 human genes (918 genes of the MitoCarta 1.0 inventory as well as 240 additional genes). This improved MitoCarta 2.0 inventory gave also evidences supporting the mitochondrial localization of various proteins and seems to provide a functional molecular framework for system-level analysis of mammalian mitochondria [9]. Importantly, retrospective analysis of various MS data showed that only the most abundant mitochondrial proteins were being detected.

More recently, the Italian Proteomics Association (ItPA) focused its attention to the "Mitochondrial Human Proteome Project (mt-HPP)", to analyze by a proteomic approach human mitochondrial proteins, encoded by both the mitochondrial DNA (mt-DNA) and the nuclear DNA (in humans, the mitochondrial genome codes for only 13 proteins; the remaining 99% of mitochondrial proteins are encoded by nuclear genes). The main goal of this project is to obtain data about the integrative role of proteins acting at the mitochondrial level [10]. It is noteworthy to stress that all these projects mainly point to characterize physiological proteome of mitochondria, however this approach can anyway give some pathophysiological indications above all permitting to compare these proteome patterns with one's obtained in specific inherited or acquired mitochondriophaties.

Finally, we must also cite the recent COST Action CA15203 called "MitoEAGLE" which aims at developing a quality management system, defining common standards for measurements in the field of mitochondrial function to obtain accurate and comparative data useful to study various aspects of mitochondrial pathophysiology [11]. In facts, a fundamental prerequisite for a careful understanding of the complex pathophysiology of mitochondria, in particular in cancer, is a clear definition of their proteome not only in qualitative and quantitative terms but also in terms of functional interrelationships.

These novel approaches to study mitochondria surely may give a new light also on some tangled aspects on cancer cell mitochondria which are yet a very debated topic of molecular oncology.

In fact, we are assisting to a profound reevaluation of the original observation of Otto Warburg about the use of glycolysis by cancer cell [12]. The so-called Warburg effect does not seem to depend on a primary mitochondrial derangement as originally hypothesized, but represents a sign of a complex metabolic reprogramming to the new aim of a subpopulation of cancer cell, i.e. uncontrolled proliferation.

In this new metabolic arrangement, the aerobic glycolysis seems to allow a rapid ATP synthesis, an improvement of metabolic flux into biosynthetic pathways, an enhancement of tissue architecture disruption and immune cell evasion and a modification of signal transduction and chromatin modulation. Finally, a significant amelioration of the solvent capacity of cells which is particularly stressed by presence of mitochondria and a reduction in metabolic requirements necessary to synthetize such complex organelles may have a role in inducing aerobic glycolysis. On the other side, mitochondria continue to have a central position in the metabolic machinery of cancer cells. In fact, in terms of cancer cell proliferation, aerobic glycolysis could facilitate precursors generation, such as aspartate, acetate, citrate and malate, used in various biosynthetic processes fundamental for cell growth [13]. Most importantly, increasing evidences seem to show a pathogenic role for mitochondrial bioenergetics in tumor initiation, cancer cell stemness, dormancy, progression and metastasis [14–16]. These fundamental pathogenetic aspects of cancer have molecular mechanisms different from ones related to cell proliferation. How to reconcile these different but concomitant pathogenetic mechanisms? And, how a proteomic approach may contribute to understand the tangled pathogenic role of mitochondria in cancer?

Proteomic has surely a promising potential, rich of innovative and unexpected evidences. However, it must be stressed that until now proteomic approach to understand molecular biology of cancer in general failed to give real translational results. This may probably depend on both too simplistic methodological approaches and, above all, on too often disregarded high cancer cell heterogeneity. Importantly, this cancer cell variability, in turn, is not only related to intrinsic cancer cell plasticity, but also strictly depends on physic-chemical microenvironmental milieu. In facts, too often it has been neglected that cancers are abnormal tissues made up of heterogeneous cells endowed with different metabolism, functions, proliferative potentials and, above all, degree of differentiation. This cancer heterogeneity may so influence the result, also any well realized proteomic approach. At present, it is well accepted that different subpopulations of cancer cells coexist in tumors, contributing to this cell heterogeneity. Just as example, in cancer, rare tumorigenic cells, the so-called cancer stem cells or tumor initiating/maintaining cells, which are in a quiescent/slow-cycling state, have a peculiar structural and functional proteome and a preferentially oxidative metabolism, coexist with a high proliferative subpopulation of cells which have instead a preferential glycolytic metabolism, as indicated by Warburg. Moreover, it must be stressed that this dichotomy cannot be absolute, but in cancer there is probably a gradual passage of cells with intermediate features between low and high proliferative potential in relation to various microenvironmental conditions. Further, also microenvironment can locally present differences in terms both of cell types (cancer associated fibroblast, immune inflammatory cells, endothelial cells and so on) and chemical mediators (prostaglandins, interleukins, growth factors, ECM proteins and ECM modifying enzymes), nutrients and oxygen concentrations [17, 18]. To further complicate the already complex picture, it should be stressed that the mitochondrial set of proteins `to local driven functions of this organelle. So differential expression of genes in various tissues, alternative splicing, post-translational modifications, turnover and spatial dynamics of proteins are all factors that influence mitochondrial proteomes increasing their versatility. This cytological, histological and functional heterogeneity push to view cancer not as an agglomerate of anaplastic cells but as a tissue with structural and functional peculiarities different from ones typical of normal tissues. Finally, the different cells interact (i.e. autophagia, metabolite exchange, and so on) each other to sustain the main aim of cancer tissue, the growth and local and distal proliferations. Importantly, to overlook these levels of complexity inevitably jeopardize any structural and functional proteomic approach on cancer in general on cancer mitochondrial metabolism.

Hence, to overcome these obstacles is fundamental not only to reduce this heterogeneity by careful methodological approaches (i.e., by addressing research towards a subpopulation of cancer cells, proliferating cancer cells or cancer maintaining cells and/or cell culture synchronization, presence of specific ECM proteins, standardization of culture methods, and so on), but also by a careful integration of the particular in vitro results with the in vivo pathophysiology of cancer disease.

Without this accurate definition of the histocytological framework, oncoproteomic approaches too often can give data related to the actual experimental settings with scarce translational applicability.

Just as example, we are going to illustrate a oncoproteomic comparative approach on human hepatocarcinoma HepG2 cell line. By inducing a drug-dependent process of cytodifferentiation, we monitored biochemical and cellular parameters, and above all, the modifications of cellular protein profiles through 2-DE and MALDI-TOF analysis. Independently by the hypothesized mechanism of action of ciglitazone, an intriguing PPARγ agonist but also a mitochondrial complex I inhibitor, results showed that this drug was a strong differentiating agent for the HepG2 cell line and, importantly, this process was associated with peculiar modifications of cell proteome. Intriguingly, the most significant modifications of intracellular protein levels affected the cell antioxidant systems (Peroxiredoxin 2, DJ-1 protein), cell proliferation (peroxiredoxin-2, Stathmin 1/Oncoprotein-18, Nucleoside diphosphate kinase A, Peptidyl-prolyl cis–trans isomerase A, RAB14, RAB1A), cellular signaling (DJ-1 protein, Nucleoside diphosphate kinase A, Peptidyl-prolyl cis–trans isomerase A, Rab1A and Rab14), cellular stress machinery (DJ-1 protein, Protein disulfide isomerase A3) and invasiveness related proteins (Nucleoside diphosphate kinase A, RAB 14, RAB1A). This particular proteome picture related to in vitro cancer cell differentiation, not only was validated by western blot analysis, but, most importantly, was confirmed in clinical specimens of liver cancer tissue and matched normal liver tissue [19]. Intriguingly, in this experiments metabolism did not display the same dramatic proteome modification. Just enolase A showed a paradoxical increase in differentiated with respect to undifferentiated cancer cells, but this data could be related to the mechanism of action of differentiating agent.

A different methodological approach carried out to analyze secretome, by using serum free media during drug induced HepG2 hepatocarcinoma differentiation, showed comparable results to the previously reported. Specifically, this analysis not only confirmed drug induced metabolic derangement but also show an intriguing activation of antioxidant defense system and HSPs signaling, highlighting peculiar molecular mechanisms activated by some extraperoxisomial activities of this class of drugs, that anyway showed a strict correlation with the grade of cancer cell proliferation [20].

Independently by the mechanism of this drug induced cancer cell differentiation, results indicate that a more careful proteomic approach can give results connected to in vivo picture.

## 1.2 The Proteome of Cancer Cell Mitochondria – Status Quo

Given the role of mitochondria in modulating many cellular functions, it is not surprising that they can play a crucial role also in etiology of cancer [21]. The discovery in recent decades of a link between cancer metabolic processes, alterations of mitochondrial DNA, oncogenes and tumour suppressors has led not only to a renaissance of interest in Warburg's pioneering work, but also to a reexamination of his original observations above all in relation to the current knowledge in cancer cell metabolism [22]. It follows that, although mitochondrial contribution to the pathogenesis of cancer has historically tended to be neglected, it is now evident that mitochondria can/must contribute to a complex bioenergetic reprogramming that sustains not only tumor formation but also its progression [23]. In fact, nuclear or mitochondrial DNA mutations which are common in cancer cells, rather than inactivate mitochondrial energy metabolism, can alter the mitochondrial bioenergetic and biosynthetic state, thereby promoting the tumor growth [21].

In recent years, proteomic technologies have been extensively employed to study mitochondria in cancer and to characterize their involvement in cancer biology. Proteomic approaches have produced a wealth of data, which have contributed, and may still now contribute, both to accurately define the mitochondrial proteome in normal and in cancer cells, and to identify all drug-induced changes in mitochondrial proteome of tumor cells, thereby generating new data about the connection between mitochondrial dysfunction, deregulation of apoptosis and tumorigenesis [24–28].

### 1.2.1 Mitochondria, Warburg Effect and Proteomics

Briefly, cancer proteomic studies focusing on various type of cancer frequently described a characteristic hyperactivation in protein expression of glycolytic patterns (HK, PKM2, LDH-A, ENO-1), which seems to confirm the Warburg's original hypothesis, regarding the cancer cell metabolism shift towards aerobic glycolysis instead of mitochondrial respiration. However, the great heterogeneity of alterations reported from different proteomic studies, above all in term of up- or down-regulation for some metabolic enzymes, generated ambiguous data, as already reviewed [29]. Really, the Warburg's findings about the key role of aerobic glycolysis in cancer cell biology is universally accepted, however the reduced dependence on mitochondrial oxidative phosphorylation for ATP generation appears now not due to a mitochondrial damage, but rather to a mitochondrial reprogramming that determines a different metabolism [30].

Together with the induction of glycolytic enzymes in different tumors, it has been reported a reduction of components of the Krebs cycle and oxidative phosphorylation, an increase in glucose transporter (GLUT) levels (i.e. GLUT-1) and an activation of pentose phosphates cycle and of anaplerotic activities producing energy and substrates for macromolecular synthesis [30]. Moreover, numerous studies have reported an induction of the HIF-1a, HIF-2a and phosphatidyl inositol 3-kinase (PI3K)/Akt [31]. Altogether, these results again demonstrated the existence of a metabolic state in cancer cells, in which various processes appear to be coordinated, in a manner still now poorly understood, to satisfy the new metabolic requests for rapidly proliferating cells.

Unwin and coworkers [24] firstly provided, by proteomic and 2-DE approach, a comprehensive evidence of both the glycolytic and mitochondrial aspects of the "Warburg effect" in renal cancer (RCC). By comparing the protein profiles of conventional RCC tissue with patient-matched normal kidney cortex, the authors showed an overall increase in proteins involved in glycolysis and a decrease in gluconeogenesis, together with a parallel downregulation of several mitochondrial enzymes involved in oxidative phosphorylation. Enzymes involved in other pathways including fatty acid and amino acid metabolism and the urea cycle were also found to be downregulated, indicating a wider role for mitochondria in tumorigenesis process. Some years later, Bi et al. [32], examining changes of protein profiles associated with the process of colorectal tumorigenesis, showed an up-regulation of several glycolytic enzymes such as aldolase A, enolase 1, GAPDH, etc., thus providing a proteomic evidence that the Warburg effect occurs in proliferating colorectal cancer cells. In addition to a significantly elevated glycolysis, results from this study revealed that alterations of the following major metabolic pathways are involved in the colorectal cancer tumorigenesis: down-regulated gluconeogenesis, decreased glucuronate metabolism, and impaired tricarboxylic acid cycle (Krebs cycle), suggesting a more complex scenario in which the aberrant metabolism, once seen just as an epiphenomenon of oncogenic reprogramming, appears now play a key role in controlling both genetic and epigenetic events in cells during oncogenesis process. Interestingly, in oncoproteomics studies 3 glycolytic enzymes often recur.

**Hexokinase (HK)** Several oncoproteomic studies reported in cancer cells an overexpression of HK enzyme, particularly in its isoform II, which is considered a pivotal player in the Warburg effect [24, 33]. Interestingly, HK bounds the outer mitochondrial membrane via the porin-like protein voltage-dependent anion channel (VDAC). Together these proteins move ATP, newly synthesized by the inner membrane located ATP synthase, to active sites on HK II. The abundant amounts of HK II bind both the ATP and the incoming glucose producing the product glucose-6-phosphate, also at an elevated rate [34].

Danial et al. [35], using a proteomic approach, demonstrated that BAD, a pro-apoptotic BCL-2 family member, forms in mitochondria a complex together with the catalytic subunits of protein kinase A and protein phosphatase 1, Wiskott-Aldrich family member WAVE-1 (an A kinase-anchoring protein) and glucokinase (HK IV). This complex is essential for modulating the activity of this HK isoform in response to glucose and to establishing a link between the molecular mechanisms which modulate glycolysis to those which regulate apoptosis. Elevated levels of HK detected in rapidly proliferating tumours could not only sustain accelerated glycolytic flux [36], but also attenuate protein oxidative damage that might compromise energy metabolism [37]. Altogether, the evidence that HK2 is highly expressed in glycolytic cancer cells and not in normal cells made it an attractive target for systemic targeting of glycolytic enzymes, providing a strong rationale for developing small-molecule inhibitors of HK2 for cancer therapy [38, 39].

**Pyruvate Kinase (PK)** Notably, among the various alterations reported in the enzymes of glycolytic pathway, a tumor-specific elevation of pyruvate kinase M2 (PKM2) have been often observed in numerous cancerous cells, and PKM2 has been investigated as a potential tumor marker for diagnostic assays [40, 41]. Interestingly, Liu et al. showed PKM2, once methylated by co-activator-associated arginine methyltransferase 1 (CARM1) may reversibly shift the balance of metabolism from oxidative phosphorylation to aerobic glycolysis breast cancer [42].

As well known, PK catalyzes the irreversible transfer of a phosphate group from phosphoenolpyruvate to ADP, to produce one molecule of pyruvate and one molecule of ATP, in the last step within the glycolytic sequence. Among the four isozymes of PK present in mammals, the dimeric form of PKM2, a splice variant of M1, is predominant in tumor cells [43] and seems to act by direct interaction with several oncoproteins [44]. It has been demonstrated that the activity of PKM2 can be regulated by tyrosine kinase signaling pathways [45]. PKM2 interaction with phosphotyrosine-containing proteins inhibits enzyme activity and increases availability of glycolytic metabolites to support cell proliferation and promote tumorigenesis. In addition, Christofk et al. [45] demonstrated that an exchange in the expression of PKM1 to PKM2 contributes to the characteristic aerobic glycolysis during tumorigenesis and a replacement of PKM2 with its splice variant PKM1 cannot efficiently support biosynthesis and tumor growth. This also suggested that selective targeting of PKM2 by small-molecule inhibitors is feasible for cancer therapy [46]. To have analytical evidence for a switch in pyruvate kinase PKM1 to PKM2 expression during tumorigenesis, Bluemlein and collaborators [47], using mass spectrometry, performed an absolute quantification of PKM1 and PKM2 splice isoforms in several cancer tissues of different origin, benign tumors and cell lines, and their tissue matched controls. Quantitative analysis of PKM1 and PKM2 expression revealed that total PKM is up-regulated in cancer, which matches the observation of a high glycolytic activity of cancer cells. However, the authors do not found evidence for an exchange of PKM1 to PKM2 expression during cancer formation, nor do these results support conclusions that PKM2 is specific for proliferating, and PKM1 for non-proliferating tissue. Being the nature of the tissue the prime determinant of the expressed PKM isoform, cancer cells appears to maintain the PKM isoform expression according to that of their tissue of origin.

In proteomic analysis of pancreatic cancer cells by mass spectrometry, Zhou et al. [48] observed that PKM2 is indeed up-regulated in cancer cells, but also that the isozyme is an abundant protein in both normal pancreatic duct cells and cancer cells, which is discrepant from some researchers' observations that PKM2 is replaced by tissue-specific isoforms during tissue differentiation in development and PKM1 is switched to PKM2 during tumorigenesis. These observations strongly do impose a carefully analytical data revision and, above all, really do suggest applying a good dose of prudence in developing the drug target to PKM2 since PKM2 has been demonstrated to be abundantly expressed also in normal cells.

**Enolase (ENO)** ENO is a metal-activated metalloenzyme that catalyzes the dehydration of 2-phospho-D-glycerate (PGA) to phosphoenolpyruvate (PEP) in the glycolytic pathway, and the reverse reaction, the hydration of PEP to PGA, in gluconeogenesis. ENO overexpression has been associated with multiple tumors and for some cancer it is considered a tumor marker [49]. Really, in many of these tumors, ENO promoted cell proliferation and tumorigenesis. Moreover, as cell surface plasminogen receptor, ENO may induce ECM degradation and cancer invasion. Proteomic studies deepened the role of ENO in cancer, confirming its upregulation and its active role in supporting the Warburg effect. Clinically, this upregulation seems to have a prognostic value because is related to cancer cell invasivity [49]. Intriguingly, some proteomic studies showed that upregulation of ENO was associated to upregulation of cathepsin D and of VDAC1 [50].

**Lactate Dehydrogenase (LDH)** Similarly, also LDH levels and its isozymes pattern seems to undergo to profound modifications during neoplastic development, as already described several decades ago [51]. Previous researches on LDH and cancer are generally focused on LDH-A, showing how the overexpression of this isozyme has been involved in the development and progression of many cancer types [52, 53]. However, recently new data revealed a more complex scenario, which seems to redraft a different history on LDH and cancer. Likewise, oncoproteomics studies describing the prevalence of one isozyme form rather than another generated some conflicting results. For example, an upregulation of LDH-A has been demonstrated in renal cancer by Unwin et al. (24), according with previous findings showing a notable change in LDH-B/LDH-A ratio, caused by a shift to the LDH-A form, which better supports the conversion of pyruvate into lactate particularly under hypoxic conditions [54]. However other data seem to suggest that loss of the LDH-B subunit may play a role both in the development and metastatic progression of human cancer, as demonstrated by the silencing of LDH-B observed in a variety of tumors [55–57]. In addition, by mass spectrometry, a lack of LDH-B expression was observed in prostate metastatic cancer [58] and, by iTRAQ proteomic analysis, also in human prostate cancer cells with different metastatic potential [59]. Proteomic analysis of pancreatic ductal adenocarcinoma cells conducted by Zhou et al. [48, 60] demonstrated that the expression level of LDH-A in pancreatic cancer cells was not significantly changed compared to normal duct cells, but LDH-B was strikingly increased in the cancer cells. Another study on LDH (A and B) isoforms in breast cancer cell lines MCF-7 and MDA-MB-231 was conducted by Hussien and Brooks [61], who compared results with those from a control, untransformed primary breast cell line. Data showed that two cancerous cell lines have differences in LDH isoform expression, with LDH-A mainly expressed in MDA-MB-231 and LDH-B mainly expressed in MCF-7. Interestingly, Zha et al. [62] identified LDH-B as an important target of mTORC1 and, being critical for hyperactive mTOR-mediated tumorigenesis, hypothesized that LDH-B may be a druggable target for the treatment of diseases associated with aberrant activation of the RTK-PI3K-AKT-mTOR signaling cascade. All these heterogeneous data, showing sometimes that LDH-A was increased in some cancers, or that lower expression of LDH-B may be more characteristic of cancer cells than is the increase in LDHA, or even that LDH-B was strikingly increased in the cancer, suggests that further investigations are needed to clarify how change in LDH-B/LDH-A ratio may contribute to tumour progression.

### 1.2.2 Mitochondrial Oxidative Metabolism and Cancer

Mitochondrial oxidative metabolism in cancer is reprogrammed to sustain cancer cell proliferation. In facts, we have already stressed that mitochondria are not simply bystanders in cancer and their function is not impaired [20, 29, 46]. On the contrary, they have significant pathogenic roles, just in terms of alteration of ROS generation which could be one of the motive force at the basis of cancer cell genomic instability and, hence cancer cell plasticity. Moreover, they may be fundamental for a more specialized role of Krebs cycle on anaplerotic and cataplerotic reactions of cancer cell.

Interestingly, several oncoproteomic studies focusing on the metabolic and structural alterations of mitochondria in cancer, frequently reported discordant results above all in terms of alteration of several mitochondrial proteins and in a variety of tumor types. Just as example, Zhou et al. [48, 60] showed interesting metabolic alterations, by a proteomic analysis of pancreatic ductal adenocarcinoma cells (PDAC) conducted by mass spectrometry using LTQ-Orbitrap. Specifically, the Authors highlighted differentially expressed metabolic enzymes and proteins involved in cytoskeleton, cell adhesion, transport, transcription, translation, and cell proliferation as well. Moreover, several enzymes of Krebs cycle, oxidative phosphorylation, and fatty acid β-oxidation pathways were showed to be down-regulated, while enzymes for pyruvate fermentation, pentose phosphate pathway, fatty acid synthesis, purine synthesis, cholesterol synthesis were demonstrated to be up-regulated. Moreover, these Authors reported that enzymes in Krebs cycle, such as pyruvate dehydrogenase, aconitase 2, isocitrate dehydrogenase 2 ($NADP^+$), succinate-CoA ligase, and succinate dehydrogenase, were down-regulated in PDAC cells, while citrate synthase was up-regulated. In addition, the components of oxidative phosphorylation, such as NADH dehydrogenase, ubiquinol-cytochrome c reductase, cytochrome c-1, cytochrome c oxidase subunit II, ATP synthase, $H^+$ transporting, mitochondrial F0 complex, subunit B1 were down-regulated as well. Moreover, mitochondrial fatty acid β-oxidation was reported to be reduced in this comparative analysis, while two enzymes involved in fatty acid synthesis, the ATP citrate lyase isoform 1 and the fatty acid synthase (FAS), were found up-regulated, indicating that FAS was extremely overexpressed in PDAC cells. This observation is remarkably concordant with previous reports on a wide variety of tumors underwent de novo biogenesis of fatty acids with FAS upregulation [63]. In addition, the comparative MS analysis unveiled anomalous metabolism of glutamine, suggesting that glutamine was largely consumed as a nitrogen donor in nucleotide and amino acid biosynthesis in PDAC cells. The data seem to demonstrate that metabolic pathways were modulated in PDAC cell line according to Warburg effect, and that the prominent energetic function of mitochondrion was altered with a glycolytic shift in energy production in PDAC cells.

An interesting and different oncoproteomic study on cancer cell metabolism [64] of pheochromocytomas and paragangliomas (PHEOs/PGLs), with distinct underlying gene defects as von Hippel-Lindau -VHL- and succinate dehydrogenase B -SDHB- mutations and aggressiveness, showed that PHEOs/PGLs metastasize rarely in VHL-, but frequently in SDHB-patients. Most importantly, in SDHB-PGLs OXPHOS complex activity was increased at complex III and, decreased at complex II. Additionally, there was an associated increase in the ratio of nuclear encoded cytochrome c oxidase subunits to mitochondrially encoded cytochrome c oxidase subunits in the tumor-derived cell lines. These results demonstrate an alteration in subunit levels of a single enzyme complex (cytochrome c oxidase) commensurate with tumor-altered metabolism. Moreover, protein and mRNA expression of all tested OXPHOS-related genes were higher in SDHB- than in VHL-derived tumors. Although there was no direct evidence for increased reactive oxygen species production, elevated superoxide dismutase 2 expression may reflect elevated oxidative stress in SDHB-derived PHEOs/PGLs.

Interestingly, a recent research of Frezza's group seems to clarify the mechanism by which a mutation of fumarate hydratase (FH), an enzyme of the tricarboxylic acid (TCA) cycle, could affect mitochondrial metabolism decreasing, accompanied by a varying degree of dysfunction of respiratory chain (RC) complex I and II. In fact, fumarate excess may induce succination of key components of the iron-sulfur cluster biogenesis family of proteins, leading to defects in the biogenesis of iron-sulfur clusters that mainly affect complex I function. Moreover, also complex II activity is impaired due to fumarate accumulation [65, 66].

Intriguingly, Gao et al. [67, 68] by a TMT method followed by mass spectrometry analysis, quantified alterations in protein abundance in mitochondria between noncancer and gastric cancer tissues. Results confirmed previous studies showing upregulation of some typical biomarkers as HSP70, HSP60, HSP90, leucine-rich pentatricopeptide repeat containing (LRPPRC), SOD2 and cathepsin B. Additionally, solute carrier family 25 and VDAC1 were strongly upregulated in gastric cancer as well. Hence, mitochondrial proteome seemed to show intriguing modifications in terms of down/up-regulation that seem to confirm a deep reprogramming of mitochondrial metabolism in cancer and not just a simple alteration/reduction.

It is important to stress that these data refer to the functional and morphological heterogeneous population of cancer cells. This heterogeneity does not only depend by intrinsic characteristics of cancer cells but also by cell culture conditions adopted, in in vitro studies. In facts, rarely proteomic studies synchronize cell cultures, and this further complicate the results in terms of accuracy and reproducibility. This heterogeneity, in fact, is associated to different metabolism and hence to different proteome picture. Generally, however, the great part of a cell culture should be formed by cell in proliferating phase. In this sense, it could be fundamental to standardize the culture conditions of cancer cell.

## 1.3 Cancer Stem Cells (Tumor Maintaining Cells, Stem Like Cancer Cells)

CSCs represent one of the most interesting topics of cancer pathophysiology studied in the last decade. Moreover, CSCs represent a fundamental component of cancer cells population.

The American Association for Cancer Research Stem Cell Workshop defined a cancer stem cell as a cell within the tumor that possesses the capacity to self-renew and, in doing so, gives rise to the heterogeneous lineages that comprise the tumor [69]. Most importantly, this subpopulation of cancer cells should permit to justify some relevant clinical aspects of cancer, above all recurrence and resistance to therapies.

From a biological point of view, CSC show the following peculiarities [70, 71]:

- Are primarily in a more quiescent cycle state;
- give rise to a more differentiated cell;
- strongly influenced by their microenvironment;
- presence of biomarkers and/or signal transduction pathways that are also important in stem cell biology;
- high levels expression of ABC transporters and DNA repair mechanisms;
- resistance to classical radio and/or chemotherapeutic protocols;
- finally, cellar metabolism does not follow the Warburg effect but mainly relies on oxidative phosphorylation (OXPHOS).

The study of proteome of cancer stem cells has been already extensively studied [72–75] and could be considered a one of the first proteomic approach to investigate the typical cancer cell heterogeneity. Moreover, recent researches tried to better characterize the mitochondria proteome and mitochondrial metabolism of this subpopulation of cancer cells. The study of Viale et al. [76] on pancreatic cancer stem cell cells represented an illustrative example of this peculiar metabolism. By adopting an inducible mouse model of mutated *KRAS*, these Authors isolated and analyzed a subpopulation of CSC characterized by a significant enrichment of protein involved in various metabolic pathways (e.g. mitochondrial electron transport chain (ETC), lysosome activity, autophagy, mitochondrial and peroxisomal β-oxidation), which confirmed an increased mitochondrial activity. Moreover, also PGC1alpha, a key regulator of mitochondrial biogenesis, was increased. Similarly, these CSCs stained intensely for MitoTracker Green, a marker of mitochondrial mass. Finally, a classical mitochondrial marker, as VDAC1, showed a significant increment both *in vitro* and *in vivo*.

Specifically, Lamb et al. [77] realized a comparative proteomic approach on mammospheres from two ER-positive breast cancer cell lines (MCF7 and T47D) and their respective suspension and monolayer cultures. 62 mitochondrial-related proteins resulted significantly upregulated in MCF7 mammospheres. 12 of these proteins (HSD17B10, BDH1, ACAT1, ACADVL, ACACA, ACLY, HADHB, SUCLG2, ACAD9, HADHA, ECHS1, ACADSB) were specifically related to beta-oxidation and ketone metabolism. 8 proteins involved in mitochondrial biogenesis (HSPA9, TIMM8A, GFM1, MRPL45, MRPL17, HSP60, TSFM, TUFM), were strongly upregulated. In addition, a lot of proteins related to electron transport (NDUFB10, COX6B1, PMPCA, COX5B, SDHA, UQCRC1), ATP synthesis (ATP5B, ATPIF1, ATP5A1, ATP5F1, ATP5H, ATP5O), ADP/ATP exchange/transport (SLC25A5), CoQ synthesis (COQ9), and, above all, ROS production (GPD2) were also increased. Finally, two proteins involved in the suppression of glycolysis, autophagy and mitophagy were also significantly increased (SOGA1, LRPPRC).

By adopting a different method to obtain CSCs (MCF7 cells stably transfected with an hTERT-promoter construct driving high GFP expression, as a surrogate marker of telomerase transcriptional activity and isolated by FACS), Lamb et al. [78] showed with respect to control, a high overexpression of more than 30 mitochondrial-related proteins, most of these proteins were related to oxidative phosphorylation, the TCA cycle, or mitochondrial biogenesis. Remarkably, MT-CO2 (encoded by mt-DNA) was upregulated by >20-fold, which testify for a significant increase in mitochondrial biogenesis. Importantly, consistent with an overall anabolic phenotype, 17 enzymes related to glycolysis and the pentose phosphate pathway were also upregulated in GFP-high cells.

Interestingly, a different approach could help to better clarify the interrelationships between mitochondria and CSCs by a proteomic point of view. In facts, metformin, the most frequently prescribed drug for type 2 diabetes, showed to be able to reduce cancer incidence. This anti-cancer activity is incompletely understood but, intriguingly, metformin inhibits mitochondrial complex I and, so it reduces NADH oxidation. Sacco et al. [79] profiled the metformin-dependent changes in the proteome and phosphoproteome of breast cancer cells using high-resolution mass spectrometry. In total, 7875 proteins and 15,813 phosphosites after metformin changes were quantified. This approach seems to suggest that metformin treatment makes cancer cells more sensitive to apoptotic stimuli and less sensitive to pro-growth stimuli. Authors tested these hypotheses in vivo and demonstrated that metformin inhibits the p70S6K-rpS6 axis in a PP2A-phosphatase dependent manner. In this study, analysis of deep proteomics reveals some molecular mechanisms related to the anti-cancer activity of metformin. Importantly, also glitazones, others well-known antidiabetic drugs showed to be able to gradually purge leukemic stem cells in chronic myeloid leukemia both in vitro and in vivo. The mechanism of action is still debated. Prost et al. [80], showed that these agonists of peroxisome proliferator-activated receptor-γ (PPARγ) can reduce expression of STAT5 and its downstream targets HIF2α and CITED2, which are key guardians of the quiescence and stemness of CML LSCs. However, also this class of drugs are well known inhibitor of NADH oxidation by deranging mitochondrial complex I activity. Noteworthy, a recent study showed that targeting mitochondrial oxidative phosphorylation eradicates therapy-resistant chronic myeloid leukemia stem cells [81].

## 1.4 Cancer Progression Proteome and Mitochondrial Metabolism

The first proteome-wide approach to study of breast cancer progression has been conducted by Choong et al. [82], who used iTRAQ-based tandem mass spectrometry to analyze the MCF10AT breast cancer model, which comprises of 4 isogenic xenograft-derived human cell lines that mimic different stages of breast cancer progression: normal, premalignant epithelium, low grade and high grade lesions. The number of proteins that showed different expression levels increased as disease progressed from AT1k pre-neoplastic cells to low grade CA1h cancer cells and high grade cancer cells. This reflects an immense degree of aberrations when cancer transit from low to high grade cancers. Importantly, the molecular changes detected suggest that cellular transformation and acquisition of aggressive phenotype is associated with a major re-programming in metabolism (42%), especially in processes associated with carbohydrate, amino acid and lipid metabolism.

To understand the molecular mechanisms associated with tumor progression and metastasis, and in particular the involvement of mitochondria, Chen et al. [83] carried out a subcellular proteomic approach by subjecting to 2D-DIGE and MALDI-TOF mass spectrometry analysis mitochondria purified from three breast cells, MCF10A, MCF7, and MDA-MB-231, respectively corresponding to the normal luminal epithelial cells, the non-invasive breast cancer cells and the invasive breast cancer cells derived from the same tissues. Numerous differentially expressed mitochondrial proteins were identified, most of which are involved in electron transport, metabolism and protein folding, including cytochrome c oxidase subunit 5B, malate dehydrogenase and elongation factor Tu, which are highly expressed in both low invasive and aggressive breast cancer cells.

This is perhaps not surprising since metabolism is an Achilles' heel in cancer biology [83]. For instance, aberrant PI3K/AKT pathway during cancer development inadvertently amplifies glucose metabolism and translational activities via glucose transporter (GLUT4), mTOR/S6K/eukaryotic initiation factor 4E-binding protein 1 (4EBP-1), respectively. Consistently, several translation initiation factors were observed to be up-regulated in this study [84].

Skvortsov et al. [85], by using 2D-DIGE in combination with laser capture microdissection (LCM) and MALDI-TOF/TOF mass spectrometry in prostate cancer, identified a number of overexpressed proteins, ten of which were associated with glycolysis and Warburg effect. Interestingly, all of the Warburg associated proteins determined in this study were also correlated with cancer progression and were upregulated in dependence of the Gleason score, a histology-based tumor grading system that is a surrogate marker of the aggressiveness of the disease. This correlation between upregulation and tumor dedifferentiation might be relevant for selection of therapeutic strategies.

Important functional perturbations in tumor metabolism were highlighted also by Pernemalm et al. [86] who analyzed quantitative proteomics profiling of primary lung adenocarcinoma tumors to elucidate the biological mechanisms behind relapse post-surgery. By measuring the bioenergetic cellular index of the tumors, the Authors detected a higher dependency of glycolysis among the tumors with poor prognosis and an up-regulation of HIF1$\alpha$ mRNA expression in tumors with early relapse.

## 1.5 Mitochondria, Oncoproteomics and Chemoresistance

In evaluating the complex phenomenon of chemoresistance, a wealth of cancer proteomics studies, focusing on mechanisms by which chemotherapeutic agents perform their antineoplastic action, showed interesting modifications in protein expression levels associated with anticancer drug-resistances. Some of these modifications interest mitochondria. More specifically, mitochondrial proteome modifications have been described in a mitoxantrone-resistant MCF-7 cells compared to its parental drug susceptible cell line [87], in adrenocortical carcinoma cell line after exposure to mitotane [88], in non-Hodgkin lymphoma following treatment with adriamycin [89] and in paclitaxel-sensitive human ovarian cancer cell lines SKOV3 and A2780 [90]. Interestingly, mitochondrial comparative proteomics of human ovarian cancer cells and their platinum-resistant sublines realized by Dai et al. [91] revealed, in platinum-resistant cell lines, a downregulation of five mitochondrial proteins involved in energy metabolism and electron transfer respiratory chain (ATP-α, PRDX3, PHB, ETF and ALDH), suggesting that dysfunctions of oxidative phosphorylation and energy production in ovarian cancer cells may be directly related to their resistance to platinum drugs, probably acting as obstacle for the apoptotic potential of the cancer cell [92].

Notably, an interesting protein whose differential expression in tumor cells has been observed in various studies on cancer chemotherapy is prohibitin (PHB), which has been reported to be upregulated in transformed cells in respect to normal [93–95]. PHB1 has been proposed to play important roles in cancer development and progression by modulation of transcription during apoptotic process [96] or by inhibition of the intrinsic apoptotic pathway [97], in addition to its potential role as a chaperone in the inner mitochondrial membrane for respiration chain proteins. In addition, results from several studies, reporting significantly elevated levels of this protein in membrane fractions isolated from paclitaxel-resistant sublines compared to their normal drug-sensitive sublines [98], and in 5-FU-resistant subline in respect to their respective drug-sensitive human colon cancer cell line [99], seem to indicate that PHB1 is closely related to the chemoresistance phenomenon. It has been however demonstrated that overexpression of PHB does not correlate with pro-tumorigenic function [100]. Intriguingly, most recent functional studies suggested instead that it is the subcellular localization of PHB, rather than the total amount of this protein in the cell, that may affect cell fate and, more importantly, hypothesized that it is through mitochondrial localization that PHB mediates its effects [101, 102]. Interestingly, Jiang and coworkers [103] recently demonstrated that PHB interacts with Akt, which, by phosphorylating PHB at Thr258, promotes its mitochondrial translocation and induces bladder cancer cell proliferation. Therefore, these findings identified the Akt/PHB signaling cascade as a novel mechanism of cancer cell proliferation, highlighting mitochondrial PHB as an important regulator during bladder cancer tumorigenesis.

## 1.6 Conclusions

The oncoproteomic approach to cancer is undoubtedly a fascinating topic of research with enormous potential as long as associated to a deep knowledge of molecular cancer pathophysiology and to a careful methodological procedure that considers also the limits of the in vivo and in vitro approaches. Independently by the results obtained, however, oncoproteomics in general and the proteomic of cancer mitochondria in particular is producing a mass of useful data which undoubtedly may promote understanding of some tangled aspect of cancer metabolism.

First of all, this methodology seems to confirm that the so called Warburg effect is an epiphenomenon of a complex metabolic reprogramming of cancer tissue capable to induce significant modifications of cell function for:

(a) not only suitable to sustain cell proliferation;
(b) but also to promote, by inducing changes in mitochondrial redox potential and in ROS generation, a DNA damaging milieu which contributes to genomic instability. Genomic instability, in turn, together microenviron-

mental influences fosters cancer cell plasticity and a mechanistic positive pressure selectivity.

In conclusion, it is well accepted that cancers are abnormal tissues made up of heterogeneous cells endowed with different functions, varying proliferative potentials and with various levels of differentiation. Consequently, cancers are also heterogeneous at the metabolic level. Even if tumors appear primarily glycolytic due to the prevalence of cells bearing glycolytic features, other metabolic programs may coexist and can become evident after perturbations such as pharmacologic treatment. Hence, various energetic programs active in different subpopulations of tumor cells renders oncometabolic proteomic approach a really hard challenge. In our opinion, an oncoproteomic approach with significant translational implications in terms of diagnosis, prognosis and therapy must be realized on carefully defined subpopulations of cancer cells in terms of differentiation level and prevalent metabolism.

**Acknowledgements** Supported by COST Action CA15203 MitoEAGLE and LINEA D.3.2 COFIN. GRASSI.

## References

1. Mootha VK, Bunkenborg J, Olsen JV et al (2003) Integrated analysis of protein composition, tissue diversity, and gene regulation in mouse mitochondria. Cell 115:629–640
2. Palmfeldt J, Bross P (2017) Proteomics of human mitochondria. Mitochondrion 33:2–14
3. Cherry C, Thompson B, Saptarshi N, Wu J, Hoh J (2016) 2016: a 'mitochondria' odyssey. Trends Mol Med 22:391–403
4. Lane RK, Hilsabeck T, Rea SL (2015) The role of mitochondrial dysfunction in age-related diseases. Biochim Biophys Acta 1847:1387–1400
5. Scatena R, Bottoni P, Giardina B (2012) Advances in mitochondrial medicine. Springer, Philadelphia
6. Singh KK, Costello L (2008) Mitochondria and cancer. Springer, Philadelphia
7. Sarangarajan R, Apte S (2008) Cellular respiration and carcinogenesis. Humana Press, Totowa, pp 45–54
8. Pagliarini DJ, Calvo SE, Chang B et al (2008) A mitochondrial protein compendium elucidates complex I disease biology. Cell 134:112–123
9. Calvo SE, Clauser KR, Mootha VK (2016) MitoCarta2.0: an updated inventory of mammalian mitochondrial proteins. Nucleic Acids Res 44:D1251–D1257
10. Alberio T, Pieroni L, Ronci M et al (2017) Toward the standardization of mitochondrial proteomics: the Italian mitochondrial human proteome project initiative. J Proteome Res 16:4319–4329
11. MitoEAGLE preprint 2017-09-21 (Version #). The protonmotive force and respiratory control: Building blocks of mitochondrial physiology Part 1. http://www.mitoeagle.org/index.php/MitoEAGLE_preprint_2017-09-21
12. Dell' Antone P (2012) Energy metabolism in cancer cells: how to explain the Warburg and Crabtree effects? Med Hypotheses 79:388–392
13. Liberti MV, Locasale JW (2016) The Warburg effect: how does it benefit Cancer cells? Trends Biochem Sci 41:211–218
14. Tran Q, Lee H, Park J et al (2016) Targeting Cancer metabolism -revisiting the Warburg effects. Toxicol Res 32:177–193
15. Vander Heiden MG (2011) Targeting cancer metabolism: a therapeutic window opens. Nat Rev Drug Discov 10:671–684
16. Vazquez A, Liu J, Zhou Y, Oltvai ZN (2010) Catabolic efficiency of aerobic glycolysis: the Warburg effect revisited. BMC Syst Biol 4:58
17. Hanahan D, Weinberg RA (2011) Hallmarks of cancer: the next generation. Cell 144:646–674
18. Räsänen K, Herlyn M (2012) Paracrine signaling between carcinoma cells and mesenchymal stem cells generates cancer stem cell niche via epithelial-mesenchymal transition. Cancer Discov 2:775–777
19. Bottoni P, Giardina B, Vitali A, Boninsegna A, Scatena R (2009) A proteomic approach to characterizing ciglitazone-induced cancer cell differentiation in Hep-G2 cell line. Biochim Biophys Acta 1794:615–626
20. Scatena R, Bottoni P, Giardina B (2014) Mitochondria, metabolism and cancer: a growing role in cancer cell differentiation and cancer cell dormancy. Cancer & Metabolism 2:P64
21. Wallace DC (2012) Mitochondria and cancer. Nat Rev Cancer 12:685–698
22. Koppenol WH, Bounds PL, Dang CV (2011) Otto Warburg's contributions to current concepts of cancer metabolism. Nat Rev Cancer 11:325–337
23. Gaude E, Frezza C (2014) Defects in mitochondrial metabolism and cancer. Cancer Metab 2:10
24. Unwin RD, Craven RA, Harnden P et al (2003) Proteomic changes in renal cancer and co-ordinate demonstration of both the glycolytic and mitochondrial aspects of the Warburg effect. Proteomics 3:1620–1632

25. Wallace DC (2005) Mitochondria and cancer: Warburg addressed. Cold Spring Harb Symp Quant Biol 70:363–374
26. Verma M, Kagan J, Sidransky D, Srivastava S (2003) Proteomic analysis of cancer-cell mitochondria. Nat Rev Cancer 3:789–795
27. Herrmann PC, Gillespie JW, Charboneau L et al (2003) Mitochondrial proteome: altered cytochrome c oxidase subunit levels in prostate cancer. Proteomics 3:1801–1810
28. Modica-Napolitano JS, Singh KK (2004) Mitochondrial dysfunction in cancer. Mitochondrion 4:755–762
29. Scatena R, Bottoni P, Pontoglio A, Giardina B (2010) Revisiting the Warburg effect in cancer cells with proteomics. The emergence of new approaches to diagnosis, prognosis and therapy. Proteomics Clin Appl 4:143–158
30. Ward PS, Thompson CB (2012) Metabolic reprogramming: a cancer hallmark even Warburg did not anticipate. Cancer Cell 21:297–308
31. Xie H, Simon MC (2017) Oxygen availability and metabolic reprogramming in cancer. J Biol Chem 292:16825–16832
32. Bi X, Lin Q, Foo TW et al (2006) Proteomic analysis of colorectal cancer reveals alterations in metabolic pathways: mechanism of tumorigenesis. Mol Cell Proteomics 5:1119–1113
33. Mathupala SP, Rempel A, Pedersen PL (1995) Glucose catabolism in cancer cells. Isolation, sequence, and activity of the promoter for type II hexokinase. J Biol Chem 270:16918–16925
34. Scatena R, Bottoni P, Pontoglio A et al (2008) Glycolytic enzyme inhibitors in cancer treatment. Expert Opin Investig Drugs 17:1533–1545
35. Danial NN, Gramm CF, Scorrano L et al (2003) BAD and glucokinase reside in a mitochondrial complex that integrates glycolysis and apoptosis. Nature 424:896–897
36. Mathupala SP, Rempel A, Pedersen PL (2001) Glucose catabolism in cancer cells: identification and characterization of a marked activation response of the type II hexokinase gene to hypoxic conditions. J Biol Chem 276:43407–43412
37. Miller DM, Thomas SD, Islam A et al (2012) C-Myc and cancer metabolism. Clin Cancer Res 18:5546–5553
38. Li S, Li J, Dai W et al (2017) Genistein suppresses aerobic glycolysis and induces hepatocellular carcinoma cell death. Br J Cancer 117:1518–1528
39. Zhang H, Du X, Sun TT et al (2017) Lectin PCL inhibits the Warburg effect of PC3 cells by combining with EGFR and inhibiting HK2. Oncol Rep 37:1765–1771
40. Hardt PD, Mazurek S, Toepler M et al (2004) Faecal tumour M2 pyruvate kinase: a new, sensitive screening tool for colorectal cancer. Brit J Cancer 91:980–984
41. Dombrauckas JD, Santarsiero BD, Mesecar AD (2005) Structural basis for tumor pyruvate kinase M2 allosteric regulation and catalysis. Biochemistry 44:9417–9429
42. Liu F, Ma F, al WY (2017) PKM2 methylation by CARM1 activates aerobic glycolysis to promote tumorigenesis. Nat Cell Biol 19:1358–1370
43. Mazurek S, Boschek CB, Hugo F, Eigenbrodt E (2005) Pyruvate kinase type M2 and its role in tumor growth and spreading. Semin Cancer Biol 15:300–308
44. Du XL, Hu H, Lin DC et al (2007) Proteomic profiling of proteins dysregulated in Chinese esophageal squamous cell carcinoma. J Mol Med 85:863–875
45. Christofk HR, Vander Heiden MG, Harris MH et al (2008) The M2 splice isoform of pyruvate kinase is important for cancer metabolism and tumour growth. Nature 452:230–233
46. Vander Heiden MG, Christofk HR, Schuman E et al (2010) Identification of small molecule inhibitors of pyruvate kinase M2. Biochem Pharmacol 79:1118–1124
47. Bluemlein K, Grüning NM, Feichtinger RG et al (2011) No evidence for a shift in pyruvate kinase PKM1 to PKM2 expression during tumorigenesis. Oncotarget 2:393–400
48. Zhou W, Capello M, Fredolini C et al (2012) Proteomic analysis reveals Warburg effect and anomalous metabolism of glutamine in pancreatic cancer cells. J Proteome Res 11:554–563
49. Isgrò MA, Bottoni P, Scatena R (2015) Neuron-specific enolase as a biomarker: biochemical and clinical aspects. Adv Exp Med Biol 867:125–143
50. Deng SS, Xing TY, Zhou HY et al (2006) Comparative proteome analysis of breast cancer and adjacent normal breast tissues in human. Genomics Proteomics Bioinformatics 4:165–172
51. Hilf R, Rector WD, Orlando RA (1976) Multiple molecular forms of lactate dehydrogenase and glucose 6-phosphate dehydrogenase in normal and abnormal human breast tissues. Cancer 37:1825–1830
52. Shim H, Dolde C, Lewis BC et al (1997) C-Myc transactivation of LDH-A: implications for tumor metabolism and growth. Proc Natl Acad Sci U S A 94:6658–6663
53. Koukourakis MI, Giatromanolaki A, Sivridis E et al (2003) Tumour and angiogenesis research group. Lactate dehydrogenase-5 (LDH-5) overexpression in non-small-cell lung cancer tissues is linked to tumour hypoxia, angiogenic factor production and poor prognosis. Br J Cancer 89:877–885
54. Fantin VR, St-Pierre J, Leder P (2006) Attenuation of LDH-A expression uncovers a link between glycolysis, mitochondrial physiology, and tumor maintenance. Cancer Cell 9:425–434
55. Kawamoto M (1994) Breast cancer diagnosis by lactate dehydrogenase isozymes in nipple discharge. Cancer 73:1836–1841
56. Koukourakis MI, Giatromanolaki A, Simopoulos C et al (2005) Lactate dehydrogenase 5 (LDH5) relates to up-regulated hypoxia inducible factor path-

way and metastasis in colorectal cancer. Clin Exp Metastasis 22:25–30
57. Koukourakis MI, Giatromanolaki A, Polychronidis A et al (2006) Endogenous markers of hypoxia/anaerobic metabolism and anemia in primary colorectal cancer. Cancer Sci 97:582–588
58. Leiblich A, Cross SS, Catto JW et al (2006) Lactate dehydrogenase-B is silenced by promoter hypermethylation in human prostate cancer. Oncogene 25:2953–2960
59. Glen A, Gan CS, Hamdy FC et al (2008) iTRAQ facilitated proteomic analysis of human prostate cancer cells identifies proteins associated with progression. J Proteome Res 7:897–907
60. Zhou W, Capello M, Fredolini C (2011) Proteomic analysis of pancreatic ductal adenocarcinoma cells reveals metabolic alterations. J Proteome Res 10:1944–1952
61. Hussien R, Brooks GA (2011) Mitochondrial and plasma membrane lactate transporter and lactate dehydrogenase isoform expression in breast cancer cell lines. Physiol Genomics 43:255–264
62. Zha X, Wang F, Wang Y et al (2011) Lactate dehydrogenase B is critical for hyperactive mTOR-mediated tumorigenesis. Cancer Res 71:13–18
63. Menendez JA, Lupu R (2007) Fatty acid synthase and the lipogenic phenotype in cancer pathogenesis. Nat Rev Cancer 7:763–777
64. Fliedner SM, Kaludercic N, Jiang XS et al (2012) Warburg effect's manifestation in aggressive pheochromocytomas and paragangliomas: insights from a mouse cell model applied to human tumor tissue. PLoS One 7:e40949
65. Costa ASH, Frezza C (2017) Metabolic reprogramming and oncogenesis: One Hallmark, Many Organelles. Int Rev Cell Mol Biol 332:213–231
66. Tyrakis PA, Yurkovich ME, Sciacovelli M et al (2017) Fumarate hydratase loss causes combined respiratory chain defects. Cell Rep 21:1036–1047
67. Gao W, Xua J, Wang F et al (2015) Mitochondrial proteomics approach reveals voltage-dependent Anion Channel 1 (VDAC1) as a potential biomarker of gastric Cancer. Cell Physiol Biochem 37:2339–2354
68. Gao W, Xu J, Wang F, Zhang L, Peng R, Shu Y, Wu J, Tang Q, Zhu Y (2015) Plasma membrane proteomic analysis of human gastric Cancer tissues: revealing flotillin 1 as a marker for gastric Cancer. BMC Cancer 15:36
69. Clarke MF, Dick JE, Dirks PB et al (2006) Cancer stem cells--perspectives on current status and future directions: AACR workshop on cancer stem cells. Cancer Res 66:9339–9344
70. Visvader JE, Lindeman GJ (2012) Cancer stem cells: current status and evolving complexities. Cell Stem Cell 10:717–728
71. Dick JE (2008) Stem cell concepts renew cancer research. Blood 112:4793–4807
72. Skvortsov S, Debbage P, Skvortsova I (2014) Proteomics of cancer stem cells. Int J Radiat Biol 90:653–658
73. Van Houdt WJ, Emmink BL, Pham TV et al (2011) Comparative proteomics of colon cancer stem cells and differentiated tumor cells identifies BIRC6 as a potential therapeutic target. Mol Cell Proteomics 10:M111.011353
74. Dormeyer W, van Hoof D, Braam SR et al (2008) Plasma membrane proteomics of human embryonic stem cells and human embryonal carcinoma cells. J Proteome Res 7:2936–2951
75. Chaerkady R, Kerr CL, Kandasamy K et al (2010) Comparative proteomics of human embryonic stem cells and embryonal carcinoma cells. Proteomics 10:1359–1373
76. Viale A, Pettazzoni P, Lyssiotis CA et al (2014) Oncogene ablation-resistant pancreatic cancer cells depend on mitochondrial function. Nature 514:628–632
77. Lamb R, Harrison H, Hulit J et al (2014) Mitochondria as new therapeutic targets for eradicating cancer stem cells: quantitative proteomics and functional validation via MCT1/2 inhibition. Oncotarget 5:11029–11037
78. Lamb R, Ozsvari B, Bonuccelli G et al (2015) Dissecting tumor metabolic heterogeneity: telomerase and large cell size metabolically define a subpopulation of stem-like, mitochondrial-rich, cancer cells. Oncotarget 6:21892–21905
79. Sacco F, Silvestri A, Posca D et al (2016) Deep proteomics of breast Cancer cells reveals that metformin rewires signaling networks away from a pro-growth state. Cell Syst 2:159–171
80. Prost S, Relouzat F, Spentchian M et al (2015) Erosion of the chronic myeloid leukaemia stem cell pool by PPARγ agonists. Nature 525:380–383
81. Kuntz EM, Baquero P, Michie AM et al (2017) Targeting mitochondrial oxidative phosphorylation eradicates therapy-resistant chronic myeloid leukemia stem cells. Nat Med 23:1234–1240
82. Choong LY, Lim S, Chong PK (2010) Proteome-wide profiling of the MCF10AT breast cancer progression model. PLoS One 5:e11030
83. Chen YW, Chou HC, Lyu PC et al (2011) Mitochondrial proteomics analysis of tumorigenic and metastatic breast cancer markers. Funct Integr Genomics 11:225–239
84. Kroemer G, Pouyssegur J (2008) Tumor cell metabolism: cancer's Achilles' heel. Cancer Cell 13:472–482
85. Skvortsov S, Schäfer G, Stasyk T et al (2011) Proteomics profiling of microdissected low- and high-grade prostate tumors identifies Lamin a as a discriminatory biomarker. J Proteome Res 10:259–268
86. Pernemalm M, De Petris L, Branca RM et al (2013) Quantitative proteomics profiling of primary lung adenocarcinoma tumors reveals functional per-

turbations in tumor metabolism. J Proteome Res 12:3934–3943
87. Wang J, Gutierrez P, Edwards N, Fenselau C (2007) Integration of 18O labeling and solution isoelectric focusing in a shotgun analysis of mitochondrial proteins. J Proteome Res 6:4601–4607
88. Stigliano A, Cerquetti L, Borro M et al (2008) Modulation of proteomic profile in H295R adrenocortical cell line induced by mitotane. Endocr Relat Cancer 15:1–10
89. Jiang YJ, Sun Q, Fang XS, Wang X (2009) Comparative mitochondrial proteomic analysis of Rji cells exposed to adriamycin. Mol Med 15:173–182
90. Chen M, Huang H, He H et al (2015) Quantitative proteomic analysis of mitochondria from human ovarian cancer cells and their paclitaxel-resistant sublines. Cancer Sci 106:1075–1083
91. Dai Z, Yin J, He H et al (2010) Mitochondrial comparative proteomics of human ovarian cancer cells and their platinum-resistant sublines. Proteomics 10:3789–3799
92. Dey R, Moraes CT (2000) Lack of oxidative phosphorylation and low mitochondrial membrane potential decrease susceptibility to apoptosis and do not modulate the protective effect of Bcl-x(L) in osteosarcoma cells. J Biol Chem 275:7087–7094
93. Jupe ER, Liu XT, Kiehlbauch JL (1996) Prohibitin in breast cancer cell lines: loss of antiproliferative activity is linked to 3′ untranslated region mutations. Cell Growth Differ 7:871–878
94. Ummanni R, Junker H, Zimmermann U et al (2008) Prohibitin identified by proteomic analysis of prostate biopsies distinguishes hyperplasia and cancer. Cancer Lett 266:171–185
95. Ross JA, Robles-Escajeda E, Oaxaca DM et al (2017) The prohibitin protein complex promotes mitochondrial stabilization and cell survival in hematologic malignancies. Oncotarget 8:65445–65456
96. Fusaro G, Dasgupta P, Rastogi S et al (2003) Prohibitin induces the transcriptional activity of p53 and is exported from the nucleus upon apoptotic signaling. J Biol Chem 278:47853–47861
97. Gregory-Bass RC, Olatinwo M, Xu W et al (2008) Prohibitin silencing reverses stabilization of mitochondrial integrity and chemoresistance in ovarian cancer cells by increasing their sensitivity to apoptosis. Int J Cancer 122:1923–1930
98. Patel N, Chatterjee SK, Vrbanac V et al (2010) Rescue of paclitaxel sensitivity by repression of Prohibitin1 in drug-resistant cancer cells. Proc Natl Acad Sci U S A 107:2503–2508
99. Kimura K, Wada A, Ueta M et al (2010) Comparative proteomic analysis of the ribosomes in 5-fluorouracil resistance of a human colon cancer cell line using the radical-free and highly reducing method of two-dimensional polyacrylamide gel electrophoresis. Int J Oncol 37:1271–1278
100. Wu TF, Wu H, Wang YW et al (2007) Prohibitin in the pathogenesis of transitional cell bladder cancer. Anticancer Res 27:895–900
101. Merkwirth C, Langer T (2009) Prohibitin function within mitochondria: essential roles for cell proliferation and cristae morphogenesis. Biochim Biophys Acta 1793:27–32
102. Sripathi SR, He W, Atkinson CL et al (2011) Mitochondrial-nuclear communication by prohibitin shuttling under oxidative stress. Biochemistry 50:8342–8351
103. Jiang L, Dong P, Zhang Z et al (2015) Akt phosphorylates Prohibitin 1 to mediate its mitochondrial localization and promote proliferation of bladder cancer cells. Cell Death Dis 6:e1660

# Mitochondrial Proteins in the Development of Parkinson's Disease

2

Mara Zilocchi, Mauro Fasano, and Tiziana Alberio

**Abstract**

Parkinson's disease (PD) is a multifactorial disorder whose etiology is not completely understood. Strong evidences suggest that mitochondrial impairment and altered mitochondrial disposal play a key role in the development of this pathology. Here we show this association in both genetic and sporadic forms of the disease. Moreover, we describe the mitochondrial dysfunctions in toxin-induced models of PD, thus highlighting the importance of environmental factors in the onset of this pathology. In particular, we focus our attention on mitochondrial dynamics, mitochondrial biogenesis, and mitophagy and explain how their impairment could have a negative impact on dopaminergic neurons function and survival. Lastly, we aim at clarifying the important role played by proteomics in this field of research, proteomics being a global and unbiased approach suitable to unravel alterations of the molecular pathways in multifactorial diseases.

**Keywords**

Parkinson's disease · Mitochondrial homeostasis · Mitochondrial impairment · Proteomics

M. Zilocchi
Department of Biochemistry, Research and Innovation Centre, University of Regina, Regina, Canada
e-mail: mara.zilocchi@uregina.ca

M. Fasano · T. Alberio (✉)
Department of Science and High Technology, University of Insubria, Busto Arsizio, VA, Italy
e-mail: mauro.fasano@uninsubria.it; tiziana.alberio@uninsubria.it

## Abbreviations

| | |
|---|---|
| CCCP | carbonyl cyanide m-chlorophenyl hydrazone |
| COR | C-terminal of ROC |
| CREB | cyclic AMP-responsive element-binding protein |
| DAT | dopamine transporter |
| DRP1 | dynamin related protein 1 |
| ERRs | estrogen-related receptors |
| ETC | electron transport chain |
| Fis1 | mitochondrial fission 1 protein |
| GCase | β-glucocerebrosidase |
| Mff | mitochondrial fission factor |
| MFNs | mitofusins |
| MiD49 | mitochondrial dynamics protein MID49 |
| MiD51 | mitochondrial dynamics protein MID51 |
| MPP | mitochondrial processing peptidase |
| $MPP^+$ | 1-methyl-4-phenylpyridinium |





A. Urbani, M. Babu (eds.), *Mitochondria in Health and in Sickness*, Advances in Experimental Medicine and Biology 1158, https://doi.org/10.1007/978-981-13-8367-0_2

| | |
|---|---|
| MPTP | 1-methyl-4-phenyl-1,2,3,6-tetrahydropyridine |
| mtDNA | mitochondrial DNA |
| MTERF1 | transcription termination factor 1 |
| mtRNAP | RNA polymerase |
| mtSSB | mitochondrial single-stranded DNA-binding protein |
| NOS | reactive nitrogen species |
| NRFs | nuclear respiratory factors |
| OMA1 | metalloendopeptidase OMA1 |
| OPA1 | dynamin-like 120 kDa protein |
| PARL | presenilins-associated rhomboid-like protein |
| PD | Parkinson's disease |
| PGC-1α | peroxisome proliferator-activated receptor gamma coactivator 1 alpha |
| PKA | protein kinase A |
| POLG | DNA polymerase γ |
| POLRMT | DNA-directed RNA polymerase |
| PPAR | peroxisome proliferator-activated receptor |
| PPP | pentose phosphate pathway |
| ROC | Ras of complex proteins |
| ROS | reactive oxygen species |
| Tfam | transcription factor A |
| TFB | dimethyladenosine transferase 1 |
| TIM | translocase of inner mitochondrial membrane |
| TOM | translocase of outer mitochondrial membrane |
| Top2α | type IIA topoisomerase |
| VDACs | voltage-dependent anion channels |
| YME1L | ATP-dependent zinc metalloprotease YME1L1 |
| α-syn | α-synuclein |

## 2.1 Mitochondria and Parkinson's Disease

Parkinson's disease (PD) is the most common neurodegenerative movement disorder in our society. It affects around 14 per 100 000 people, with an increasing prevalence in the aging population [2]. The progressive loss of dopaminergic neurons in the substantia nigra pars compacta that characterizes this pathology causes the appearance of several and typical motor symptoms (i.e., bradykinesia, rigidity, postural instability and resting tremor) [93]. Although the cause of PD is currently unknown, strong evidence indicates that a complex interplay between several factors including genetic susceptibility, environmental factors, abnormal protein handling and oxidative stress could be involved [65, 196, 215]. Many of the molecular pathways implicated in PD etiology converge on mitochondria, resulting in their dysfunction, which could affect neuronal survival (Fig. 2.1).

In this chapter, we will explore the tight relation between mitochondrial dysfunction and PD pathogenesis, in order to clarify the role of these organelles in triggering neurodegeneration.

## 2.2 Autosomal Dominant Forms of PD and Mitochondria

### 2.2.1 SNCA (PARK1)

SNCA (PARK1) was the first gene identified to be associated with familial PD [178, 179]. In particular, three different missense mutations (A53T, A30P and E46K) and duplication or triplication of the SNCA gene are associated with autosomal dominant forms of this pathology [110, 179, 210, 257]. This gene encodes for α-synuclein (α-syn), a presynaptic protein formed by 140 amino acids that plays several roles in neuronal cells, and is particularly involved in endocytosis and synaptic vesicles trafficking [7, 231]. α-Syn protein is prone to form fibrillar aggregates that are the major components of Lewy bodies [231], thus emphasizing the common pathogenetic traits in sporadic and familial forms of PD.

It is well known that aggregation of α-syn causes several cellular damages, such as disruption of lysosomal and proteasomal functions [32, 238, 248], axonal and synaptic transport impairment [30, 33]. α-Syn aggregation directly affects also the mitochondrial shape and function. Indeed, it has been demonstrated that overexpression of mutant or wild type forms of α-syn causes the association of this protein to the mitochondrial membrane, thus determining the release of cytochrome c and the increase of reactive oxygen

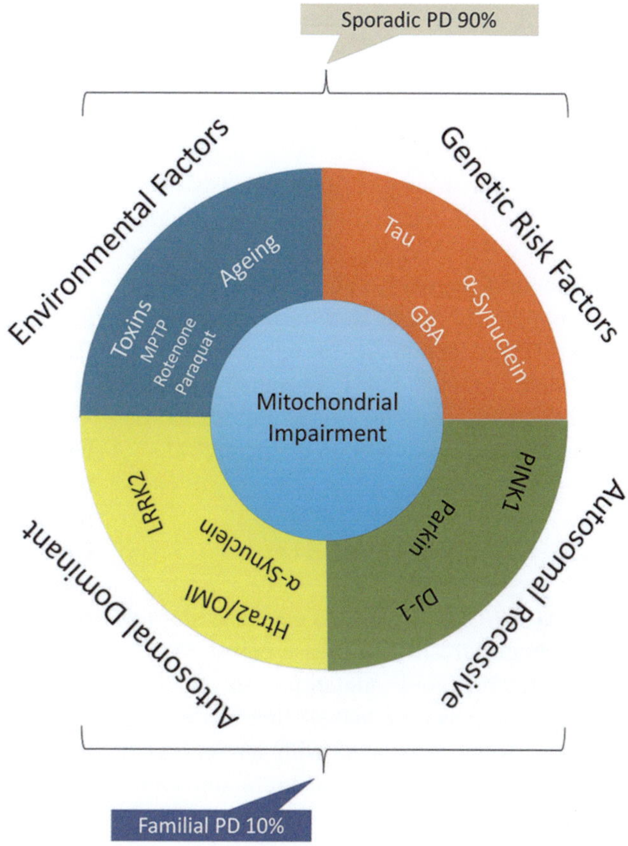


**Fig. 2.1** Schematic representation of PD etiology and mitochondrial impairment relation. The complex interplay between genetic susceptibility and environmental factors plays a key role in the development of sporadic PD. Both familial and sporadic forms of the disease are characterized by mitochondrial impairment, a common pathogenetic mechanism

species (ROS) production [168]. Moreover, α-syn aggregates directly inhibit complex I, leading to mitochondrial impairment [187, 231]. Mutations in the SNCA gene were related to mitochondrial fragmentation because of a decreased association between mitochondria and the endoplasmic reticulum and to the cleavage of OPA1 into its shorter form [77]. Lastly, the expression level of voltage-dependent anion channel 1 (VDAC1) protein is decreased following the accumulation and aggregation of α-syn, either in the sporadic and in the SNCA familial forms of the disease [34].

### 2.2.2 LRRK2 (PARK8)

Mutations in PARK8 gene are associated with autosomal dominantly inherited late onset PD [43, 263]. LRRK2 mutations constitute around 10% of the familial PD cases [107]. Mutations in this gene are associated also with 1% of the sporadic form of the disease [71, 120]. To date, it has been proven that seven sequence variants (N1437H, R1441H, R1441C, R1441G, Y1699C, I2020T and G2019S) cause the development of PD. Indeed, these mutations are located within the functional domains of the protein as well as in evolutionary conserved regions [107].

LRRK2 gene encodes a large multidomain protein with GTP-regulated serine/threonine kinase activity, whose catalytic core is characterized by the presence of tandem ROC-COR and kinase domains [129]. Most of the LRRK2 mutations are located in the GTPase and kinase domain and for several of them an increased kinase activity was demonstrated [37, 129]. It has been shown that LRRK2 is located not only in the cytoplasm and the nucleus, but also in mitochondria [141, 245], thus highlighting the role of this protein in mitochondrial function, morphology and dynamics. Indeed, LRRK2 is implicated in the translocation of dynamin related protein 1 (DRP1), a mitochondrial fission protein, from the cytosol to the mitochondria [242]. This role of LRRK2 protein in mitochondrial morphology maintenance was confirmed, for example, in fibroblasts from LRRK2 G2019S mutated PD patients, characterized by a more elongated and interconnected mitochondrial network [141]. However, these skin fibroblasts were also characterized by lower ATP production and lower mitochondrial membrane potential [141]. Eventually, inhibition of LRRK2 activity causes mitochondrial fission and increased ROS production [190].

### 2.2.3 HTRA2/OMI (PARK13)

The HTRA2/OMI (PARK13) gene encodes for a mitochondrial serine protease called HtrA2/OMI. Heterozygous mutations in HTRA2 gene, which cause the loss of function of this protease, were associated with an increased susceptibility to the development of autosomal dominant Parkinson disease [214]. Nevertheless, following studies failed to confirm this association [209]. Mitochondrial dysfunction and mutations in the mitochondrial DNA (mtDNA) are a hallmark of the loss of HtrA2/OMI protease activity [73]. Moreover, it has been suggested that PINK1 acts upstream of HtrA2/OMI, thus facilitating its phosphorylation and the resulting cellular resistance to mitochondrial stress.

## 2.3 Autosomal Recessive Forms of PD and Mitochondria

### 2.3.1 Parkin (PARK2)

The PARK2 gene encodes for the Parkin protein, an E3 ubiquitin ligase, and mutations in this gene have been linked to autosomal recessive juvenile Parkinsonism [106, 206]. Age at onset is typically between childhood and 40 years [216]. Moreover, around half of the confirmed cases of autosomal recessive forms of PD are caused by mutations in this gene [123]. Most of the PARK2 point mutations cause alterations in the cellular localization or in the solubility of this protein. Other mutations, including insertions and deletions, result in Parkin loss-of-function [216]. Parkin plays a role in proteins and mitochondrial degradation via the proteasome: for this reason, the loss-of-function of Parkin determines the accumulation of its substrates, which may have a toxic effect on dopaminergic neurons [70, 259]. Parkin is also strongly involved in the regulation of mitochondrial dynamics and turnover [69]. The role of Parkin in the mitophagic process will be described in detail in paragraph 2.8. Moreover, it has been proposed that Parkin also plays an indirect role in mitochondrial biogenesis. Indeed, Parkin loss-of-function causes the increase of PARIS protein levels, a repressor of peroxisome proliferator-activated receptor gamma coactivator 1 alpha (PGC-1$\alpha$) [207] and a central regulator of mitochondrial encoded gene expression [79]. These data are further supported by the finding of PARIS accumulation in brains of PD patients [207].

### 2.3.2 PINK1 (PARK6)

Mutations in PARK6 gene were found in familial cases of PD but also in patients suffering from sporadic PD [226], and, for this reason, it was proposed that heterozygous mutations in PINK1 might be a risk factor for PD [227]. Most of the mutations reside in the kinase domain of PINK1 causing a loss of function of the protein [208]. PINK1 is a serine/threonine kinase that acts upstream to Parkin in the regulation of mitochondrial disposal: indeed, following mitochondrial depolarization, PINK1 accumulates on the outer mitochondrial membrane, thus recruiting Parkin [107, 147]. Moreover, the PINK1/Parkin pathway is involved in mitochondrial traffic, mitochondrial dynamics and complex I activity. Indeed, when mitochondria are damaged and lose their membrane potential, PINK1 phosphorylates Miro (a mitochondrial GTPase involved in the mitochondrial trafficking and in the subcellular distribution of these organelles) to induce a

Parkin and proteasomal-dependent degradation of this protein. This process leads to the inhibition of mitochondrial motility [239]. Following mitochondrial depolarization, fusion and fission proteins (e.g. mitofusins and DRP1) are ubiquitinated by Parkin, thus suggesting a role of PINK1/Parkin pathway also in mitochondrial dynamics [25, 67, 240]. Lastly, it has been demonstrated that PINK1 deficiency causes lower complex I activity and increased sensitivity to apoptotic stress [140].

### 2.3.3 DJ-1 (PARK7)

Mutations in DJ-1 (designated as PARK7) cause the development of early onset familial PD [13]. PARK7 mutations are quite rare and the typical age of onset is between 20 and 30 years [211]. DJ-1 is a multifunctional protein involved in several cellular processes, e.g., anti-oxidative stress and chaperone activity [199, 218]. Following oxidative stress, DJ-1, normally localized in the cytosol, translocates to the mitochondria, thus playing its protecting role [18, 103]. It has been demonstrated that DJ-1 deficiency causes not only an increased oxidative stress, but also decreased complex I activity and mitochondrial respiration [81, 82]. Moreover, DJ-1 seems to be involved in the maintenance of the endoplasmic reticulum and mitochondria communication. Indeed, decreased DJ-1 levels cause mitochondrial fragmentation due to the reduction of contact sites with the endoplasmic reticulum [17, 162].

## 2.4 Sporadic PD and Mitochondria

Strong evidence suggests the tight relation between mitochondrial dysfunctions and PD pathogenesis. The first hint in this direction was the parkinsonian syndrome induced by 1-methyl-4-phenyl-1,2,3,6-tetrahydropyridine (MPTP), a neurotoxin that selectively inhibits the complex I activity in dopaminergic neurons [112]. MPTP-induced parkinsonism results from an acute toxic insult and therefore differs from the slow and progressive disease process that characterizes the sporadic PD. Nevertheless, the reduction of the complex I activity is one of the main hallmark of PD. Indeed, mitochondrial complex I deficiency was reported not only in the substantia nigra [197], but also in other cells and tissues of PD patients, e.g., fibroblasts [144], skeletal muscle [10] and lymphocytes [78]. Moreover, the exposure to pesticides that inhibit complex I activity, such as rotenone and paraquat, represents a crucial risk factor for PD [200].

Oxidative stress is another typical feature of PD pathogenesis [94]. Normally the production and the detoxification of ROS are well balanced [237]. However, when mitochondrial complexes of the electron transport chain (ETC) are inhibited, the production of free radicals increases, thus allowing ROS to activate specific signaling pathways, eventually inducing cell death [61, 95]. This mechanism has been confirmed by the presence of oxidized proteins and of high level of mtDNA deletions in substantia nigra of PD patients [6, 94]. Neurons of the substantia nigra are more sensitive to ROS production due to the presence of dopamine, which is able to produce reactive quinones at neutral pH. This process leads to the formation of endogenous toxins, which cause mitochondrial dysfunction and oxidative damage [143]. Moreover, it has been demonstrated that the inhibition of pentose phosphate pathway (PPP) leads to the selective death of dopaminergic neurons and to the onset of typical parkinsonian features. Since neurons are able to maintain their antioxidant status primarily by metabolizing glucose via the PPP, decreased levels of PPP enzymes, which in turn cause a reduction in the antioxidant defenses, may be a leading event in idiopathic PD pathogenesis [14, 47].

The most known genetic risk factor involved in the development of sporadic PD is the GBA gene. GBA encodes for β-glucocerebrosidase (GCase), a lysosomal enzyme with an important role in glycolipid metabolism. Mutations in this gene have been found to increase the risk of developing sporadic PD and are found in 8%-14% of autopsy-proven diagnoses of PD [232]. Some GBA mutations result in misfolded protein, which might induce lysosomal insufficiency and autophagic pathways impairment, thus causing neurodegeneration [5]. The pathogenic mechanism leading to mitochondrial dysfunction is probably the inhibition of proper mitochondrial disposal (i.e., mitophagy) [66]. Moreover, it has been demonstrated that reduction in GCase determines an increase in α-syn and conversely, overexpression of α-syn reduces GCase levels [127, 196]. It was demonstrated that GCase inhibition leads to decreased ADP phosphorylation, reduced mitochondrial membrane potential and increased free radical formation and damage [36].

Lastly, mitochondrial dysfunctions may be triggered or intensified by ubiquitin-proteasome-system and/or autophagy-lysosomal pathway impairments [121, 135, 139].

## 2.5 Mitochondrial Toxins and PD

All mitochondrial toxins used to reproduce some aspects of the PD pathogenesis are environmental contaminants. The discovery of their relation to PD pathogenesis not only highlighted the role of mitochondria in neurodegeneration but also originated the environmental etiological hypothesis for PD. These toxins are able to increase ROS cellular levels due either to the direct inhibition of complex I (i.e., 1-methyl-4-phenylpyridinium - $MPP^+$, and rotenone) or to the conversion of free radicals to reactive species (i.e., paraquat). Increased ROS levels lead to mitochondrial alterations and consequently to the dysfunction of several cellular pathways, thus causing dopaminergic cell death. In this paragraph we will clarify the role of each toxin in the PD pathogenesis.

### 2.5.1 MPTP

The discovery that the administration of MPTP toxin causes parkinsonism was done in the late 1970s [40, 112]. The patients that developed PD due to the MPTP exposure were heroin users. Indeed, this compound was found to be a by-product in the synthesis of a meperidine analog [112]. In particular, these subjects showed degeneration of dopaminergic neurons and the consequent development of several parkinsonian symptoms, but without Lewy bodies formation [40, 114]. For this reason, the MPTP toxin became of common use in research in various animal species to recapitulate PD pathology.

MPTP is a molecule able to cross the blood-brain barrier. It is transformed in the $MPP^+$ toxin in glial cells, due to the action of monoamine oxidase [113, 128]. Then, $MPP^+$ accumulates in the mitochondria of dopaminergic neurons thanks to the dopamine transporter (DAT) [151, 186]. At mitochondrial level, this toxin inhibits the activity of the complex I of the ETC, thus leading to lower ATP production, increased ROS generation and neuronal cell death (Fig. 2.2) [21, 38, 57, 171, 172, 182].

MPTP treatment in neuronal cells causes several mitochondrial alterations. Indeed, expression levels of several proteins, e.g., chaperones, VDAC1, metabolic enzymes and mitofilin, are altered following the inhibition of complex I [16]. An *in vivo* proteomics study described alterations caused in the striatum and/or in the substantia nigra by this toxin. The majority of altered proteins are involved in dopamine signaling, ubiquitin-proteasome system, calcium signaling, apoptosis and mitochondrial maintenance [260]. The proteome alterations after MPTP administration can be explained by the damage at the DNA level and by the reduction of several mitochondrial proteins. Indeed, it has been demonstrated that MPTP exposure causes the loss of genome integrity due to the appearance of several double or single DNA strand breaks [85]. Moreover, mRNA levels of several mitochondrial-related genes were demonstrated to be drastically reduced after $MPP^+$ treatment. In particular, the mRNA levels of several subunits of the ETC (e.g., ND1, ND2, ND3, ND4, ND4L, and ND6 of complex I) and of two proteins involved in mitochondrial biogenesis (i.e., Tfam, and NRF1) were affected by $MPP^+$ [175]. However, it must be kept in mind that MPTP exposure does not reproduce all the clinical and pathological symptoms of the PD in non-human primates [62, 80, 230]. Moreover, MPTP toxin has no effect on the rat model. As a matter of fact, none of the symptoms

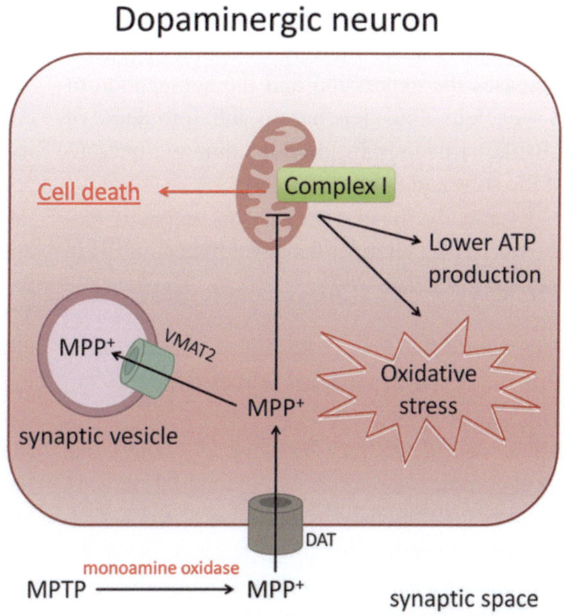


**Fig. 2.2** Schematic representation of $MPP^+$ neurotoxicity. MPTP is converted by a monoamine oxidase (i.e., MAOB) to $MPP^+$. $MPP^+$ is then taken up by the dopamine transporter (DAT). Afterwards, it can be accumulated in mitochondria, leading to complex I inhibition and the generation of free radicals, or stored in synaptic vesicles, thus limiting its toxic effects

could be reproduced in rats [29, 48]. The clinical and pathological differences between sporadic PD patients and MPTP animal models can explain the ineffectiveness of several compounds in clinical studies [87].

### 2.5.2 Rotenone

Rotenone is an organic compound that was widely used for several decades as a pesticide. Rotenone is a lipophilic molecule that is able to cross the blood-brain barrier and biological membranes without using a specific receptor or transporter. This toxin, like $MPP^+$, inhibits the complex I of the mitochondrial ETC, thus causing high ROS generation, lower ATP production and apoptotic cell death [101]. The use of rotenone also causes proteolytic stress due to the inhibition of the proteasome activity [31]. The selective death of dopaminergic neurons was observed in rats chronically treated with rotenone [8]. This finding was followed by the discovery that *in vivo* rotenone administration causes the formation of LB-like cytoplasmic inclusions in dopaminergic neurons, and an increase in oxidative damage, enhanced iron deposits and microgliosis [8, 201, 202]. Moreover, rotenone-related symptoms better resemble typical parkinsonian motor features, such as bradykinesia, rigidity, postural instability and tremor. It has also been demonstrated that people exposed to this toxin show an increased risk in developing PD [222].

### 2.5.3 Paraquat

Paraquat is a commonly used herbicide and it has a molecular structure similar to that of the $MPP^+$. Paraquat is another compound commonly used in research to reproduce some clinical and pathological features of PD. Despite this toxin is not able to cross the blood-brain barrier, it enters the brain through the neutral amino acid transporter [132, 204]. Paraquat accumulates in mitochondria, where it exerts its toxic action. Actually, paraquat is able to increase the cellular oxidative stress due to its ability to convert free radicals to superoxide and other ROS [104, 255]. In contrast to $MPP^+$ and rotenone, it has weak inhibitory effects on the mitochondrial complex I activity [188] but leads to glutamate-mediated excitotoxicity and reactive nitrogen species (RNS) generation [205].

It has been demonstrated that administration of paraquat causes the selective loss of dopaminergic neurons, with a consequent decrease in motor functions [15, 133, 159]. Administration of paraquat causes lipid peroxidation [181] and increases the expression and the aggregation of α-synuclein, thus leading to the formation of LB-like structures in the dopaminergic neurons of the substantia nigra.

Eventually, the exposure to this herbicide represents a risk factor for the development of PD in humans [145, 222, 247].

## 2.6 Mitochondrial Dynamics

Mitochondria are very dynamic organelles that form complex networks. Mitochondrial network changes continuously in response to the activation of a specific signaling pathway or to the presence of a particular metabolic stimulus.

Two different processes are involved in the mitochondrial network dynamics, i.e., fission and fusion. Fusion of the outer and inner mitochondrial membranes is controlled by three different proteins, known as mitofusin 1 (MFN1), mitofusin 2 (MFN2) and dynamin-like 120 kDa protein (OPA1), while the recruitment of DRP1 protein is responsible for the activation of the fission process.

Mitochondrial fusion allows the exchange of components between mitochondria, thus enabling the maintenance of their functional state [24]. Fusion of the outer mitochondrial membranes is controlled by two homologous proteins, MFN1 and MFN2 (Fig. 2.3) [47]. MFN1 protein consists of an N-terminal GTPase followed by a predicted helix bundle segment, two transmembrane domains, and the C-terminal tail, a heptad-repeat domain and a GTP-binding domain [109, 184]. The heptad-repeat domain mediates the first step of fusion due to its role in the formation of a dimeric antiparallel coiled-coil structure, which can be homotypic (MFN1-MFN1 or MFN2-MFN2) or heterotypic (MFN1-MFN2) [26, 109]. MFN1 protein has a higher GTPase activity than MFN2. For this reason, mitochondria that own only MFN1 protein show a better tethering efficiency than mitochondria with only MFN2 protein [89]. Moreover, MFN1 partially rescues the defects caused by MFN2 mutation, thus suggesting that the function of these proteins is the same [42].

Like MFN1, MFN2 is an integral outer mitochondrial membrane protein with GTPase activity, which is necessary for the fusion process [26, 55]. This protein exposes both terminal ends to the cytosol [190] and exerts a protective function against neurodegeneration [28] and in particular in dopaminergic neurons [116, 174]. Moreover, MFN2 is involved in the tethering between mitochondria and the endoplasmic reticulum and in the maintenance of the $Ca^{2+}$ signaling between these organelles [41]. MFN2 plays also a role in the mitochondrial motility, thanks to its connection with the Miro/Milton transport complex [137]. It has been demonstrated that deletion of MFN1 or MFN2 causes mitochondrial fragmentation and mitochondrial dysfunction [23, 27].

MFN1 and MFN2 turnover depends on the recruitment of the AAA-ATPase p97, which permits the degradation of the ubiquitinated mitofusins through the proteasome [166, 221].

OPA1 is a GTPase protein that is responsible for the fusion of the inner mitochondrial membrane (Fig. 2.3) [19]. There are eight possible isoforms of OPA1 protein that derive from alternative splicing [212]. Moreover, once OPA1 protein is imported in the intermembrane space, this protein can be processed by several proteases, e.g., presenilins-associated rhomboid-like protein (PARL), ATP-dependent zinc metalloprotease YME1L1 (YME1L), and metalloendopeptidase OMA1 (OMA1) [35, 86, 90, 126], thus leading to the generation of long (L-OPA1) and short (S-OPA1) protein forms. In physiological conditions, a correct balance between L-OPA1 and S-OPA1 is maintained in order to preserve the mitochondrial network morphology [90, 212]. Following mitochondrial depolarization, L-OPA1 is completely converted in its short form, thus causing the inhibition of the fusion process and mitochondrial fragmentation [212, 225]. Eventually, it has been demonstrated that OPA1 is involved in the protection against apoptosis. Indeed, this protein has a key role in the maintenance of the cristae structure and junctions, thus preventing the release of cytochrome c from

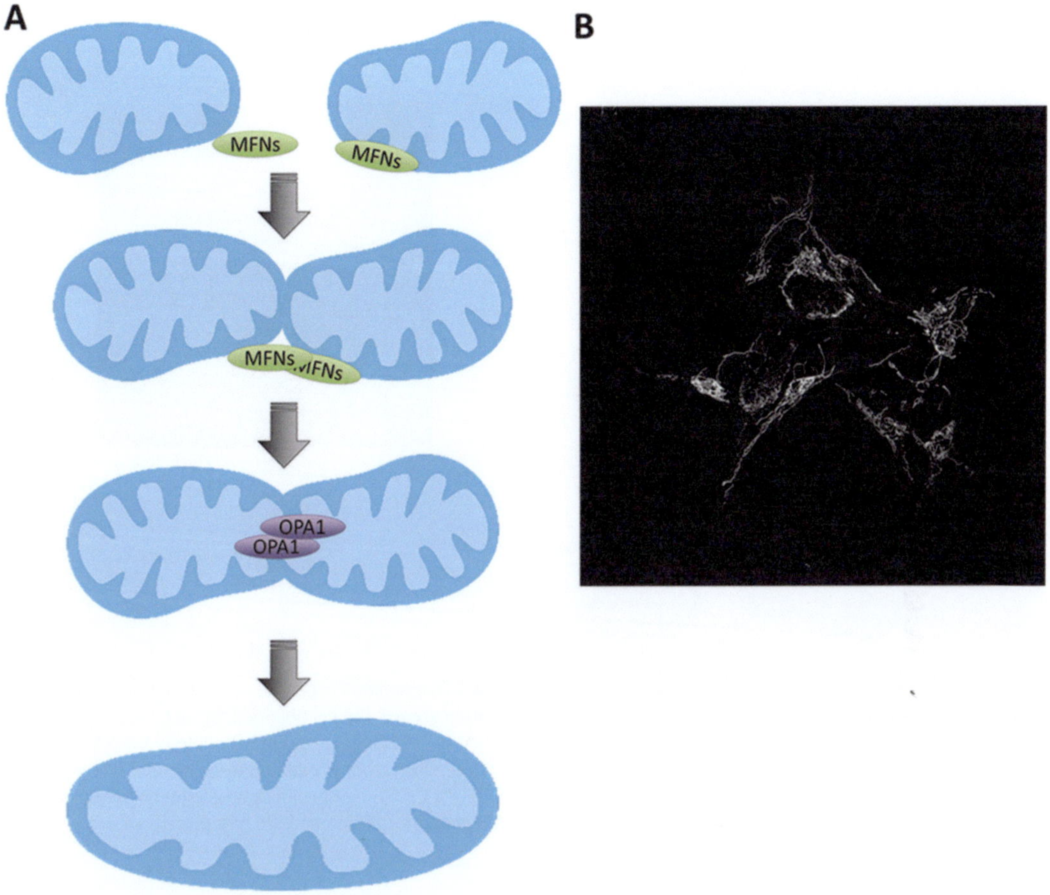


**Fig. 2.3** (**a**) Schematic representation of the mitochondrial fusion process. Both MFN1 and MFN2 are responsible for the fusion of the outer mitochondrial membranes, while OPA1 plays a key role in the fusion of the inner mitochondrial membrane. (**b**) Representative immunofluorescence image of mitochondrial networks with filamentous shape (immunofluorescence image obtained by anti-ATP synthase β antibody staining in undifferentiated SH-SY5Y cells)

mitochondria [116, 185, 258]. In this view, OPA1 knock down leads to Bax translocation, release of cytochrome c and caspase activation [155, 198].

Mitochondrial fission is important to allow the proper cellular distribution of these organelles and the degradation of damaged mitochondria through the mitophagy process [160]. This process is made possible by the cytosolic protein DRP1 that is recruited to the outer mitochondrial membrane thanks to specific post-translational modifications. Once this GTPase protein localizes on mitochondria, it oligomerizes into ring-like structures and constricts mitochondria thanks to the GTP hydrolysis (Fig. 2.4) [88]. The fission process occurs when calcineurin dephosporilates DRP1 protein, while it is prevented when protein kinase A (PKA) phosphorylates the conserved residue Ser-637. Therefore, the phosphorylation status of DRP1 determines its mitochondrial localization even if also sumoylation and ubiquitination have an impact on the fission machinery [160].

The recruitment of DRP1 protein on the outer mitochondrial membrane requires the presence of specific mitochondrial receptors. This is due to the absence in DRP1 of a specific domain involved in the mitochondrial membrane binding [169]. In particular, mitochondrial fission 1 protein

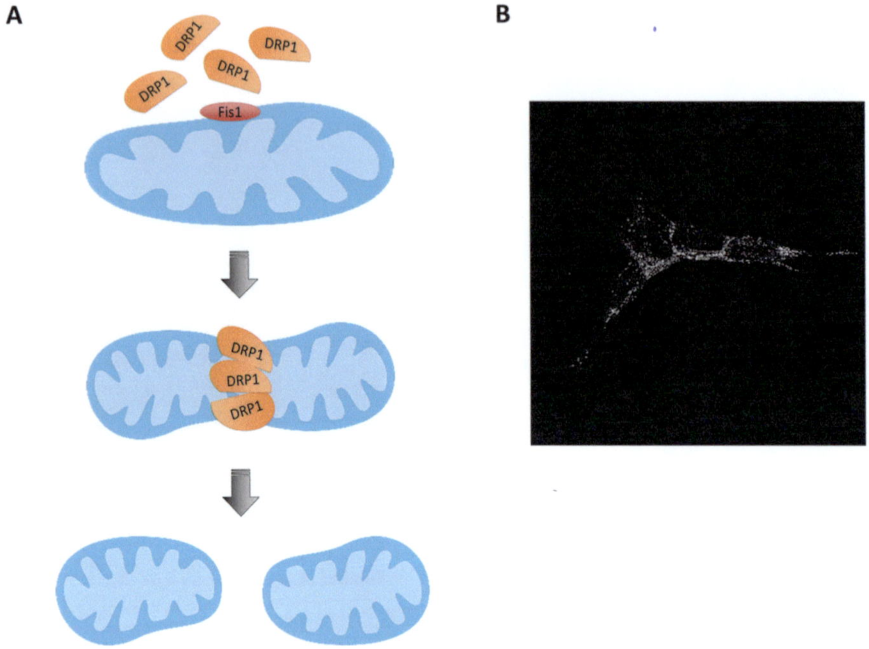


**Fig. 2.4** (**a**) Schematic representation of the mitochondrial fission process. Fis1 is a mitochondrial membrane receptor involved in the recruitment of DRP1 protein. Once DRP1 localizes and oligomerizes on the outer mitochondrial membrane, it constricts mitochondria thanks to the hydrolysis of GTP molecules. (**b**) Representative immunofluorescence image of mitochondrial networks with a fragmented shape, due to fission induction (immunofluorescence image obtained by anti-ATP synthase β antibody staining in undifferentiated SH-SY5Y cells treated with carbonyl cyanide m-chlorophenyl hydrazone (CCCP). CCCP is a protonophore widely used to induce mitochondrial fission and the mitophagic process)

(Fis1) is a protein of the outer mitochondrial membrane that interacts with DRP1, thus forming the fission complex (Fig. 2.4) [104, 118, 142, 252]. Fis1 protein consists of several domains, but the first N-terminal alpha helix seems to be the most important one for the oligomerization of this protein and its fission activity [100]. Other outer mitochondrial membrane proteins are involved in the mitochondrial recruitment of DRP1. For example, mitochondrial fission factor (Mff) protein can recruit DRP1 independently of Fis1 and its overexpression causes mitochondrial fragmentation [161]. Mitochondrial dynamics protein MID49 (MiD49) and mitochondrial dynamics protein MID51 (MiD51) are also involved in the fission process through their capacity to recruit DRP1 on mitochondria [167].

### 2.6.1 Alteration of Mitochondrial Dynamics in PD

Several studies demonstrated that mutations in the genes associated with early-onset PD affect the mitochondrial network morphology. It has been proposed, for example, that PINK1 and Parkin proteins act in a common pathway to modulate mitochondrial shape [56, 180, 251]. Indeed, PINK1/Parkin pathway is pro-fusion as overexpression of PINK1 causes the formation of more elongated and interconnected mitochondria, whereas knockdown of PINK1 increases the mitochondrial fragmentation [125, 254]. Moreover, mitochondrial network morphology is not altered in PARK2-mutated fibroblasts, but mutations in this gene cause the appearance of "chain-like" networks [256].

Strong evidences suggest that LRRK2 protein interacts with DRP1. Indeed, LRRK2 phosphorylates DRP1 thus causing the translocation of DRP1 from the cytosol to mitochondria [153, 242]. In addition, A30P and A53T α-syn mutations have an impact on mitochondrial morphology. Indeed, the increase in the cleavage of OPA1 was observed in M17 cells overexpressing these types of mutated α-syn [77]. A fragmented mitochondrial phenotype was also observed in cells transfected with DJ-1 mutated protein [241]. A recent study demonstrated that dopamine treatment in SH-SY5Y cells, used to reproduce the altered dopamine homeostasis, causes the formation of spheroidal mitochondrial aggregates. In this case, MFN1 protein level was not affected, while both forms of OPA1 were downregulated [12], well recapitulating what happens in the substantia nigra of sporadic PD patients [262].

## 2.7 Mitochondrial Biogenesis

The life cycle of mitochondria includes both the growth, and the consequent division, of preexisting organelles (mitochondrial biogenesis) and the degradation of older or damaged organelles (mitophagy). The balance between these two processes is indispensable to optimize mitochondrial functions. Several energy and growth stimuli control the mitochondrial mass and function.

The mtDNA replication is made possible by the DNA polymerase γ (POLG). This protein is a heterotrimer made by two subunits, the catalytic and the homodimeric accessory subunit [76]. The replication of mtDNA needs also the activity of other replisome components, such as the topoisomerase, the Twinkle mtDNA helicase, the mitochondrial single-stranded DNA-binding protein (mtSSB), and the mitochondrial DNA ligase III [108, 136, 261].

The mtDNA transcription requires a single DNA-directed RNA polymerase (POLRMT), two stimulatory transcription factors (Transcription factor A - Tfam, and dimethyladenosine transferase 1 - TFB) and the transcription termination factor 1 (MTERF1) [11, 63, 193]. Transcription occurs bi-directionally from two different promoters, known as LSP and HSP, which are inside the D-loop regulatory region [22, 149]. The mitochondrial genetic system is regulated by several nuclear-encoded proteins that play a key role in mitochondria: Tfam, for example, is a protein that stimulates mtDNA transcription through specific promoter recognition [60, 149]. It has been demonstrated that Tfam knockout mouse shows embryonic lethality due to oxidative phosphorylation impairment and to a severe reduction of the mtDNA content, thus suggesting the importance of this protein in the mitochondrial biogenesis process [115]. Moreover, it has been observed that transgenic mice overexpressing human Tfam have increased mtDNA copy number [52].

The nuclear respiratory factors NRF1 and NRF2 are two important nuclear transcription factors that are able to bind several promoters and to positively regulate the expression of some key genes involved in various processes. In particular, NRF1 recognizes and activates the promoter of Tfam and TFB genes, thus enabling their transcription [71, 235]. Moreover, NRF1 positively regulates the transcription of other genes that codify for mitochondrial proteins, such as cytochrome c, TOMM20 and COX17 [9, 194, 219]. NRF2 was first identified as the transcriptional activator of the COXIV promoter [192]. NRF2 has also a regulatory function in the expression of all the ten cytocrome oxidase subunits [155]. This finding, obtained by chromatin immunoprecipitation, was further confirmed by the discovery that dominant negative suppression of NRF2 by RNA interference causes a reduced expression of all ten nucleus-encoded COX subunits [156]. Moreover, NRF2 is a transcriptional activator of other important genes, including Tfam, TFB1M and TFB2M and three different succinate dehydrogenase (complex II) subunit genes [3, 54, 72, 86, 235].

Estrogen-related receptor (ERR) α, ERRβ, and ERRγ are other important receptors that are involved in the positive transcriptional regulation of several nuclear genes encoding mitochondrial proteins involved in fatty-acid oxidation (FAO),

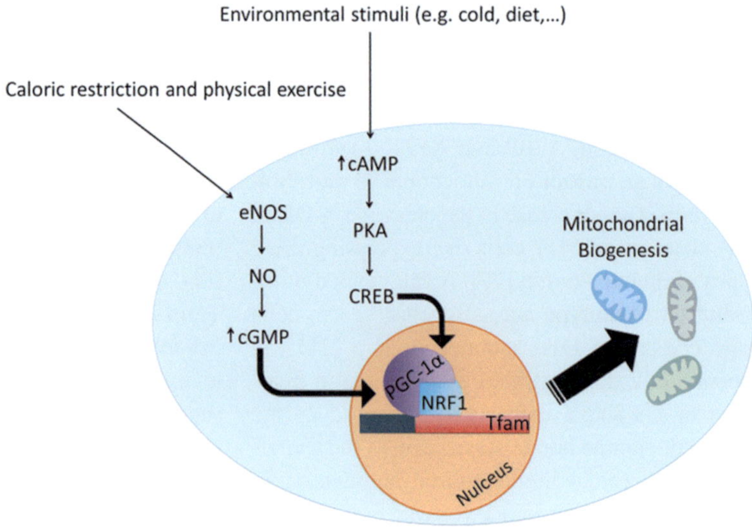


**Fig. 2.5** Schematic representation of the mitochondrial biogenetic process. Both cGMP and cAMP mediate the increased expression of PGC-1α, which in turn positively regulates the transcription of Tfam. Eventually, Tfam stimulates the mtDNA transcription through specific promoter recognition

tricarboxylic acid (TCA) cycle and respiratory chain [50].

Eventually, the peroxisome proliferator-activated receptors (PPARα, PPARβ and PPARγ) are involved in the transcriptional control of several mitochondrial FAO enzymes, thus suggesting a role for these nuclear receptors in the regulation of the mitochondrial metabolism [46].

PGC-1α is an important transcriptional coactivator that plays a key role in the mitochondrial biogenesis process. Indeed, PGC-1α is able to bind both NRF1 and NRF2 proteins, thus activating the transcription of their target genes (e.g. Tfam, TFB1M and TFB2M) [72, 246]. This process leads to the increase in mitochondrial genes transcription and to the expression of proteins of the respiratory chain, thus positively influencing the mitochondrial functions. Moreover, it has been demonstrated that PGC-1α controls the mitochondrial density in all neuronal subtypes [243].

The c-AMP and c-GMP dependent signaling are able to induce the transcription of PGC-1α at physiological level. When cAMP cellular levels are elevated, PKA phosphorylates the cyclic AMP-responsive element-binding protein (CREB) leading to its activation. CREB, in turn, induces the expression of PGC-1α, thus increasing mitochondrial biogenesis [83, 183]. Moreover, c-GMP dependent signaling, activated following an increase of nitric oxide cellular levels, induces the mitochondrial biogenesis process through the transcpritional activation of PGC-1α along with Tfam and NRF-1 (Fig. 2.5) [150, 151].

Recently, the involvement of the PINK1/Parkin-mediated mitophagy pathway has been also related to the activation of mitochondrial biogenesis. Following the activation of this mitophagy pathway and upon oxidative stress, NRF2 is able to translocate to the nucleus, thus inducing the activation of the mitochondrial biogenesis and the transcription of genes with an antioxidant response element in the regulatory sequence. Also the activation of TFEB and PGC-1α upon mitophagy induction can contribute to the increased mitochondrial biogenesis [92]. The involvement of Parkin protein in the mitochondrial biogenesis process was previously suggested also from Kuroda and colleagues. Indeed, they found that overexpression of Parkin protein in proliferating cells was associated with increased mtDNA transcription and replication through Tfam activity, wheras this process was suppressed by Parkin siRNA [111].

### 2.7.1 Altered Mitochondrial Biogenesis and PD

Sporadic PD is associated with the presence of mtDNA deletions. These mtDNA deletions seems to be due to high ROS levels in the dopaminergic neurons of the substantia nigra [44]. Moreover, mice with a conditional knockout of Tfam show both lower mtDNA expression, respiratory chain impairment and slowly progressive levodopa-responsive motor deficits, thus suggesting that an impairment in the mtDNA expression may be a leading event in PD pathogenesis [53]. It has been demonstrated that mutations in the gene that encodes for the mitochondrial protein POLG, which is responsible for the synthesis and the proofreading of mtDNA, is associated with the development of a more complex syndrome, characterized by levodopa-responsive parkinsonism [39, 124].

The mitochondrial dysfunctions that occur in α-syn-transfected cells or in transgenic mice expressing A53T α-syn can be rescued by the overexpression of PGC-1α. Indeed, this protein is able to increase the mitochondrial mass, thus compensating the mitochondrial dysfunctions typical of PD [243]. Downstream targets of PGC-1α (e.g., Tfam, NRF1 and NRF2) were observed to be downregulated in PARK2 patients, confirming the important role of Parkin in the biogenetic process [164]. The same conclusion can be driven by the observation that the major repressor of PGC-1α, PARIS, accumulates in Parkin knockout mice and in PD brains [79, 207]. Eventually, it has been proposed that the mitochondrial protease HtrA2/OMI is involved in the regulation of the mitochondrial biogenetic process. Indeed, the loss of HtrA2/OMI protease activity causes the downregulation of PGC-1α and increased expression of the GSK3β kinase, which is responsible for PGC-1α degradation [249].

## 2.8 Mitophagy

Mitophagy is an important quality control process that allows the recognition and the consequent lysosomal degradation of damaged mitochondria [253]. The improper mitochondrial disposal causes the accumulation of dysfunctional mitochondria [49, 220] that can lead to cell death [75], the activation of inflammation [76] and participate in disease pathogenesis [20, 91, 236].

The most important and studied mitophagy mechanism is the PINK1/Parkin pathway. Under basal conditions, the precursor of PINK1 is synthesized in the cytosol and is imported into the OMM via translocase of outer mitochondrial membrane complex (TOM). Full-length PINK1 is further transferred into the inner membrane of mitochondria through the translocase of inner mitochondrial membrane (TIM) complex in a membrane potential dependent manner. Here, PINK1 is processed by the mitochondrial processing peptidase (MPP), that cleaves the MTS sequence, resulting in a ~60-kDa PINK1 form. This MPP-cleaved form of PINK1, which spans the inner mitochondrial membrane, is then further cleaved by PARL, to give rise to the 52-kDa mature form. PINK1 is then degraded through the proteasome (Fig. 2.6). This pathway keeps endogenous PINK1 levels very low in polarized mitochondria to prevent mitophagy of healthy mitochondria [51, 99, 131].

However, when mitochondria are damaged and lose their mitochondrial membrane potential, the PINK1 import and its consequent proteasomal degradation are inactivated, thus causing the accumulation of this protein onto the outer mitochondrial membrane. This process leads to the activation and the recruitment of the E3-ubiquitin ligase Parkin to the outer mitochondrial membrane, where it catalyzes the covalent attachment of ubiquitin moieties onto specific mitochondrial proteins [147, 176, 191]. The recruitment of Parkin to mitochondria depends on the phosphorylation of MFN2 by PINK1 [25] and on VDAC proteins [68, 217]. Moreover, the activation of its E3 ligase activity relies on the phosphorylation of the ubiquitin-like domain by PINK1 [158, 203]. Once the OMM proteins are ubiquitinated, several autophagy receptors are recruited to damage mitochondria in order to bind the ubiquitin-tagged outer mitochondrial membrane proteins. In particular, p62/SQSTM1 is an autophagy receptor involved in the mitochondrial clustering during

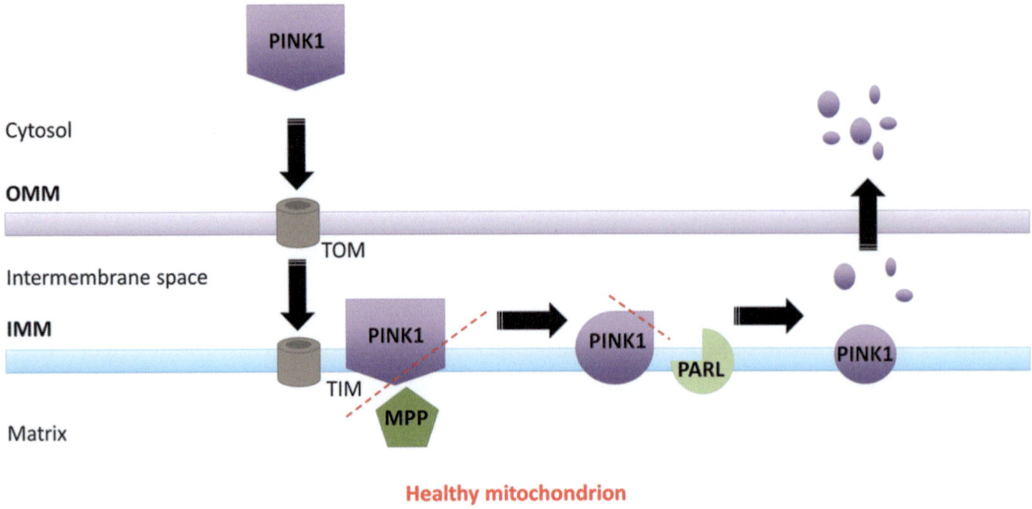


**Fig. 2.6** Schematic representation of PINK1 degradation in healthy mitochondria. In healthy mitochondria, the precursor of PINK1 (64 kDa) is imported into the outer mitochondrial membrane via TOM and further transferred into the inner mitochondrial membrane through TIM complex in a membrane potential dependent manner. Here, 64-kDa PINK1 is processed by the MPP, that remove the MTS sequence. This form of PINK1 is then cleaved by PARL, to give rise to the 52-kDa mature form, subsequently degraded by other mitochondrial peptidases and by the proteasome

mitophagy and it is required for mitochondrial disposal (Fig. 2.7) [68, 146, 154].

Moreover, Parkin protein is able to bind AMBRA1, an autophagy-promoting protein that determines the local formation of autophagosomes, thus causing the PINK1/Parkin-mediated mitophagy [59, 213, 228].

An impairment of PINK1/Parkin mitophagy axis could lead to the accumulation of dysfunctional mitochondria that may contribute to dopaminergic cell death by an increased production of ROS and an enhanced release of mitochondrial apoptogenic factors [58, 233].

### 2.8.1 Impaired Mitophagy and PD

Giving the important role of PINK1 and Parkin in the mitophagic process, it appears clear that mutations in the genes that encodes for these proteins cause an alteration of the correct mitochondrial disposal and the accumulation of damaged organelles, thus leading to dopaminergic cell death [70, 224, 259]. Moreover, as described above, loss of function of PINK1 or Parkin determines several alterations of mitochondrial functions and dynamics.

Recently, it has been demonstrated that mitophagy impairment is not only a hallmark of PARK2 and PARK6 PD pathogenesis, but it is also a common feature of the sporadic form and of the toxins-induced models of PD. Indeed, the specific inhibition of complex I by $MPP^{+}$ induces mitophagy dysfunction through a specific pathway, linked to BNIP3L degradation, decrease in protein ubiquitination and p62 inactivation [64, 148], thus suggesting that $MPP^{+}$ causes the impairment of the ubiquitin-proteasome system and of the autophagic pathway. Moreover, a cellular model of altered dopamine homeostasis revealed that dopamine oxidation causes the lack of PINK1 accumulation and a failed recruitment of Parkin onto outer mitochondrial membrane [12], again recapitulating what happens in the substantia nigra of sporadic PD patients [262] . Therefore, these two treatments cause mitophagy alteration and accumulation of damaged mitochondria, a process that may lead to neuronal cell death.

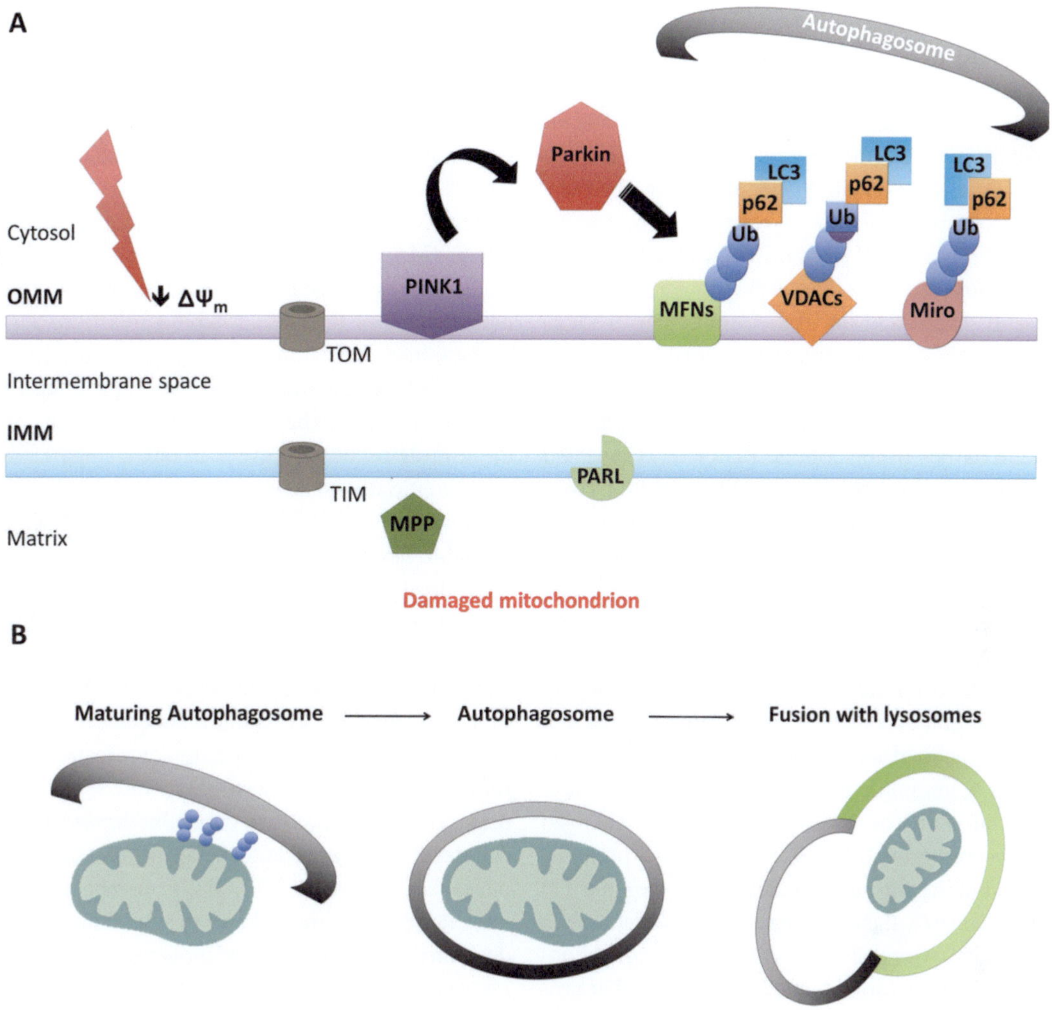


**Fig. 2.7** (**a**) Schematic representation of PINK1/Parkin mitophagy pathway. When mitochondria are depolarized, PINK1 accumulates onto the outer mitochondrial membrane, thus recruiting Parkin. Parkin catalyzes the covalent attachment of ubiquitin moieties onto specific mitochondrial proteins. This process leads to the selective engulfment of damaged organelles by the autophagosome. (**b**) The autophagosome, with its cargo, fuses with a lysosome, forming the acidic autophagolysosome. Within this compartment, mitochondria and proteins are degraded by lysosomal hydrolysis

As already described, GCase is an important protein involved in the autophagy-lysosome pathway. Loss of GCase activity leads to mitochondrial dysfunctions, fragmentation of mitochondrial network and increased ROS production. Therefore, GCase loss of function results in the accumulation of damaged macromolecules and dysfunctional mitochondria, which in turn cause the activation of apoptosis [66].

## 2.9 Proteomics as a Global and Unbiased Approach to Study Mitochondrial Proteome in PD

PD is a multifactorial disorder whose exact etiology remains unknown. Given the involvement of several pathogenic mechanisms in the development of PD, such as mitochondrial impairment, oxidative and nitrative stress, and autophagy dysfunction, the identification of specific molecular factors and pathways altered by the disease would help in the understanding of its etiology and in identifying new therapeutic targets. To achieve this goal, the –OMICS techniques are particularly suitable to obtain an overview of molecular alterations induced by multifactorial pathological conditions. In particular, proteomics represents a useful global and integrative approach to focus on proteins as the emergent properties of the system.

Proteomics analysis performed on human brain tissues from sporadic PD patients revealed an increased expression of mitochondrial complex III and ATP synthase. Since the mitochondrial activity is compromised in PD, their increased expression suggests a possible compensatory mechanism [4]. A proteomic study conducted on mitochondrial-enriched fractions obtained from substantia nigra of sporadic PD patients and of control subjects demonstrated that 119 mitochondrial proteins linked to signal transduction, ubiquitin proteasomal system, or regulation of oxidative stress quantitatively changed between the two groups. Among these proteins, mortalin significantly decreased in sporadic PD patients [97], thus causing a lower resistance to oxidative stress and the consequent mitochondrial and proteasomal dysfunctions. Another proteomics study on substantia nigra specimens from PD patients suggested a major vulnerability against neurotoxins as a possible pathogenetic mechanism. Indeed, a protein involved in the detoxification of aldehydes, the cytosolic aldehyde dehydrogenase (ALDH1A1), was found to be reduced in PD patients. This result can explain the up-regulation of annexin V in idiopathic PD patients, a protein involved in the apoptotic pathway. In the same study β-tubulin cofactor A and co-actosin-like protein were increased in human brain tissues of PD patients, witnessing a cytoskeletal rearrangement [244].

Since altered dopamine homeostasis might be a key factor in the early steps of PD pathogenesis, several proteomics studies were conducted on cellular models of dopamine toxicity in order to understand the tight relation between dopamine oxidation and mitochondrial dysfunctions. In a study conducted by Van Laar and colleagues, it has been demonstrated that the exposure of isolated mitochondria to reactive dopamine quinone caused the dysregulation of several mitochondrial proteins. In particular, mitochondrial creatine kinase (MtCK) and mitofilin levels were significantly decreased. Since MtCK is involved in the maintenance of the cellular ATP equilibrium and of the mitochondrial morphology, and mitofilin is responsible for the maintenance of mitochondrial cristae structure, the decreased expression of these proteins might have a negative impact on mitochondrial structure and function. Also the 75 KDa subunit of NADH dehydrogenase, VDAC2, mortalin, and superoxide dismutase 2 (SOD2) were drastically reduced in these mitochondria. The altered proteins expression revealed by this study suggested that dopamine oxidation might have a detrimental effect on several mitochondrial functions, such as structural maintenance, transport, and metabolism [229]. Another research performed on mitochondria isolated from SH-SY5Y cells revealed that altered dopamine homeostasis can deeply affect the mitochondrial proteome. In particular, several HSP70 family members were altered by dopamine. Moreover, the same study described that mitochondrial proteome alterations induced by $MPP^+$ toxin and compared the different pattern of changes with respect to dopamine. Indeed, 27 proteins were altered only by the selective inhibition of complex I by $MPP^+$. Therefore, this study suggested that both treatments used to reproduce some PD molecular mechanisms have a general negative impact on the mitochondrial proteome, although altering different key pathways [1].

Several proteomics studies were conducted on the MPTP-induced model of PD. One of these researches demonstrated that more than 100 proteins showed a different expression profiles in the substantia nigra of mice treated chronically with MPTP when compared to control mice. For example, mortalin is differently regulated in treated mice [96], thus confirming that the activity of this chaperone is altered in different PD models. Another study conducted on the MPTP-induced PD mouse model revealed that 32 proteins associated with mitochondrial functions were differently expressed in striatum and other brain regions after MPTP exposure. In particular, several of these identified proteins belonged to the mitochondrial ETC, thus suggesting that the lower complex I activity and the lower ATP production are due to the reduced expression of these subunits [260]. Recently, it was conducted a meta-analysis of the literature that took into account all the proteomics studies conducted on the $MPP^+$-induced model of PD. In this way, it was possible to identify the molecular pathways affected by the selective inhibition of complex I, increasing the power of every single study and considering results as a whole. The main pathways affected were the synapse functionality, the production of ATP, the mitochondrial unfolded stress response, the autophagic and the apoptotic pathways. Moreover, this analysis showed the central role of HSP60 protein in counteracting $MPP^+$ toxicity [138].

Rotenone is also able to alter the mitochondrial proteome. Indeed, a research performed by Jin and colleagues on mitochondrial-enriched fractions obtained from MES cells evidenced that rotenone quantitatively changed the expression of 110 mitochondrial proteins. Moreover, the functional classification of these proteins highlighted their involvement in several important molecular processes, i.e., folding degradation stability, metabolism, morphology, respiratory chain, protein synthesis, signaling, and transport [98], thus suggesting that this mitochondrial toxin damages the cell at different molecular levels. A similar outcome was obtained also for paraquat. Indeed, this mitochondrial toxin induced a change in the expression of several proteins involved in different molecular pathways, i.e., cytoskeleton organization, redox signaling, and mitochondrial function and metabolism [119].

These results suggest that these toxins (i.e. $MPP^+$, rotenone, and paraquat) deeply affect the cellular functions at various levels, acting directly on mitochondria. Therefore, the selective mitochondrial damage caused by these toxins has several secondary effects on cells due to the central role of mitochondria in influencing cell fate.

Regarding the genetic forms of the disease, several proteomics studies were conducted in order to highlight the molecular mechanisms altered by mutations in specific PD related genes. It has been demonstrated that the loss of PINK1 protein causes several mitochondrial dysfunctions. Indeed, the expression levels of several ETC complex subunits, numerous enzymes of the glycolytic pathway and proteins involved in the mitochondrial dynamics and trafficking were affected in PINK1 knockout rats [234]. Moreover, evaluation of the brain proteome and phosphoproteome in PINK1-deficient mice demonstrated that the expression of 23 proteins was altered by the loss of PINK1 (e.g., V-type proton ATPase subunit F, heat-shock-related 70 kDa protein 2, malate dehydrogenase and proteasome subunit alpha type-2) [223]. These proteomics differences determined a negative impact on several molecular pathways, for example energy production, mitochondrial metabolism and protein homeostasis. Since, as described above, PINK1 plays a key role in the mitophagy process by targeting the damaged mitochondria for the subsequent degradation [107, 147], it appears clear that the loss of function of this protein may determine the accumulation of impaired organelles that contribute to higher ROS production and activation of the apoptotic pathway.

The cellular role of Parkin was also largely studied and described using the proteomics approach. For example, Martinez and colleagues used a Drosophila model that co-expressed biotinylated ubiquitin together with FLAG-tagged Parkin. In this way, it was possible to identify 37 Parkin substrates, among which several mitochondrial proteins (e.g. Cisd2/CISD1,2, VDAC1,2,3 and Tom70) [130], thus confirming its role in mitochondrial disposal. However, many identified Parkin substrates were cytosolic and endosomal proteins. Therefore, Parkin has broader roles beyond the mitophagic pathway [130]. A research conducted on the SH-SY5Y cell line expressing a mutant form of Parkin (i.e., two missense mutations, Q311R and A371T) revealed that this protein was subjected to post-translational modifications, without affecting its stability and localization. The 2D-DIGE analysis highlighted also the proteome alterations of these cells. In particular, nine proteins were differently expressed in Parkin mutant cells, such as UCHL-1, a protein involved both in the processing of ubiquitin precursors and of ubiquitinated proteins, and 14-3-3, which is involved in the regulation of a large spectrum of both general and specialized signaling pathways [163]. Moreover, studies conducted on Parkin knockout mice revealed a general impairment of the mitochondrial function, antioxidant system, energy metabolism, protein folding and degradation, signal transduction pathways, and vesicle trafficking [165, 173]. A different expression of proteins involved in the stress response, redox balance and protein processing were also found in skin fibroblasts from PARK2-mutated patients [122].

Eventually, the proteomics approach was also used to characterize the dominant forms of PD. Regarding α-syn protein, the A53T mutation caused a different expression of 24 proteins associated with different cellular components, i.e., membrane, endoplasmic reticulum, cytoskeleton, mitochondria, and ribosome [250]. In particular, this paper revealed that two mitochondrial proteins were increased in A53T α-syn flies, i.e., Mn-SOD, an antioxidant enzyme, and β-ATPase, which plays an important role in the production of ATP. The increased expression of Mn-SOD might be a compensatory mechanism in order to counteract the high levels of ROS, while the alterations in β-ATPase expression might represent a malfunction of the ATP synthase enzyme [250]. Another study conducted on differentiated SH-SY5Y cells overexpressing WT α-syn revealed that this protein altered the expression of 17 mitochondrial proteins. Among these, the most interesting were the increased levels of ATP synthase subunits and the decreased levels of NDUFS1, DLAT, an enzyme that links the glycolytic pathway to the tricarboxylic cycle, and HSP70, a molecular chaperone [170]. Eventually, a study conducted on the phosphorylated form of α-syn, which is associated to PD, revealed a several number of proteins interacting with this post-translational modified α-syn, such as cytoskeletal proteins, proteins involved in synaptic vesicle endocytosis, and chaperone proteins [134].

Since PD is a multifactorial disorder characterized by the alteration of several molecular pathways, proteomics approaches represent a useful method to fill the gaps in the field of PD research. Indeed, proteomics is able to give a global vision of the dysfunctions that occur in organelles, cells, or tissues. Moreover, systems biology approach can merge the proteomics data and, in this way, highlight the molecular pathways altered by the disease.

In conclusion, mitochondria may be considered the central hub of the PD pathogenesis. Indeed, both sporadic and genetic forms of the disease and the toxins-induced PD models are related to mitochondrial impairment and altered mitochondrial disposal. In all cases, mitochondria are damaged at various levels (e.g. structure, energy production, and metabolism) and the permanence of these dysfunctional organelles due to the mitophagy impairment cause the alteration of several molecular pathways, which in turn aggravates the mitochondrial status. This process leads to the activation of the apoptotic pathway, thus causing neurodegeneration and the consequent onset of the typical motor symptoms.

## References

1. Alberio T, Bondi H, Colombo F et al (2014) Mitochondrial proteomics investigation of a cellular model of impaired dopamine homeostasis, an early step in Parkinson's disease pathogenesis. Mol Biosyst 10:1332–1344. https://doi.org/10.1039/c3mb70611g
2. Ascherio A, Schwarzschild MA (2016) The epidemiology of Parkinson's disease: risk factors and prevention. Lancet Neurol 15:1257–1272. https://doi.org/10.1016/S1474-4422(16)30230-7
3. Au HC, Scheffler IE (1998) Promoter analysis of the human succinate dehydrogenase iron-protein gene. Both nuclear respiratory factors NRF-1 and NRF-2 are required. Eur J Biochem 251:164–174
4. Basso M, Giraudo S, Corpillo D et al (2004) Proteome analysis of human substantia nigra in Parkinson's disease. Proteomics 4:3943–3952. https://doi.org/10.1002/pmic.200400848
5. Bekris LM, Mata IF, Zabetian CP (2010) The genetics of Parkinson disease. J Geriatr Psychiatry Neurol 23:228–242. https://doi.org/10.1177/0891988710383572

6. Bender A, Krishnan KJ, Morris CM et al (2006) High levels of mitochondrial DNA deletions in substantia nigra neurons in aging and Parkinson disease. Nat Genet 38:515–517. https://doi.org/10.1038/ng1769
7. Benskey MJ, Perez RG, Manfredsson FP (2016) The contribution of alpha synuclein to neuronal survival and function - Implications for Parkinson's disease. J Neurochem 137:331–359. https://doi.org/10.1111/jnc.13570
8. Betarbet R, Sherer TB, MacKenzie G et al (2000) Chronic systemic pesticide exposure reproduces features of Parkinson's disease. Nat Neurosci 3:1301–1306. https://doi.org/10.1038/81834
9. Blesa JR, Prieto-Ruiz JA, Hernandez JM et al (2007) NRF-2 transcription factor is required for human TOMM20 gene expression. Gene 391:198–208. https://doi.org/10.1016/j.gene.2006.12.024
10. Blin O, Desnuelle C, Rascol O et al (1994) Mitochondrial respiratory failure in skeletal muscle from patients with Parkinson's disease and multiple system atrophy. J Neurol Sci 125:95–101
11. Bonawitz ND, Clayton DA, Shadel GS (2006) Initiation and beyond: multiple functions of the human mitochondrial transcription machinery. Mol Cell 24:813–825. https://doi.org/10.1016/j.molcel.2006.11.024
12. Bondi H, Zilocchi M, Mare MG et al (2016) Dopamine induces mitochondrial depolarization without activating PINK1-mediated mitophagy. J Neurochem 136:1219–1231. https://doi.org/10.1111/jnc.13506
13. Bonifati V, Rizzu P, van Baren MJ et al (2003) Mutations in the DJ-1 gene associated with autosomal recessive early-onset parkinsonism. Science 299:256–259. https://doi.org/10.1126/science.1077209
14. Bouzier-Sore AK, Bolanos JP (2015) Uncertainties in pentose-phosphate pathway flux assessment underestimate its contribution to neuronal glucose consumption: relevance for neurodegeneration and aging. Front. Aging Neurosci 7:89. https://doi.org/10.3389/fnagi.2015.00089
15. Brooks AI, Chadwick CA, Gelbard HA et al (1999) Paraquat elicited neurobehavioral syndrome caused by dopaminergic neuron loss. Brain Res 823:1–10
16. Burte F, De Girolamo LA, Hargreaves AJ et al (2011) Alterations in the mitochondrial proteome of neuroblastoma cells in response to complex 1 inhibition. J Proteome Res 10:1974–1986. https://doi.org/10.1021/pr101211k
17. Cali T, Ottolini D, Brini M (2013) Calcium and endoplasmic reticulum-mitochondria tethering in neurodegeneration. DNA Cell Biol 32:140–146. https://doi.org/10.1089/dna.2013.2011
18. Canet-Aviles RM, Wilson MA, Miller DW et al (2004) The Parkinson's disease protein DJ-1 is neuroprotective due to cysteine-sulfinic acid-driven mitochondrial localization. Proc. Natl. Acad. Sci. USA 101:9103–9108. https://doi.org/10.1073/pnas.0402959101
19. Chan DC (2006a) Dissecting mitochondrial fusion. Dev Cell 11:592–594. https://doi.org/10.1016/j.devcel.2006.10.009
20. Chan DC (2006b) Mitochondria: dynamic organelles in disease, aging, and development. Cell 125:1241–1252. https://doi.org/10.1016/j.cell.2006.06.010
21. Chan P, DeLanney LE, Irwin I et al (1991) Rapid ATP loss caused by 1-methyl-4-phenyl-1,2,3,6-tetrahydropyridine in mouse brain. J Neurochem 57:348–351
22. Chang DD, Clayton DA (1985) Priming of human mitochondrial DNA replication occurs at the light-strand promoter. Proc Natl Acad Sci U S A. 82:351–355
23. Chen H, Chan DC (2005) Emerging functions of mammalian mitochondrial fusion and fission. Hum Mol Genet 14 Spec 2:R283–R289. https://doi.org/10.1093/hmg/ddi270
24. Chen H, Chan DC (2006) Critical dependence of neurons on mitochondrial dynamics. Curr Opin Cell Biol 18:453–459. https://doi.org/10.1016/j.ceb.2006.06.004
25. Chen Y, Dorn GW 2nd (2013) PINK1-phosphorylated mitofusin 2 is a Parkin receptor for culling damaged mitochondria. Science 340:471–475. https://doi.org/10.1126/science.1231031
26. Chen H, Detmer SA, Ewald AJ et al (2003) Mitofusins Mfn1 and Mfn2 coordinately regulate mitochondrial fusion and are essential for embryonic development. J Cell Biol 160:189–200. https://doi.org/10.1083/jcb.200211046
27. Chen H, Chomyn A, Chan DC (2005) Disruption of fusion results in mitochondrial heterogeneity and dysfunction. J Biol Chem 280:26185–26192. https://doi.org/10.1074/jbc.M503062200
28. Chen H, McCaffery JM, Chan DC (2007) Mitochondrial fusion protects against neurodegeneration in the cerebellum. Cell 130:548–562. https://doi.org/10.1016/j.cell.2007.06.026
29. Chiueh CC, Markey SP, Burns RS et al (1984) Neurochemical and behavioral effects of 1-methyl-4-phenyl-1,2,3,6-tet- rahydropyridine (MPTP) in rat, guinea pig, and monkey. Psychopharmacol Bull 20:548–553
30. Choi BK, Kim JY, Cha MY et al (2015) beta-Amyloid and alpha-synuclein cooperate to block SNARE-dependent vesicle fusion. Biochemistry. 54:1831–1840. https://doi.org/10.1021/acs.biochem.5b00087
31. Chou AP, Li S, Fitzmaurice AG et al (2010) Mechanisms of rotenone-induced proteasome inhibition. Neurotoxicology 31:367–372. https://doi.org/10.1016/j.neuro.2010.04.006
32. Chu Y, Dodiya H, Aebischer P et al (2009) Alterations in lysosomal and proteasomal markers in Parkinson's disease: relationship to alpha-synuclein inclusions. Neurobiol Dis. 35:385–398. https://doi.org/10.1016/j.nbd.2009.05.023
33. Chu Y, Morfini GA, Langhamer LB et al (2012) Alterations in axonal transport motor proteins in sporadic and experimental Parkinson's disease.

Brain 135:2058–2073. https://doi.org/10.1093/brain/aws133
34. Chu Y, Goldman JG, Kelly L et al (2014) Abnormal alpha-synuclein reduces nigral voltage-dependent anion channel 1 in sporadic and experimental Parkinson's disease. Neurobiol Dis. 69:1–14. https://doi.org/10.1016/j.nbd.2014.05.003
35. Cipolat S, Rudka T, Hartmann D et al (2006) Mitochondrial rhomboid PARL regulates cytochrome crelease during apoptosis via OPA1-dependent cristae remodeling. Cell 126:163–175. https://doi.org/10.1016/j.cell.2006.06.021
36. Cleeter MW, Chau KY, Gluck C et al (2013) Glucocerebrosidase inhibition causes mitochondrial dysfunction and free radical damage. Neurochem Int 62:1–7. https://doi.org/10.1016/j.neuint.2012.10.010
37. Cookson MR (2010) The role of leucine-rich repeat kinase 2 (LRRK2) in Parkinson's disease. Nat Rev Neurosci 11:791–797. https://doi.org/10.1038/nrn2935
38. Dauer W, Przedborski S (2003) Parkinson's disease: mechanisms and models. Neuron 39:889–909
39. Davidzon G, Greene P, Mancuso M et al (2006) Early-onset familial parkinsonism due to POLG mutations. Ann Neurol 59:859–862. https://doi.org/10.1002/ana.20831
40. Davis GC, Williams AC, Markey SP et al (1979) Chronic Parkinsonism secondary to intravenous injection of meperidine analogues. Psychiatry Res 1:249–254
41. de Brito OM, Scorrano L (2008) Mitofusin 2 tethers endoplasmic reticulum to mitochondria. Nature 456:605–610. https://doi.org/10.1038/nature07534
42. Detmer SA, Chan DC (2007) Complementation between mouse Mfn1 and Mfn2 protects mitochondrial fusion defects caused by CMT2A disease mutations. J Cell Biol 176:405–414. https://doi.org/10.1083/jcb.200611080
43. Di Fonzo A, Rohé CF, Ferreira J et al (2005) A frequent LRRK2 gene mutation associated with autosomal dominant Parkinson's disease. Lancet. 365:412–415. https://doi.org/10.1016/S0140-6736(05)17829-5
44. Dias V, Junn E, Mouradian MM (2013) The role of oxidative stress in Parkinson's disease. J Parkinsons Dis 3:461–491. https://doi.org/10.3233/JPD-130230
45. Dimmer KS, Scorrano L (2006) (De)constructing mitochondria: what for? Physiology (Bethesda). 21:233–241. https://doi.org/10.1152/physiol.00010.2006
46. Djouadi F, Bastin J (2008) PPARs as therapeutic targets for correction of inborn mitochondrial fatty acid oxidation disorders. J Inherit Metab Dis 31:217–225. https://doi.org/10.1007/s10545-008-0844-7
47. Dunn L, Allen GF, Mamais A et al (2014) Dysregulation of glucose metabolism is an early event in sporadic Parkinson's disease. Neurobiol Aging 35:1111–1115. https://doi.org/10.1016/j.neurobiolaging.2013.11.001
48. Duty S, Jenner P (2011) Animal models of Parkinson's disease: a source of noveltreatments and clues to the cause of the disease. Br J Pharmacol 164:1357–1391. https://doi.org/10.1111/j.1476-5381.2011.01426.x
49. Egan DF, Shackelford DB, Mihaylova MM et al (2011) Phosphorylation of ULK1 (hATG1) by AMP-activated protein kinase connects energy sensing to mitophagy. Science 331:456–461. https://doi.org/10.1126/science.1196371
50. Eichner LJ, Giguère V (2011) Estrogen related receptors (ERRs): a new dawn in transcriptional control of mitochondrial gene networks. Mitochondrion 11:544–552. https://doi.org/10.1016/j.mito.2011.03.121
51. Eiyama A, Okamoto K (2015) PINK1/Parkin-mediated mitophagy in mammalian cells. Curr Opin Cell Biol 33:95–101. https://doi.org/10.1016/j.ceb.2015.01.002
52. Ekstrand MI, Falkenberg M, Rantanen A et al (2004) Mitochondrial transcription factor A regulates mtDNA copy number in mammals. Hum Mol Genet 13:935–944. https://doi.org/10.1093/hmg/ddh109
53. Ekstrand MI, Terzioglu M, Galter D et al (2007) Progressive parkinsonism in mice with respiratory-chain-deficient dopamine neurons. Proc Natl Acad Sci U S A 104:1325–1330. https://doi.org/10.1073/pnas.0605208103
54. Elbehti-Green A, Au HC, Mascarello JT et al (1998) Characterization of the human SDHC gene encoding of the integral membrane proteins of succinate-quinone oxidoreductase in mitochondria. Gene 213:133–140
55. Eura Y, Ishihara N, Yokota S et al (2003) Two mitofusin proteins, mammalian homologues of FZO, with distinct functions are both required for mitochondrial fusion. J Biochem 134:333–344
56. Exner N, Treske B, Paquet D et al (2007) Loss-of-function of human PINK1 results in mitochondrial pathology and can be rescued by parkin. J Neurosci. 27:12413–12418. https://doi.org/10.1523/JNEUROSCI.0719-07.2007
57. Fabre E, Monserrat J, Herrero A et al (1999) Effect of MPTP on brain mitochondrial H2O2 and ATP production and on dopamine and DOPAC in the striatum. J Physiol Biochem 55:325–331
58. Fernández-Moriano C, González-Burgos E, Gómez-Serranillos MP (2015) Mitochondria-Targeted Protective Compounds in Parkinson's and Alzheimer's Diseases. Oxid Med Cell Longev. 2015:408927. https://doi.org/10.1155/2015/408927
59. Fimia GM, Stoykova A, Romagnoli A et al (2007) Ambra1 regulates autophagy and development of the nervous system. Nature 447:1121–1125. https://doi.org/10.1038/nature05925
60. Fisher RP, Clayton DA (1988) Purification and characterization of human mitochondrial transcription factor 1. Mol Cell Biol 8:3496–3509

61. Fiskum G, Starkov A, Polster BM et al (2003) Mitochondrial mechanisms of neural cell death and neuroprotective interventions in Parkinson's disease. Ann N Y Acad Sci 991:111–119
62. Forno LS, DeLanney LE, Irwin I et al (1993) Similarities and differences between MPTP-induced parkinsonsim and Parkinson's disease. Neuropathologic considerations. Adv Neurol 60:600–608
63. Fusté JM, Wanrooij S, Jemt E et al (2010) Mitochondrial RNA polymerase is needed for activation of the origin of light-strand DNA replication. Mol Cell 37:67–78. https://doi.org/10.1016/j.molcel.2009.12.021
64. Gao F, Chen D, Si J et al (2015) The mitochondrial protein BNIP3L is the substrate of PARK2 and mediates mitophagy in PINK1/PARK2 pathway. Hum Mol Genet 24:2528–2538. https://doi.org/10.1093/hmg/ddv017
65. Gasser T (2009) Molecular pathogenesis of Parkinson disease: insights from genetic studies. Expert Rev Mol Med. 11:e22. https://doi.org/10.1017/S1462399409001148
66. Gegg ME, Schapira AH (2016) Mitochondrial dysfunction associated with glucocerebrosidase deficiency. Neurobiol Dis 90:43–50. https://doi.org/10.1016/j.nbd.2015.09.006
67. Gegg ME, Cooper JM, Chau KY et al (2010) Mitofusin 1 and mitofusin 2 are ubiquitinated in a PINK1/parkin-dependent manner upon induction of mitophagy. Hum Mol Genet 19:4861–4870. https://doi.org/10.1093/hmg/ddq419
68. Geisler S, Holmström KM, Skujat D et al (2010a) PINK1/Parkin-mediated mitophagy is dependent on VDAC1 and p62/SQSTM1. Nat Cell Biol 12:119–131. https://doi.org/10.1038/ncb2012
69. Geisler S, Holmström KM, Treis A et al (2010b) The PINK1/Parkin-mediated mitophagy is compromised by PD-associated mutations. Autophagy 6:871–878
70. Giasson BI, Lee VM (2001) Parkin and the molecular pathways of Parkinson's disease. Neuron. 31:885–888
71. Gilks WP, Abou-Sleiman PM, Gandhi S et al (2005) A common LRRK2 mutation in idiopathic Parkinson's disease. Lancet. 365:415–416. https://doi.org/10.1016/S0140-6736(05)17830-1
72. Gleyzer N, Vercauteren K, Scarpulla RC (2005) Control of mitochondrial transcription specificity factors (TFB1M and TFB2M) by nuclear respiratory factors (NRF-1 and NRF-2) and PGC-1 family coactivators. Mol Cell Biol 25:1354–1366. https://doi.org/10.1128/MCB.25.4.1354-1366.2005
73. Goo HG, Jung MK, Han SS et al (2013) HtrA2/Omi deficiency causes damage and mutation of mitochondrial DNA. Biochim Biophys Acta. 1833:1866–1875. https://doi.org/10.1016/j.bbamcr.2013.03.016
74. Graziewicz MA, Longley MJ, Copeland WC (2006) DNA polymerase gamma in mitochondrial DNA replication and repair. Chem Rev 106:383–405. https://doi.org/10.1021/cr040463d
75. Green DR, Kroemer G (2004) The pathophysiology of mitochondrial cell death. Science 305:626–629. https://doi.org/10.1126/science.1099320
76. Green DR, Galluzzi L, Kroemer G (2011) Mitochondria and the autophagy-inflammation-cell death axis in organismal aging. Science 333:1109–1112. https://doi.org/10.1126/science.1201940
77. Guardia-Laguarta C, Area-Gomez E, Rüb C et al (2014) α-Synuclein is localized to mitochondria-associated ER membranes. J Neurosci. 34:249–259. https://doi.org/10.1523/JNEUROSCI.2507-13.2014
78. Haas RH, Nasirian F, Nakano K et al (1995) Low platelet mitochondrial complex I and complex II/III activity in early untreated Parkinson's disease. Ann Neurol. 37:714–722
79. Handschin C (2009) The biology of PGC-1α and its therapeutic potential. Trends Pharmacol Sci. 30:322–329. https://doi.org/10.1016/j.tips.2009.03.006
80. Hantraye P, Varastet M, Peschanski M et al (1993) Stable parkinsonian syndrome and uneven loss of striatal dopamine fibres following chronic MPTP administration in baboons. Neuroscience 53:169–178
81. Hayashi T, Ishimori C, Takahashi-Niki K et al (2009) DJ-1 binds to mitochondrial complex I and maintains its activity. Biochem Biophys Res Commun 390:667–672. https://doi.org/10.1016/j.bbrc.2009.10.025
82. Heo JY, Park JH, Kim SJ et al (2012) DJ-1 null dopaminergic neuronal cells exhibit defects in mitochondrial function and structure: involvement of mitochondrial complex I assembly. PLoS One 7:e32629. https://doi.org/10.1371/journal.pone.0032629
83. Herzig S, Long FX, Jhala US et al (2001) CREB regulates hepatic gluconeogenesis through the coactivator PGC-1. Nature 413:179–183. https://doi.org/10.1038/35093131
84. Hirawake H, Taniwaki M, Tamura A et al (1999) Characterization of the human SDHD gene encoding the small subunit of cytochrome b (cybS) in mitochondrial succinate- ubiquinone oxidoreductase. Biochim Biophys Acta 1412:295–300
85. Hoang T, Choi DK, Nagai M et al (2009) Neuronal NOS and cyclooxygenase-2 contribute to DNA damage in a mouse model of Parkinson disease. Free Radic Biol Med 47:1049–1056. https://doi.org/10.1016/j.freeradbiomed.2009.07.013
86. Hoppins S, Lackner L, Nunnari J (2007) The machines that divide and fuse mitochondria. Annu Rev Biochem 76:751–780. https://doi.org/10.1146/annurev.biochem.76.071905.090048
87. Hung AY, Schwarzschild MA (2007) Clinical trials for neuroprotection in Parkinson's disease: overcoming angst and futility? Curr Opin Neurol 20:477–483. https://doi.org/10.1097/WCO.0b013e32826388d6
88. Ingerman E, Perkins EM, Marino M et al (2005) Dnm1 forms spirals that are structurally tailored to fit mitochondria. J Cell Biol 170:1021–1027. https://doi.org/10.1083/jcb.200506078

89. Ishihara N, Eura Y, Mihara K (2004) Mitofusin 1 and 2 play distinct roles in mitochondrial fusion reactions via GTPase activity. J Cell Sci 117:6535–6546. https://doi.org/10.1242/jcs.01565
90. Ishihara N, Fujita Y, Oka T et al (2006) Regulation of mitochondrial morphology through proteolytic cleavage of OPA1. EMBO J 25:2966–2977. https://doi.org/10.1038/sj.emboj.7601184
91. Itoh K, Nakamura K, Iijima M et al (2013) Mitochondrial dynamics in neurodegeneration. Trends Cell Biol 23:64–71. https://doi.org/10.1016/j.tcb.2012.10.006
92. Ivankovic D, Chau KY, Schapira AH et al (2016) Mitochondrial and lysosomal biogenesis are activated following PINK1/parkin-mediated mitophagy. J Neurochem 136:388–402. https://doi.org/10.1111/jnc.13412
93. Jankovic J (2008) Parkinson's disease: clinical features and diagnosis. J Neurol Neurosurg Psychiatry 79:368–376. https://doi.org/10.1136/jnnp.2007.131045
94. Jenner P (2003) Oxidative stress in Parkinson's disease. Ann Neurol 53(Suppl 3):S26–S36. https://doi.org/10.1002/ana.10483
95. Jenner P, Olanow CW (2006) The pathogenesis of cell death in Parkinson's disease. Neurology. 66:S24–S36
96. Jin J, Meredith GE, Chen L et al (2005) Quantitative proteomic analysis of mitochondrial proteins: relevance to Lewy body formation and Parkinson's disease. Brain Res Mol Brain Res 134:119–138. https://doi.org/10.1016/j.molbrainres.2004.10.003
97. Jin J, Hulette C, Wang Y et al (2006) Proteomic identification of a stress protein, mortalin/mthsp70/GRP75: relevance to Parkinson disease. Mol Cell Proteomics. 5:1193–1204. https://doi.org/10.1074/mcp.M500382-MCP200
98. Jin J, Davis J, Zhu D et al (2007) Identification of novel proteins affected by rotenone in mitochondria of dopaminergic cells. BMC Neurosci 8:67. https://doi.org/10.1186/1471-2202-8-67
99. Jin SM, Lazarou M, Wang C et al (2010) Mitochondrial membrane potential regulates PINK1 import and proteolytic destabilization by PARL. J Cell Biol 191:933–942. https://doi.org/10.1083/jcb.201008084
100. Jofuku A, Ishihara N, Mihara K (2005) Analysis of functional domains of rat mitochondrial Fis1, the mitochondrial fission-stimulating protein. Biochem Biophys Res Commun 333(2):650–659. https://doi.org/10.1016/j.bbrc.2005.05.154
101. Johnson ME, Bobrovskaya L (2015) An update on the rotenone models of Parkinson's disease: their ability to reproduce the features of clinical disease and model gene-environment interactions. Neurotoxicology 46:101–116. https://doi.org/10.1016/j.neuro.2014.12.002
102. Jones GM, Vale JA (2000) Mechanisms of toxicity, clinical features, and management of diquat poisoning: a review. J Toxicol Clin Toxicol 38:123–128
103. Junn E, Jang WH, Zhao X et al (2009) Mitochondrial localization of DJ-1 leads to enhanced neuroprotection. J Neurosci Res 87:123–129. https://doi.org/10.1002/jnr.21831
104. Karren MA, Coonrod EM, Anderson TK et al (2005) The role of Fis1p-Mdv1p interactions in mitochondrial fission complex assembly. J Cell Biol 171:291–301. https://doi.org/10.1083/jcb.200506158
105. Kim I, Rodriguez-Enriquez S, Lemasters JJ (2007) Selective degradation of mitochondria by mitophagy. Arch Biochem Biophys 462:245–253. https://doi.org/10.1016/j.abb.2007.03.034
106. Kitada T, Asakawa S, Hattori N et al (1998) Mutations in the parkin gene cause autosomal recessive juvenile parkinsonism. Nature 392:605–608
107. Klein C, Westenberger A (2012) Genetics of Parkinson's disease. Cold Spring Harb Perspect Med 2:a008888. https://doi.org/10.1101/cshperspect.a008888
108. Korhonen JA, Gaspari M, Falkenberg M (2003) TWINKLE Has 5' -> 3' DNA helicase activity and is specifically stimulated by mitochondrial single-stranded DNA-binding protein. J Biol Chem 278:48627–48632. https://doi.org/10.1074/jbc.M306981200
109. Koshiba T, Detmer SA, Kaiser JT (2004) Structural basis of mitochondrial tethering by mitofusin complexes. Science 305:858–862. https://doi.org/10.1126/science.1099793
110. Kruger R, Kuhn W, Muller T et al (1998) Ala30Pro mutation in the gene encoding alphasynuclein in Parkinson's disease. Nature genetics 18:106–108
111. Kuroda Y, Mitsui T, Kunishige M et al (2006) Parkin enhances mitochondrial biogenesis in proliferating cells. Hum Mol Genet 15:883–895. https://doi.org/10.1093/hmg/ddl006
112. Langston JW, Ballard PA Jr (1983) Parkinson's disease in a chemist working with 1-methyl-4-phenyl-1,2,5,6-tetrahydropyridine. N Engl J Med 309:310
113. Langston JW, Irwin I, Langston EB et al (1984) 1-Methyl-4-phenylpyr- idinium ion (MPP+): identification of a metabolite of MPTP, a toxin selective to the substantia nigra. Neuroscience Letters 48:87–92
114. Langston JW, Forno LS, Tetrud J et al (1999) Evidence of active nerve cell degeneration in the substantia nigra of humans years after 1-methyl-4-phenyl-1,2,3,6-tetrahydropyridine exposure. Ann Neurol 46:598–605
115. Larsson NG, Wang J, Wilhelmsson H et al (1998) Mitochondrial transcription factor A is necessary for mtDNA maintenance and embryogenesis in mice. Nat Genet 18:231–236. https://doi.org/10.1038/ng0398-231
116. Lee YJ, Jeong SY, Karbowski M et al (2004) Roles of the mammalian mitochondrial fission and fusion mediators Fis1, Drp1, and Opa1 in apoptosis. Mol Biol Cell 15:5001–5011. https://doi.org/10.1091/mbc.E04-04-0294
117. Lee S, Sterky FH, Mourier A et al (2012) Mitofusin 2 is necessary for striatal axonal projections of mid-

brain dopamine neurons. Hum Mol Genet 21:4827–4835. https://doi.org/10.1093/hmg/dds352
118. Legesse-Miller A, Massol RH, Kirchhausen T (2003) Constriction and Dnm1p recruitment are distinct processes in mitochondrial fission. Mol Biol Cell 14:1953–1963. https://doi.org/10.1091/mbc.E02-10-0657
119. Lei S, Zavala-Flores L, Garcia-Garcia A et al (2014) Alterations in energy/redox metabolism induced by mitochondrial and environmental toxins: a specific role for glucose-6-phosphate-dehydrogenase and the pentose phosphate pathway in paraquat toxicity. ACS Chem Biol 9:2032–2048. https://doi.org/10.1021/cb400894a
120. Lesage S, Brice A (2009) Parkinson's disease: from monogenic forms to genetic susceptibility factors. Hum Mol Genet 18:R48–R59. https://doi.org/10.1093/hmg/ddp012
121. Lim KL, Tan JM (2007) Role of the ubiquitin proteasome system in Parkinson's disease. BMC Biochem 8(Suppl 1):S13. https://doi.org/10.1186/1471-2091-8-S1-S13
122. Lippolis R, Siciliano RA, Pacelli C et al (2015) Altered protein expression pattern in skin fibroblasts from parkin-mutant early-onset Parkinson's disease patients. Biochim Biophys Acta 1852:1960–1970. https://doi.org/10.1016/j.bbadis.2015.06.015
123. Lücking CB, Dürr A, Bonifati V et al (2000) Association between early-onset Parkinson's disease and mutations in the parkin gene. N Engl J Med 342:1560–1567. https://doi.org/10.1056/NEJM200005253422103
124. Luoma P, Melberg A, Rinne JO et al (2004) Parkinsonism, premature menopause, and mitochondrial DNA polymerase gamma mutations: clinical and molecular genetic study. Lancet 364:875–882. https://doi.org/10.1016/S0140-6736(04)16983-3
125. Lutz AK, Exner N, Fett ME et al (2009) Loss of parkin or PINK1 function increases Drp1-dependent mitochondrial fragmentation. J Biol Chem 284:22938–22951. https://doi.org/10.1074/jbc.M109.035774
126. MacVicar T, Langer T (2016) OPA1 processing in cell death and disease - the long and short of it. J Cell Sci 129:2297–2306. https://doi.org/10.1242/jcs.159186
127. Manning-Bog AB, Schüle B, Langston JW (2009) Alpha-synuclein-glucocerebrosidase interactions in pharmacological Gaucher models: a biological link between Gaucher disease and parkinsonism. Neurotoxicology. 30:1127–1132. https://doi.org/10.1016/j.neuro.2009.06.009
128. Markey SP, Johannessen JN, Chiueh CC et al (1984) Intraneuronal generation of a pyridinium metabolite may cause drug-induced parkinsonism. Nature 311:464–467
129. Martin I, Kim JW, Dawson VL et al (2014) LRRK2 pathobiology in Parkinson's disease. J Neurochem 131:554–565. https://doi.org/10.1111/jnc.12949
130. Martinez A, Lectez B, Ramirez J et al (2017) Quantitative proteomic analysis of Parkin substrates in Drosophila neurons. Mol Neurodegener 12:29. https://doi.org/10.1186/s13024-017-0170-3
131. Matsuda N, Sato S, Shiba K et al (2010) PINK1 stabilized by mitochondrial depolarization recruits Parkin to damaged mitochondria and activates latent Parkin for mitophagy. J Cell Biol 189:211–221. https://doi.org/10.1083/jcb.200910140
132. McCormack AL, Di Monte DA (2003) Effects of L-dopa and other amino acids against paraquat-induced nigrostriatal degeneration. J Neurochem 85:82–86
133. McCormack AL, Thiruchelvam M, Manning-Bog AB et al (2002) Environmental risk factors and Parkinson's disease: selective degeneration of nigral dopaminergic neurons caused by the herbicide paraquat. Neurobiol Dis 10:119–127
134. McFarland MA, Ellis CE, Markey SP et al (2008) Proteomics analysis identifies phosphorylation-dependent alpha-synuclein protein interactions. Mol Cell Proteomics 7:2123–2137. https://doi.org/10.1074/mcp.M800116-MCP200
135. Migdalska-Richards A, Schapira AH (2016) The relationship between glucocerebrosidase mutations and Parkinson disease. J Neurochem 139 Suppl 1:77–90. https://doi.org/10.1111/jnc.13385
136. Miralles Fusté J, Shi Y, Wanrooij S et al (2014) In vivo occupancy of mitochondrial single-stranded DNA binding protein supports the strand displacement mode of DNA replication. PLoS Genet 10:e1004832. https://doi.org/10.1371/journal.pgen.1004832
137. Misko A, Jiang S, Wegorzewska I et al (2010) Mitofusin 2 is necessary for transport of axonal mitochondria and interacts with the Miro/Milton complex. J Neurosci 30:4232–4240. https://doi.org/10.1523/JNEUROSCI.6248-09.2010
138. Monti C, Bondi H, Urbani A et al (2015) Systems biology analysis of the proteomic alterations induced by MPP(+), a Parkinson's disease-related mitochondrial toxin. Front Cell Neurosci 9:14. https://doi.org/10.3389/fncel.2015.00014
139. Moors T, Paciotti S, Chiasserini D et al (2016) Lysosomal dysfunction and α-synuclein aggregation in Parkinson's disease: diagnostic links. Mov Disord 31:791–801. https://doi.org/10.1002/mds.26562
140. Morais VA, Verstreken P, Roethig A et al (2009) Parkinson's disease mutations in PINK1 result in decreased Complex I activity and deficient synaptic function. EMBO Mol Med 1:99–111. https://doi.org/10.1002/emmm.200900006
141. Mortiboys H, Johansen KK, Aasly JO et al (2010) Mitochondrial impairment in patients with Parkinson disease with the G2019S mutation in LRRK2. Neurology 75:2017–2020. https://doi.org/10.1212/WNL.0b013e3181ff9685
142. Mozdy AD, McCaffery JM, Shaw JM (2000) Dnm1p GTPase-mediated mitochondrial fission is a multi-

step process requiring the novel integral membrane component Fis1p. J Cell Biol 151:367–380
143. Muñoz P, Huenchuguala S, Paris I et al (2012) Dopamine oxidation and autophagy. Parkinsons Dis 2012:920953. https://doi.org/10.1155/2012/920953
144. Mytilineou C, Werner P, Molinari S et al (1994) Impaired oxidative decarboxylation of pyruvate in fibroblasts from patients with Parkinson's disease. J Neural Transm Park Dis Dement Sect. 8:223–228
145. Nandipati S, Litvan I (2016) Environmental Exposures and Parkinson's Disease. Int J Environ Res Public Health. 13:pii: E881. https://doi.org/10.3390/ijerph13090881
146. Narendra D, Kane LA, Hauser DN et al (2010a) p62/SQSTM1 is required for Parkin-induced mitochondrial clustering but not mitophagy; VDAC1 is dispensable for both. Autophagy 6:1090–1106
147. Narendra DP, Jin SM, Tanaka A et al (2010b) PINK1 is selectively stabilized on impaired mitochondria to activate Parkin. PLoS Biol 8:e1000298. https://doi.org/10.1371/journal.pbio.1000298
148. Navarro-Yepes J, Anandhan A, Bradley E et al (2016) Inhibition of protein ubiquitination by paraquat and 1-methyl-4-phenylpyridinium impairs ubiquitin-dependent protein degradation pathways. Mol Neurobiol 53:5229–5251. https://doi.org/10.1007/s12035-015-9414-9
149. Neupert W (2016) Mitochondrial gene expression: a playground of evolutionary tinkering. Annu Rev Biochem 85:65–76. https://doi.org/10.1146/annurev-biochem-011116-110824
150. Nicklas WJ, Vyas I, Heikkila RE (1985) Inhibition of NADH-linked oxidation in brain mitochondria by 1-methyl-4-phenyl-pyridine, a metabolite of the neurotoxin, 1-methyl-4-phenyl-1,2,5,6-tetrahydropyridine. Life Sciences 36:2503–2508
151. Nisoli E, Clementi E, Paolucci C et al (2003) Mitochondrial biogenesis in mammals: the role of endogenous nitric oxide. Science 299:896–899. https://doi.org/10.1126/science.1079368
152. Nisoli E, Falcone S, Tonello C et al (2004) Mitochondrial biogenesis by NO yields functionally active mitochondria in mammals. Proc Natl Acad Sci USA 101:16507–16512. https://doi.org/10.1073/pnas.0405432101
153. Niu J, Yu M, Wang C et al (2012) Leucine-rich repeat kinase 2 disturbs mitochondrial dynamics via Dynamin-like protein. J Neurochem 122:650–658. https://doi.org/10.1111/j.1471-4159.2012.07809.x
154. Okatsu K, Saisho K, Shimanuki M et al (2010) p62/SQSTM1 cooperates with Parkin for perinuclear clustering of depolarized mitochondria. Genes Cells 15:887–900. https://doi.org/10.1111/j.1365-2443.2010.01426.x
155. Olichon A, Baricault L, Gas N et al (2003) Loss of OPA1 perturbates the mitochondrial inner membrane structure and integrity, leading to cytochrome c release and apoptosis. J Biol Chem 278:7743–7746. https://doi.org/10.1074/jbc.C200677200
156. Ongwijitwat S, Wong-Riley MT (2005) Is nuclear respiratory factor 2 a master transcriptional coordinator for all ten nuclear-encoded cy- tochrome c oxidase subunits in neurons? Gene 360:65–77. https://doi.org/10.1016/j.gene.2005.06.015
157. Ongwijitwat S, Liang HL, Graboyes EM et al (2006) Nuclear respiratory factor 2 senses changing cellular energy demands and its silencing down-regulates cytochrome oxidase and other target gene mRNAs. Gene 374:39–49. https://doi.org/10.1016/j.gene.2006.01.009
158. Ordureau A, Sarraf SA, Duda DM et al (2014) Quantitative proteomics reveal a feedforward mechanism for mitochondrial PARKIN translocation and ubiquitin chain synthesis. Mol Cell 56:360–375. https://doi.org/10.1016/j.molcel.2014.09.007
159. Ossowska K, Wardas J, Smiałowska M et al (2005) A slowly developing dysfunction of dopaminergic nigrostriatal neurons induced by long-term paraquat administration in rats: an animal model of preclinical stages of Parkinson's disease? Eur J Neurosci 22:1294–1304. https://doi.org/10.1111/j.1460-9568.2005.04301.x
160. Otera H, Mihara K (2011) Molecular mechanisms and physiologic functions of mitochondrial dynamics. J Biochem 149:241–251. https://doi.org/10.1093/jb/mvr002
161. Otera H, Wang C, Cleland MM et al (2010) Mff is an essential factor for mitochondrial recruitment of Drp1 during mitochondrial fission in mammalian cells. J Cell Biol 191:1141–1158. https://doi.org/10.1083/jcb.201007152
162. Ottolini D, Calì T, Negro A et al (2013) The Parkinson disease-related protein DJ-1 counteracts mitochondrial impairment induced by the tumour suppressor protein p53 by enhancing endoplasmic reticulum-mitochondria tethering. Hum Mol Genet 22:2152–2168. https://doi.org/10.1093/hmg/ddt068
163. Ozgul S, Kasap M, Akpinar G et al (2015) Linking a compound-heterozygous Parkin mutant (Q311R and A371T) to Parkinson's disease by using proteomic and molecular approaches. Neurochem Int 85-86:1–13. https://doi.org/10.1016/j.neuint.2015.03.007
164. Pacelli C, De Rasmo D, Signorile A et al (2011) Mitochondrial defect and PGC-1α dysfunction in parkin-associated familial Parkinson's disease. Biochim Biophys Acta 1812:1041–1053. https://doi.org/10.1016/j.bbadis.2010.12.022
165. Palacino JJ, Sagi D, Goldberg MS et al (2004) Mitochondrial dysfunction and oxidative damage in parkin-deficient mice. J Biol Chem 279:18614–18622
166. Pallanck LJ (2010) Culling sick mitochondria from the herd. J Cell Biol 191:1225–1227. https://doi.org/10.1083/jcb.201011068
167. Palmer CS, Osellame LD, Laine D et al (2011) MiD49 and MiD51, new components of the mitochondrial fission machinery. EMBO Rep 12:565–573. https://doi.org/10.1038/embor.2011.54

168. Parihar MS, Parihar A, Fujita M et al (2008) Mitochondrial association of alpha-synuclein causes oxidative stress. Cell Mol Life Sci 65:1272–1284. https://doi.org/10.1007/s00018-008-7589-1
169. Parone PA, Da Cruz S, Tondera D et al (2008) Preventing mitochondrial fission impairs mitochondrial function and leads to loss of mitochondrial DNA. PLoS One 3:e3257. https://doi.org/10.1371/journal.pone.0003257
170. Pennington K, Peng J, Hung CC et al (2010) Differential effects of wild-type and A53T mutant isoform of alpha-synuclein on the mitochondrial proteome of differentiated SH-SY5Y cells. J Proteome Res 9:2390–23401. https://doi.org/10.1021/pr901102d
171. Perier C, Tieu K, Guegan C et al (2005) Complex I deficiency primes Bax- dependent neuronal apoptosis through mitochondrial oxidative damage. Proc Natl Acad Sci U S A 102:19126–19131. https://doi.org/10.1073/pnas.0508215102
172. Perier C, Bove J, Wu DC et al (2007) Two molecular pathways initiate mitochondria-dependent dopaminergic neurodegeneration in experimental Parkinson's disease. Proc Natl Acad Sci U S A 104:8161–8166. https://doi.org/10.1073/pnas.0609874104
173. Periquet M, Corti O, Jacquier S et al (2005) Proteomic analysis of parkin knockout mice: alterations in energy metabolism, protein handling and synaptic function. J Neurochem 95:1259–1276. https://doi.org/10.1111/j.1471-4159.2005.03442.x
174. Pham AH, Meng S, Chu QN et al (2012) Loss of Mfn2 results in progressive, retrograde degeneration of dopaminergic neurons in the nigrostriatal circuit. Hum Mol Genet 21:4817–4826. https://doi.org/10.1093/hmg/dds311
175. Piao Y, Kim HG, Oh MS et al (2012) Overexpression of TFAM, NRF-1 and myr-AKT protects the MPP(+)-induced mitochondrial dysfunctions in neuronal cells. Biochim Biophys Acta 1820:577–585. https://doi.org/10.1016/j.bbagen.2011.08.007
176. Pickrell AM, Youle RJ (2015) The roles of PINK1, parkin, and mitochondrial fidelity in Parkinson's disease. Neuron 85:257–273. https://doi.org/10.1016/j.neuron.2014.12.007
177. Plun-Favreau H, Klupsch K, Moisoi N et al (2007) The mitochondrial protease HtrA2 is regulated by Parkinson's disease-associated kinase PINK1. Nat Cell Biol 9:1243–1252. https://doi.org/10.1038/ncb1644
178. Polymeropoulos MH, Higgins JJ, Golbe LI et al (1996) Mapping of a gene for Parkinson's disease to chromosome 4q21-q23. Science 274:1197–1199
179. Polymeropoulos MH, Lavedan C, Leroy E et al (1997) Mutation in the alpha-synuclein gene identified in families with Parkinson's disease. Science 276:2045–2047
180. Poole AC, Thomas RE, Andrews LA et al (2008) The PINK1/parkin pathway regulates mitochondrial morphology. Proc Natl Acad Sci U S A 105:1638–1643. https://doi.org/10.1073/pnas.0709336105
181. Prasad K, Winnik B, Thiruchelvam MJ et al (2007) Prolonged toxicokinetics and toxicodynamics of paraquat in mouse brain. Environ Health Perspect 115:1448–1453. https://doi.org/10.1289/ehp.9932
182. Przedborski S, Jackson-Lewis V, Yokoyama R et al (1996) Role of neuronal nitric oxide in 1-methyl-4-phenyl-1,2,3,6-tetrahydropyridine (MPTP)-induced dopaminergic neurotoxicity. Proc Natl Acad Sci U S A 93:4565–4571
183. Puigserver P, Spiegelman BM (2003) Peroxisome proliferator-activated receptor-gamma coactivator 1 alpha (PGC-1 alpha): transcriptional coactivator and metabolic regulator. Endocr Rev 24:78–90. https://doi.org/10.1210/er.2002-0012
184. Qi Y, Yan L, Yu C et al (2016) Structures of human mitofusin 1 provide insight into mitochondrial tethering. J Cell Biol 215:621–629. https://doi.org/10.1083/jcb.201609019
185. Ramonet D, Perier C, Recasens A et al (2013) Optic atrophy 1 mediates mitochondria remodeling and dopaminergic neurodegeneration linked to complex I deficiency. Cell Death Differ 20:77–85. https://doi.org/10.1038/cdd.2012.95
186. Ramsay RR, Salach JI, Dadgar J et al (1986) Inhibition of mitochondrial NADH dehydrogenase by pyridine derivatives and its possible relation to experimental and idiopathic parkinsonism. Biochem Biophys Res Commun 135:269–275
187. Reeve AK, Ludtmann MH, Angelova PR et al (2015) Aggregated α-synuclein and complex I deficiency: exploration of their relationship in differentiated neurons. Cell Death Dis 6:e1820. https://doi.org/10.1038/cddis.2015.166
188. Richardson JR, Quan Y, Sherer TB et al (2005) Paraquat neurotoxicity is distinct from that of MPTP and rotenone. Toxicol Sci 88:193–201. https://doi.org/10.1093/toxsci/kfi304
189. Rojo M, Legros F, Chateau D et al (2002) Membrane topology and mitochondrial targeting of mitofusins, ubiquitous mammalian homologs of the transmembrane GTPase Fzo. J Cell Sci 115:1663–1674
190. Saez-Atienzar S, Bonet-Ponce L, Blesa JR et al (2014) The LRRK2 inhibitor GSK2578215A induces protective autophagy in SH-SY5Y cells: involvement of Drp-1-mediated mitochondrial fission and mitochondrial-derived ROS signaling. Cell Death Dis 5:e1368. https://doi.org/10.1038/cddis.2014.320
191. Sarraf SA, Raman M, Guarani-Pereira V et al (2013) Landscape of the PARKIN-dependent ubiquitylome in response to mitochondrial depolarization. Nature 496:372–376. https://doi.org/10.1038/nature12043
192. Scarpulla RC (1999) Nuclear Transcription Factors in Cytochrome *c* and Cytochrome Oxidase Expression. In: Papa S, Guerrieri F, Tager JM (eds) Frontiers of Cellular Bioenergetics. Springer, Boston, MA, pp 553–591

193. Scarpulla RC (2008) Transcriptional paradigms in Mammalian mitochondrial biogenesis and function. Physiol Rev 88:611–638. https://doi.org/10.1152/physrev.00025.2007
194. Scarpulla RC (2011) Nucleus-encoded regulators of mitochondrial function: integration of respiratory chain expression, nutrient sensing and metabolic stress. Biochim Biophys Acta 1819:1088–1097. https://doi.org/10.1016/j.bbagrm.2011.10.011
195. Schapira AH (2015) Glucocerebrosidase and Parkinson disease: recent advances. Mol Cell Neurosci 66:37–42. https://doi.org/10.1016/j.mcn.2015.03.013
196. Schapira AH, Jenner P (2011) Etiology and pathogenesis of Parkinson's disease. Mov Disord. 26:1049–1055. https://doi.org/10.1002/mds.23732
197. Schapira AH, Cooper JM, Dexter D et al (1990) Mitochondrial complex I deficiency in Parkinson's disease. J Neurochem 54:823–827
198. Scorrano L, Ashiya M, Buttle K et al (2002) A distinct pathway remodels mitochondrial cristae and mobilizes cytochrome c during apoptosis. Dev Cell 2:55–67
199. Shendelman S, Jonason A, Martinat C et al (2004) DJ-1 is a redox-dependent molecular chaperone that inhibits alpha-synuclein aggregate formation. PLoS Biol 2:e362. https://doi.org/10.1371/journal.pbio.0020362
200. Sherer TB, Betarbet R, Greenamyre JT (2002a) Environment, mitochondria, and Parkinson's disease. Neuroscientist. 8:192–197. https://doi.org/10.1177/1073858402008003004
201. Sherer TB, Betarbet R, Stout AK et al (2002b) An in vitro model of Parkinson's disease: linking mitochondrial impairment to altered alpha-synuclein metabolism and oxidative damage. J Neurosci 22:7006–7015. doi: 20026721
202. Sherer TB, Kim JH, Betarbet R et al (2003) Subcutaneous rotenone exposure causes highly selective dopaminergic degeneration and alpha-synuclein aggregation. Exp Neurol 179:9–16
203. Shiba-Fukushima K, Imai Y, Yoshida S et al (2012) PINK1-mediated phosphorylation of the Parkin ubiquitin-like domain primes mitochondrial translocation of Parkin and regulates mitophagy. Sci Rep 2:1002. https://doi.org/10.1038/srep01002
204. Shimizu K, Ohtaki K, Matsubara K et al (2001) Carrier-mediated processes in blood—brain barrier penetration and neural uptake of paraquat. Brain Res 906:135–142
205. Shimizu K, Matsubara K, Ohtaki K et al (2003) Paraquat induces long-lasting dopamine overflow through the excitotoxic pathway in the striatum of freely moving rats. Brain Res 976:243–252
206. Shimura H, Hattori N, Kubo S et al (2000) Familial Parkinson disease gene product, parkin, is a ubiquitin-protein ligase. Nat Genet 25:302–305. https://doi.org/10.1038/77060
207. Shin JH, Ko HS, Kang H et al (2011) PARIS (ZNF746) repression of PGC-1α contributes to neurodegeneration in Parkinson's disease. Cell 144:689–702. https://doi.org/10.1016/j.cell.2011.02.010
208. Sim CH, Lio DS, Mok SS et al (2006) C-terminal truncation and Parkinson's disease-associated mutations down-regulate the protein serine/threonine kinase activity of PTEN-induced kinase-1. Hum Mol Genet 365:412–415. https://doi.org/10.1093/hmg/ddl398
209. Simon-Sanchez J, Singleton AB (2008) Sequencing analysis of OMI/HTRA2 shows previously reported pathogenic mutations in neurologically normal controls. Hum Molec Genet 17:1988–1993. https://doi.org/10.1093/hmg/ddn096
210. Singleton AB, Farrer M, Johnson J et al (2003) alpha-Synuclein locus triplication causes Parkinson's disease. Science. 302:841. https://doi.org/10.1126/science.1090278
211. Singleton AB, Farrer MJ, Bonifati V (2013) The genetics of Parkinson's disease: progress and therapeutic implications. Mov Disord 28:14–23. https://doi.org/10.1002/mds.25249
212. Song Z, Ghochani M, McCaffery JM et al (2009) Mitofusins and OPA1 mediate sequential steps in mitochondrial membrane fusion. Mol Biol Cell 20:3525–3532. https://doi.org/10.1091/mbc.E09-03-0252
213. Strappazzon F, Vietri-Rudan M, Campello S et al (2011) Mitochondrial BCL-2 inhibits AMBRA1-induced autophagy. EMBO J 30:1195–1208. https://doi.org/10.1038/emboj.2011.49
214. Strauss KM, Martins LM, Plun-Favreau H et al (2005) Loss of function mutations in the gene encoding Omi/HtrA2 in Parkinson's disease. Hum Mol Genet 14:2099–2111. https://doi.org/10.1093/hmg/ddi215
215. Subramaniam SR, Chesselet MF (2013) Mitochondrial dysfunction and oxidative stress in Parkinson's disease. Prog Neurobiol 106-107:17–32. https://doi.org/10.1016/j.pneurobio.2013.04.004
216. Sun M, Latourelle JC, Wooten GF et al (2006) Influence of heterozygosity for parkin mutation on onset age in familial Parkinson disease: the GenePD study. Arch Neurol 63:826–832. https://doi.org/10.1001/archneur.63.6.826
217. Sun Y, Vashisht AA, Tchieu J et al (2012) Voltage-dependent anion channels (VDACs) recruit Parkin to defective mitochondria to promote mitochondrial autophagy. J Biol Chem 287:40652–40660. https://doi.org/10.1074/jbc.M112.419721
218. Taira T, Saito Y, Niki T et al (2004) DJ-1 has a role in antioxidative stress to prevent cell death. EMBO Rep 5:213–218. https://doi.org/10.1038/sj.embor.7400074
219. Takahashi Y, Kako K, Arai H et al (2002) Characterization and identification of promoter elements in the mouse COX17 gene. Biochim Biophys Acta 1574:359–364
220. Tal MC, Sasai M, Lee HK et al (2009) Absence of autophagy results in reactive oxygen species-dependent amplification of RLR signaling. Proc

Natl Acad Sci USA 106:2770–2775. https://doi.org/10.1073/pnas.0807694106
221. Tanaka A, Cleland MM, Xu S et al (2010) Proteasome and p97 mediate mitophagy and degradation of mitofusins induced by Parkin. J Cell Biol 191:1367–1380. https://doi.org/10.1083/jcb.201007013
222. Tanner CM, Kamel F, Ross GW et al (2011) Rotenone, paraquat, and Parkinson's disease. Environ Health Perspect 119:866–872. https://doi.org/10.1289/ehp.1002839
223. Triplett JC, Zhang Z, Sultana R et al (2015) Quantitative expression proteomics and phosphoproteomics profile of brain from PINK1 knockout mice: insights into mechanisms of familial Parkinson's disease. J Neurochem 133:750–765. https://doi.org/10.1111/jnc.13039
224. Truban D, Hou X, Caulfield TR et al (2017) PINK1, Parkin, and mitochondrial quality control: what can we learn about Parkinson's disease pathobiology? J Parkinsons Dis 7:13–29. https://doi.org/10.3233/JPD-160989
225. Twig G, Hyde B, Shirihai OS (2008) Mitochondrial fusion, fission and autophagy as a quality control axis: the bioenergetic view. Biochim Biophys Acta 1777:1092–1097. https://doi.org/10.1016/j.bbabio.2008.05.001
226. Valente EM, Abou-Sleiman PM, Caputo V et al (2004a) Hereditary early-onset Parkinson's disease caused by mutations in PINK1. Science 304:1158–1160. https://doi.org/10.1126/science.1096284
227. Valente EM, Salvi S, Ialongo T et al (2004b) PINK1 mutations are associated with sporadic early-onset parkinsonism. Ann Neurol 56:336–341. https://doi.org/10.1002/ana.20256
228. Van Humbeeck C, Cornelissen T, Hofkens H et al (2011) Parkin interacts with Ambra1 to induce mitophagy. J Neurosci 31:10249–10261. https://doi.org/10.1523/JNEUROSCI.1917-11.2011
229. Van Laar VS, Dukes AA, Cascio M et al (2008) Proteomic analysis of rat brain mitochondria following exposure to dopamine quinone: implications for Parkinson disease. Neurobiol Dis 29:477–489. https://doi.org/10.1016/j.nbd.2007.11.007
230. Varastet M, Riche D, Maziere M (1994) Chronic MPTP treatment reproduces in baboons the differential vulnerability of mesencephalic dopaminergic neurons observed in Parkinson's disease. Neuroscience 63:47–56
231. Vekrellis K, Xilouri M, Emmanouilidou E et al (2011) Pathological roles of α-synuclein in neurological disorders. Lancet Neurol 10:1015–10125. https://doi.org/10.1016/S1474-4422(11)70213-7
232. Velayati A, Yu WH, Sidransky E (2010) The role of glucocerebrosidase mutations in Parkinson disease and Lewy body disorders. Curr Neurol Neurosci Rep 10:190–198. https://doi.org/10.1007/s11910-010-0102-x
233. Vila M, Przedborski S (2003) Targeting programmed cell death in neurodegenerative diseases. Nat Rev Neurosci 4:365–375. https://doi.org/10.1038/nrn1100
234. Villeneuve LM, Purnell PR, Boska MD et al (2016) Early Expression of Parkinson's Disease-Related Mitochondrial Abnormalities in PINK1 Knockout Rats. Mol Neurobiol 53:171–186. https://doi.org/10.1007/s12035-014-8927-y
235. Virbasius JV, Scarpulla RC (1994) Activation of the human mitochondrial transcription factor A gene by nuclear respiratory factors: a potential regulatory link between nuclear and mitochondrial gene expression in organelle biogenesis. Proc Natl Acad Sci U S A 91:1309–1313
236. Wallace DC (1999) Mitochondrial diseases in man and mouse. Science 283:1482–1488
237. Wang X, Hai C (2016) Novel insights into redox system and the mechanism of redox regulation. Mol Biol Rep 43:607–628. https://doi.org/10.1007/s11033-016-4022-y
238. Wang G, Mao Z (2014) Chaperone-mediated autophagy: roles in neurodegeneration. Transl Neurodegener 3:20. https://doi.org/10.1186/2047-9158-3-20
239. Wang X, Winter D, Ashrafi G et al (2011a) PINK1 and Parkin target Miro for phosphorylation and degradation to arrest mitochondrial motility. Cell 147:893–906. https://doi.org/10.1016/j.cell.2011.10.018
240. Wang H, Song P, Du L et al (2011b) Parkin ubiquitinates Drp1 for proteasome-dependent degradation: implication of dysregulated mitochondrial dynamics in Parkinson disease. J Biol Chem 286:11649–11658. https://doi.org/10.1074/jbc.M110.144238
241. Wang X, Petrie TG, Liu Y et al (2012a) Parkinson's disease-associated DJ-1 mutations impair mitochondrial dynamics and cause mitochondrial dysfunction. J Neurochem 121:830–839. https://doi.org/10.1111/j.1471-4159.2012.07734.x
242. Wang X, Yan MH, Fujioka H et al (2012b) LRRK2 regulates mitochondrial dynamics and function through direct interaction with DLP1. Hum Mol Genet 21:1931–1944. https://doi.org/10.1093/hmg/dds003
243. Wareski P, Vaarmann A, Choubey V et al (2009) PGC-1{alpha} and PGC-1{beta} regulate mitochondrial density in neurons. J Biol Chem 284:21379–21385. https://doi.org/10.1074/jbc.M109.018911
244. Werner CJ, Heyny-von Haussen R, Mall G et al (2008) Proteome analysis of human substantia nigra in Parkinson's disease. Proteome Sci 6:8. https://doi.org/10.1186/1477-5956-6-8
245. West AB, Moore DJ, Biskup S et al (2005) Parkinson's disease-associated mutations in leucine-rich repeat kinase 2 augment kinase activity. Proc Natl Acad Sci U S A 102:16842–16847. https://doi.org/10.1073/pnas.0507360102
246. Wu Z, Puigserver P, Andersson U et al (1999) Mechanisms controlling mitochondrial biogenesis and respiration through the thermogenic coactivator PGC-1. Cell 98(1):115–124

247. Wu B, Song B, Tian S et al (2012) Central nervous system damage due to acute paraquat poisoning: a neuroimaging study with 3.0 T MRI. Neurotoxicology 33:1330–1337. https://doi.org/10.1016/j.neuro.2012.08.007
248. Xilouri M, Brekk OR, Stefanis L (2016) Autophagy and alpha-synuclein: relevance to Parkinson's disease and related synucleopathies. Mov Disord 31:178–192. https://doi.org/10.1002/mds.26477
249. Xu R, Hu Q, Ma Q et al (2014) The protease Omi regulates mitochondrial biogenesis through the GSK3β/PGC-1α pathway. Cell Death Dis 5:e1373. https://doi.org/10.1038/cddis.2014.328
250. Xun Z, Sowell RA, Kaufman TC et al (2008) Quantitative proteomics of a presymptomatic A53T alpha-synuclein Drosophila model of Parkinson disease. Mol Cell Proteomics 7:1191–1203. https://doi.org/10.1074/mcp.M700467-MCP200
251. Yang Y, Ouyang Y, Yang L et al (2008) Pink1 regulates mitochondrial dynamics through interaction with the fission/fusion machinery. Proc Natl Acad Sci U S A 105:7070–7075. https://doi.org/10.1073/pnas.0711845105
252. Yoon Y, Krueger EW, Oswald BJ et al (2003) The mitochondrial protein hFis1 regulates mitochondrial fission in mammalian cells through an interaction with the dynamin-like protein DLP1. Mol Cell Biol 23:5409–5420
253. Youle RJ, Narendra DP (2011) Mechanisms of mitophagy. Nat Rev Mol Cell Biol. 12:9–14. https://doi.org/10.1038/nrm3028
254. Yu W, Sun Y, Guo S et al (2011) The PINK1/Parkin pathway regulates mitochondrial dynamics and function in mammalian hippocampal and dopaminergic neurons. Hum Mol Genet 20:3227–3240. https://doi.org/10.1093/hmg/ddr235
255. Yumino K, Kawakami I, Tamura M et al (2002) Paraquat- and diquat-induced oxygen radical generation and lipid peroxidation in rat brain microsomes. J Biochem 131:565–570
256. Zanellati MC, Monti V, Barzaghi C (2015) Mitochondrial dysfunction in Parkinson disease: evidence in mutant PARK2 fibroblasts. Front Genet 6:78. https://doi.org/10.3389/fgene.2015.00078
257. Zarranz JJ, Alegre J, Gomez-Esteban JC et al (2004) The new mutation, E46K, of alpha-synuclein causes Parkinson and Lewy body dementia. Annals of neurology 55:164–173. https://doi.org/10.1093/hmg/ddr235
258. Zhang Y, Chan DC (2007) New insights into mitochondrial fusion. FEBS Lett 581:2168–2173. https://doi.org/10.1016/j.febslet.2007.01.095
259. Zhang Y, Gao J, Chung KK et al (2000) Parkin functions as an E2-dependent ubiquitin–protein ligase and promotes the degradation of the synaptic vesicle-associated protein, CDCrel-1. Proc Natl Acad Sci U S A 97:13354–13359. https://doi.org/10.1073/pnas.240347797
260. Zhang X, Zhou JY, Chin MH et al (2010) Region-specific protein abundance changes in the brain of MPTP-induced Parkinson's disease mouse model. J Proteome Res 9:1496–1509. https://doi.org/10.1021/pr901024z
261. Zhang H, Zhang YW, Yasukawa T et al (2014) Increased negative supercoiling of mtDNA in TOP1mt knockout mice and presence of topoisomerases IIα and IIβ in vertebrate mitochondria. Nucleic Acids Res 42:7259–7267. https://doi.org/10.1093/nar/gku384
262. Zilocchi M, Finzi G, Lualdi M et al (2018) Mitochondrial alterations in Parkinson's disease human samples and cellular models. Neurochem Int 118:61–72. https://doi.org/10.1016/j.neuint.2018.04.013
263. Zimprich A, Biskup S, Leitner P et al (2004) Mutations in LRRK2 cause autosomal- dominant parkinsonism with pleomorphic pathology. Neuron 44:601–607. https://doi.org/10.1016/j.neuron.2004.11.005

# Role of Mitochondria in Host-Pathogen Interaction

# 3

Alessio Soggiu, Paola Roncada, Luigi Bonizzi, and Cristian Piras

**Abstract**

The centrality of the mitochondrion in the evolution and control of the cellare now supported by many experimental studies. Not only with regard to the energy metabolism but also and especially with regard to the other functions indispensable for the cell such as apoptosis and the control of innate immunity through different complex cell signaling pathways. All this makes them one of the main targets during infections supported by pathogenic microorganisms. The interaction and control of these organelles by pathogens results, from the latest experimental evidence, of fundamental importance in the fate of the host cell and in the progression of infectious diseases.

**Keywords**

Mitochondria · Programmed cell death · Mitophagy · Apoptosis · Necrosis · Pyroptosis interaction

A. Soggiu (✉) · L. Bonizzi · C. Piras
Department of Veterinary Medicine, University of Milan, Milan, Italy
e-mail: alessio.soggiu@unimi.it

P. Roncada
Department of Health Sciences, University 'Magna Græcia' of Catanzaro, Catanzaro, Italy

## Abbreviations

| | |
|---|---|
| AC | Adenylate cyclase |
| ADP | Adenosine diphosphate |
| AT | Alpha toxin |
| ATP | Adenosine triphosphate |
| Bax | Bcl-2-associated X protein |
| BimEL | Bcl-2 interacting mediator extra long |
| cAMP | Cyclic adenosine monophosphate |
| Cyt-c | Cytochrome C |
| HKII | Mitochondrial hexokinase II |
| PCD | Programmed cell death |
| PDPK1 | 3-phosphoinositide dependent protein kinase 1 |
| T3SS | Type 3 secretion system |
| VDAC | Voltage dependent anion channel |

## 3.1 Mitochondria and Role in Infectious Diseases

Many of the cellular activities require an input of energy that is provided by the ATP hydrolysis to ADP. One of the main functions of mitochondria is to carry out the energy transformations that lead to the production of ATP. Moreover, are responsible for the synthesis of heme, control calcium levels, regulate cell signaling and can, if necessary, induce cell death. Having a decisive role in metabolism and therefore in cellular respiration, mitochondria are the major cause of



A. Urbani, M. Babu (eds.), *Mitochondria in Health and in Sickness*, Advances in Experimental Medicine and Biology 1158, https://doi.org/10.1007/978-981-13-8367-0_3

cellular production of waste products such as reactive oxygen and nitrogen species which in turn can activate inflammatory processes and cell damage. So mitochondria have emerged as organelles with many cellular functions other than energy production and many of them are related to innate and adaptive immune responses [65]. A pivotal role for mitochondria in signaling and cell fate control during infection with pathogenic microorganism has been summarized mainly for bacteria [55] and viruses [5].

### 3.1.1 Host Mitochondria and Bacteria

The relationship between host and microorganisms is, generally, a mutualistic symbiosis with advantages for the host (vitamins and protection from pathogen) and the microorganism (nutrients and habitat). When the equilibrium between the host and microorganisms is modified by environmental and metabolic changes, pathogenic microorganisms can take control of the host cell using different mechanisms. Parasitic gram-positive and gram-negative bacteria initially lived outside the host cell (extracellular pathogens), but subsequently evolved mechanisms to ensure intracellular survival (intracellular pathogens) by evading the digestion/destruction mechanisms inside the host cell. The evidence that some bacteria have strongly integrated their biology with that of the host cell confirms the success of such strategy. The highest level of this integration is represented by the mitochondria of eukaryotic cells and maybe is due to this functional similarity if the actual pathogenic bacteria can control the fate of the host cell targeting several mitochondrial signaling pathways.

### 3.1.2 Mitochondrial Programmed Cell Death (PCD) Targeting by Gram-Negative Bacteria

In this paragraph, we will describe the different strategies followed by several pathogenic gram-negative bacteria during the infection of the host cell. Manipulation of different types of programmed cell death (PCD) in immune cells has recently emerged as one of the main strategies of pathogenic bacteria to overthrow the host defense machinery [55]. Since mitochondria are central organelles in the control of PCD [26], we will focus our interest on mitochondrial PCD targeting by bacterial toxins and effectors protein (Tables 3.1 and 3.2). Bordetella pertussis, the etiological agent of whooping cough disease, during the establishment of the infection produce a bifunctional toxin, CyaA, an hemolysin/adenylate cyclase (AC) that specifically target myeloid cells. At higher concentration, exert cytotoxic activities mediated by uncontrolled cAMP signaling and hemolytic activity, at low concentration Ahmad and colleagues showed the activation of a pro-apoptotic signaling in THP1 cells via mitochondrial route [1]. More precisely, the authors showed that apoptosis was triggered by CyaA-produced cAMP with the rapid accumulation of BimEL protein and Bax activation that associates with mitochondria and permeabilizes the outer mitochondrial membrane.

Mitochondrial down-regulation of protein synthesis and transport, electron transfer, and small-molecule transfer as well as pathways involved in cytochrome c release has been demonstrated in a murine macrophage-like J774.A1 cell line infected with smooth Brucella melitensis 16M. In the early post-infection stage a marked inhibition of apoptosis via cyt-c release has been detected that lead to extensive Brucella replication after the initial killing [35]. As described by other authors [16], *Brucella abortus* RB51 is able to induce both apoptotic and necrotic cell death in murine RAW264.7 macrophages via caspase 2 and activation of the mitochondrial cell death pathway [16]. Recently RB51-induced cell death driven by the caspase-2/mitochondrial axis having features of apoptosis and pyroptosis has been described [11]. As seen for Brucella melitensis 16 M also Chlamydia trachomatis is able to prevent apoptosis during host cell infection. A complex signalling axes that involve PDPK1 activation, MYC accumulation and the subsequent association of hexokinase II with mitochondrial VDAC leads to a stabilized

**Table 3.1** Gram-negative protein effectors and host protein involved into the control of mitochondrial cell death and related pathways

| Pathogen | Bacterial protein effector | Host proteins | Death type | Modality | References |
|---|---|---|---|---|---|
| Bordetella pertussis | CyaA | BimEL/Bax | Apoptosis | Induction | [1] |
| Brucella melitensis 16m | Many | Many | Apoptosis | Inhibition | [35] |
| *Brucella abortus* RB51 | Many | Casp2/cyt-c | Apoptosis/necrosis | Induction | [16] |
| Brucella abortus RB51 | Many | Casp2/cyt-c | Apoptosis/pyroptosis | Induction | [11] |
| Chlamydia trachomatis | Many | PDPK1/MYC/HKII | Apoptosis | Inhibition | [2] |
| Francisella tularensis | Many | Bax/Bid/casp-8 | Apoptosis | Inhibition | [50] |
| *Pseudomonas aeruginosa* | Pyoverdin | bZIPs/ESRE | Mitophagy | Activation | [64] |
| Vibrio cholerae | VopE | Miro1–2 | Na | Na | [62] |
| *Salmonella enterica* | SopB | TRAF6 | Apoptosis | Inhibition | [54] |
| EPEC | EspC | Procaspase-3/calpain | Apoptosis/necrosis | Activation | [56] |
| Shigella flexneri | IpaD | Caspase 2,8,9 | Apoptosis | Activation | [4] |
| Shigella flexneri | VirA-IpgD | BID,SMAC,XIAP | Apoptosis | Inhibition | [3, 6] |

**Table 3.2** Gram positive protein effectors and host protein involved into the control of mitochondrial death and related pathways

| Pathogen | Pathogenic protein effector | Host proteins | Death type | Modality | References |
|---|---|---|---|---|---|
| S.aureus RN6390 | PVL | Caspase-9, SMAC | Apoptosis (low PVL) | Activation | [29] |
| | | Cytochrome c | | | |
| S.aureus RN6390 | PVL | Caspase-9, SMAC | Necrosis (high PVL) | Activation | [29] |
| | | Cytochrome c | | | |
| S.aureus newman | AT | NRLP3 | Necrosis (high AT) | Activation | [38] |
| S.aureus USA300 strain SF8300 | AT | NRLP3 | Pyroptosis (low AT) | Activation | [19] |
| S.aureus ATCC 25293 | Many | NRF2 | Mitophagy | Activation | [14] |
| *S. aureus* 25,293 (Seattle 45) | Many | P62, Pink1,Parkin | Mitophagy | Activation | [61] |
| *S. epidermidis* | Many | P62, Pink1,Parkin,Drp1 | Mitophagy | Activation | [32] |

mitochondrial energetics and in-vitro infected cell survival [2]. A different strategy is followed by Francisella tularensis (FT) to inhibit PMN apoptosis and permit dissemination of this pathogen. During 24–48 h of infection, FT significantly inhibits both Bax expression and translocation to the mitochondria and conversion of Bid into active tBid via caspase 8 [50]. Pyoverdine released by *Pseudomonas aeruginosa* is at the basis of the bacterial virulence during the interaction with the experimental host *c. elegans*. This siderophore damages mitochondria and initiate the mitophagic turnover activating a hypoxia-like defense response. During the infection, the disrupted mitochondrial function activate the conserved Ethanol and Stress Response Element (ESRE) via the bZIP family of transcription factors leading mainly to the transcription of small HSPs in *c. elegans* during the infection [64]. In this report, the authors highlight the existence of relatively new conserved mitochondrial surveillance network that link stress response and innate immune function. The same network, during Vibrio cholera infection, is targeted by a Type 3 secretion system (T3SS) effector, VopE, which acts as a GTPase-activating protein that interacts with the activity of mitochondrial Rho GTPases Miro1 and Miro2. In this way the VopE interfere with host-specific mitochondrial proteins involved in the cellular surveillance-activated detoxification and defenses (cSADD) leading to an altered innate immune signalling [62]. In a similar fashion, *Salmonella enterica* serovar typhimurium inject via the T3SS effector proteins that facilitate the infection inhibiting the apoptosis of the infected cells. In particular SopB encoded by the Salmonella pathogenicity island 1 (SP1) attenuate the production of ROS in mitochondria binding the cytosolic tumor necrosis factor receptor associated factor 6 (TRAF6) that is the switch of ROS generation that leads to intrinsic apoptosis into the host epithelial cells [54]. Another bacterium that target the intestinal epithelium through the T3SS injectosome is the Enteropathogenic *Escherichia coli* (EPEC). During the infection a multitude of effectors are injected and several target the mitochondrion leading to the modulation of the host apoptotic process. A particular effector, EspC, has been investigated recently and authors reported for the first time the induction of intrinsic apoptosis by the catalytic activity of an EPEC protein. In particular, the serine protease activity is at the basis of the direct cleavage of the procaspase-3, moreover the same effector is able to activate calpain that increases cellular calcium leading to an enhanced cell death [56]. As described for other pathogenic gram-negative bacteria, also Shigella flexneri is able to trigger apoptosis in macrophages through activation of caspases accompanied by mitochondrial disruption. This is accomplished by a structural component of the T3SS, IpaD, that, once secreted, acts on multiple initiator caspases (2, 8 and 9) and trigger a loss of mitochondrial membrane potential thereby demonstrating the central role of mitochondria in IpaD mediated pathogenesis [4]. When intestinal epithelial cells are infected shigella do not induce apoptosis like in macrophages because the bacterium have to keep alive the host cells to sustain the infection rate. To control the cell death machinery shigella inject several effectors via the T3SS. VirA lead to p53 degradation and IpgD increasing PI5P promote PI3K/Akt-dependent survival pathways [6]. VirA induce a survival process that start with BID (BH3 interacting domain death agonist) cleavage by bacterial activated calpain, this leads to the release of SMAC (second mitochondria-derived activator of caspases) from mitochondria and the inhibition of apoptosis and immune response [3] mediated by XIAP (X-linked inhibitor of apoptosis protein). The activation of a PCD program remains one of the principal host defence mechanisms during the infection, and using different effectors, bacteria can promote or stop the PCD depending on the cellular type infected leading to a specific infection dynamics. Due to the multiplicity of effectors released by microorganisms, they can simultaneously induce an activation or inhibition of PCD depending of the host tissue invaded and this strategy can be followed potentially by each gram-negative pathogen.

### 3.1.3 Mitophagy and Related Pathway Targeting by Gram-Positive Bacteria

*Staphylococcus aureus* is one of the main etiological agents of hospital infections and has become a major health problem, due to the spread of strains resistant to different antibiotics, known as methicillin-resistant (MRSA). MRSA is a very serious problem both for human health and, given its presence in animals, for animal health. New clones have emerged in the food chain that can be transmitted from food-producing animals to humans [67, 68].

*Staphylococcus aureus* also has the ability to survive in immunocompetent cells [69, 70] through modulation of mitochondrial protein expression (Jiang et al. 2012). It can also alter, by acting on the host mitochondrial signalling, fundamental physiological responses such as the intrinsic pathway of apoptosis [29]. In the latter paper, authors highlighted the role of Panton-Valentine leukocidin (PVL) on human peripheral PMN, in particular the involvement of this exotoxin in the control of cellular death through apoptosis or necrosis. It has been shown how, depending on the concentration of the toxin, PMN undergone apoptosis (low PVL) or necrosis (high PVL) and that the preferential activation of caspase specific of the mitochondrial pathway (caspase 9) was a feature of the apoptotic phenotype. Additionally PVL was able, during the experimental infection, to induce the release of Smac/DIABLO and cytochrome c from isolated mitochondria [29] so fully confirming the central role of this organelle in the infection dynamics. Many other virulence factors can be used by S.aureus to escape from immune system. In particular, the alpha toxin (AT) is one of the more relevant toxins and his presence has been detected in 99% of isolates, compared to the 20% of PVL. Moreover AT expression in respiratory infections has been correlated with worse clinical outcome [60] and the associated mortality [9] and morbidity.

As described for PVL, also AT is able to elicit different molecular responses depending to the quantity released at the cellular membrane. At high concentration it causes the lysis of the target cell, at low concentration it forms a discrete number of membrane pores that activate several pathways among them the calpain signaling [20] and the NRLP3 inflammasome [22, 25, 38] both involved in innate immune defences. In a recent paper the activation of the NLRP3 inflammasome in response to AT blocked mitochondrial localization to phagocytized bacteria leading to a limited killing by human monocytes. In particular, mitochondria preferentially bind NRLP3 and the resulting complex activate caspase 1 with production of IL1b and IL18. The NRLP3-AT mediated uncoupling of the organelle from endosome prevented caspase 1 activation, mitochondrial ROS production against internalized bacteria and initiation of phagosome acidification [19]. The inflammatory response during *S. aureus* infection is closely linked to oxidative stress and ROS production. It has been demonstrated recently in a murine model of *S. aureus* sepsis the importance of a redox switch in mitochondrial quality control. In particular, data highlighted the main role of ROS-activated Nrf2 in the simultaneous modulation of mitochondrial biogenesis and mitophagy in specific tissues [14]. A similar molecular response has been described in an *s. aureus* pneumonia infection model in mice, where in mitochondria-rich alveolar type II (AT2) cells cell death was limited while both mitochondrial biogenesis and mitophagy were activated. Particularly the overexpression of nuclear mitochondrial biogenesis regulators nuclear respiratory factor-1 (NRF-1) and peroxisome proliferator- activated receptor-γ coactivator-1α (PGC-1α) indicated an augmentated mitochondrial biogenesis. Moreover a simultaneous activation of mitophagy mediated by p62, Pink 1 and Parkin has been detected leading to an activated mitochondrial QC and a promotion of cell survival in s.aureus pneumonia damaged AT2 cells [61]. Also known commensal bacteria are able to induce a similar molecular response at the host level. This phenomenon has been demonstrated in bone marrow mesenchymal stromal cells (BMSCs) challenged with *S. epidermidis*. The bacterial challenge activated, by the PRE/DAMP response, several stress-responsive, autolysosomal and mitophagy-related genes and the related proteins. Globally, the infection of BMSCs with *S. epidermidis* lead mainly to the over-expression of mitophagy related proteins (Parkin, p62, LC3) that interacted with damaged mitochondria which were subjected to autophagosomal clearance [32]. Was therefore observed a potential role in suppressing the intrinsic apoptotic pathway linked to the up-regulation of the mitophagic pathway by s.epidermidis challenged BMSCs in view of survival mechanism linked to the innate defence response.

## 3.2 Host Mitochondria and Viruses

Viruses are obligate parasites that completely rely on the host machinery for their survivor and replication. For this reasons, they have to strongly interact with the cellular structures in order to shape metabolism and architecture for their replication. Since mitochondria play a key role in cellular homeostasis and energy production, the successful replication and completion of viral cycles is guaranteed by their interaction with this mechanisms [39].

From this perspective, it has to be mentioned that mitochondria can be either directly targeted by viruses or, what can happen is that their homeostasis will be just disrupted by the viral modification of other cellular pathways.

Viruses, during their evolution, developed many different methods to suppress mitochondrial antiviral action, and this argument, still represents a field of intense investigation because of the extremely blurry knowledge [57].

### 3.2.1 Viral Targeting of Mitochondrial Ca++ Homeostasis and Membrane Potential

One of the major ways to negatively interact with mitochondrial homeostasis is related with the deregulation of ions in the cell. Among all ions, Ca2+ represents one of the most important elements specially if considering the fact that it is as well an intra-cellular messenger [8]. Outer mitochondrial membrane has voltage-dependent anion channels (VDAC) that are able to control the entrance of Ca++ ions into the mitochondrial intermembrane [47]. Moreover, Ca++ concentration in the intermembrane is as well regulated through the inner membrane by the mitochondrial Ca2+ uniporter (MCU) which is as well selective for Ca++ ions [42]. These represent just two examples of the multiple regulation mechanisms that finely regulate calcium homeostasis inside mitochondria.

Viral infections can interact with calcium metabolism and, in turn with mitochondria in many different ways according to the viral family or specie.

It has been demonstrated how Herpes simplex type (HSV) 1 can completely take over mitochondrial Ca++ concentration control by reducing it of 65% leaving constant the overall cellular calcium concentration. This different homeostasis helps it cycle in viral replication [48].

On the contrary, other types of viruses, as hepatitis C virus (HCV) cause mitochondrial increase of Ca++ through the action of its core protein or NS5A [45]. The p7 protein of HCV is responsible for the formation of pores in Ca++ storage organelles that lead to the overall perturbation of cellular homeostasis [21, 34].

Tat HIV protein is strictly related to an impaired uptake of mitochondrial Ca++, which strongly disturbs the electrophysiological activity of cardiomyocytes. This could represent the reason linked to the onset of cardiac disease linked to HIV infection [63]. The mitochondrial membrane potential (MMP) corresponds to $\cong$ 180 mV across the inner membrane and is generated by the gradient of H+ and the electron transport chain. There are many viral species capable to alter MMP to enhance bacterial survival and replication. Among virus capable to interact with MMP it can be mentioned HCV, HIV, myxoma poxvirus, influenza A virus and many others. All of them are produce proteins that are capable to localize in the mitochondria and to directly interact with mitochondrial membrane permeability [7, 23, 24, 31]. A stronger confirmation of this mechanism has been described in case of (HIV-1). In this specific case, viral protein R (Vpr) is responsible for the loss of membrane potential (MMP) that at the end brings to cellular death [36].

### 3.2.2 Viral Modulation of Mitochondrial Oxidative Stress and Apoptosis

Electron transport chain represents the major source of ROS in the cell. Therefore, every perturbation of the mitochondrial membrane could lead to the release of these compounds in the cytosol producing cellular damage. There are several viruses that are capable to induce the formation or the release of ROS. Among them Human-Adenovirus- (HAdV-) 5 is capable to break endosomal membrane causing the leak of lysosomal cathepsins. Cathepsins, in turn, breaks mitochondrial membrane and leads to the release of ROS from mitochondria [51]. The core protein of HCV is responsible for the production of oxidative stress in the cell and it interferes with the regulation of apoptotic pathways [49, 53]. Both these mechanisms are positively related and favorable to viral replication. Among other viral species, both HIV and EBV are causative agents for ROS production [43, 46].

Chronic hepatitis B virus (HBV) is capable of the synthesis of HBV X protein (HBx) which is responsible for oxidative-stress dependent apoptosis. More precisely, it is involved in the opening of mitochondrial permeability transition pore (MPTP) that leads to the disruption of mitochondrial membrane potential an to the production of oxidative stress [27].

One of the first mechanisms of cell defense against viruses is linked to the activation of pro-apoptotic signals when some perturbation is detected. This will ensure the controlled death of infected cells and the avoidance of viral particles replication and spread. For this reasons, viruses evolved several mechanisms capable of controlling and regulating apoptotic pathways. One common mechanism is related to the viral synthesis of Bcl-2 homologous protein (vBcl-2) that is involved in the regulation of pro-apoptotic and anti-apoptotic pathways. The capability of viruses to regulate this process gives the possibility to complete their life cycle and to control the cellular disruption and the release of viral particles [10, 41, 59]. There are some species of virus that can also encode for viral mitochondrial inhibitors of apoptosis (vMIA) that are responsible for protecting the cell [30]. An important mechanism of cellular protection to enhance viral survival has been recently described in Hepatitis B virus. More precisely, this virus, is capable to silence Parkin signaling and, in this way, promote viral persistence [40].

### 3.2.3 Viral Modulation of Mitochondrial Antiviral Immunity

Mitochondrial antiviral signaling protein (MAVS) has an important role in the cellular defense against viruses. This protein has a trans membrane region (TM) and it has been show that its expression is linked to the activation of NFkB that, in turn, leads to the production of type I interferon [57]. Several viruses developed some mechanisms to cleave MAVs domains that are placed out of mitochondrial membrane reducing the cellular ability to induce interferon response [44, 52, 66]. The main mechanism is related to the synthesis of a protein called NS3/4A that co-localizes with MAVS leading to its cleavage and to the consequent block of its defensive role against HCV [17, 52].

Dengue virus nonstructural protein (NS)4B is involved in the alteration of mitochondrial morphology and, this mechanism is linked to the protection of the virus from mitochondrial immunity transforming the environment into being favorable for viral replication [15].

## 3.3 Host Mitochondria and Fungi

Several authors reported the role of host mitochondria in fungi infections [13]. Mitochondrial activity has been involved in several aspects of fungal cell biology. These aspects are fundamental for the understanding of several human pathogens, like *candida albicans*, Cryptococcus spp. and Aspergillus spp. Within fungal pathogenesis, mitochondria play a central role in the development of morphological switch, as hyphal differentiation, biofilm formation, stress adaptation, cell wall biosynthesis, drug resistance and host-pathogen interaction. Fungi had their particular protein repertoire that mitochondria are specific to exert the function described [12, 58]. In particular, a pivotal role is given to the complex system ERMES, ER-mitochondria encounter structure. This complex is constituted by four subunits named as Mmm1, Mdm10, Mdm12, and Mdm34. Mmm1 is the only one that is reported as integral ER proteins, while the others. Mdm10, 12 and 34 are placed in or at the mitochondrial outer membrane. Moreover, it is described the regulator of ERMES, the mitochondrial Rho (Miro) GTPase Gem1. Metazoans lack entirely in ERMES; so ERMES could be used as target in antifungal therapy; since authors demonstrated that in *S. cerevisiae* mutants lacking core subunits of ERMES typically provoke alteration in mitochondrial morphology with respiratory defect growth till to complete absence, that is essential for the growth of Aspergillus fumigatus [28]. Interestingly, Cohelo and collaborators focused the attention on infection of *C. neoformans* in macrophage phagocytosis. In general, authors reported a role for mitochondria in activating the inflammasome pathway, but their data instead showed a clear position of NO in the modulation of mitochondria. In fact, it is well known that more extended inhibition NO attenuate mitochondria depolarisation, In this way, *C. neoformans* is not able to be virulent because it is necessary efficient glycolytic pathway together full mitochondria function [18]. About sugar metabolism, Huang et al. identified a novel role of mitochondrial complex I (CI). In particular, CI is essential for mannitol-induced biofilm. They use a combination of in vitro and in vivo methods to show for the first time the explanation of the mechanism related to regulation of alternative carbon assimilation affecting *C. albicans* hyphal morphology, biofilm development and intestinal commensalism. Their data established the influences of CI in mannitol cascade (mannitol dehydrogenase) that is essential for both biofilm formation and hyphal growth in *Candida albicans* [37]. These effects, and in particular the CI dysfunction, have a role in gut colonization by Candida. In conclusion, in the field of fungal pathogenesis, the mitochondria exert a fundamental role to design a new drug. Mitochondrial inhibitors can act new therapeutic target. Lastly, considering the limited choice of antifungal drugs, the toxicity and the long-time treatment, mitochondria fungal-specific targeting drug can represent a challenge for antifungal therapies [33].

## 3.4 Conclusions

The study of mitochondrial physiology in infectious bacterial, viral and fungal diseases is rapidly expanding to clarify in particular how autophagy and programmed cell death are activated during viral infection and its role in modulating the replication of some microorganisms. A better understanding of these mechanisms will be essential in developing new strategies to control the growth and spread of pathogens.

## References


1. Ahmad JN, Cerny O, Linhartova I, Masin J, Osicka R, Sebo P (2016) cAMP signalling of Bordetella adenylate cyclase toxin through the SHP-1 phosphatase activates the BimEL-Bax pro-apoptotic cascade in phagocytes. Cell Microbiol 18(3):384–398. https://doi.org/10.1111/cmi.12519
2. Al-Zeer MA, Xavier A, Abu Lubad M, Sigulla J, Kessler M, Hurwitz R, Meyer TF (2017) Chlamydia trachomatis prevents apoptosis via activation of PDPK1-MYC and enhanced mitochondrial binding of hexokinase II. EBioMedicine 23:100–110. https://doi.org/10.1016/j.ebiom.2017.08.005

3. Andree M, Seeger JM, Schull S, Coutelle O, Wagner-Stippich D, Wiegmann K, Wunderlich CM, Brinkmann K, Broxtermann P, Witt A, Fritsch M, Martinelli P, Bielig H, Lamkemeyer T, Rugarli EI, Kaufmann T, Sterner-Kock A, Wunderlich FT, Villunger A, Martins LM, Kronke M, Kufer TA, Utermohlen O, Kashkar H (2014) BID-dependent release of mitochondrial SMAC dampens XIAP-mediated immunity against Shigella. EMBO J 33(19):2171–2187. https://doi.org/10.15252/embj.201387244
4. Arizmendi O, Picking WD, Picking WL (2016) Macrophage apoptosis triggered by IpaD from Shigella flexneri. Infect Immun 84(6):1857–1865. https://doi.org/10.1128/IAI.01483-15
5. Arnoult D, Carneiro L, Tattoli I, Girardin SE (2009) The role of mitochondria in cellular defense against microbial infection. Semin Immunol 21(4):223–232. https://doi.org/10.1016/j.smim.2009.05.009
6. Ashida H, Mimuro H, Sasakawa C (2015) Shigella manipulates host immune responses by delivering effector proteins with specific roles. Front Immunol 6:219. https://doi.org/10.3389/fimmu.2015.00219
7. Azuma A, Matsuo A, Suzuki T, Kurosawa T, Zhang X, Aida Y (2006) Human immunodeficiency virus type 1 Vpr induces cell cycle arrest at the G1 phase and apoptosis via disruption of mitochondrial function in rodent cells. Microbes Infect 8(3):670–679
8. Berridge MJ, Bootman MD, Lipp P (1998) Calcium-a life and death signal. Nature 395(6703):645
9. Bonesso MF, Yeh AJ, Villaruz AE, Joo HS, McCausland J, Fortaleza CM, Cavalcante RS, Sobrinho MT, Ronchi CF, Cheung GY, Cunha ML, Otto M (2016) Key role of alpha-toxin in fatal pneumonia caused by *Staphylococcus aureus* sequence type 398. Am J Respir Crit Care Med 193(2):217–220. https://doi.org/10.1164/rccm.201506-1225LE
10. Borutaite V (2010) Mitochondria as decision-makers in cell death. Environ Mol Mutagen 51(5):406–416
11. Bronner DN, O'Riordan MX, He Y (2013) Caspase-2 mediates a *Brucella abortus* RB51-induced hybrid cell death having features of apoptosis and pyroptosis. Front Cell Infect Microbiol 3:83. https://doi.org/10.3389/fcimb.2013.00083
12. Brown GD, Denning DW, Gow NA, Levitz SM, Netea MG, White TC (2012) Hidden killers: human fungal infections. Sci Transl Med 4(165):165rv113. https://doi.org/10.1126/scitranslmed.3004404
13. Calderone R, Li D, Traven A (2015) System-level impact of mitochondria on fungal virulence: to metabolism and beyond. FEMS Yeast Res 15(4):fov027. https://doi.org/10.1093/femsyr/fov027
14. Chang AL, Ulrich A, Suliman HB, Piantadosi CA (2015) Redox regulation of mitophagy in the lung during murine *Staphylococcus aureus* sepsis. Free Radic Biol Med 78:179–189. https://doi.org/10.1016/j.freeradbiomed.2014.10.582
15. Chatel-Chaix L, Cortese M, Romero-Brey I, Bender S, Neufeldt CJ, Fischl W, Scaturro P, Schieber N, Schwab Y, Fischer B (2016) Dengue virus perturbs mitochondrial morphodynamics to dampen innate immune responses. Cell Host Microbe 20(3):342–356
16. Chen F, He Y (2009) Caspase-2 mediated apoptotic and necrotic murine macrophage cell death induced by rough *Brucella abortus*. PLoS One 4(8):e6830. https://doi.org/10.1371/journal.pone.0006830
17. Chen Z, Benureau Y, Rijnbrand R, Yi J, Wang T, Warter L, Lanford RE, Weinman SA, Lemon SM, Martin A (2007) GB virus B disrupts RIG-I signaling by NS3/4A-mediated cleavage of the adaptor protein MAVS. J Virol 81(2):964–976
18. Coelho C, Souza AC, Derengowski Lda S, de Leon-Rodriguez C, Wang B, Leon-Rivera R, Bocca AL, Goncalves T, Casadevall A (2015) Macrophage mitochondrial and stress response to ingestion of *Cryptococcus neoformans*. J Immunol 194(5):2345–2357. https://doi.org/10.4049/jimmunol.1402350
19. Cohen TS, Boland ML, Boland BB, Takahashi V, Tovchigrechko A, Lee Y, Wilde AD, Mazaitis MJ, Jones-Nelson O, Tkaczyk C, Raja R, Stover CK, Sellman BR (2018) *S. aureus* evades macrophage killing through NLRP3-dependent effects on mitochondrial trafficking. Cell Rep 22(9):2431–2441. https://doi.org/10.1016/j.celrep.2018.02.027
20. Cohen TS, Hilliard JJ, Jones-Nelson O, Keller AE, O'Day T, Tkaczyk C, DiGiandomenico A, Hamilton M, Pelletier M, Wang Q, Diep BA, Le VT, Cheng L, Suzich J, Stover CK, Sellman BR (2016) *Staphylococcus aureus* alpha toxin potentiates opportunistic bacterial lung infections. Sci Transl Med 8(329):329ra331. https://doi.org/10.1126/scitranslmed.aad9922
21. Cook GA, Opella SJ (2010) NMR studies of p7 protein from hepatitis C virus. Eur Biophys J 39(7):1097–1104
22. Craven RR, Gao X, Allen IC, Gris D, Bubeck Wardenburg J, McElvania-Tekippe E, Ting JP, Duncan JA (2009) *Staphylococcus aureus* alpha-hemolysin activates the NLRP3-inflammasome in human and mouse monocytic cells. PLoS One 4(10):e7446. https://doi.org/10.1371/journal.pone.0007446
23. Danishuddin M, Khan SN, Khan AU (2010) Molecular interactions between mitochondrial membrane proteins and the C-terminal domain of PB1-F2: an in silico approach. J Mol Model 16(3):535–541
24. Everett H, Barry M, Lee SF, Sun X, Graham K, Stone J, Bleackley RC, McFadden G (2000) M11L: a novel mitochondria-localized protein of myxoma virus that blocks apoptosis of infected leukocytes. J Exp Med 191(9):1487–1498
25. Ezekwe EA Jr, Weng C, Duncan JA (2016) ADAM10 cell surface expression but not activity is critical for *Staphylococcus aureus* alpha-Hemolysin-mediated activation of the NLRP3 inflammasome in human monocytes. Toxins 8(4):95. https://doi.org/10.3390/toxins8040095

26. Galluzzi L, Vitale I, Aaronson SA, Abrams JM, Adam D, Agostinis P, Alnemri ES, Altucci L, Amelio I, Andrews DW, Annicchiarico-Petruzzelli M, Antonov AV, Arama E, Baehrecke EH, Barlev NA, Bazan NG, Bernassola F, Bertrand MJM, Bianchi K, Blagosklonny MV, Blomgren K, Borner C, Boya P, Brenner C, Campanella M, Candi E, Carmona-Gutierrez D, Cecconi F, Chan FK, Chandel NS, Cheng EH, Chipuk JE, Cidlowski JA, Ciechanover A, Cohen GM, Conrad M, Cubillos-Ruiz JR, Czabotar PE, D'Angiolella V, Dawson TM, Dawson VL, De Laurenzi V, De Maria R, Debatin KM, DeBerardinis RJ, Deshmukh M, Di Daniele N, Di Virgilio F, Dixit VM, Dixon SJ, Duckett CS, Dynlacht BD, El-Deiry WS, Elrod JW, Fimia GM, Fulda S, Garcia-Saez AJ, Garg AD, Garrido C, Gavathiotis E, Golstein P, Gottlieb E, Green DR, Greene LA, Gronemeyer H, Gross A, Hajnoczky G, Hardwick JM, Harris IS, Hengartner MO, Hetz C, Ichijo H, Jaattela M, Joseph B, Jost PJ, Juin PP, Kaiser WJ, Karin M, Kaufmann T, Kepp O, Kimchi A, Kitsis RN, Klionsky DJ, Knight RA, Kumar S, Lee SW, Lemasters JJ, Levine B, Linkermann A, Lipton SA, Lockshin RA, Lopez-Otin C, Lowe SW, Luedde T, Lugli E, MacFarlane M, Madeo F, Malewicz M, Malorni W, Manic G, Marine JC, Martin SJ, Martinou JC, Medema JP, Mehlen P, Meier P, Melino S, Miao EA, Molkentin JD, Moll UM, Munoz-Pinedo C, Nagata S, Nunez G, Oberst A, Oren M, Overholtzer M, Pagano M, Panaretakis T, Pasparakis M, Penninger JM, Pereira DM, Pervaiz S, Peter ME, Piacentini M, Pinton P, Prehn JHM, Puthalakath H, Rabinovich GA, Rehm M, Rizzuto R, Rodrigues CMP, Rubinsztein DC, Rudel T, Ryan KM, Sayan E, Scorrano L, Shao F, Shi Y, Silke J, Simon HU, Sistigu A, Stockwell BR, Strasser A, Szabadkai G, Tait SWG, Tang D, Tavernarakis N, Thorburn A, Tsujimoto Y, Turk B, Vanden Berghe T, Vandenabeele P, Vander Heiden MG, Villunger A, Virgin HW, Vousden KH, Vucic D, Wagner EF, Walczak H, Wallach D, Wang Y, Wells JA, Wood W, Yuan J, Zakeri Z, Zhivotovsky B, Zitvogel L, Melino G, Kroemer G (2018) Molecular mechanisms of cell death: recommendations of the nomenclature committee on cell death 2018. Cell Death Differ. https://doi.org/10.1038/s41418-017-0012-4
27. Gao W-Y, Li D, Cai D-E, Huang X-Y, Zheng B-Y, Huang Y-H, Chen Z-X, Wang X-Z (2017) Hepatitis B virus X protein sensitizes HL-7702 cells to oxidative stress-induced apoptosis through modulation of the mitochondrial permeability transition pore. Oncol Rep 37(1):48–56
28. Geissel B, Penka M, Neubauer M, Wagener J (2017) The ER-mitochondria encounter structure contributes to hyphal growth, mitochondrial morphology and virulence of the pathogenic mold Aspergillus fumigatus. Int J Med Microbiol: IJMM 307(1):37–43. https://doi.org/10.1016/j.ijmm.2016.11.005
29. Genestier AL, Michallet MC, Prevost G, Bellot G, Chalabreysse L, Peyrol S, Thivolet F, Etienne J, Lina G, Vallette FM, Vandenesch F, Genestier L (2005) *Staphylococcus aureus* Panton-Valentine leukocidin directly targets mitochondria and induces Bax-independent apoptosis of human neutrophils. J Clin Invest 115(11):3117–3127. https://doi.org/10.1172/JCI22684
30. Goldmacher VS, Bartle LM, Skaletskaya A, Dionne CA, Kedersha NL, Vater CA, J-w H, Lutz RJ, Watanabe S, McFarland EDC (1999) A cytomegalovirus-encoded mitochondria-localized inhibitor of apoptosis structurally unrelated to Bcl-2. Proc Natl Acad Sci 96(22):12536–12541
31. Gonzalez ME, Carrasco L (2003) Viroporins. FEBS Lett 552(1):28–34
32. Gorbunov NV, McDaniel DP, Zhai M, Liao PJ, Garrison BR, Kiang JG (2015) Autophagy and mitochondrial remodelling in mouse mesenchymal stromal cells challenged with *Staphylococcus epidermidis*. J Cell Mol Med 19(5):1133–1150. https://doi.org/10.1111/jcmm.12518
33. Green DR, Galluzzi L, Kroemer G (2014) Cell biology. Metabolic control of cell death. Science 345(6203):1250256. https://doi.org/10.1126/science.1250256
34. Griffin SD, Harvey R, Clarke DS, Barclay WS, Harris M, Rowlands DJ (2004) A conserved basic loop in hepatitis C virus p7 protein is required for amantadine-sensitive ion channel activity in mammalian cells but is dispensable for localization to mitochondria. J Gen Virol 85(2):451–461
35. He Y, Reichow S, Ramamoorthy S, Ding X, Lathigra R, Craig JC, Sobral BW, Schurig GG, Sriranganathan N, Boyle SM (2006) Brucella melitensis triggers time-dependent modulation of apoptosis and down-regulation of mitochondrion-associated gene expression in mouse macrophages. Infect Immun 74(9):5035–5046. https://doi.org/10.1128/IAI.01998-05
36. Huang C-Y, Chiang S-F, Lin T-Y, Chiou S-H, Chow K-C (2012) HIV-1 Vpr triggers mitochondrial destruction by impairing Mfn2-mediated ER-mitochondria interaction. PLoS One 7(3):e33657
37. Huang X, Chen X, He Y, Yu X, Li S, Gao N, Niu L, Mao Y, Wang Y, Wu X, Wu W, Wu J, Zhou D, Zhan X, Chen C (2017) Mitochondrial complex I bridges a connection between regulation of carbon flexibility and gastrointestinal commensalism in the human fungal pathogen *Candida albicans*. PLoS Pathog 13(6):e1006414. https://doi.org/10.1371/journal.ppat.1006414
38. Kebaier C, Chamberland RR, Allen IC, Gao X, Broglie PM, Hall JD, Jania C, Doerschuk CM, Tilley SL, Duncan JA (2012) *Staphylococcus aureus* alpha-hemolysin mediates virulence in a murine model of severe pneumonia through activation of the NLRP3

inflammasome. J Infect Dis 205(5):807–817. https://doi.org/10.1093/infdis/jir846
39. Khan M, Syed GH, Kim S-J, Siddiqui A (2015) Mitochondrial dynamics and viral infections: a close nexus. Biochim Biophys Acta (BBA)-Mol Cell Res 1853(10):2822–2833
40. Kim S-J, Khan M, Quan J, Till A, Subramani S, Siddiqui A (2013) Hepatitis B virus disrupts mitochondrial dynamics: induces fission and mitophagy to attenuate apoptosis. PLoS Pathog 9(12):e1003722
41. Kim Y-J, Jung JK, Lee SY, Jang KL (2010) Hepatitis B virus X protein overcomes stress-induced premature senescence by repressing p16INK4a expression via DNA methylation. Cancer Lett 288(2):226–235
42. Kirichok Y, Krapivinsky G, Clapham DE (2004) The mitochondrial calcium uniporter is a highly selective ion channel. Nature 427(6972):360
43. Lassoued S, Gargouri B, El Feki AF, Attia H, Van Pelt J (2010) Transcription of the Epstein–Barr virus lytic cycle activator BZLF-1 during oxidative stress induction. Biol Trace Elem Res 137(1):13–22
44. Li X-D, Sun L, Seth RB, Pineda G, Chen ZJ (2005) Hepatitis C virus protease NS3/4A cleaves mitochondrial antiviral signaling protein off the mitochondria to evade innate immunity. Proc Natl Acad Sci U S A 102(49):17717–17722
45. Li Y, Boehning DF, Qian T, Popov VL, Weinman SA (2007) Hepatitis C virus core protein increases mitochondrial ROS production by stimulation of Ca2+ uniporter activity. FASEB J 21(10):2474–2485
46. Lin W, Wu G, Li S, Weinberg EM, Kumthip K, Peng LF, Méndez-Navarro J, Chen W-C, Jilg N, Zhao H (2011) HIV and HCV cooperatively promote hepatic fibrogenesis via induction of reactive oxygen species and NFκB. J Biol Chem 286(4):2665–2674
47. Liu Y, Gao L, Xue Q, Li Z, Wang L, Chen R, Liu M, Wen Y, Guan M, Li Y (2011) Voltage-dependent anion channel involved in the mitochondrial calcium cycle of cell lines carrying the mitochondrial DNA A4263G mutation. Biochem Biophys Res Commun 404(1):364–369
48. Lund K, Ziola B (1985) Cell sonicates used in the analysis of how measles and herpes simplex type 1 virus infections influence Vero cell mitochondrial calcium uptake. Can J Biochem Cell Biol 63(11):1194–1197
49. Machida K, McNamara G, Cheng KT-H, Huang J, Wang C-H, Comai L, Ou J-HJ, Lai MM (2010) Hepatitis C virus inhibits DNA damage repair through reactive oxygen and nitrogen species and by interfering with the ATM-NBS1/Mre11/Rad50 DNA repair pathway in monocytes and hepatocytes. J Immunol 185(11):6985–6998
50. McCracken JM, Kinkead LC, McCaffrey RL, Allen LA (2016) Francisella tularensis modulates a distinct subset of regulatory factors and sustains mitochondrial integrity to impair human neutrophil apoptosis. J Innate Immun 8(3):299–313. https://doi.org/10.1159/000443882
51. McGuire KA, Barlan AU, Griffin TM, Wiethoff CM (2011) Adenovirus type 5 rupture of lysosomes leads to cathepsin B-dependent mitochondrial stress and production of reactive oxygen species. J Virol 85(20):10806–10813
52. Meylan E, Curran J, Hofmann K, Moradpour D, Binder M, Bartenschlager R, Tschopp J (2005) Cardif is an adaptor protein in the RIG-I antiviral pathway and is targeted by hepatitis C virus. Nature 437(7062):1167
53. Ming-Ju H, Yih-Shou H, Tzy-Yen C, Hui-Ling C (2011) Hepatitis C virus E2 protein induce reactive oxygen species (ROS)-related fibrogenesis in the HSC-T6 hepatic stellate cell line. J Cell Biochem 112(1):233–243
54. Ruan H, Zhang Z, Tian L, Wang S, Hu S, Qiao JJ (2016) The Salmonella effector SopB prevents ROS-induced apoptosis of epithelial cells by retarding TRAF6 recruitment to mitochondria. Biochem Biophys Res Commun 478(2):618–623. https://doi.org/10.1016/j.bbrc.2016.07.116
55. Rudel T, Kepp O, Kozjak-Pavlovic V (2010) Interactions between bacterial pathogens and mitochondrial cell death pathways. Nat Rev Microbiol 8(10):693–705. https://doi.org/10.1038/nrmicro2421
56. Serapio-Palacios A, Navarro-Garcia F (2016) EspC, an autotransporter protein secreted by Enteropathogenic *Escherichia coli*, causes apoptosis and necrosis through caspase and calpain activation, including direct procaspase-3 cleavage. mBio 7(3). doi:https://doi.org/10.1128/mBio.00479-16
57. Seth RB, Sun L, Ea C-K, Chen ZJ (2005) Identification and characterization of MAVS, a mitochondrial antiviral signaling protein that activates NF-κB and IRF3. Cell 122(5):669–682
58. Shingu-Vazquez M, Traven A (2011) Mitochondria and fungal pathogenesis: drug tolerance, virulence, and potential for antifungal therapy. Eukaryot Cell 10(11):1376–1383. https://doi.org/10.1128/EC.05184-11
59. Soriano ME, Scorrano L (2010) The interplay between BCL-2 family proteins and mitochondrial morphology in the regulation of apoptosis. In: BCL-2 protein family. Springer, pp 97–114
60. Stulik L, Malafa S, Hudcova J, Rouha H, Henics BZ, Craven DE, Sonnevend AM, Nagy E (2014) Alpha-Hemolysin activity of methicillin-susceptible *Staphylococcus aureus* predicts ventilator-associated pneumonia. Am J Respir Crit Care Med 190(10):1139–1148. https://doi.org/10.1164/rccm.201406-1012OC
61. Suliman HB, Kraft B, Bartz R, Chen L, Welty-Wolf KE, Piantadosi CA (2017) Mitochondrial quality

control in alveolar epithelial cells damaged by *S. aureus* pneumonia in mice. Am J Phys Lung Cell Mol Phys 313(4):L699–L709. https://doi.org/10.1152/ajplung.00197.2017

62. Suzuki M, Danilchanka O, Mekalanos JJ (2014) Vibrio cholerae T3SS effector VopE modulates mitochondrial dynamics and innate immune signaling by targeting Miro GTPases. Cell Host Microbe 16(5):581–591. https://doi.org/10.1016/j.chom.2014.09.015
63. Tahrir FG, Shanmughapriya S, Ahooyi TM, Knezevic T, Gupta MK, Kontos CD, McClung JM, Madesh M, Gordon J, Feldman AM (2018) Dysregulation of mitochondrial bioenergetics and quality control by HIV-1 Tat in cardiomyocytes. J Cell Physiol 33(2):748–758
64. Tjahjono E, Kirienko NV (2017) A conserved mitochondrial surveillance pathway is required for defense against *Pseudomonas aeruginosa*. PLoS Genet 13(6):e1006876. https://doi.org/10.1371/journal.pgen.1006876
65. Weinberg SE, Sena LA, Chandel NS (2015) Mitochondria in the regulation of innate and adaptive immunity. Immunity 42(3):406–417. https://doi.org/10.1016/j.immuni.2015.02.002
66. Xu Y, Zhong H, Shi W (2010) MAVS protects cells from apoptosis by negatively regulating VDAC1. Mol Cell Biochem 1–8
67. van Cleef BA, Monnet DL, Voss A, Krziwanek K, Allerberger F, Struelens M, Zemlickova H, Skov RL, Vuopio-Varkila J, Cuny C, Friedrich AW, Spiliopoulou I, Pászti J, Hardardottir H, Rossney A, Pan A, Pantosti A, Borg M, Grundmann H, Mueller-Premru M, Olsson-Liljequist B, Widmer A, Harbarth S, Schweiger A, Unal S, Kluytmans JA (2011 Mar) Livestock-associated methicillin-resistant *Staphylococcus aureus* in humans, Europe. Emerg Infect Dis 17(3):502–505. https://doi.org/10.3201/eid1703.101036. PubMed PMID: 21392444; PubMed Central PMCID: PMC3166010
68. Hanselman BA, Kruth SA, Rousseau J, Low DE, Willey BM, McGeer A, Weese JS (2006 Dec) Methicillin-resistant *Staphylococcus aureus* colonization in veterinary personnel. Emerg Infect Dis 12(12):1933–1938. PubMed PMID: 17326947; PubMed Central PMCID: PMC3291342
69. Kubica M, Guzik K, Koziel J, Zarebski M, Richter W, Gajkowska B, Golda A, Maciag-Gudowska A, Brix K, Shaw L, Foster T, Potempa J (2008 Jan 9) A potential new pathway for *Staphylococcus aureus* dissemination: the silent survival of *S. aureus* phagocytosed by human monocyte-derived macrophages. PLoS One 3(1):e1409. https://doi.org/10.1371/journal.pone.0001409. PubMed PMID: 18183290; PubMed Central PMCID: PMC2169301
70. Thwaites GE, Gant V (2011 Mar) Are bloodstream leukocytes Trojan horses for the metastasis of *Staphylococcus aureus*? Nat Rev Microbiol 9(3):215–222. https://doi.org/10.1038/nrmicro2508. Epub 2011 Feb 7. PubMed PMID: 21297670

# Mitochondria Lysine Acetylation and Phenotypic Control

**4**

Federica Ciregia

**Abstract**

Mitochondria have a central role in cellular metabolism and reversible post-translational modifications regulate activity of mitochondrial proteins. Thanks to advances in proteomics, lysine acetylation has arisen as an important post-translational modification in the mitochondrion. During acetylation an acetyl group is covalently attached to the epsilon amino group in the side chain of lysine residues using acetyl-CoA as the substrate donor. Therefore the positive charge is neutralized, and this can affect the function of proteins thereby regulating enzyme activity, protein interactions, and protein stability. The major deacetylase in mitochondria is SIRT3 whose activity regulates many mitochondrial enzymes. The method of choice for the analysis of acetylated proteins foresees the combination of mass spectrometry-based proteomics with affinity enrichment techniques. Beyond the identification of lysine-acetylated proteins, many studies are moving towards the characterization of acetylated patterns in different diseases. Indeed, modifications in lysine acetylation status can directly alter mitochondrial function and, therefore, be linked to human diseases such as metabolic diseases, cancer, myocardial injury and neurodegenerative diseases. Despite the progress in the characterization of different lysine acetylation sites, additional studies are needed to differentiate the specific changes with a significant biological relevance.

**Keywords**

Mitochondria · Lysine acetylation · Sirtuin3 · Post-translational modifications · Proteomics

## 4.1 Introduction

Mitochondria have a central role in cellular metabolism considering that they harbour the main metabolic pathways. Urea cycle, tricarboxylic acid cycle, oxidative phosphorylation and fatty acid oxidation take place in these organelles. Therefore, since mitochondria provide the major source of ATP for cellular activity, defects in mitochondrial function contribute to pathologies. Indeed, mitochondrial dysfunctions have been associated with the pathogenesis of many diseases including cardiac diseases [45, 86], neurodegenerative diseases [51, 98], cancer [71, 79], and metabolic disorders such as obesity [23, 44] and type 2 diabetes [62, 84].

It is known that reversible post-translational modifications (PTMs) regulate activity not only

F. Ciregia (✉)
Department of Rheumatology, GIGA Research, Centre Hospitalier Universitaire (CHU) de Liège, University of Liège, Liège, Belgium
e-mail: federica.ciregia@uliege.be





A. Urbani, M. Babu (eds.), *Mitochondria in Health and in Sickness*, Advances in Experimental Medicine and Biology 1158, https://doi.org/10.1007/978-981-13-8367-0_4

of cytosolic and nuclear proteins, but also of mitochondrial proteins. Indeed, PTMs are involved in many cellular processes such as enzyme regulation, signal transduction, mediation of protein localization, interactions and stability [61]. Since genomic cannot be used for the analysis of modified proteins, proteomics is the method of choice in this area. Thanks to advances in mass spectrometry-based proteomics and affinity enrichment techniques, lysine acetylation has arisen as an important PTM that rivals phosphorylation and ubiquitination in its prevalence [15, 88]. Proteomics revealed that acetylation occurs in more than 2000 proteins involved in different cellular processes [15]. Moreover, lysine acetylation is a PTM more frequent than phosphorylation in bacteria. Thus, considering the endosymbiotic theory, it is not surprising that lysine acetylation is also more abundant in mitochondria of eukaryotic cells than other PTMs [17, 56].

Acetylation of proteins on lysine residues is a reversible PTM that was discovered more than 50 years ago on histones [2, 56] where it reduces their affinity with DNA. p53 was the first non-histone protein identified to be lysine acetylated [26]. Subsequently, in 2006, an extensive proteomics survey revealed, for the first time, that acetylation is a common posttranslational modification in the mitochondrion [42], and it has been assessed that nearly 35% of all proteins in mitochondria have one acetylation site, with an average of 5.6 sites per protein [4, 50]. Moreover, Lundby et al., by combining lysine-acetylated peptide immunoprecipitation and high-accuracy tandem mass spectrometric, discovered that, in muscle, mitochondria is the cells compartment which had the highest level of acetylation and the fraction of lysine-acetylated proteins is approximately three-fold greater than phosphorylated proteins [50].

## 4.2 Lysine Acetylation

### 4.2.1 General Aspects

The lysine acetylation levels are reflective of the balance between lysine acetyl-transferase (KATs) and lysine deacetylase (KDACs) activity, the specific enzymes which regulate acetylation and deacetylation, respectively (Fig. 4.1a). However, non-enzymatic acetylation can occur in mitochondria (Fig. 4.1b), where the physiologic pH (pH $\geq$7.5) in the matrix is sufficient to cause enzyme-independent acetylation [85]. During enzymatic acetylation an acetyl group is covalently attached to the epsilon amino group in the side chain of lysine residues using acetyl-CoA as the substrate donor. Therefore the positive charge is neutralized, and this can affect the function of proteins thereby regulating enzyme activity, protein interactions, and protein stability [35]. The proteins with acetyl-lysine residues can be specifically recognized by bromodomain-containing proteins which are mostly nuclear and cytosolic. Even if bromodomains have not yet been identified in mitochondria, unknown proteins might have similar role. Conversely, KDACs catalyze deacetylation by using $Zn^{2+}$ as a cofactor (Fig. 4.1c), and are mainly localized in the nucleus and cytosol, whereas sirtuins (SIRTs) are a family of $NAD^+$-dependent deacetylases which release *O*-acetyl-ribose and nicotinamide (NAM) as by-products of the deacetylation [35] (Fig. 4.1d). This specific group includes seven members (SIRT 1-7) which are involved in the regulation of many cellular activities. Sirtuins 3, 4 and 5 are the three SIRTs located in mitochondria. In particular, SIRT 3 is the main deacetylase in mitochondria whose deregulation has been implicated in the development of diseases such as diabetes, myocardial injury, and cancer [90].

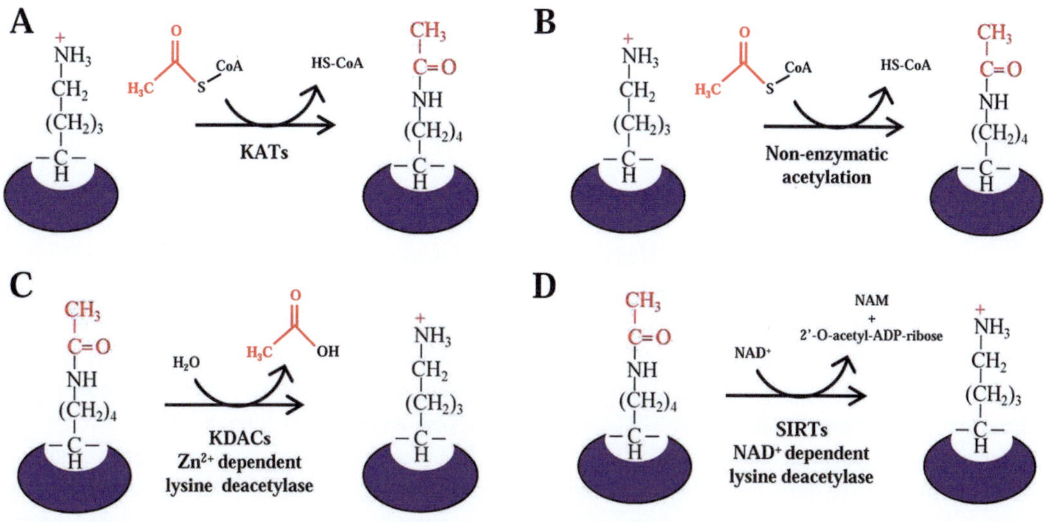


**Fig. 4.1** Scheme of acetylation and deacetylation. (**a**) Enzymatic transfer of an acetyl group from acetyl-CoA to the ε-amino group of a lysine residue, catalyzed by lysine acetyl-transferase (KATs). (**b**) Non-enzymatic acetylation of the ε-amino group. (**c**) Deacetylation of acetyl-lysine residues catalyzed by lysine deacetylases (KDACs) by using $Zn^{2+}$ as a cofactor. (**d**) Deacetylation of acetyl-lysine residues catalyzed by sirtuins (SIRTs), $NAD^+$ dependent deacetylases which release *O*-acetyl-ribose and nicotinamide (NAM)

### 4.2.2 Proteomics Analysis

So far, phosphorylation and ubiquitylation have been the most studied PTMs which are involved in the regulation of many cellular processes. Indeed, acetylation is generally less abundant, and its low stoichiometry nature complicates the identification of this modification [6]. Recently, analysis of acetylation has been propelled by progresses in proteomics approaches and it has been demonstrated that the regulatory role of lysine acetylation is wide and comparable with that of other major PTMs [15]. The method of choice for the analysis of acetylated proteins foresees that proteins are extracted from the biological samples. The isolation of subcellular fractions, such as mitochondria, enables the enrichment of acetylated peptides specifically present in different cell compartments. By this way it can be enhanced the sensitivity of the analysis when focused on a specific subset. Subsequently, proteins are digested into peptides, typically using trypsin. Peptides can be fractionated prior to enrichment, usually by ZIC®-Hydrophilic Interaction Liquid Chromatography (HILIC) or Strong cation exchange (SCX) based HPLC. Thereafter, the enrichment of acetylated peptides is required to decrease sample complexity since not all the proteins are acetylated. For this purpose, specific anti-acetyl-lysine antibodies are used for the immunoaffinity purification. Seeing the low specificity of the most commercially available antibodies, the use of at least two different antibodies is suggested [75]. Sample complexity can be further reduced by peptide fractionation methods (*e.g.* isoelectric focusing, reversed-phase and SCX chromatography) [35].

Fractions are then analyzed by nano-liquid chromatography tandem mass spectrometry (LC-MS/MS) using reversed-phase chromatography under acidic conditions. MS and MS/MS spectra are then computationally processed to acquire peptide sequences, including the presence and location of PTMs [25]. Top-down proteomics might be suitable but the previous described bottom-up approach is the most common for the analysis of lysine-acetylated proteins.

## 4.3 Roles of Lysine Acetylation in Mitochondria

The three mitochondrial SIRTs (SIRT3, SIRT4 and SIRT5) mediate mitochondrial protein acetylation levels. However, SIRT4 and SIRT5 only display weak deacetylase activity [83], besides, SIRT5 seems to have demalonylase and desuccinylase activities instead of deacetylase activity [18]. Thence, SIRT3 is recognized as the major deacetylase in mitochondria. Indeed, when it is absent, mitochondrial proteins become hyperacetylated, whereas the lack of either SIRT4 or SIRT5 does not entail significant changes in acetylation [1, 48]. Interestingly, mitochondrial acetyltransferases (MATs) have not yet been characterized, therefore the issue how mitochondrial proteins can be acetylated has emerged [4]. One hypothesis is that high acetyl-CoA levels in mitochondria could promote a non-enzymatic acetylation mechanism seeing that non-enzymatic acetylation of histones with acetyl-CoA can occur in vitro [55]. Otherwise, MATs are probably acetyltransferases dissimilar from the known enzymes present in nucleus and cytosol that might be awaiting discovery [4].

As expected, the majority of mitochondrial proteins that have been identified as acetylated take part to some aspects of energy metabolism: e.g. TCA cycle (malate dehydrogenase, isocitrate dehydrogenase, fumarate hydratase, succinate dehydrogenase); antioxidant system (superoxide dismutase); lipid pathway (long-chain acyl-CoA dehydrogenase, enoyl-coA hydratase/3-hydroxyacyl-CoA dehydrogenase acetyl-CoA synthetase); amino acid metabolism (glutamate dehydrogenase, aldehyde dehydrogenase); ketone body metabolism (3-hydroxy-3-methylglutaryl-CoA synthase 2, 3-hydroxybutyrate dehydrogenase); urea cycle (carbamoyl phosphate synthetase 1, ornithine transcarbamoylase); oxidative phosphorylation (NADH dehydrogenase, ubiquinol-cytochrome c reductase, ATP synthase) [30, 42, 48, 53, 72, 96].

As a consequence, it was predictable and to date it is well known, that lysine acetylation on mitochondrial protein is sensitive to metabolic states and dietary conditions. Indeed, high-fat diet, fasting and feeding have been linked to acetylation of mitochondrial proteins, even if data are controversial. In order to characterized the proteins which are acetylated under diverse dietary and metabolic conditions, proteomic studies have been performed [15, 42, 66, 87, 96].

The proteomics survey by Kim et al. was the first analysing lysine acetylation of proteins in liver mitochondria from fasted and fed mice. After tryptic digestion, peptides from samples were purified by using agarose beads bearing immobilized anti-acetyl lysine antibody. The enriched peptides were then analyzed by nano-HPLC/MS/MS in an LTQ mass spectrometer. Among the mitochondrial acetylated proteins, 62% were identified in both fractions, 14% were specific to fed mice, and 24% were unique to fasted mice [42]. This could suggest the increase of acetylation during fasting. Actually, with another large-scale mass spectrometry screening, it was described an increase in mitochondrial protein acetylation in mice fed a calorie-restricted diet [66] similar to the acetylation patterns observed during fasting. But with the raise of researches on lysine acetylation, a paradoxical data came out.

SIRT3 expression is upregulated during fasting, modulating mitochondrial intermediary metabolism and fatty acid utilization. For examples the fatty acid oxidation enzymes long-chain acylCoA dehydrogenase (LCAD) and medium-chain acylCoA dehydrogenase (MCAD) are deacetylated and therefore activated by SIRT3 [8, 33]. Another study [67] showed that SIRT3 deacetylates and increases the enzymatic activity of 3-hydroxy-3-methylglutaryl-CoA synthase 2 (HMGCS2). This mitochondrial liver enzyme catalyses the conversion of acetoacetyl-CoA and acetyl-CoA into 3-hydroxy-3-methylglutaryl-CoA. This is a step in ketone body synthesis, the pathway which is up-regulated during starvation.

To corroborate this interpretation, under dietary restricted conditions, many fatty acid oxidation enzymes show increased lysine residue deacetylation in control mice versus SIRT3 knockout, and SIRT3 null mice have increased accumulation of acylcarnitines, a finding consistent with reduced fat oxidation [30, 49]. On the other hand, although there is this pervasive observation that SIRT3 is down-regulated in high fat feeding [41], the rate of mitochondrial protein acetylation and fat oxidation are invariably enhanced in this dietary milieu [3, 49]. These apparently controversial remarks suggest that any altered metabolic state, such as nutrient lack or excess, lead to mitochondrial protein acetylation, considering that hyperacetylation of mitochondrial protein is also detected with the ethanol supplementation in diet [57].

Up to now, although thousands of mitochondrial acetylation sites have been characterized, the role of acetylation on the majority of these proteins is still unknown. The first work that described the functional role for acetylation on a mitochondrial protein was by Schwer et al. [65]. They reported that mitochondrial acetyl-CoA synthetase 2 (AceCS2) was reversibly acetylated at Lys-642 in the active site of the enzyme. SIRT3 deacetylates Lys-642 of AceCS2 both in vitro and in vivo activating its acetyl-CoA synthetase activity. Since then, some findings have been reached. Still and collaborators applied multiplexed quantitative mass spectrometry for detecting alterations in the mouse liver mitochondrial acetylproteome after acute and chronic alterations in diet [72]. Their findings suggested that SIRT3 usually tends to enhance the activity of its target enzymes, especially in the refed state and early during the onset of obesity, showing an adjustable target preference under different nutritional states. For example, among the mitochondrial proteins with dynamic acetylation sites, they demonstrated that acetylation of acetyl-CoA acetyltransferase 1 (Acat1) inhibits its activity by disrupting CoA binding. Another important mitochondrial substrate of SIRT3 is superoxide dismutase 2 (SOD2) which, once deacetylated, has increased activity and as a result increases the detoxification of ROS [76, 77]. The ROS reduction is also fostered through the increase of the expression of SOD2 and of catalase via the activation of the transcription factor Foxo3a [59]. This is trigger by the interaction of SIRT3 with FoxO3a in mitochondria which support FoxO3a DNA-binding to the SOD2 promoter [37].

**Table 4.1** Mitochondrial enzymes whose activity is regulated by SIRT3

| Protein | Effect | References |
|---|---|---|
| Aldehyde dehydrogenase 2 | Inhibition | [91] |
| Aconitase | Inhibition | [19] |
| F1F0-ATPase subunit | Activation | [7] |
| Malate DH | Activation | [32] |
| NADH dehydrogenase 1a subcomplex 9 | Activation | [1] |
| Very long chain acyl-coa DH | Activation | [94] |
| ATP synthase | Activation | [81] |
| Succinate DH | Activation | [16, 20] |
| AceCS2 acetyl-CoA synthetase 2 | Activation | [65, 29] |
| Complex I | Activation | [1] |
| Mitoribosome | Inhibition | [92] |
| Complex IIa | Activation | [16, 20] |
| Complex V ATP5E and ATP5O | Activation | [7, 89] |
| HMGCS2 (3-hydroxy-3-methylglutaryl-CoA synthase 2 | Activation | [67] |
| Long-chain acyl-CoA dehydrogenase | Activation | [33, 8] |
| Pyruvate dehydrogenase | Activation | [39, 54] |
| Glutamate dehydrogenase | Activation | [48, 64] |
| Isocitrate dehydrogenase 2 | Activation | [70] |
| Superoxide dismutase 2 | Activation | [13, 58, 76] |
| MPTP cyclophilin D | Inhibition | [27] |
| 8-oxoguanine-DNA glycosylase | Activation | [14] |

Table 4.1 lists mitochondrial enzymes whose activity has been found to be regulated by SIRT3.

## 4.4 Mitochondrial Protein Acetylation in Human Diseases

Beyond the identification of lysine-acetylated proteins, many studies are moving towards the characterization of acetylated patterns in different diseases. Modifications in lysine acetylation status can directly alter mitochondrial function and, therefore, be linked to human diseases. In this section, we will focus on the description of lysine acetylation in metabolic diseases, cancer, myocardial injury and neurodegenerative diseases.

### 4.4.1 Metabolic Diseases

The two major features of type 2 diabetes are insulin resistance and the impaired mitochondrial function in muscle. Altered mitochondrial lipid oxidation and glycolytic capacity have also been detected in subjects with type 1 diabetes obesity [38]. Jing and collaborators have demonstrated that the expression of SIRT3 is modified both in type 1 and type 2 diabetes models. This change regulates mitochondrial metabolism and production of ROS, which finally alters insulin signaling. This proves the importance of the balance between acetylation and deacetylation levels in mitochondria but also their probable relation with insulin resistance and metabolic disorders. Moreover, this could suggest that factors increasing SIRT3 activity might potentially reverse some of the adverse effects of type 2 diabetes [38].

An adverse effect of both type 1 and 2 diabetes, is cardiomyopathy. Diabetic cardiomyopathy is characterized by decreased glucose oxidation, mitochondrial bioenergetics, and cardiac function and increased fatty acid oxidation, lipid storage, and myocardial fibrosis [10, 82]. The quantitative analysis, by proteomics approach, of mitochondrial lysine acetylation revealed that the extent of this PTM is higher in diabetes compared with the control [82]. Deficiencies of SIRT3 and compromised acetylation balance have been found to be related to the development of the diseases of the metabolic syndrome. The metabolic syndrome is defined by metabolic abnormalities, including obesity, insulin resistance, hyperlipidemia, hyperglycemia, and hypertension [34, 60]. Sedentary lifestyles and high-fat diets are implicated in the increase of metabolic syndrome. It has been discovered that SIRT3 deficiency and the associated hyperacetylation of mitochondrial proteins cause mitochondrial dysfunction that leads to metabolic syndrome [34].

Therefore, specific deacetylation/acetylation regulation of mitochondrial proteins could be used as a therapeutic tool in diabetes and insulin resistance, as well as other mitochondrial diseases.

### 4.4.2 Cancer

Cancer is one of the leading causes of mortality, characterized by cellular metabolic alteration (e.g. the Warburg effect) and cells with excessive proliferation, resistant to apoptosis. The origin of these cellular behaviors has always been the target of extensive studies. PTMs are well known mechanisms involved in tumorigenesis, by which signaling and pathways are fine regulated. In particular, SIRT3 exhibits a tumor suppressor role. This protective role against cancer is related to the depletion of ROS. Indeed, SIRT3 reduces ROS levels by activating, via deacetylation, the mitochondrial SOD 2 [58, 76, 90], isocitrate dehydrogenase 2 [70], and FoxO3a [73]. To corroborate this role, a study from Haigis et al., with mice lacking SIRT3, provided a model connecting aberrant ROS, the Warburg effect, and carcinogenesis [28]. Murine without Sirt3 had abnormal levels of elevated ROS which lead to genomic instability and causes carcinogenesis in various cell types [21, 28, 90]. Programmed cell death or apoptosis is another well-known hallmark of cancer cells, and SIRT3 modulates pro-apoptotic but also antiapoptotic members of the Bcl-2 family. It is worth to mention that some studies have reached different conclusion, foreseeing a role of SIRT3 in promoting tumorigenesis. In esophageal cancer, the high expression of SIRT3 is associated with a poor outcome [97]. In addition, elevated levels of SIRT3 expression are related with poor prognosis in patients with grade 3 breast cancer [78]. However, it must be considered that different works can obtained diverse results since SIRT3 has a role in numerous cellular conditions on many substrates.

### 4.4.3 Myocardial Injury

Cardiac hypertrophy is a usual myocytes response to different pathologic and physiologic stimuli [73]. After birth, mammalian cardiomyocytes lose their proliferation ability; hence the only way for them to deal with an increased workload on the heart is to undergo hypertrophy [31]. However, prolonged hypertrophy and a continuous growth signal lead to malfunction and sudden cell death. By this way, the workload of the remaining cells increases which further cause heart failure and sudden death due to arrhythmias [22]. The molecular mechanism of myocyte death during heart failure is not yet fully understood [40, 74]. During stress, SIRT3 is highly expressed, probably because the modification in cellular NAD/NADH ratio may contribute to its increase. Therefore, increased NAD content elevates the deacetylase activity of SIRT3, contributing to protect cells against stress-mediated cell death. It has been demonstrated that SIRT3 is capable of blocking the cardiac hypertrophic response by reducing cellular ROS levels. This effect is mediated by the activation of Foxo-dependent antioxidants, catalase and manganese SOD, but also by suppressing ROS-mediated Ras activation and the downstream MAPK/ERK and PI3K/Akt signaling pathways [73]. Moreover, Ku70 has been identified as a new target of SIRT3. Deacetylation of Ku70 by SIRT3 promotes Ku70/Bax interaction, and this makes cells resistant to Bax-mediated cell damage [73].

Mitochondria are also believed to be the key organelle for cardioprotection against ischemia/reperfusion injury. Some works have demonstrated that caloric restriction (CR) restores mitochondrial dysfunction and attenuates oxidative damage in mitochondria [52, 69, 80]. A comprehensive proteomics analysis was undertaken to clarify the effect of aging and CR on mitochondrial proteome; and the findings showed that CR has a minor effect on age-related changes in proteins [12]. Subsequently, the DIGE system was used to identify changes in the expression levels and acetylated state of mitochondrial proteins [68]. Authors identified many deacetylated proteins, among them they found NDUFS1 and Rieske subunit of cytochrome bc1 complex, which belong to complexes I and III, respectively. These proteins were deacetylated in the CR heart, reflecting the decrease in mitochondrial ROS production after ischemia/ reperfusion in the CR heart. Hence, they proposed that the beneficial effect of CR on mitochondrial function is mediated by deacetylating specific mitochondrial proteins [68].

### 4.4.4 Neurodegenerative Diseases

Neurons consume about 20% of the body's energy, therefore they require the delicate maintenance of mitochondrial function [24]. Considering the critical role of SIRT3 as a regulator of mitochondrial protein function, it is not surprising that this enzyme is most probably involved in neurodegenerative diseases.

Alzheimer's Disease (AD) and Parkinson's disease (PD) are the most common neurodegenerative pathology in adult population.

AD is characterised by dementia, starting initially with the loss of short term memory and damaged cognitive abilities. The aggregation of abnormally folded amyloid β (Aβ) in amyloid plaques and tau proteins (hyper-phosphorylated tau; p-tau) in neuronal tangles, are directly related to neurodegenerative processes in patients' brains. Deficiency in mitochondrial metabolism, in particular, defects in Complex I and Complex IV activity, have been found in the early stage of AD [36]. Mitochondrial damage may derive from the accumulation of amyloid β in these organelles, which could lead to apoptotic cell death [11]. Another hypothesis is that the lack of SIRT3 can play a key role in AD. Expression pattern and mRNA of SIRT3 are altered in the cortex of double transgenic APP/PS1 mice [93]. Physical training of these animals decreases DNA damages and ROS production while improved mitochondrial metabolism increasing the activity of Complex I, Complex IV and ATPase. All of these effects may resulted from the exercise-induced expression of SIRT3 which implies the activation, via deacetylation, of SOD2 [9, 63]. Actually, SIRT3 expression has been found increased in both human and mouse AD pathology. Since ROS marker analysis revealed the prevalence of oxidative stress in AD, SIRT3 level may increase in response to increased ROS synthesis [46].

PD is a gradual neurodegenerative disease whose symptoms aggravate with time. These comprise bradykinesia, resting tremor, postural instability. Whereas the mechanisms underlying the clinical and pathological features of PD remain to be defined, the characteristic sign is the aggregation of α-synuclein, a protein with dopaminergic neurotoxicity. It has been demonstrated that SIRT3 null mice do not have motor and non-motor defects respect to wild-type controls. Moreover, SIRT3 deficiency intensified the degeneration of nigrostriatal dopaminergic neurons in 1-methyl-4-phenyl-1,2,3,6-tetrahydropyridine (MPTP)-induced PD mice [47]. This effect is probably due to the role of SIRT3 in preserving free radical scavenging capacity in mitochondria. In line with these observations, Zhang et al. demonstrated that Sirt3 protects against DAergic neuronal damage. They used immunoprecipitation and LC-MS/MS to identify the interacting proteins of SIRT3 [95]. With this study, they found that SIRT3 reduction leads to increased mitochondrial protein acetylation and dysregulation of two critical acetylation substrates: SOD2 and ATP synthase β, which are implicated in the regulation of ROS elimination and ATP production [95]. Hence, SIRT3 prevents ROS accumulation and ATP depletion, contributing in reducing DAergic neuronal death upon MPTP treatment.

Amyotrophic lateral sclerosis (ALS) is a rare, lethal, progressive neurodegenerative disease which perturbs corticospinal tract and conducts to the death of motor neuron in spinal cord, cortex, and brainstem. The presence of SIRT3 together with peroxisome proliferator-activated receptor-c coactivator-1a (PGC-1a) defends from neuronal cell death and mitochondrial fragmentation [5, 43]. SIRT3 promoter region has an estrogen-related receptor (ERR)-binding element (ERRE) [43]. PGC-1a recruits ERRa to the ERRE when there is an increase of oxidative stress. The overexpression of SIRT3 lowers ROS levels by deacetylating SOD2.

Thus, considering that SIRT3 is deeply involved in many aspects of mitochondrial metabolism, it can be expected its role in neurodegenerative disorders. However, further studies are desirable in order to clarify the putative role of SIRT3 as a mediator of neuronal damage.

## 4.5 Concluding Remarks

In the last years protein lysine acetylation has emerged as an important PTM in mitochondria. The improvements in the sensitivity, mass accuracy and mass resolution of mass analyzer instruments have helped in the characterization of different lysine acetylation sites. Despite this progress, the subsequent step is to differentiate the specific changes with a significant biological relevance from adventitious modifications. The improvement of affinity-based quantitative acetyl-proteomics together with optimized alternative specific anti-acetyl lysine antibodies, in a robust workflow, can help to differentiate specific changes in the acetylation status. Indeed, the balance between acetylation and deacetylation may be critical in metabolic pathways and additional studies are needed to unveil the fine molecular mechanisms that regulate this balance in physiological and pathological conditions.

## References

1. Ahn BH, Kim HS, Song S et al (2008) A role for the mitochondrial deacetylase Sirt3 in regulating energy homeostasis. Proc Natl Acad Sci U S A 105(38):14447–14452. https://doi.org/10.1073/pnas.0803790105
2. Allfrey VG, Faulkner R, Mirsky AE (1964) Acetylation and methylation of histones and their possible role in the regulation of RNA synthesis. Proc Natl Acad Sci U S A 51:786–794
3. Alrob OA, Sankaralingam S, Ma C et al (2014) Obesity-induced lysine acetylation increases cardiac fatty acid oxidation and impairs insulin signalling. Cardiovasc Res 103(4):485–497. https://doi.org/10.1093/cvr/cvu156
4. Anderson KA, Hirschey MD (2012) Mitochondrial protein acetylation regulates metabolism. Essays Biochem 52:23–35. https://doi.org/10.1042/bse0520023

5. Ansari A, Rahman MS, Saha et al (2017) Function of the SIRT3 mitochondrial deacetylase in cellular physiology, cancer, and neurodegenerative disease. Aging Cell 16(1):4–16. https://doi.org/10.1111/acel.12538
6. Baeza J, Dowell JA, Smallegan MJ et al (2014) Stoichiometry of site-specific lysine acetylation in an entire proteome. J Biol Chem 289:21326–21338. https://doi.org/10.1074/jbc.M114.581843
7. Bao J, Scott I, Lu Z et al (2010) SIRT3 is regulated by nutrient excess and modulates hepatic susceptibility to lipotoxicity. Free Radic Biol Med 49:1230–1237. https://doi.org/10.1016/j.freeradbiomed.2010.07.009
8. Bharathi SS, Zhang Y, Mohsen AW et al (2013) Sirtuin 3 (SIRT3) protein regulates long-chain acyl-CoA dehydrogenase by deacetylating conserved lysines near the active site. J Biol Chem 288(47):33837–33847. https://doi.org/10.1074/jbc.M113.510354
9. Bo H, Kang W, Jiang N et al (2014) Exercise-induced neuroprotection of hippocampus in APP/PS1 transgenic mice via upregulation of mitochondrial 8-oxoguanine DNA glycosylase. Oxidative Med Cell Longev 2014:834502. https://doi.org/10.1155/2014/834502
10. Bugger H, Abel ED (2009) Rodent models of diabetic cardiomyopathy. Dis Model Mech 2:454–466. https://doi.org/10.1242/dmm.001941
11. Cha MY, Han SH, Son SM et al (2012) Mitochondria-specific accumulation of amyloid β induces mitochondrial dysfunction leading to apoptotic cell death. PLoS One 7(4):e34929. https://doi.org/10.1371/journal.pone.0034929
12. Chang J, Cornell JE, Van Remmen H et al (2007) Effect of aging and caloric restriction on the mitochondrial proteome. J Gerontol A Biol Sci Med Sci 62:223–234
13. Chen Y, Zhang J, Lin Y et al (2011) Tumour suppressor SIRT3 deacetylates and activates manganese superoxide dismutase to scavenge ROS. EMBO Rep 12:534–541. https://doi.org/10.1038/embor.2011.65
14. Cheng Y, Ren X, Gowda AS et al (2013) Interaction of Sirt3 with OGG1 contributes to repair of mitochondrial DNA and protects from apoptotic cell death under oxidative stress. Cell Death Dis 4:e731. https://doi.org/10.1038/cddis.2013.254
15. Choudhary C, Kumar C, Gnad F et al (2009) Lysine acetylation targets protein complexes and co-regulates major cellular functions. Science 325(5942):834–840. https://doi.org/10.1126/science.1175371
16. Cimen H, Han MJ, Yang Y et al (2010) Regulation of succinate dehydrogenase activity by SIRT3 in mammalian mitochondria. Biochemistry 49(2):304–311. https://doi.org/10.1021/bi901627u
17. Drazic A, Myklebust LM, Ree R et al (2016) The world of protein acetylation. Biochim Biophys Acta 1864(10):1372–1401. https://doi.org/10.1016/j.bbapap.2016.06.007
18. Du J, Zhou Y, Su X et al (2011) Sirt5 is a NAD-dependent protein lysine demalonylase and desuccinylase. Science 334(6057):806–809. https://doi.org/10.1126/science.1207861
19. Fernandes J, Weddle A, Kinter CS et al (2015) Lysine acetylation activates mitochondrial aconitase in the heart. Biochemistry 54(25):4008–4018. https://doi.org/10.1021/acs.biochem.5b00375
20. Finley LW, Haas W, Desquiret-Dumas V et al (2011) Succinate dehydrogenase is a direct target of sirtuin 3 deacetylase activity. PLoS One 6(8):e23295. https://doi.org/10.1371/journal.pone.0023295
21. Finley LWS, Haigis MC (2012) Metabolic regulation by SIRT3: implications for tumorigenesis. Trends Mol Med 18(9):516–523. https://doi.org/10.1016/j.molmed.2012.05.004
22. Frey N, Katus HA, Olson EN et al (2004) Hypertrophy of the heart: a new therapeutic target? Circulation 109(13):1580–1589. https://doi.org/10.1161/01.CIR.0000120390.68287.BB
23. Fucho R, Casals N, Serra D et al (2016) Ceramides and mitochondrial fatty acid oxidation in obesity. FASEB J 31(4):1263–1272. https://doi.org/10.1096/fj.201601156R
24. Gao J, Wang L, Liu J et al (2017) Abnormalities of mitochondrial dynamics in neurodegenerative diseases. Antioxidants (Basel) 6(2). pii: E25). https://doi.org/10.3390/antiox6020025
25. Gil J, Ramírez-Torres A, Encarnación-Guevara S (2017) Lysine acetylation and cancer: a proteomics perspective. J Proteome 150:297–309. https://doi.org/10.1016/j.jprot.2016.10.003
26. Gu W, Roeder RG (1997) Activation of p53 sequence-specific DNA binding by acetylation of the p53 C-terminal domain. Cell 90(4):595–606
27. Hafner AV, Dai J, Gomes AP et al (2010) Regulation of the mPTP by SIRT3-mediated deacetylation of CypD at lysine 166 suppresses age-related cardiac hypertrophy. Aging (Albany NY) 2(12):914–923. https://doi.org/10.18632/aging.100252
28. Haigis MC, Deng CX, Finley LW et al (2012) SIRT3 is a mitochondrial tumor suppressor: a scientific tale that connects aberrant cellular ROS, the Warburg effect, and carcinogenesis. Cancer Res 72(10):2468–2472. https://doi.org/10.1158/0008-5472.CAN-11-3633
29. Hallows WC, Lee S, Denu JM (2006) Sirtuins deacetylate and activate mammalian acetyl-CoA synthetases. Proc Natl Acad Sci U S A 103(27):10230–10235. https://doi.org/10.1073/pnas.0604392103
30. Hallows WC, Yu W, Smith BC et al (2011) Sirt3 promotes the urea cycle and fatty acid oxidation during dietary restriction. Mol Cell 41(2):139–149. https://doi.org/10.1016/j.molcel.2011.01.002
31. Heineke J, Molkentin JD (2006) Regulation of cardiac hypertrophy by intracellular signalling pathways. Nat Rev Mol Cell Biol 7:589–600. https://doi.org/10.1038/nrm1983
32. Hebert AS, Dittenhafer-Reed KE, Yu W et al (2013) Calorie restriction and SIRT3 trigger global reprogramming of the mitochondrial protein acetylome. Mol Cell 49(1):186–199. https://doi.org/10.1016/j.molcel.2012.10.024
33. Hirschey MD, Shimazu T, Goetzman E et al (2010) SIRT3 regulates mitochondrial fatty-acid oxida-

tion by reversible enzyme deacetylation. Nature 464(7285):121–125. https://doi.org/10.1038/nature08778
34. Hirschey MD, Shimazu T, Jing E, Grueter CA et al (2011) SIRT3 deficiency and mitochondrial protein hyperacetylation accelerate the development of the metabolic syndrome. Mol Cell 44(2):177–190. https://doi.org/10.1016/j.molcel.2011.07.019
35. Hosp F, Lassowskat I, Santoro V et al (2017) Lysine acetylation in mitochondria: from inventory to function. Mitochondrion 33:58–71. https://doi.org/10.1016/j.mito.2016.07.012
36. Hroudova J, Singh N, Fisar Z (2014) Mitochondrial dysfunctions in neurodegenerative diseases: relevance to Alzheimer's disease. Biomed Res Int 2014:175062. https://doi.org/10.1155/2014/175062
37. Jacobs KM, Pennington JD, Bisht KS et al (2008) SIRT3 interacts with the daf-16 homolog FOXO3a in the mitochondria, as well as increases FOXO3a dependent gene expression. Int J Biol Sci 4(5):291–299
38. Jing E, Emanuelli B, Hirschey MD et al (2011) Sirtuin-3 (Sirt3) regulates skeletal muscle metabolism and insulin signaling via altered mitochondrial oxidation and reactive oxygen species production. Proc Natl Acad Sci U S A 108(35):14608–14613. https://doi.org/10.1073/pnas.1111308108
39. Jing E, O'Neill BT, Rardin MJ et al (2013) Sirt3 regulates metabolic flexibility of skeletal muscle through reversible enzymatic deacetylation. Diabetes 62(10):3404–3417. https://doi.org/10.2337/db12-1650
40. Kang PM, Izumo S (2000) Apoptosis and heart failure: a critical review of the literature. Circ Res 86(11):1107–1113
41. Kendrick AA, Choudhury M, Rahman SM et al (2011) Fatty liver is associated with reduced SIRT3 activity and mitochondrial protein hyperacetylation. Biochem J 433(3):505–514. https://doi.org/10.1042/BJ20100791
42. Kim SC, Sprung R, Chen Y et al (2006) Substrate and functional diversity of lysine acetylation revealed by a proteomics survey. Mol Cell 23(4):607–618. https://doi.org/10.1016/j.molcel.2006.06.026
43. Kong X, Wang R, Xue Y et al (2010) Sirtuin 3, a new target of PGC-1a, plays an important role in the suppression of ROS and mitochondrial biogenesis. PLoS One 5(7):e11707. https://doi.org/10.1371/journal.pone.0011707
44. Lahera V, de Las Heras N, López-Farré A et al (2017) Role of mitochondrial dysfunction in hypertension and obesity. Curr Hypertens Rep 19(2):11. https://doi.org/10.1007/s11906-017-0710-9
45. Lesnefsky EJ, Moghaddas S, Tandler B et al (2001) Mitochondrial dysfunction in cardiac disease: ischemia – reperfusion, aging, and heart failure. J Mol Cell Cardiol 33(6):1065–1089. https://doi.org/10.1006/jmcc.2001.1378
46. Lin MT, Beal MF (2006) Mitochondrial dysfunction and oxidative stress in neurodegenerative diseases. Nature 443(7113):787–795. https://doi.org/10.1038/nature05292
47. Liu L, Peritore C, Ginsberg J et al (2015) SIRT3 attenuates MPTP-induced nigrostriatal degeneration via enhancing mitochondrial antioxidant capacity. Neurochem Res 40(3):600–608. https://doi.org/10.1007/s11064-014-1507-8
48. Lombard DB, Alt FW, Cheng HL et al (2007) Mammalian Sir2 homolog SIRT3 regulates global mitochondrial lysine acetylation. Mol Cell Biol 27(24):8807–8814. https://doi.org/10.1128/MCB.01636-07
49. Lu Z, Chen Y, Aponte AM et al (2015) Prolonged fasting identifies heat shock protein 10 as a Sirtuin 3 substrate: elucidating a new mechanism linking mitochondrial protein acetylation to fatty acid oxidation enzyme folding and function. J Biol Chem 290(4):2466–2476. https://doi.org/10.1074/jbc.M114.606228
50. Lundby A, Lage K, Weinert BT et al (2012) Proteomic analysis of lysine acetylation sites in rat tissues reveals organ specificity and subcellular patterns. Cell Rep 2(2):419–431. https://doi.org/10.1016/j.celrep.2012.07.006
51. Martinez-Vicente M (2017) Neuronal mitophagy in neurodegenerative diseases. Front Mol Neurosci 10:64. https://doi.org/10.3389/fnmol.2017.00064. eCollection 2017
52. Merry BJ (2004) Oxidative stress and mitochondrial function with aging: the effects of calorie restriction. Aging Cell 3(1):7–12
53. Nakagawa T, Lomb DJ, Haigis MC et al (2009) SIRT5 deacetylates carbamoyl phosphate synthetase 1 and regulates the urea cycle. Cell 137(3):560–570. https://doi.org/10.1016/j.cell.2009.02.026
54. Ozden O, Park SH, Wagner BA et al (2014) SIRT3 deacetylates and increases pyruvate dehydrogenase activity in cancer cells. Free Radic Biol Med 76:163–172. https://doi.org/10.1016/j.freeradbiomed.2014.08.001
55. Paik WK, Pearson D, Lee HW et al (1970) Nonenzymatic acetylation of histones with acetyl-CoA. Biochim Biophys Acta 213:513–522
56. Phillips DM (1963) The presence of acetyl groups of histones. Biochem J 87:258–263
57. Picklo MJ (2008) Ethanol intoxication increases hepatic N-lysyl protein acetylation. Biochem Biophys Res Commun 376(3):615–619. https://doi.org/10.1016/j.bbrc.2008.09.039
58. Qiu X, Brown K, Hirschey MD et al (2010) Calorie restriction reduces oxidative stress by SIRT3-mediated SOD2 activation. Cell Metab 12(6):662–667. https://doi.org/10.1016/j.cmet.2010.11.015
59. Rangarajan P, Karthikeyan A, Lu J et al (2015) Sirtuin 3 regulates Foxo3a-mediated antioxidant pathway in microglia. Neuroscience 311:398–414. https://doi.org/10.1016/j.neuroscience.2015.10.048
60. Reaven GM (1988) Banting lecture 1988. Role of insulin resistance in human disease. Diabetes 37:1595–1607

61. Reinders J, Sickmann A (2007) Modificomics: post-translational modifications beyond protein phosphorylation and glycosylation. Biomol Eng 24(2):169–177. https://doi.org/10.1016/j.bioeng.2007.03.002
62. Rovira-Llopis S, Bañuls C, Diaz-Morales N et al (2017) Mitochondrial dynamics in type 2 diabetes: pathophysiological implications. Redox Biol 11:637–645. https://doi.org/10.1016/j.redox.2017.01.013
63. Salvatori I, Valle C, Ferri A et al (2017) SIRT3 and mitochondrial metabolism in neurodegenerative diseases. Neurochem Int pii S0197-0186(17):30141–30149. https://doi.org/10.1016/j.neuint.2017.04.012
64. Schlicker C, Gertz M, Papatheodorou P et al (2008) Substrates and regulation mechanisms for the human mitochondrial sirtuins Sirt3 and Sirt5. J Mol Biol 382(3):790–801. https://doi.org/10.1016/j.jmb.2008.07.048
65. Schwer B, Bunkenborg J, Verdin RO et al (2006) Reversible lysine acetylation controls the activity of the mitochondrial enzyme acetyl-CoA synthetase 2. Proc Natl Acad Sci U S A 103(27):10224–10229. https://doi.org/10.1073/pnas.0603968103
66. Schwer B, Eckersdorff M, Li Y et al (2009) Calorie restriction alters mitochondrial protein acetylation. Aging Cell 8(5):604–606. https://doi.org/10.1111/j.1474-9726.2009.00503.x
67. Shimazu T, Hirschey M, Hua L et al (2010) SIRT3 deacetylates mitochondrial 3-hydroxy-3-methylglutaryl CoA synthase 2, increases its enzymatic activity and regulates ketone body production. Cell Metab 12(6):654–661. https://doi.org/10.1016/j.cmet.2010.11.003
68. Shinmura K, Tamaki K, Sano M et al (2011) Caloric restriction primes mitochondria for ischemic stress by deacetylating specific mitochondrial proteins of the electron transport chain. Circ Res 109(4):396–406. https://doi.org/10.1161/CIRCRESAHA.111.243097
69. Sohal RS, Weindruch R (1996) Oxidative stress, caloric restriction, and aging. Science 273(5271):59–63
70. Someya S, Yu W, Hallows WC et al (2010) Sirt3 mediates reduction of oxidative damage and prevention of age-related hearing loss under caloric restriction. Cell 143(5):802–812. https://doi.org/10.1016/j.cell.2010.10.002
71. Srinivasan S, Guha M, Kashina A et al (2017) Mitochondrial dysfunction and mitochondrial dynamics-the cancer connection. Biochim Biophys Acta. pii: S0005-2728(17)30005-1. https://doi.org/10.1016/j.bbabio.2017.01.004
72. Still AJ, Floyd BJ, Hebert AS et al (2013) Quantification of mitochondrial acetylation dynamics highlights prominent sites of metabolic regulation. J Biol Chem 288(36):26209–26219. https://doi.org/10.1074/jbc.M113.483396
73. Sundaresan NR, Gupta M, Kim G et al (2009) Sirt3 blocks the cardiac hypertrophic response by augmenting Foxo3a-dependent antioxidant defense mechanisms in mice. J Clin Invest 119(9):2758–2771. https://doi.org/10.1172/JCI39162
74. Sundaresan NR, Samant SA, Pillai VB et al (2008) SIRT3 is a stress-responsive deacetylase in cardiomyocytes that protects cells from stress-mediated cell death by deacetylation of Ku70. Mol Cell Biol 28(20):6384–6401. https://doi.org/10.1128/MCB.00426-08
75. Svinkina T, Gu H, Silva JC et al (2015) Deep, quantitative coverage of the lysine acetylome using novel anti-acetyl-lysine antibodies and an optimized proteomic workflow. Mol Cell Proteomics 14(9):2429–2440. https://doi.org/10.1074/mcp.O114.047555
76. Tao R, Coleman MC, Pennington JD et al (2010) Sirt3-mediated deacetylation of evolutionarily conserved lysine 122 regulates MnSOD activity in response to stress. Mol Cell 40(6):893–904. https://doi.org/10.1016/j.molcel.2010.12.013
77. Tao R, Vassilopoulos A, Parisiadou L et al (2014) Regulation of MnSOD enzymatic activity by Sirt3 connects the mitochondrial acetylome signaling networks to aging and carcinogenesis. Antioxid Redox Signal 20(10):1646–1654. https://doi.org/10.1089/ars.2013.5482
78. Torrens-Mas M, Pons DG, Sastre-Serra J et al (2017) SIRT3 silencing sensitizes breast cancer cells to cytotoxic treatments through an increment in ROS production. J Cell Biochem 118(2):397–406. https://doi.org/10.1002/jcb.25653
79. Trotta AP, Chipuk JE (2017) Mitochondrial dynamics as regulators of cancer biology. Cell Mol Life Sci 74(11):1999–2017. https://doi.org/10.1007/s00018-016-2451-3
80. Ungvari Z, Parrado-Fernandez C, Csiszar A et al (2008) Mechanisms underlying caloric restriction and lifespan regulation: implications for vascular aging. Circ Res 102(5):519–528. https://doi.org/10.1161/CIRCRESAHA.107.168369
81. Vassilopoulos A, Pennington JD, Andresson T et al (2014) SIRT3 deacetylates ATP synthase F1 complex proteins in response to nutrient- and exercise-induced stress. Antioxid Redox Signal 21(4):551–564. https://doi.org/10.1089/ars.2013.5420
82. Vazquez EJ, Berthiaume JM, Kamath V et al (2015) Mitochondrial complex I defect and increased fatty acid oxidation enhance protein lysine acetylation in the diabetic heart. Cardiovasc Res 107(4):453–465. https://doi.org/10.1093/cvr/cvv183
83. Verdin E, Hirschey MD, Finley LW et al (2010) Sirtuin regulation of mitochondria: energy production, apoptosis, and signaling. Trends Biochem Sci 35(12):669–675. https://doi.org/10.1016/j.tibs.2010.07.003
84. Wada J, Nakatsuka A (2016) Mitochondrial dynamics and mitochondrial dysfunction in diabetes. Acta Med Okayama 70(3):151–158
85. Wagner GR, Payne RM (2013) Widespread and enzyme-independent Nε-acetylation and Nε-succinylation of proteins in the chemical conditions of the mitochondrial matrix. J Biol Chem 288(40):29036–29045. https://doi.org/10.1074/jbc.M113.486753

86. Walters AM, Porter GA Jr, Brookes PS (2012) Mitochondria as a drug target in ischemic heart disease and cardiomyopathy. Circ Res Oct 111(9):1222–1236. https://doi.org/10.1161/CIRCRESAHA.112.265660
87. Wang Q, Zhang Y, Yang C et al (2010) Acetylation of metabolic enzymes coordinates carbon source utilization and metabolic flux. Science 327(5968):1004–1007. https://doi.org/10.1126/science.1179687
88. Weinert BT, Wagner SA, Horn H et al (2011) Proteome-wide mapping of the Drosophila acetylome demonstrates a high degree of conservation of lysine acetylation. Sci Signal 4(183):ra48. https://doi.org/10.1126/scisignal.2001902
89. Wu YT, Lee HC, Liao CC et al (2013) Regulation of mitochondrial F(o)F(1) ATPase activity by Sirt3-catalyzed deacetylation and its deficiency in human cells harboring 4977 bp deletion of mitochondrial DNA. Biochim Biophys Acta 1832(1):216–227. https://doi.org/10.1016/j.bbadis.2012.10.002
90. Xiong Y, Wang M, Zhao J et al (2016) Sirtuin 3: a Janus face in cancer (review). Int J Oncol 49(6):2227–2235. https://doi.org/10.3892/ijo.2016.3767
91. Xue L, Xu F, Meng L et al (2012) Acetylation-dependent regulation of mitochondrial ALDH2 activation by SIRT3 mediates acute ethanol-induced eNOS activation. FEBS Lett 586(2):137–142. https://doi.org/10.1016/j.febslet.2011.11.031
92. Yang Y, Cimen H, Han MJ et al (2010) NAD+-dependent deacetylase SIRT3 regulatesmitochondrial protein synthesis by deacetylation of the ribosomal protein MRPL10. J Biol Chem 285(10):7417–7429. https://doi.org/10.1074/jbc.M109.053421
93. Yang W, Nagasawa K, Münch C et al (2016) Mitochondrial sirtuin network reveals dynamic SIRT3-dependent deacetylation in response to membrane depolarization. Cell 167(4):985–1000.e21. https://doi.org/10.1016/j.cell.2016.10.016
94. Zhang Y, Bharathi SS, Rardin MJ et al (2015) SIRT3 and SIRT5 regulate the enzyme activity and cardiolipin binding of very long-chain acyl-CoA dehydrogenase. PLoS One 10(3):e0122297. https://doi.org/10.1371/journal.pone.0122297. eCollection 2015
95. Zhang X, Ren X, Zhang Q et al (2016) PGC-1α/ERRα- Sirt3 pathway regulates DAergic neuronal death by directly deacetylating SOD2 and ATP synthase β. Antioxid Redox Signal 24(6):312–328. https://doi.org/10.1089/ars.2015.6403
96. Zhao S, Xu W, Jiang W et al (2010) Regulation of cellular metabolism by protein lysine acetylation. Science 327(5968):1000–1004. https://doi.org/10.1126/science.1179689
97. Zhao Y, Yang H, Wang X et al (2013) Sirtuin-3 (SIRT3) expression is associated with overall survival in esophageal cancer. Ann Diagn Pathol 17(6):483–485. https://doi.org/10.1016/j.anndiagpath.2013.06.001
98. Zsurka G, Kunz WS (2013) Mitochondrial involvement in neurodegenerative diseases. IUBMB Life 65(3):263–272. https://doi.org/10.1002/iub.1126

# Crosstalk Between Oxidative Stress and Mitochondrial Damage: Focus on Amyotrophic Lateral Sclerosis

5

Viviana Greco, Patrizia Longone, Alida Spalloni, Luisa Pieroni, and Andrea Urbani

**Abstract**

Proteins oxidation by reactive species is implicated in the aetiology or progression of a panoply of disorders and diseases such as neurodegenerative disorders. It is becoming increasingly evident that redox imbalance in the brain mediates neurodegeneration. Free radicals, as reactive species of oxygen (ROS) but also reactive nitrogen species (RNS) and reactive sulfur species (RSS), are generated in vivo from several sources. Within the cell the mitochondria represent the main source of ROS and mitochondrial dysfunction is both the major contributor to oxidative stress (OS) as well its major consequence.

To date there are no doubts that a condition of OS added to other factors as mitochondrial damage in mtDNA or mitochondrial respiratory chain, may contribute to trigger or amplify mechanisms leading to neurodegenerative disorders.

In this chapter, we aim at illustrate the molecular interplay occurring between mitochondria and OS focusing on Amyotrophic Lateral Sclerosis, describing a phenotypic reprogramming mechanism of mitochondria in complex neurological disorder.

**Keywords**

ALS · Mitochondrial damage · Reactive species · Oxidative stress · Redox proteomics

## 5.1 Introduction

Oxidative stress (OS) induced by free radicals plays a critical role in the setting of and pathophysiology of a wide range of diseases including aging and neurodegeneration [32].

Free radicals such as reactive oxygen species (ROS), reactive nitrogen species (RNS) and reactive sulfur species (RSS) are always present in cells at basal levels and are often involved in several signal transduction processes [58].

To counteract the increase of free radicals levels, the body has developed several antioxidant

V. Greco · A. Urbani (✉)
Institute of Biochemistry and Clinical Biochemistry, Università Cattolica del Sacro Cuore, Rome, Italy

Fondazione Policlinico Universitario A. Gemelli – IRCCS, Rome, Italy
e-mail: andrea.urbani@unicatt.it

P. Longone · A. Spalloni
Molecular Neurobiology Unit, Fondazione Santa Lucia-IRCCS, Rome, Italy

L. Pieroni
Proteomics and Metabonomics Unit, Fondazione Santa Lucia-IRCCS, Rome, Italy


A. Urbani, M. Babu (eds.), *Mitochondria in Health and in Sickness*, Advances in Experimental Medicine and Biology 1158, https://doi.org/10.1007/978-981-13-8367-0_5

defences. When the antioxidant mechanisms fail, a condition of redox imbalance or OS occurs.

The brain is one of the most metabolically active organs in the body; in addition it is also one of the most susceptible to OS, due to the activity of less efficient antioxidant systems in comparison to those of other organs [25].

Cellular free radicals are generated in vivo from several sources. However, they are mainly produced as a by-product of aerobic metabolism, resulting from the oxidative phosphorylation process. Therefore, mitochondria represent the main source of ROS in physiological conditions (1–5%) [73].

To date, there is no doubt that oxidative species and mitochondria are playing important roles in neurodegenerative processes. Alzheimer's disease (AD), Parkinson's disease (PD), Huntington's disease (HD) and Amyotrophic lateral sclerosis (ALS) are associated to increased levels of free radicals and oxidative damage as well mitochondrial dysfunction, which contribute to disease progression [13, 58].

This chapter discusses the close interaction between OS and mitochondrial damage, focusing on ALS as example of neurological disorder in whose aetiology these factors are extremely involved and interconnected.

## 5.2 Reactive Species, Oxidative Stress and Mitochondria

In biological systems, the term free radicals coincides with ROS because a large percentage of biologically relevant. As the name indicates, ROS are oxygen-centred radicals.

Several organelles and physiological processes contribute to the generation of ROS, in particular the mitochondria are the major contributors.

Mitochondria represent the "key organelle" contributing to different several cellular processes including homeostasis of intracellular calcium, phospholipid biogenesis, energy metabolism and regulation of apoptotic processes. Anomalies in mitochondrial morphology and function occur in most of neurodegenerative disorders, including ALS.

The main function of mitochondria is the ATP production being the sites of aerobic respiration via the oxidative phosphorylation system (OXPHOS). In the mitochondrial electron transport chain, during energy transduction, a small number of electrons 'leak' to oxygen prematurely resulting in incomplete reduction of molecular oxygen to produce ROS such as the superoxide radical anion ($O_2(^{-})$) and hydrogen peroxide ($H_2O_2$) [47].

As it is well known, the OXPHOS system is composed by five multiprotein complexes (complex I–V).

Briefly, electrons move from NADH to complex I (NADH coenzyme Q reductase) and then to coenzyme Q (ubiquinone), which also takes electrons from complex II (succinate dehydrogenase). From Coenzyme Q the electrons move towards to complex III (cytochrome bc1), and then to cytochrome C, which transfers them to complex IV (cytochrome C oxidase). In turn cytochrome C oxidase uses the electrons to reduce molecular oxygen to water.

Under physiological conditions, mitochondrial antioxidant mechanisms exist to maintain ROS levels. Among these, principally the superoxide dismutase (SOD1) catalyses the dismutation of the superoxide anion $O_2^{(-)}$ to oxygen ($O_2$) and hydrogen peroxide ($H_2O_2$) according to the reaction $O_2^- + 2H^+ \rightarrow H_2O_2$ (superoxide dismutation). Hydrogen peroxide is then converted by peroxidase and glutathione peroxidases to oxygen and water. $O_2$ is generated predominantly by complexes I and III [47].

OS arises from the unbalance between the production and the removal of ROS and RNS species [6, 16].

Superoxide and hydrogen peroxide are not highly reactive per se, however each may undergo further reactions.

When ROS levels increase, superoxide anion can interact with other molecules generating primary and secondary reactive species. These may occur either by direct or metabolic processes, as through enzyme or metal-catalyzed processes.

Superoxide anion reacts rapidly with nitric oxide, a weakly oxidizing producted by nitric oxide synthase (NOS), to produce peroxynitrite (ONOO-). Likewise, hydrogen peroxide slowly decomposes into reactive hydroxyl radicals (·OH) in a process that is catalysed by metal ions reduced such as $Fe^{2+}$ in the Fenton reaction [53].

Although it has always been thought that redox imbalance was mainly due to ROS, there is a growing awareness that RNS species such as Nitric oxide or Peroxynitrite contribute in a significant way to changes in the redox homeostasis [9].

ROS and RNS are highly reactive and may damage proteins, lipids and nucleic acid. Similarly, at mitochondrial level, an excessive accumulation of ROS may induce damage to proteins and lipids and to mtDNA as well. In particular, this is due to defects in the complex I or III and in the DNA repair mechanisms resulting in a high rate of mutations. On the other side, mitochondrial dysfunction (MD) may lead to ATP depletion, increased overload of superoxide anions and ROS, and release of pro-apoptotic factors including cytochrome C. Therefore, MD is both the main contributor to OS and, on the other side, one of the major consequences of this.

It is well known how the brain is characterized by the highest metabolic rate and energy demands. The mitochondrial production of ATP supports both metabolism and brain processes, such as synaptic signal transmission, synaptic plasticity and growth and assembly of neurons.

In addition, more recent advances in the field of sulfur biology and hydrogen sulfide ($H_2S$) have also proposed reactive sulfur species (RSS) as signaling molecules. Since 2001, RSS, such as $H_2O_2$, $^1O_2$, $ONOO^-$ and $O_2^{\bullet-}$ generated from thiols by reaction with oxidizing agents, have been suggested as new oxidizing agents together with ROS and RNS [27].

The mitochondria are also a rich source of RSS especially during sulfide oxidation; sulfur species such as glutathione persulfide (GSSH) are generated during mitochondrial functions [40, 46].

As well, sulfur amino acids are a main target of redox modifications and thus the identification of an increasing number of sulphur-derived radicals adds additional layers of complexity in the redox signaling processes. Moreover, sulfydration or persulfydration mediated by $H_2S$ is one of the mechanisms through which $H_2S$ can modulate several features of cellular physiology and pathology, also in the brain [68].

Because of this close link between brain and mitochondria, there are no doubts that a condition of OS added to other factors such as damage in mtDNA and in mitochondrial respiratory chain, and cellular excitotoxicity, may contribute to trigger, amplify or worsen mechanisms affecting neurodegenerative processes.

The mitochondria-OS-neurodegenerative disease interaction is well evident in Amyotrophic Lateral Sclerosis (Fig. 5.1).

## 5.3 Amyotrophic Lateral Sclerosis

Amyotrophic lateral sclerosis (ALS) is a fatal neurodegenerative disease characterized by a progressive degeneration of upper and lower motorneurons in the spinal cord, brainstem and motor cortex [36, 37].

Characteristic signs are muscle weakness and paralysis leading to dysphagia and orthopnea and ultimately death within 2–5 years from diagnosis. To date, there are no effective pharmacological treatments. Riluzole and the antioxidant edavarone, currently approved by the FDA, can prolong life, but only by a few months.

ALS onset can be spinal or bulbar with an average age of onset between 40 and 60 years [49]. ALS is a sporadic disease without clear genetic basis. Approximately, a minor fraction of ALS cases (5–10%) are familial (fALSL) related to an autosomal dominant inheritance. To date, more than 13 genes and loci related to a disease clinical phenotype have been identified: superoxide dismutase 1 (SOD1), TAR-DNA binding protein (TAR-DBP, TDP-43), Fused in Sarcoma (FUS), angiogenin (ANG) and C9orf72 [55].

Despite the research progresses over the years, the ALS aetiology remains unclear. The relevance

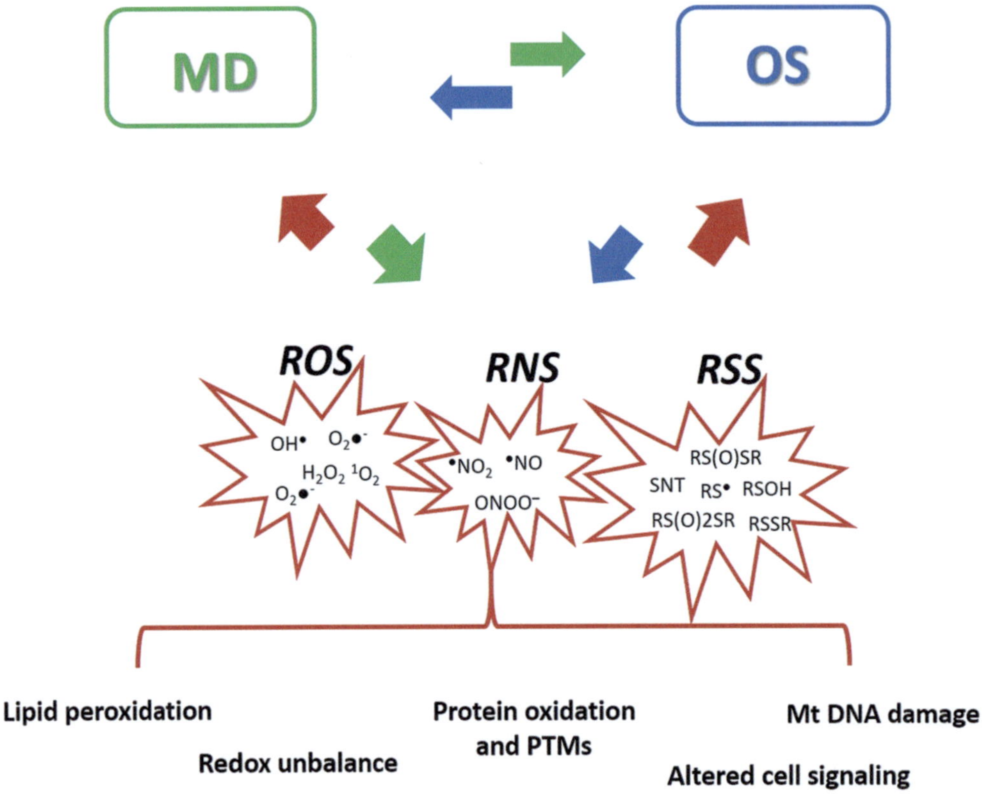


**Fig. 5.1** Cross-talk mitochondrial dysfunction-oxidative stress

of some pathogenic mechanisms come from animal model carrying these mutations, in particular SOD1 or TDP-43.

Several processes have been identified as causing and/or contributing to the disease pathogenesis. These include defect in axonal transport, glutamatergic excitotoxicity, dysfunctions in proteasome and endoplasmic reticulum (ER), mitochondrial dysfunction and damage and OS.

However, it is well accepted that ALS is a multifactorial disease in which a severe MD remarkably contributes to an unavoidable neuronal death.

A wide literature already exists on the role of mitochondria in the ALS (reviewed in [63]). In this chapter, we have chosen to delve into the interaction mitochondria-OS and examine how these factors contribute to compromise the onset and progression of such disease.

### 5.3.1 Cross Talk Between Mitochondria and Oxidative Stress in ALS

Motorneurons are characterized by high metabolic demand and consequently by high request of $O_2$ becoming more sensitive to ROS over the years. MD associated with ALS occurs in many guises, which include primarily faulty oxidative phosphorylation, production of ROS, reduced calcium buffering capacity, and defective mitochondrial dynamics [63]. Furthermore, MD appears to be directly or indirectly related to a "non-mitochondrial" toxicity in ALS, due to excitotoxicity, loss of protein homeostasis and defects in axonal transport [63].

Anomalies in mitochondrial morphology affect most of neurodegenerative disorders, including ALS [63]. The alteration of the mitochondrial structure with aggregated and vacuolated mitochondria is an evident and common feature in motorneurons of ALS patients.

In the ALS animal model altered structure and networks and the resulting functional impairment have been shown at an early stage of the disease [70]. For this reason, this aspect seems to contribute to the onset of upstream degeneration rather than being a consequence.

Evidences of such abnormal mitochondrial morphology and accumulation of neurofilaments have also been reported in sporadic ALS [56] and ALS animal models. Aggregated and vacuolated mitochondria have been shown in ALS TDP43 mice; on the contrary, mitochondria appear to be more spherical and less elongated in SOD1G93A by aggregating in clusters along the axons [44].

To date, there is a significant evidence that OS is a key mechanism by which motorneurons undergo death. However, it is not still clear if OS is a cause of ALS degeneration rather than a consequence of some other etiologic factors.

Increased levels of OS biomarkers have been shown in ALS post-mortem tissues: elevated levels of carbonyl proteins in spinal cord [60] and motor cortex [22], markers of lipid peroxidation [62] and DNA damage [23]. Significant increase of both oxidized DNA biomarkers and lipid oxidation biomarkers have been also found in ALS CSF and sera [30]. Moreover transgenic ALS animal model and cell culture show evident sign of OS as we will discuss below.

In addition, an involvement of Fe has been demonstrated. An increase in iron levels due to the Fenton reaction have been reported in ALS spinal cord [31] as well the accumulation of Fe due to increased uptake by lactotransferrin in ALS motorneuron [39]. Some fALS show mutations in copper/zinc (Cu/Zn) superoxide mutase (SOD1), component of the oxidant defence system. More recently, evidence of such crosstalk has also been found in non-SOD1 linked ALS. We will discuss this aspect in the following paragraphs.

### 5.3.2 Role of SOD1

#### 5.3.2.1 Non Mutated SOD1

Human SOD1 gene maps on the 21q22 chromosome and encodes for the main antioxidant enzyme of the eukaryote cell, SOD1.

SOD1 is a cytosolic enzyme, with only a portion (less than 5%) [69] present in the mitochondrial inner membrane space (IMS) [64]. Moreover, the eukaryotic cell has two additional superoxide dismutases, SOD2 and SOD3, which are found respectively within the mitochondrial matrix and in the extracellular milieu.

The mitochondria, through the mitochondrial respiratory chain, located within the inner mitochondrial membrane (IMM), produces elevated amounts of ROS, as a bypass product of the electron transfer reactions, occurring mainly in the complex I and complex III [47].

SOD2, within the mitochondrial matrix, and SOD1, within the IMS, are the two enzymes dedicated to the detoxification of the superoxide anions produced during the respiratory reactions. The two enzymes catalyze the superoxide dismutation leading to the production of $H_2O_2$. $H_2O_2$ is a dangerous ROS that can react with cytochrome C to produce highly reactive oxygen species. Therefore, $H_2O_2$ must be carefully neutralized. The cells express, along with the SOD enzymes, also catalase, glutathione peroxidase and peroxiredoxins to properly process $H_2O_2$ [69]. In conditions of mitochondrial stress, a significant gain in SOD1 activity could lead to an increased production of toxic amounts of ROS and $H_2O_2$ within IMS, ultimately leading to cytochrome peroxidation [1].

To work properly, SOD1 requires post-translational modifications (PTMs), which are necessary for both its function and subcellular localization. These PTMs include the formation of intramolecular disulfide bonds for a correct folding, the binding of zinc and copper metal ions, and the exposition of a hydrophobic region for the organization of SOD1 dimers. In an unfolded state, SOD1 traslocates from cytosol into the mitochondria through the OMM translocator TOM. Within the mitochondria the SOD1 apoenzyme results from the establishment of disulfide bonds and by the insertion of copper metal promoted by the copper chaperone for superoxide dismutase (CCS) [18]. Hence, SOD1 reaches its mature form through metals and disulfide bonds that prevent its leakage back to the cytosol [35].

The intracellular distribution of SOD1 is strictly related to the localization of CCS. In fact, CCS acts as redox sensor for SOD1 subcellular localization. When the cytosolic levels of oxygen are high, CCS is maintained in the cytosol to fold apoenzyme SOD1, when hypoxia occurs, CCS translocates in the mitochondrial IMS, promotes the formation of sulfur bound and metal copper leading to the mature form of SOD. In addition, hypoxia induces a boost of the mitochondrial respiratory chain, leading to the increases of the concentration of superoxide anion inside the mitochondria. Hence, the enhanced translocation of CCS, promoting the maturation of SOD1, could be seen as a compensatory antioxidation strategy. Because CCS lacks a signalling sequence for its mitochondrial localization, its import in mitochondria occurs through the so-called disulfide relay system (DRS) that normally is used by small cysteine-rich IMS proteins. Therefore, apCCS behaves as substrate of DRS and enters IMS. Here, the IMS proteins MIA-40 mediates its oxidation involving the shifting between specific cysteine pairs. In this way, on one side the oxided-apCCS remains inside IMS, and on the other side MIA40 is reduced. This cycle reduced-MIA40 and oxided-apoCCS is regulated by the involvement of the respiratory chain. In fact, the sulfydril oxidase ERV/ALR1 transfers electrons to cytochrome C and reoxides MIA-40 [65].

#### 5.3.2.2 Mutated SOD1 and Oxidative Stress

In the normal nervous system, only a low amount of SOD1 are found in or on the mitochondria. Mitochondrial SOD1 accumulation together with mitochondrial dysfunction are key features leading to the motorneuron death [65].

To date, more than 150 SOD1 mutations have been identified in fASL patients [2], most of them characterized by mutation of single amino acid, such as principally SOD1$^{G93A}$ and SOD1$^{G37R}$ or SOD1$^{G85E}$. These different mutations show different properties depending on the residual affected and with the ability to accumulate and aggregate in high molecular weight complexes [65].

Mutant SOD1 (mSOD1) aggregates are a well-known hallmark of ALS. The exact mechanism underlying the basis of SOD1 toxicity are not yet fully understood [67]. Insoluble SOD1 aggregates have been described in ALS rodent mouse models [72]. Soluble SOD1 may to be toxic as well [74].

Likewise, accumulation of SOD1 in mitochondria has been proposed as a particular hallmark of motorneuron degeneration. In fact, signs of mutant SOD1 accumulation and, consequently, evident mitochondrial dysfunction have been demonstrated in spinal cord [57] and neurons [29, 33]. Moreover, SOD1 accumulation increases with disease progression [65].

ALS related-SOD1 mutations occur as misfolding and intracellular mislocalization. SOD1 misfolded seems to be principally due to PTMs involving disulphide bridge [34], metal ion binding [4] and incorrect C6 and C111 bond leading to instable SOD 1 dimerization and tertiary structure. Mutant SOD1 enters IMS through TOM and precipitates in insoluble aggregates accumulating in the IMS overriding the activity of CCS. In addition, mSOD1might deposit onto the outer mitochondrial membrane clumping the transport across the mitochondrial membrane and engaging the mitochondrial cell apoptosis.

Misfolded SOD1 interacts with OMM proteins such as Bcl2 and voltage-dependent anion channel (VDAC) leading to the activation of pro-apoptotic pathways and ultimately cell death [50]. Localization of misfolded SOD1 into mitochondria can depend on tissue, cells, and cytosol composition.

The macrofage migration inhibitory factor (MIF) mediates the interaction SOD1-OMM proteins. At high cytosolic levels, soluble MIF is able to prevent SOD1 mislocalization. In contrast, low levels favour SOD1 deposition.

Moreover, while both mSOD1 and wtSOD1 have been shown to be in mitochondria, only mSOD1 seems to have a major selectivity for OMM and interacts with the voltage dependent ion channel (VDAC) [66].

Bcl-2 and VDAC are both OMM proteins involved in the intrinsic apoptotic process. In particular Bcl-2 regulates the release of cytocrome c by VDAC inhibiting the proapoptotic factor Bax. mSOD1 interacts with Bcl-2 inducing the exposition of the BH3 death domain resulting into toxic effect for mitochondria. In fact, the interaction with Bcl2 induces the release of Cyt c and changes in the mitochondrial morphology [50].

This toxic association SOD1-Bcl2 and the exposition of BH3 domain contribute to induce the bond between Bcl-2-VDAC. VADC is a general diffusion pore for anions and cations characterized by two different dynamic conformation status: in its open conformation, it is an anionic pore that allows ATP and ions flow from mitochondria to cytosol, in closed conformation it acts as small cation. Due to the VDAC closure, the link mutant SOD1-Bcl2 VDAC results in the alteration of OMM polarity and permeability and in the consequent drop of ATP/ADP ratio leading to the increase of ROS production. Indeed, it has been demonstrated that mutant SOD can bind directly to the cytoplasmic surface of VDAC.

SOD1 mutations reveal a strong link with OS. It is widely demonstrated that SOD1 toxicity is mediated by a gain of toxic characteristics according two different possible mechanisms [14]. The first one hypothesis of the SOD1 toxicity is related to the ability of mSOD1 to aggregate [14]. The second one is closely related to the protein misfolding due to the Cu/Zn sites [8, 41]. Therefore, misfolding at Cu, Zn or SOD level may induce an increase of ROS and the consequence of an OS condition within motorneuron [54]. This last mechanism is considered as the real "oxidative damage hypothesis".

All the several mechanisms by which the toxicity of SOD1 occur lead to an alteration of the dismutase activity. A loss of function could increase superoxide levels, which can react with nitric oxide for the peroxynitrite production; likewise also amplified SOD1 activity may increase levels of hydrogen peroxide and hydroxyl radicals.

The toxic function of mutant SOD1 is also mediated by several aberrant oxidative reactions. In fact, mSOD1 has a more open conformation than wtSOD1. As a result, substrates other than superoxide enter the active site and react with the copper and zinc ions it contains.

In addition, SOD1 cysteine residues have been implicated in the aggregation [17]. Human SOD1 has four cysteine residues. Among these, Cys57 and Cys146, highly conserved in other species, form an intramolecular disulfide bond, while Cys6 and Cys111, that are not conserved, do not form bridge [15, 58]. It has been demonstrated that Cys111 is modified by a persulfide group being prone to modifications by a wide variety of reactive cysteine interacting compounds [21].

Cys111 mediates both SOD1 aggregation and localization. The aggregation occurs when SOD1 proteins are in the metal-free form, through oxidation of Cys6 and Cys111, and are stabilized by hydrogen bonds.

Furthermore, in order for SOD1 to be targeted to mitochondria, interaction with CCS, that is also redox sensitive, is required, as we discussed before [35].

Furthermore, mitochondrial SOD1 Oligomers cause a shift in the redox state leading an impairment of respiratory complexes and mitochondrial dynamics anomalies in the motorneuron cell line NSC34 [45]. Association of mSOD1s with mitochondria decreases the reduced/oxidized glutathione ratio (GSH/GSSG).

#### 5.3.2.3 Effects on Non-SOD1 ALS Mitochondria

Several evidences indicate that MD and OS play a critical role also in non-SOD1 ALS. Mitochondrial damage may be related to an alteration of other proteins than SOD1. The involvement of such other proteins in OS-MD interaction supports the idea that these factors can represent the common denominator of ALS motorneuron degeneration.

Here we report just a few examples.

CHCHD10 is a mitochondrial protein located at contact sites between the inner and outer mitochondrial membranes (OMM). ALS-CHCHD10 mutations have been shown to have a significant effect on mitochondrial structure such as the disruption of mitochondrial cristae, as reported by Bannwarth et al. [5]. Recently a novel CHCHD10 mutation has been shown to be related to an ALS deficit of the Mia40-dependent mitochondrial import [38].

Another protein, valosin containing protein (VCP), which is an ATPase involved in mitochondrial quality control and mitophagy, is mutated in about 2% of familial ALS. Moreover, Decreased levels of VCP are related to a global changes in mitochondrial homeostasis [7].

In addition, mutation in TDP-43 amplify OS and MD through the nuclear accumulation of nuclear factor E2 related factor 2 (Nrf2) [7].

## 5.4 Studies Based on Proteomics Tools: Redox Proteomics

### 5.4.1 Protein Oxidation and Redox Proteomics

Protein oxidation is associated with numerous conditions characterized by oxidative stress.

Being the major non-water components of most biological systems, proteins represent the major targets of the action of radical species. The extent of oxidation damage depends on several factors including: the target protein concentration, the target protein localization and oxidant localization, and the constant rate for oxidative reaction. Additional contributory factors may include concomitant events of secondary damage and the effectiveness of the repair mechanisms mediated by the oxidising scavenger [19].

Modifications in several aminoacid residues, such as cysteine, methionine, lysine and proline, can result in protein unfolding, alteration of conformation and turnover, toxic aggregation due to hydrophobic interactions, increased side chain hydrophilicity, alteration of cell signalling activating cell death pathways [12, 14].

Among the oxidative PTMs induced by ROS/RNS, proteins carbonyl, protein nitration and protein bound 4-hydroxy-2-tras-nonenal (HNE) are the most common [13]. In addition, Cysteine residues are extremely sensitive to cellular redox. The three oxidation states of sulfur atom of the Cys thiol group can change in response to numerous alterations in the cell redox state [13].

As a result, Cys residues are the main targets of redox modification which affect the activity of target proteins in determining susceptibility to oxidative damage and neurodegeneration [52].

Thus, the redox proteomics supported by the advances in Mass spectrometry, is the branch of the proteomics aimed at detecting the oxidized proteins in order to investigate the proteome adaptation due to mechanisms of redox unbalance and oxidative stress [14, 61].

### 5.4.2 Application of Redox Proteomics to ALS

Redox proteomics studies have clarified some of the molecular mechanisms related to the ALS pathology.

In the SOD1$^{G93A}$ mouse lipid peroxidation and the resulting increased production of hydroxyl radicals and/or radical species has supported the hypothesis that SOD1 itself could be a key protein target, and that these modifications could compromise its antioxidant function [3]. Redox proteomics studies have also highlighted the crucial role of the Hsp70 chaperone in the correct folding of SOD1 and ultimately in the ALS pathogenesis. Hsp70 is a chaperone localized in the mitochondrial matrix and functions in the folding and ATP-dependent transport into the mitochondrial matrix of proteins that have been just synthetized. The mSOD1 forms aggregates with Hsp70, Hsp40 a key chaperone protein important in the regulation of Hsp70, and crystallins, a family of heat shock proteins involved in the prevention of heat- and oxidative stress-induced aggregation of proteins. When Hsp70is overexpressed -SOD1 aggregates decrease [11]. Moreover, the importance of Hsp70 has grown, over the years, with the ever growing knowledge about its link to the mitochondrial quality control system [42]. The mitochondrial quality control system (MQC) is fundamental in maintaining the proper mitochondrial homeostasis through a synchronized regulation of mitochondrial proteins biogenesis and degradation [71].

Mitochondrial heme-containing proteins are essential for normal mitochondrial functions such as oxidative phosphorylation and mitochondrial biogenesis. In fact, both events, oxygen-sensitive accumulation and degradation of heme proteins in mitochondria might affect mitochondrial functions, such as bioenergetic and oxygen-sensing pathways. Cystathionine β-synthase (CBS) is a cytosolic heme-containing protein, central in the homocysteine and cysteine metabolism and in the synthesis of endogenous $H_2S$. Its presence and accumulation within the mitochondria are regulated by the interaction of the heme group with the oxygen molecule under the control of the Lon protease.

Lon protease, a member of the MQC family of proteases, is of utmost importance, within the mitochondrial matrix, in controlling mitochondrial proteins degradation, to prevent their toxic aggregation, by regulating many mitochondrial processes such as chaperone activity and the assembly of the respiratory complexes [10]. Indeed, Teng et al. (2013) have shown how a failing Lon protease triggers the mitochondrial accumulation of CBS.

Recently, we have shown an involvement of $H_2S$ in the ALS pathogenesis [20] and a toxicity mediated by aberrant $H_2S$-levels in motorneurons [28]. Regarding CBS we have demonstrated its accumulation in the mitochondria of the $SOD1^{G93A}$ mice compared to the control group mitochondria. With regard to the aforementioned evidences, we may infer that the poisonous levels of $H_2S$ might induce modification in the thiol groups or in the sulphydration processes that could be involved in the disease. The SOD-$H_2S$ interaction has been known for some time [59]. Recently it has been shown that SOD oxidizes $H_2S$ to form persulfides. This study suggests that SOD contributes directly to sulphur metabolism and RSS production [48].

Moreover, it has been widely demonstrated that a correlation between mitochondria and $H_2S$ does exists in several disorders [24, 26], and sulphydration is one of the PTMs involved in several diseases [43] including the neurological disorders [51].

## 5.5 Conclusions

ALS is a terrible neurodegenerative disorder, which also represents a complex pathogenic paradigm on the bidirectional mechanism of mitochondrial homeostasis within a cell. The mechanisms of oxidative damages are a central hallmark but do not explain the whole picture in particular the triggering and recursive feedback interactions. In this context, redox proteomics is an attractive tool to identify oxidated proteins as putative targets of such complex mechanisms consequent to OS-MD interaction.

## References

1. Ahtoniemi T, Jaronen M, Keksa-Goldsteine V, Goldsteins G, Koistinaho J (2008) Mutant SOD1 from spinal cord of G93A rats is destabilized and binds to inner mitochondrial membrane. Neurobiol Dis 32(3):479–485
2. Andersen PM (2000) Genetic factors in the early diagnosis of ALS. Amyotroph Lateral Scler Other Motor Neuron Disord 1(Suppl 1):S31–S42
3. Andrus PK, Fleck TJ, Gurney ME, Hall ED (1998) Protein oxidative damage in a transgenic mouse model of familial amyotrophic lateral sclerosis. J Neurochem 71(5):2041–2048
4. Banci L, Bertini I, Cantini F, Kozyreva T, Massagni C, Palumaa P, Rubino JT, Zovo K (2012) Human superoxide dismutase 1 (hSOD1) maturation through interaction with human copper chaperone for SOD1 (hCCS). Proc Natl Acad Sci 109(34):13555–13560
5. Bannwarth S, Ait-El-Mkadem S, Chaussenot A, Genin EC, Lacas-Gervais S, Fragaki K, Berg-Alonso L, Kageyama Y, Serre V, Moore DG (2014) A mitochondrial origin for frontotemporal dementia and amyotrophic lateral sclerosis through CHCHD10 involvement. Brain 137(8):2329–2345
6. Barber SC, Shaw PJ (2010) Oxidative stress in ALS: key role in motor neuron injury and therapeutic target. Free Radic Biol Med 48(5):629–641

7. Bartolome F, Wu H-C, Burchell VS, Preza E, Wray S, Mahoney CJ, Fox NC, Calvo A, Canosa A, Moglia C (2013) Pathogenic VCP mutations induce mitochondrial uncoupling and reduced ATP levels. Neuron 78(1):57–64
8. Beckman JS, Chen J, Crow JP, Ye YZ (1994) Reactions of nitric oxide, superoxide and peroxynitrite with superoxide dismutase in neurodegeneration. Prog Brain Res., Elsevier 103:371–380
9. Bian K, Murad F (2003) Nitric oxide (NO) – biogeneration, regulation, and relevance to human diseases. Front Biosci 8:d264–d278
10. Bota DA, Davies KJ (2002) Lon protease preferentially degrades oxidized mitochondrial aconitase by an ATP-stimulated mechanism. Nat Cell Biol 4(9):674
11. Bruening W, Roy J, Giasson B, Figlewicz DA, Mushynski WE, Durham HD (1999) Up-regulation of protein chaperones preserves viability of cells expressing toxic Cu/Zn-superoxide dismutase mutants associated with amyotrophic lateral sclerosis. J Neurochem 72(2):693–699
12. Butterfield DA, Hensley K, Cole P, Subramaniam R, Aksenov M, Aksenova M, Bummer PM, Haley BE, Carney JM (1997) Oxidatively induced structural alteration of glutamine synthetase assessed by analysis of spin label incorporation kinetics: relevance to Alzheimer's disease. J Neurochem 68(6):2451–2457
13. Butterfield DA, Perluigi M (2017) Redox proteomics: a key tool for new insights into protein modification with relevance to disease. Mary Ann Liebert, New Rochelle
14. Butterfield DA, Perluigi M, Reed T, Muharib T, Hughes CP, Robinson RA, Sultana R (2012) Redox proteomics in selected neurodegenerative disorders: from its infancy to future applications. Antioxid Redox Signal 17(11):1610–1655
15. Calabrese L, Federici G, Rotilio G, Finazzi-Agrò A, Bannister WH, Bannister JV (1975) Labile sulfur in human superoxide dismutase. FEBS J 56(1):305–309
16. Carrì MT, Valle C, Bozzo F, Cozzolino M (2015) Oxidative stress and mitochondrial damage: importance in non-SOD1 ALS. Front Cell Neurosci 9:41
17. Cozzolino M, Ferri A, Teresa Carri M (2008) Amyotrophic lateral sclerosis: from current developments in the laboratory to clinical implications. Antioxid Redox Signal 10(3):405–444
18. Culotta VC, Klomp LW, Strain J, Casareno RLB, Krems B, Gitlin JD (1997) The copper chaperone for superoxide dismutase. J Biol Chem 272(38):23469–23472
19. Davies MJ (2016) Protein oxidation and peroxidation. Biochem J 473(7):805–825
20. Davoli A, Greco V, Spalloni A, Guatteo E, Neri C, Rizzo GR, Cordella A, Romigi A, Cortese C, Bernardini S (2015) Evidence of hydrogen sulfide involvement in amyotrophic lateral sclerosis. Ann Neurol 77(4):697–709
21. de Beus MD, Chung J, Colón W (2004) Modification of cysteine 111 in Cu/Zn superoxide dismutase results in altered spectroscopic and biophysical properties. Protein Sci 13(5):1347–1355
22. Ferrante RJ, Browne SE, Shinobu LA, Bowling AC, Baik MJ, MacGarvey U, Kowall NW, Brown RH, Beal MF (1997) Evidence of increased oxidative damage in both sporadic and familial amyotrophic lateral sclerosis. J Neurochem 69(5):2064–2074
23. Fitzmaurice P, Shaw I, Kleiner H, Miller R, Monks T, Lau S, Mitchell J, Lynch P (1996) Evidence for DNA damage in amyotrophic lateral sclerosis. Muscle Nerve 19(6):797–798
24. Fu M, Zhang W, Wu L, Yang G, Li H, Wang R (2012) Hydrogen sulfide (H2S) metabolism in mitochondria and its regulatory role in energy production. Proc Natl Acad Sci 109(8):2943–2948
25. Gadoth N, Göbel HH (2011) Oxidative stress and free radical damage in neurology. Springer, New York
26. Gerő D, Torregrossa R, Perry A, Waters A, Le-Trionnaire S, Whatmore JL, Wood M, Whiteman M (2016) The novel mitochondria-targeted hydrogen sulfide (H2S) donors AP123 and AP39 protect against hyperglycemic injury in microvascular endothelial cells in vitro. Pharmacol Res 113:186–198
27. Giles GI, Tasker KM, Jacob C (2001) Hypothesis: the role of reactive sulfur species in oxidative stress. Free Radic Biol Med 31(10):1279–1283
28. Greco V, Spalloni A, Corasolla Carregari V, Pieroni L, Persichilli S, Mercuri N, Urbani A, Longone P (2018) Proteomics and toxicity analysis of spinal-cord primary cultures upon hydrogen sulfide treatment. Antioxidants 7(7):87
29. Higgins CM, Jung C, Ding H, Xu Z (2002) Mutant Cu, Zn superoxide dismutase that causes motoneuron degeneration is present in mitochondria in the CNS. J Neurosci 22(6):RC215
30. Ihara Y, Nobukuni K, Takata H, Hayabara T (2005) Oxidative stress and metal content in blood and cerebrospinal fluid of amyotrophic lateral sclerosis patients with and without a Cu, Zn-superoxide dismutase mutation. Neurol Res 27(1):105–108
31. Ince P, Shaw P, Candy J, Mantle D, Tandon L, Ehmann W, Markesbery W (1994) Iron, selenium and glutathione peroxidase activity are elevated in sporadic motor neuron disease. Neurosci Lett 182(1):87–90
32. Islam MT (2017) Oxidative stress and mitochondrial dysfunction-linked neurodegenerative disorders. Neurol Res 39(1):73–82
33. Jaarsma D, Rognoni F, van Duijn W, Verspaget HW, Haasdijk ED, Holstege JC (2001) CuZn superoxide dismutase (SOD1) accumulates in vacuolated mitochondria in transgenic mice expressing amyotrophic lateral sclerosis-linked SOD1 mutations. Acta Neuropathol 102(4):293–305
34. Karch CM, Prudencio M, Winkler DD, Hart PJ, Borchelt DR (2009) Role of mutant SOD1 disulfide oxidation and aggregation in the pathogenesis of familial ALS. Proc Natl Acad Sci 106(19):7774–7779
35. Kawamata H, Manfredi G (2008) Different regulation of wild-type and mutant Cu, Zn superoxide dismutase

localization in mammalian mitochondria. Hum Mol Genet 17(21):3303–3317
36. Khalil B, Liévens J-C (2017) Mitochondrial quality control in amyotrophic lateral sclerosis: towards a common pathway? Neural Regen Res 12(7):1052
37. Kiernan MC, Vucic S, Cheah BC, Turner MR, Eisen A, Hardiman O, Burrell JR, Zoing MC (2011) Amyotrophic lateral sclerosis. Lancet 377(9769):942–955
38. Lehmer C, Schludi MH, Ransom L, Greiling J, Junghänel M, Exner N, Riemenschneider H, van der Zee J, Van Broeckhoven C, Weydt P (2018) A novel CHCHD10 mutation implicates a Mia40-dependent mitochondrial import deficit in ALS. EMBO Mol Med;10(6)
39. Leveugle B, Spik G, Perl DP, Bouras C, Fillit HM, Hof PR (1994) The iron-binding protein lactotransferrin is present in pathologic lesions in a variety of neurodegenerative disorders: a comparative immunohistochemical analysis. Brain Res 650(1):20–31
40. Libiad M, Yadav PK, Vitvitsky V, Martinov M, Banerjee R (2014) Organization of the human mitochondrial hydrogen sulfide oxidation pathway. J Biol Chem 289(45):30901–30910
41. Liochev SI, Fridovich I (2000) Copper-and zinc-containing superoxide dismutase can act as a superoxide reductase and a superoxide oxidase. J Biol Chem 275(49):38482–38485
42. Liu Q, D'silva P, Walter W, Marszalek J, Craig EA (2003) Regulated cycling of mitochondrial Hsp70 at the protein import channel. Science 300(5616):139–141
43. Longen S, Richter F, Köhler Y, Wittig I, Beck K-F, Pfeilschifter J (2016) Quantitative persulfide site identification (qPerS-SID) reveals protein targets of H 2 S releasing donors in mammalian cells. Sci Rep 6:29808
44. Magrane J, Cortez C, Gan W-B, Manfredi G (2013) Abnormal mitochondrial transport and morphology are common pathological denominators in SOD1 and TDP43 ALS mouse models. Hum Mol Genet 23(6):1413–1424
45. Magrané J, Hervias I, Henning MS, Damiano M, Kawamata H, Manfredi G (2009) Mutant SOD1 in neuronal mitochondria causes toxicity and mitochondrial dynamics abnormalities. Hum Mol Genet 18(23):4552–4564
46. Mishanina TV, Libiad M, Banerjee R (2015) Biogenesis of reactive sulfur species for signaling by hydrogen sulfide oxidation pathways. Nat Chem Biol 11(7):457
47. Murphy MP (2009) How mitochondria produce reactive oxygen species. Biochem J 417(1):1–13
48. Olson KR, Gao Y, Arif F, Arora K, Patel S, DeLeon ER, Sutton TR, Feelisch M, Cortese-Krott MM, Straub KD (2018) Metabolism of hydrogen sulfide (H2S) and production of reactive sulfur species (RSS) by superoxide dismutase. Redox Biol 15:74–85
49. Orsini M, Oliveira AB, Nascimento OJ, Reis CHM, Leite MAA, de Souza JA, Pupe C, de Souza OG, Bastos VH, de Freitas MR (2015) Amyotrophic lateral sclerosis: new perpectives and update. Neurol Int 7(2):5885
50. Pasinelli P, Belford ME, Lennon N, Bacskai BJ, Hyman BT, Trotti D, Brown RH Jr (2004) Amyotrophic lateral sclerosis-associated SOD1 mutant proteins bind and aggregate with Bcl-2 in spinal cord mitochondria. Neuron 43(1):19–30
51. Paul BD, Snyder SH (2017) Gasotransmitter hydrogen sulfide signaling in neuronal health and disease. Biochem Pharmacol 149:101–109
52. Pieragostino D, Del Boccio P, Di Ioia M, Pieroni L, Greco V, De Luca G, D'Aguanno S, Rossi C, Franciotta D, Centonze D (2013) Oxidative modifications of cerebral transthyretin are associated with multiple sclerosis. Proteomics 13(6):1002–1009
53. Raha S, Robinson BH (2000) Mitochondria, oxygen free radicals, disease and ageing. Trends Biochem Sci 25(10):502–508
54. Rakhit R, Cunningham P, Furtos-Matei A, Dahan S, Qi X-F, Crow JP, Cashman NR, Kondejewski LH, Chakrabartty A (2002) Oxidation-induced misfolding and aggregation of superoxide dismutase and its implications for amyotrophic lateral sclerosis. J Biol Chem 277(49):47551–47556
55. Rowland LP, Shneider NA (2001) Amyotrophic lateral sclerosis. N Engl J Med 344(22):1688–1700
56. Sasaki S, Iwata M (2007) Mitochondrial alterations in the spinal cord of patients with sporadic amyotrophic lateral sclerosis. J Neuropathol Exp Neurol 66(1):10–16
57. Sasaki S, Warita H, Murakami T, Shibata N, Komori T, Abe K, Kobayashi M, Iwata M (2005) Ultrastructural study of aggregates in the spinal cord of transgenic mice with a G93A mutant SOD1 gene. Acta Neuropathol 109(3):247–255
58. Sbodio JI, Snyder SH, Paul BD (2018) Redox mechanisms in neurodegeneration: from disease outcomes to therapeutic opportunities. Antioxid Redox Signal
59. Searcy DG, Whitehead JP, Maroney MJ (1995) Interaction of Cu, Zn superoxide dismutase with hydrogen sulfide. Arch Biochem Biophys 318(2):251–263
60. Shaw PJ, Ince PG, Falkous G, Mantle D (1995) Oxidative damage to protein in sporadic motor neuron disease spinal cord. Ann Neurol 38(4):691–695
61. Sheehan D, McDonagh B, Bárcena JA (2010) Redox proteomics. Expert Rev Proteomics 7(1):1–4
62. Shibata N, Nagai R, Uchida K, Horiuchi S, Yamada S, Hirano A, Kawaguchi M, Yamamoto T, Sasaki S, Kobayashi M (2001) Morphological evidence for lipid peroxidation and protein glycoxidation in spinal cords from sporadic amyotrophic lateral sclerosis patients. Brain Res 917(1):97–104
63. Smith EF, Shaw PJ, De Vos KJ (2017) The role of mitochondria in amyotrophic lateral sclerosis. Neurosci Lett Jun 30
64. Sturtz LA, Diekert K, Jensen LT, Lill R, Culotta VC (2001) A fraction of yeast cu, zn-superoxide dismutase and its metallochaperone, ccs, localize to the

intermembrane space of mitochondria a physiological role for sod1 in guarding against mitochondrial oxidative damage. J Biol Chem 276(41):38084–38089
65. Tafuri F, Ronchi D, Magri F, Comi GP, Corti S (2015) SOD1 misplacing and mitochondrial dysfunction in amyotrophic lateral sclerosis pathogenesis. Front Cell Neurosci 9:336
66. Tan W, Pasinelli P, Trotti D (2014) Role of mitochondria in mutant SOD1 linked amyotrophic lateral sclerosis. Biochim Biophys Acta (BBA)-Mol Basis Dis 1842(8):1295–1301
67. Taylor JP, Hardy J, Fischbeck KH (2002) Toxic proteins in neurodegenerative disease. Science 296(5575):1991–1995
68. Vandiver MS, Snyder SH (2012) Hydrogen sulfide: a gasotransmitter of clinical relevance. J Mol Med 90(3):255–263
69. Vehviläinen P, Koistinaho J, Gundars G (2014) Mechanisms of mutant SOD1 induced mitochondrial toxicity in amyotrophic lateral sclerosis. Front Cell Neurosci 8:126
70. Velde CV, McDonald KK, Boukhedimi Y, McAlonis-Downes M, Lobsiger CS, Hadj SB, Zandona A, Julien J-P, Shah SB, Cleveland DW (2011) Misfolded SOD1 associated with motor neuron mitochondria alters mitochondrial shape and distribution prior to clinical onset. PLoS One 6(7):e22031
71. Voos W (2013) Chaperone–protease networks in mitochondrial protein homeostasis. Biochim Biophys Acta (BBA)-Mol Cell Res 1833(2):388–399
72. Wang J, Xu G, Borchelt DR (2002) High molecular weight complexes of mutant superoxide dismutase 1: age-dependent and tissue-specific accumulation. Neurobiol Dis 9(2):139–148
73. Wei Y-H, Lu C-Y, Wei C-Y, Ma Y-S, Lee H-C (2001) Oxidative stress in human aging and mitochondrial disease-consequences of defective mitochondrial respiration and impaired antioxidant enzyme system. Chin J Physiol 44(1):1–12
74. Zetterström P, Stewart HG, Bergemalm D, Jonsson PA, Graffmo KS, Andersen PM, Brännström T, Oliveberg M, Marklund SL (2007) Soluble misfolded subfractions of mutant superoxide dismutase-1s are enriched in spinal cords throughout life in murine ALS models. Proc Natl Acad Sci 104(35):14157–14162

# A Tag-Based Affinity Purification Mass Spectrometry Workflow for Systematic Isolation of the Human Mitochondrial Protein Complexes

6

Zhuoran Wu, Ramy Malty, Mohamed Taha Moutaoufik, Qingzhou Zhang, Matthew Jessulat, and Mohan Babu

**Abstract**

Mitochondria (mt) are double-membraned, dynamic organelles that play an essential role in a large number of cellular processes, and impairments in mt function have emerged as a causative factor for a growing number of human disorders. Given that most biological functions are driven by physical associations between proteins, the first step towards understanding mt dysfunction is to map its protein-protein interaction (PPI) network in a comprehensive and systematic fashion. While mass-spectrometry (MS) based approaches possess the high sensitivity ideal for such an endeavor, it also requires stringent biochemical purification of bait proteins to avoid detecting spurious, non-specific PPIs. Here, we outline a tagging-based affinity purification coupled with mass spectrometry (AP-MS) workflow for discovering new mt protein associations and providing novel insights into their role in mt biology and human physiology/pathology. Because AP-MS relies on the creation of proteins fused with affinity tags, we employ a versatile-affinity (VA) tag, consisting of 3× FLAG, 6 × His, and Strep III epitopes. For efficient delivery of affinity-tagged open reading frames (ORF) into mammalian cells, the VA-tag is cloned onto a specific ORF using Gateway recombinant cloning, and the resulting expression vector is stably introduced in target cells using lentiviral transduction. In this chapter, we show a functional workflow for mapping the mt interactome that includes tagging, stable transduction, selection and expansion of mammalian cell lines, mt extraction, identification of interacting protein partners by AP-MS, and lastly, computational assessment of protein complexes/PPI networks.

**Keywords**

Affinity purification · Protein complex · Versatile-affinity tagging · Lentivirus · Mass spectrometry · Mitochondria · Protein-protein interaction

Authors Zhuoran Wu, Ramy Malty, Mohamed Taha Moutaoufik, and Qingzhou Zhang are contributed equally to this chapter

Z. Wu · R. Malty · M. T. Moutaoufik · Q. Zhang
M. Jessulat · M. Babu (✉)
Department of Biochemistry, University of Regina, Regina, Saskatchewan, Canada
e-mail: zwo674@uregina.ca; mohan.babu@uregina.ca


A. Urbani, M. Babu (eds.), *Mitochondria in Health and in Sickness*, Advances in Experimental Medicine and Biology 1158, https://doi.org/10.1007/978-981-13-8367-0_6

## 6.1 Introduction

Mitochondria (mt) are intimately involved in many biological processes including energy metabolism, apoptosis, and cell signalling [1]. Therefore, it is unsurprising that many important human pathologies, especially those of neurodegenerative [2, 3], metabolic [4], cardiovascular [5] and malignant [6] origins, often have mt dysfunction as a common hallmark. Given its disease relevance, mt research has become a "hot spot" in different fields, driving the discovery of novel physiological and pathological mechanisms as well as the development of new techniques and paradigms.

In recent years, proteomic approaches have moved to the forefront of the exploration of molecular mechanisms involved in mt function and disease [1, 7–9]. Unlike the classical reductionist approach focusing exclusively on one particular protein or process, proteomics is the large-scale study of complete groups of proteins sharing a common characteristic, such as subcellular, nuclear, or membrane localization, in a living organism, to gather systems level insights in protein expression, physical and consequently functional relationships, and potential role in pathogenesis in an unbiased manner [1]. More importantly, given that many biological processes are mediated through the physical association between proteins, the systematic mapping of protein-protein interaction (PPI) networks is becoming an increasingly powerful means of studying how molecular events drive biological systems. In the context of mt health and disease, mapping PPIs can be especially valuable in the development of new diagnostic biomarkers and elucidating drug targets. As well, the differences in the PPI network between healthy and diseased states enable researchers to design models and make predictions about the underlying biological processes contributing to pathogenesis.

The construction of large-scale human PPI networks were traditionally conducted by systematically examining pair-wise interactions identified through several experimental means, including yeast-two hybrid [10, 11] and biochemical fractionation coupled with mass spectrometry (MS) [12, 13]. However, affinity-purification using the tag-based approach or immunoprecipitation (IP) using a protein-specific antibody coupled with MS [7, 8, 14, 15] has become the method of choice for purification of a single protein and interactors at a time. These methods are well-suited to identify and characterize transiently or stably interacting proteins.

The recent successful purification of mt proteins using a lentivirus and cell culture-based AP-MS framework appeared to be a promising approach to isolate stable protein complexes [7, 8]. This flexible and robust AP-MS procedure allows the tagged proteins to be expressed at a level similar to the native physiological state and then purified via a two-step enrichment procedure. Although its high sensitivity makes MS-based approach crucial for large-scale PPI research [13, 15, 16], great care must be exercised to minimize non-specific (spurious) PPIs and contaminants [17]. Thus, the AP procedure provides a comprehensive yet selective strategy of protein purification to minimize false negative and positives, allowing for accurate and sensitive MS identification of distinct complexes and PPIs.

As an alternative to AP-MS, immunoprecipitation coupled with MS (IP-MS) has been commonly used to purify mammalian protein complexes [18]; however, the scale of such studies is often limited by the availability of validated antibodies against specific bait targets. Therefore, for large-scale, systematic identification of PPI by AP-MS, a more suitable strategy that can be integrated into a high-throughput workflow would be necessary. To this end, since the main requirement for AP-MS is the creation of proteins fused with affinity tags, rather than relying on protein-specific antibodies, transgenic cells could be generated with an affinity tag introduced on the C-terminus (to avoid interference with mt protein import signal peptides, typically located at the N-terminal) of the open reading frame (ORF) of interest. In this manner, the tagged protein can still be expressed at a near physiological level.

The purification method described here uses the versatile-affinity (VA) tag [19], a 12-kDa triple affinity tag consisting of 3× FLAG, 6 × His, and Strep III epitopes. Here, the 3× FLAG is spaced from the rest of the sequence by a dual tobacco etch virus (TEV) protease cleavage site, thus allowing for its proteolytic cleavage, and the use of two other tags, permitting a highly selective two-stage protein purification and enrichment. A widely used epitope for purification, 3× FLAG is not only small in size, which reduces the chance of functional perturbation of the tagged proteins, but also highly amenable to sensitive immunodetection with commercially available high-affinity anti-FLAG antibodies. In addition, the inclusion of 6 × His and Strep III adds considerable versatility to the VA-tag. The widely used 6 × His gives researchers the option of purifying proteins under denaturing conditions, while Strep III is highly selective, tightly binds to Strep-tactin and is efficiently displaced (competed) with biotin for elution.

The overall workflow proceeds as outlined in Fig. 6.1. Briefly, a Gateway cloning procedure is first used to introduce the affinity tag coding sequence on the C-terminus of the mt protein of interest (also called "baits") on a lentiviral expression backbone. The resulting expression constructs are packaged into lentiviral particles and used to transduce target mammalian cells at multiplicity of infection (MOI) of ≈0.3, thereby generating cell lines stably expressing the VA-tagged mt protein at near-physiological expression level. Transgenic cells are cross-linked, lysed, and from the lysates derived from these expanded cell lines the tagged bait proteins can be isolated by tandem affinity purification and then processed for analysis by MS. The identification of co-purifying proteins (also called "preys") is achieved by searching the resulting mass/charge spectra against peptide sequence databases, which then enables computational filtration, scoring and clustering algorithms to be used to generate a mt interaction network.

In the rest of this chapter, we describe detailed step-by-step procedures for the entire workflow outlined above. It is noteworthy that while the overall context for methodologies recorded here deals primarily with mt, the basic approach can be adapted for analysis of the proteomes of other organelles and subcellular compartments or whole cells.

## 6.2 Materials

### 6.2.1 Vector Construction and Generation

- pLD-puro-Cc-VA (Addgene# 24588) maintained in *ccdA*-containing strains such as one shot *ccdB* survival or DB3.1 *E. coli.*
- Gateway LR Clonase II Enzyme Mix (Thermo Fisher Scientific, cat# 11791-020)
- Stbl3 (Thermo Fisher Scientific, cat# C737303) or NEB Stable (NEB, cat# C3040H)
- Expression clones from Harvard Plasmid ID repository
- Roche High Pure RNA Isolation kit (Sigma-Aldrich, cat# 12033674001)
- iScript cDNA Synthesis kit (BioRad, cat# 1708890).

### 6.2.2 Lentiviral Transduction

#### 6.2.2.1 Lentiviral Production

- HEK 293T cells (ATCC, cat# CRL-3216)
- Lentiviral coat protein-encoding plasmid, e.g. pMD2.G (Addgene# 12259)
- Lentiviral packaging plasmid: the third-generation packaging plasmids pMDLg/pRRE (Plasmid # 12251) and pRSV-Rev (Plasmid# 12253)
- Lipofectamine LTX with Plus reagent (Thermo Fisher Scientific, cat# 15338100)
- Opti-MEM I (Thermo Fisher Scientific, cat# 31985070)
- Low-protein binding 0.4 μm membrane syringe filter (e.g. polyether sulfone, Sarstedt, cat# 83.1826)
- 0.2 μm membrane syringe filter (Sarstedt, cat# 83.1826.001)

#### 6.2.2.2 Lentiviral Infection

- Six-well plate (Sarstedt, cat# 83.3920.005)
- Hexadimethrine bromide (Sigma-Aldrich, cat# H9268)

#### 6.2.2.3 Antibiotic Selection

- Microtest Plate 96 Well (Sarstedt, cat# 82.1581.001)
- Blasticidin S (Fisher Scientific, cat# BP2647-25)

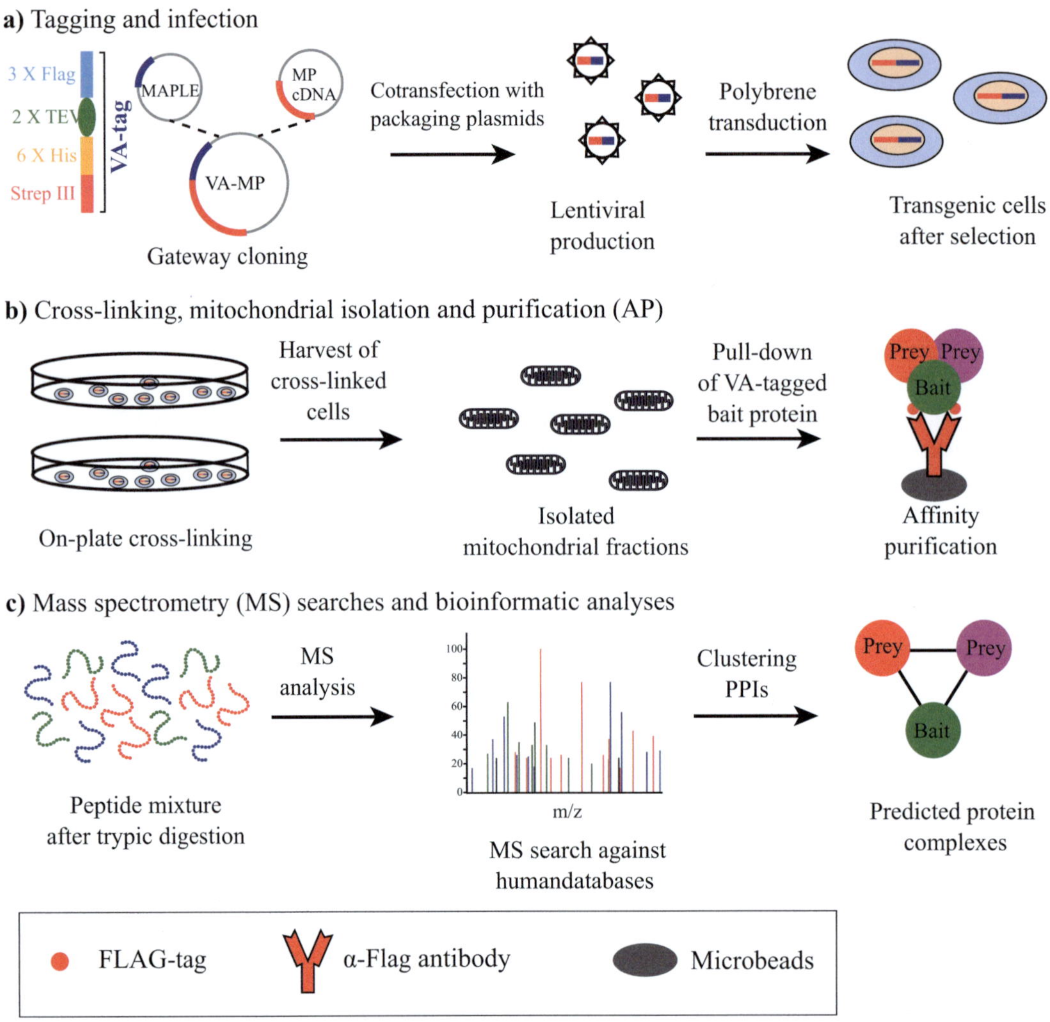


**Fig. 6.1** Outline of the AP-MS workflow. Schematic of Lentiviral-based versatile-affinity (VA) tagging of MP using the MAPLE lentiviral transfer vector (**a**), followed by purification of VA-tagged proteins and their interactors from mt fractions isolated from transgenic HEK293 cells (**b**). Proteins co-purified with the target bait protein identified using MS and analyzed by clustering algorithm to predict protein complexes (**c**). MAPLE mammalian affinity purification and lentiviral expression, MP mitochondrial protein, AP affinity purification, VA versatile-affinity, PPIs protein-protein interactions

- Puromycin (Fisher Scientific, cat# BP2956-100)
- Presto Blue (Thermo Fisher Scientific, A13261)

### 6.2.3 Confirmation of Tagging by Immunoblotting

- Trans-Blot ® Cell system (Bio-Rad, cat# 170-3853)
- Nitrocellulose membrane (Bio-Rad, cat# 162-0115)
- Transfer stock buffer (10X): 120 g Tris -HCl and 576 g glycine in distilled water
- 1X transfer buffer: 400 mL of 10X transfer stock buffer with 800 mL methanol and 2800 mL of sterile distilled water. Store at room temperature.
- TBS stock buffer (5X): 48.44 g Tris and 584.4 g NaCl in distilled water
- Wash buffer: 1X TBS buffer containing 0.05% Tween-20

- Blocking buffer: 5% (w/v) non-fat dry milk in 1X TBS buffer containing 0.1% Tween-20
- Primary antibody: M2 antibody (Sigma-Aldrich, cat# F3165)
- Primary antibody buffer: TBS buffer supplemented with 1% gelatin, 0.05% Tween-20, and 0.02% sodium azide
- Secondary antibody: Anti-mouse IgG conjugated to horseradish peroxidase (HRP; SantaCruz, cat# sc-516102)
- Secondary antibody buffer: 5% (w/v) non-fat dry milk in TBS buffer containing 0.1% Tween-20
- Chemiluminescence reagent (PIERCE, cat# 1856136), prepared (0.125 mL of chemiluminescence reagent per $cm^2$ of membrane) by mixing equal volumes of the enhanced luminol reagent and the oxidizing reagent
- Kodak X-OMAT autoradiography film (Clonex Corp., cat# CLEC810)
- Stripping buffer: 62.5 mM Tris-HCl (pH 6.7), 2% (w/v) SDS, and 100 mM β- mercaptoethanol. Store at room temperature.

### 6.2.4 Freezing, Reviving and Differentiating Mammalian Cells

#### 6.2.4.1 Cryogenic Preservation of Successfully Tagged Mammalian Cells

- DMSO (Sigma-Aldrich, cat# C6164)
- Trypsin-EDTA solution (Thermo Fisher Scientific, cat# 15400054)
- CryoPure Storage System (Sarstedt, cat# 72.377)
- Mr. Frosty™ Freezing Container (Thermo Fisher Scientific, cat# 5100-0001)

#### 6.2.4.2 Reviving and Large-Scale Culturing of Mammalian Cells

- One hundred millimeter (Sarstedt, cat# 83.3902) and 150 mm (Sarstedt, cat# 83.3903) polystyrene culture dishes.
- DMEM complete: DMEM high glucose (Thermo Fisher Scientific, cat# 11995073), supplemented with 10% (v/v) fetal bovine serum (Thermo Fisher Scientific, cat# 12483020), 100 units/mL of penicillin and 100 μg/mL of streptomycin (Thermo Fisher Scientific, cat# 15140122).

#### 6.2.4.3 Differentiation of SH-SY5Y Cells

- Human neuroblastoma SH-SY5Y (ATCC, cat# CRL-2266)
- All-trans retinoic acid (at-RA, Sigma-Aldrich; cat# R2625)
- Brain-derived neurotrophic factor (BDNF, Thermo Fisher Scientific, cat# 14-8366-80)

### 6.2.5 Cell Harvesting, Mt Fractionation and Protein Purification

#### 6.2.5.1 Cell Preparation

- Phosphate-Buffered Saline (PBS, pH 7.4) (Thermo Fisher Scientific, cat# 10010023)

#### 6.2.5.2 Cross-Linking with DSP

- Dithiobis succinimidyl propionate (DSP, Thermo Fisher Scientific, cat# PG82081)
- Dimethyl sulfoxide (DMSO, Sigma-Aldrich, cat# D8418)
- One molar Tris-HCl (pH 7.5) stock solution: 121.1 g of Tris base in 800 mL of distilled water, adjusted to pH 7.5 with concentrated HCl and brought to a final volume of 1 L.
- Quenching solution: 100 mM Tris-HCl (pH 7.5) supplemented with 2 mM EDTA.

#### 6.2.5.3 Mt Purification

- Homogenization buffer: 10 mM NaCl, 1.5 mM $MgCl_2$ and 10 mM Tris-HCl.
- Mitochondria stabilization buffer: 210 mM mannitol, 70 mM sucrose, 5 mM Tris-HCl (pH 7.5), and 1 mM EDTA.

#### 6.2.5.4 Tandem Affinity Purification Using VA-Tag

- Mitochondria lysis buffer: 30 mM Tris-HCl (pH 7.5), 150 mM NaCl, 1% digitonin (Sigma-Aldrich, cat# D140), and supplemented with

protease and phosphatase inhibitors (Thermo Fisher Scientific, cat# 78440).
- Bio-Spin chromatography columns (Bio-Rad, cat# 7326008)
- Anti-Flag M2 agarose beads (Sigma-Aldrich, cat# M8823)
- TBS buffer: 50 mM Tris (pH 7.5) and 150 mM NaCl
- TEV buffer: 10 mM Tris-HCl (pH 7.9), 125 mM NaCl, and 0.1% digitonin (Sigma-Aldrich, cat# D140)
- TEV protease (Sigma-Aldrich, cat# T4455), Dilute to working concentration of 2 mg/mL
- 3× FLAG peptide (Sigma-Aldrich, cat# F4799)
- Strep-Tactin-Sepharose bead (IBA, cat# 2-1201-010)
- Ammonium bicarbonate (Sigma-Aldrich, cat # 09830)
- D-biotin (Sigma-Aldrich, B4501)
- Tris (2-carboxyethyl) phosphine (TCEP-HCl, Sigma-Aldrich, cat# C4706)

### 6.2.6 Sample Preparation for MS Analysis

- Digestion buffer: 50 mM $NH_4HCO_3$ and 1 mM $CaCl_2$
- Alkylating solution: 10 mM iodoacetamide (Sigma-Aldrich, cat# I1149)
- Trypsin gold (Promega, cat# V5280)
- C18 Zip-Tip (Millipore, cat# Z720046)
- HPLC-grade $H_2O$ in 0.1% formic acid (Fisher Scientific, cat# LS118)
- Elution solution (1% acetic acid, 65% acetonitrile, in HPLC $H_2O$ in 0.1% formic acid)
- Washing solution (1% acetic acid, 2% acetonitrile in HPLC $H_2O$ in 0.1% formic acid)

### 6.2.7 MS for Identifying Interacting Proteins

- EASY nano liquid chromatography (nLC) 1000 (Proxeon, Mississauga, ON, Canada)
- Velos Pro Ion Trap and Orbitrap Elite Mass Spectrometer (Thermo Fisher Scientific)
- P-2000 Laser-based Micropipette Puller System (Sutter Instruments)
- Luna 3 μm C18 100A from Phenomenex (cat# 04A-4251)
- Fused-silica capillary tubing (ID 75.4 mm) from Polymicro technologies (cat# 1068150019)
- Ninety-six-well plate from Thermo Fisher Scientific (AB-0800)
- Methanol from Thermo Fisher Scientific (A456-4)
- Formic acid from Thermo Fisher Scientific (117-50)
- Water from Thermo Fisher Scientific (W6-4)
- Sample buffer: 1% formic acid in water
- MS Buffer A: 0.1% formic acid in water from Thermo Fisher Scientific (LS118-1)
- MS Buffer B: 0.1% formic acid in acetonitrile from Thermo Fisher Scientific (LS120-1)

## 6.3 Methods

### 6.3.1 Vector Construction and Generation

In this section, we will specifically address cloning of Gateway-compatible entry clone into pLD-puro-Cc-VA (Addgene# 24588). The protocol can be modified and applied to a variety of other destination vectors. ***Note:*** *(1) pLD-puro-Cc-VA is maintained in either One shot ccdB survival or DB3.1 Escherichia coli since it contains gene ccdB; (2) since pLD-puro-Cc-VA is a lentiviral vector, host bacteria is grown at 30 °C for no longer than 16 h and in case of liquid media at rotating speed of no more than 180 RPM; (3) The recipe below can be downscaled to save ingredients.*

**Procedure:**

1. In a 1.5 mL Eppendorf tube, 25–75 ng entry plasmid is combined with 75 ng pLD-puro-Cc-VA, and made the volume up to 4 μL with TE buffer (pH 8.0).
2. LR Clonase II is vortexed briefly and 2 μL is added to the mixture above and the reaction is incubated at 25 °C for 1 h.

3. Proteinase K (0.5 μL) is added and mixed prior to incubation at 37 °C for 10 min.

**Transformation:**

1. Of the above mixture, 1–5 μL is pipetted into 50 μL of competent bacteria suitable for propagation of lentiviral plasmids (have low potential to mutate or recombine long terminal repeats (LTR)) such as Stbl3 or NEB Stable. The tube is mixed by gentle flicking a few times.
2. Tube is incubated on ice for 30 min followed by heat shock at 42 °C for 20–30 s.
3. After heat shock, tube is returned to ice briefly and 950 μL SOC media is added.
4. The tube is shaken at 37 °C for 1 h at 225 RPM.
5. With the above mixture, 50–100 μL is plated on ampicillin-containing Luria-Broth agar plate and incubated at 30 °C for 16 h.
6. Individual colonies are picked and grown in liquid media with ampicillin at 30 °C for 16 h with shaking at 180 RPM.
7. Plasmid is extracted and sanger-sequenced in the forward direction using CMV-F and reverse using the primer 5′-CTTTCCCCTGCACTGTACCC-3′, to validate cloning.

#### 6.3.1.1 Vector Generation

1. Where available, we obtained lentiviral expression vectors from Harvard Plasmid ID repository. If not available, Gateway-compatible vectors cloned into pLD-puro-Cc-VA vector using Gateway cloning.
2. If neither lentiviral nor Gateway-compatible clones of the coding sequence (CDS) of choice were available, the target cDNA can be cloned as follows.
3. Cloning of target CDS from cDNA library:
   (A) cDNA library generation: Total RNA is extracted from human cells using a compatible kit such as Roche High Pure RNA Isolation kit as per manufacturer's instructions. ***Note:*** *Care must be taken to avoid contaminating RNA sample with RNAse, which can lead to RNA degradation and poor cDNA library quality, resulting in failure of cDNA synthesis.*
   (B) Next, cDNA is synthesized from RNA using iScript cDNA Synthesis kit. Using Primer3Plus tool, gene-specific primers are designed to amplify from the mRNA sequence. ***Note:*** *The following rules should be taken into account while designing primers. Use default parameters whenever possible. Following Primer3Plus instructions, the CDS sequence is flanked to ensure that the designed primers amplify the full sequence. In general settings, under "Mispriming/Repeat Library" menu, choose Human. In the advanced settings, it is preferable to choose primers with a 1 or 2 CG clamp nucleotides at the 3′ end of the primer under "Advanced Settings" menu. If default settings failed to identify selective primers, adjust the "Product Size" setting under "Advanced Settings" menu so that the Min size matches the desired CDS length, and the Opt and Max is set reasonably above that. For example, if the CDS sequence is 1000 nucleotides, then Min size should be set to 1000, Opt to 1100, and Max to 1200. The Opt and Max sizes can be further adjusted accordingly until appropriate primers are identified.*
   (C) If the above primers flank the specific CDS, then a second should be manually designed to amplify the exact desired CDS with the STOP codon at its 3′ end removed to allow in-frame fusion with the VA-tag in the pLD-puro-Cc-VA vector. This second PCR can be performed using the former set of primers designed using Primer3Plus. ***Note:*** *Primers designed manually as above should be prepended on its 5′ end with either restriction sites of extra 6 nucleotides (on the 5′ end of the primer) that match restriction sites on the MAPLE vector. Alternately, the manually-designed primers can be used when cloning into a*

*donor (pDONR) vector, rendering it into a suitable entry (pENTR) vector that is Gateway-compatible with MAPLE. For details, review Gateway® Technology Guide from Thermo Fisher Scientific.*

### 6.3.2 Lentiviral Transduction

#### 6.3.2.1 Lentiviral Production

1. A fresh vial of healthy frozen 293T cells is thawed, plated, and regularly passaged in media composed of DMEM complete media.
2. After three passages, 293T cells are plated in standard 100 mm diameter dishes with 30–40% confluence at the time of transfection.
3. Media is replaced with antibiotic-free DMEM or Opti-MEM I 1 h before transfection.
4. Before transfection, individual plasmids are decontaminated by heating in a thermal cycler at 85 °C for 30 min followed by slow cooling at the rate of 1 °C/min. ***Note:*** *Alternatively, plasmid extraction can be performed under aseptic condition and eluted from the plasmid mini-prep column in sterile tubes using filter-sterilized nuclease-free water or elution buffer.*
5. After 8–12 h post plating, cells are transfected with a mixture of the following plasmids: 7.5 µg lentiviral expression (transfer) plasmid, 1.9 µg lentiviral coat protein encoding plasmid (e.g. pMD2.G), and third-generation lentiviral packaging plasmids (3.75 µg pMDLg/pRRE; 1.9 µg pRSV-Rev). Using the transfection reagent Lipofectamine LTX with Plus reagent, DNA is mixed according to the following recipe: (Tube A: DNA variable, 15 µl Plus reagent, up to 750 µl Opti-MEM I; Tube B: 52.5 µl Lipofectamine LTX, up to 750 µl Opti-MEM I). Tube A is added drop-wise to Tube B with frequent shaking and DNA-transfection reagent mixture is incubated at room temperature for 15 min. Media is replaced again with antibiotic-free DMEM or Opti-MEM I and DNA-transfection reagent mixture is added drop-wise to the cells and mixed gently. After 6–12 h, media is changed to complete media containing antibiotic and 30% FBS and returned to the incubator. Media is harvested after 24 h and again 48 h post-transfection. Media from both harvests are pooled and filtered through a low-protein binding 0.4 µm membrane syringe filter. Hexadimethrine bromide is freshly dissolved at 10 mg/mL, filter-sterilized through 0.2 µm syringe membrane filter and added to the lentiviral supernatant obtained in the previous step to a final concentration of 10 µg/mL.

#### 6.3.2.2 Lentiviral Infection

**Adherent Cells:**

1. Target cells are plated in six-well plate at 10–20% confluence and allowed to fully adhere prior to exposure to lentiviral supernatant.
2. Determination of multiplicity of infection (MOI):
   (A) Target cells are plated at ~10 to 20% confluence
   (B) Serially diluted lentiviral supernatant in complete media is added to target cells
   (C) Cells are centrifuged at 140 *xg* for 0.5–1 h at 30 °C.
   (D) After 24–48 h of infection, selection antibiotic (in case of pLD-puro-Cc-VA, puromycin 2 µg/ml) is added and a kill curve is constructed by plotting percent viability at different lentiviral dilutions versus control.
   (E) Lentiviral supernatant dilution that results in 30–50% cell viability provides highest probability that any given cell is infected with a single lentiviral particle and thus ensures near-physiological level of expression.
3. Media replaced with lentiviral supernatant (0.3–0.5 MOI) is centrifuged at 140 *xg*, for 0.5–1 h at 30 °C, and returned to the incubator.
4. After 24 h, lentiviral supernatant is replaced with fresh media, and antibiotic selection can be carried out 24–48 h after infection.
5. If toxicity is evident, lentiviral supernatant can be mixed with complete medium at 1:1

ratio. Exposure duration can be also minimized anywhere from 6 to 16 h. These measures will reduce the number of stably transduced cells, and will limit toxicity in difficult-to-work with cell lines.

**Suspension Cells:**

1. Approximately $6 \times 10^6$ cells are centrifuged at 140 *xg* for 5 min at room temperature.
2. Media is aspirated and replaced with lentiviral supernatant at 0.3–0.5 MOI as defined above.
3. Cells are centrifuged at 140 *xg* for 0.5–1 h at 30 °C, and resuspended cells are then returned to the incubator for 24 h.
4. After 24 h, cells are again centrifuged at 140 *xg* for 5 min at room temperature and lentiviral supernatant is replaced with fresh media, followed by antibiotic selection 24–48 h after infection.

#### 6.3.2.3 Antibiotic Selection

Because mammalian cell lines differ in their antibiotics sensitivity, prior to antibiotic selection, a preliminary experiment should be performed to determine the lowest antibiotic concentration and duration of exposure required to efficiently kill 100% of uninfected cells and thus selection of successfully lentivirus-transduced cells. Here, we provide general guideline for establishing a kill curve using puromycin or blasticidin. The same principle can be extended to other mammalian selection antibiotics, such as zeocin, hygromycin, and neomycin with adaptation to ensure longer incubation as the latter antibiotics have slower effect on non-transduced cells.

**Establish Antibiotic Kill Curve:**

**Day 0**

1. Seed cells at 50% confluence per well, in 24-well plate with blasticidin or puromycin selection, respectively.
2. Incubate cells overnight.

**Day 1**

1. Make three sets of eight concentrations of antibiotic in growth media. For puromycin: 0, 0.06, 0.2, 0.6, 2, 6 and 20 μg/mL. For blasticidin: 0, 0.15, 0.5, 1.5, 5, 15 and 50 μg/mL. Each dilution should have at least 110 μL as the final volume.
2. Remove cells from incubator, aspirate media, and replace with selection media supplemented at varying concentrations of antibiotics.

**Day 3–15**

1. Cell viability is determined using Presto Blue (Thermo Fisher Scientific, A13261). Optimum effectiveness is typically reached within 1–2 days for puromycin and 5–7 days for blasticidin.
2. Change media every day with the appropriate selection media. If cells become too confluent, they should be split and allowed to adhere and recover before resuming the antibiotic selection.
3. The minimum concentration resulting in complete cell death after selection period (depending on the antibiotic) should be used for antibiotic selection of transduced cells.
4. Validate transgene expression using immunoblotting.

### 6.3.3 Confirmation of Successful Tagging with Immunoblotting

#### 6.3.3.1 SDS-Polyacrylamide Gel Electrophoresis

1. The following protocol uses the Mini Protean 3 Cell (Bio-Rad cat# 165-3301) gel system. Before use, clean the glass plates, spacers, and combs to ensure they are free of dried gel fragments, grease, and dust.
2. For each gel, assemble one small and one large glass plate so that they are separated by spacers and an alignment card. Make sure to have the small glass plate on top of the spacers so that both sides and the bottom of the plates and the spacers are even.
3. Slide glass plates into the holder without tightening the screws. Make sure the glass plates are pushed all the way to the bottom before tightening the screws. If the assembly is correct, then the whole setup should snap into place when placed above the gasket.

4. Take out the alignment card, slide in the comb and mark a line at 2–3 cm from the bottom of the comb. The comb should be of the same thickness as that of the spacers.
5. Gently remove the comb. Pour 12% polyacrylamide resolving gel up to the marked line. Pipette a thin layer of ethanol on top of the gel and allow to polymerize for about 30 min.
6. Pour 4% polyacrylamide stacking layer to fill up the remaining volume between glass plates. Insert the comb and leave the gel to polymerize for another 30 min.
7. Remove the comb once the stacking gel has set, then use a 3 mL syringe fitted with a 22-gauge needle to rinse the wells with running buffer.
8. Place the gel in the electrophoresis unit and add running buffer.
9. Load each well with 10–20 μl samples in 1X SDS sample buffer. Reserve one lane for precision plus protein dual color standards (Bio-Rad cat#1610374). Complete the assembly of the electrophoresis unit and attach the power cords to the power supply.
10. Turn the power supply on and run the gel at 120 V. Stop running when the blue dye front reaches the bottom.
11. Once the electrophoresis is completed, remove the gel plate assembly and place on a paper towel. Carefully pry the glass plates apart using a spatula.
12. Immediately transfer the gel to the nitrocellulose membrane and follow the transfer protocol described below.

#### 6.3.3.2 Immunoblotting Using Chemiluminescence

***Note:** These instructions assume the use of a Trans-Blot ® Cell system. Wear gloves at all times when handling the membrane.*

1. Fill the cooling unit with water and store at −20 °C until ready to use.
2. Cut the nitrocellulose membrane to be slightly larger than the gel.
3. Equilibrate the gel by soaking it in transfer buffer for 15 min.
4. Briefly soak the membrane, four pieces of Quick Draw TM blotting paper (Sigma Aldrich cat#P7796), and two scotch-brite® fiber pads in transfer buffer.
5. Mark one side of the membrane for future reference.
6. Preparing the gel sandwich:
   (a) On a clean, even surface, place the cassette with the gray side down
   (b) Place one pre-wetted fiber pad on the gray side of the cassette.
   (c) Then put two pieces of wet blotting paper on top of the fiber pad.
   (d) Place the equilibrated gel on top of the filter paper.
   (e) Carefully place the pre-wetted membrane on top of the gel.
   (f) Place the other two pieces of blotting paper over the membrane, creating a 'sandwich'.
   (g) Remove any air bubbles trapped between layers by rolling a pipet over the sandwich.
7. Complete the assembly of the sandwich with the second fiber pad.
8. Close and lock the cassette firmly with minimal disturbance to the gel sandwich.
9. Insert the sandwich into the transfer apparatus. Make sure the membrane is positioned between the gel and anode as most polypeptides eluted from the gels are anions.
10. Add transfer buffer into the transfer apparatus, pour slowly to avoid forming bubbles. Once filled, put the lid on and turn on the power supply set at either 30 V overnight or 100 V for 1 h. To maintain even buffer temperature and ion distribution, add one stir bar.
11. After the transfer, unclamp the apparatus and remove the membrane, and allow it to air dry at room temperature. Mark the side of the membrane that is facing the gel as well as the positions of the pre-stained markers.

#### 6.3.3.3 Confirm Tagging with Anti-FLAG M2 Antibody

1. After transferring the proteins to the membrane, block the membrane by incubating in

blocking buffer for 1 h at room temperature on a C2 platform rocking shaker (New Brunswick Scientific, Edison, NJ, USA).

2. Wash the membrane twice for 5 min with wash buffer.
3. Dilute the anti-FLAG M2 primary antibody at 1:5000 in primary antibody buffer, and incubate with the membrane for 1 h on the rocking shaker.
4. Wash the membrane with wash buffer for 10 min three times.
5. Dilute HRP-labelled secondary antibody at 1:20,000 in secondary antibody buffer, and incubate with the membrane for 45 min at room temperature on the rocking shaker.
6. Wash the membrane with wash buffer three times with 10 min for each wash.

#### 6.3.3.4 Imaging Blots

1. Transfer the membrane to a shallow tray and incubate in the Western lightning™ chemiluminescence reagent plus for 5 min by gently shaking in the dark.
2. Drain off the excess chemiluminescence reagent.
3. Cover the membrane with Saran wrap.
4. Expose the membrane to Kodak X-OMAT blue autoradiography film for 30 s. If necessary, expose the membrane for up to another 30 min.

#### 6.3.3.5 Stripping and Re-probing the Blots

1. After imaging, wash the membrane for 15 min in TBS buffer.
2. Incubate the membrane for 30 min at 50 °C in stripping buffer.
3. Wash the membrane for 20 min in TBS buffer.
4. Incubate the membrane for 1 min in the Western lightning ™ chemiluminescence reagent plus. Expose the membrane to Kodak X-OMAT blue autoradiography film for 30 min to 1 h to ensure that the original signal is removed.
5. Wash membrane again for 20 min in TBS buffer. The membrane is now ready for reuse.
6. If successful tagging of the target bait protein is confirmed by immunoblotting, frozen stocks of mammalian cells can then be made and preserved in liquid nitrogen following the steps in the next section.

### 6.3.4 Freezing, Reviving, and Differentiating Mammalian Cells

***Note:*** *Successful use of cell line models is pivotal for studying the structure and function of the mt. While mt fractions can be obtained from other sources such as animal or human tissues, cell line models continue to be the mainstay for studying mt as they not only provide an abundant source of mt samples, but also allow for easy transfection and tagging of ORFs. Here, we will cover how to freeze and grow mammalian cells. In addition, given the link between mt and neurodegenerative diseases, we will cover the steps needed to differentiate neuronal-like cells from the model human SH-SY5Y neuroblastoma cell line.*

#### 6.3.4.1 Cryogenic Preservation of Successfully-Tagged Mammalian Cells

1. Take culture dishes from the incubator, aspirate media, and wash twice with 10 mL PBS.
2. Add Versene solution to gently detach cells (3 mL Versene for 150 mm cell culture dish). Put the culture dishes back into the incubator for 10–15 min.
3. Meanwhile, make freezing media by supplementing complete media with 10% DMSO.
4. Take out the dishes, gently tap on the sides to dislodge the cells.
5. Transfer the cell suspension into a 15 mL falcon tube.
6. Rinse the dish with another 6 mL of PBS, add this to the 15 mL falcon tube as well.
7. Centrifuge the cell suspension at 200 *xg* for 2–5 min, aspirate the supernatant while being careful not to disturb the pellet.

8. Resuspend the cells in freezing media by gently flicking the tube and slowly pipetting up and down.
9. Gently aliquot the cell suspension in cryogenic tubes.
10. Gradual freezing at the rate of 1 °C per minute can be achieved by using a Mr. Frosty freezing container in a − 80 °C freezer overnight.
11. Transfer to liquid nitrogen when feasible.

#### 6.3.4.2 Reviving and Large-Scale Culturing of Mammalian Cells

1. Transfer 10 mL of complete growth medium in a 100 mm diameter tissue culture dish, place it in a 37 °C incubator with 5% $CO_2$, and allow to equilibrate for 15 min.
2. Retrieve one vial containing the frozen cells from liquid nitrogen storage.
3. Immediately thaw the cells by placing it in the 37 °C water bath until only small silver of ice remains in the vial.
4. Dry the vial thoroughly, wipe its outside with 70% ethanol, and then place it into a biological safety cabinet.
5. Retrieve the pre-warmed 100 mm diameter tissue culture dish from the incubator, transfer the thawed cells to the dish by gentle pipetting.
6. Rock the plate gently a few times to distribute the cells evenly, and put back the plate into the incubator for overnight incubation.
7. Change media the next day, and afterwards change every 3 days to maintain the culture.
8. Passage into 150 mm tissue culture dishes when cells have reached ~80% confluence.

#### 6.3.4.3 Differentiation of SH-SY5Y Cells

***Note:*** *The SH-SY5Y cell line can be differentiated into neuron-like cells by the addition of all-trans-retinoic acid (at-RA), and brain-derived neurotrophic factor (BDNF) essentially as previously described with minor modifications* [20]. *Undifferentiated cells can be tagged with lentiviral transduction first and then differentiated.*

1. For differentiation, plate cells at a density of 4 × $10^4$ cells/$cm^2$ in DMEM complete media containing 5% FBS.
2. The next day, add differentiation medium containing 5% FBS and 10 μM *at*-RA to the cells.
3. Change the differentiation medium every day for 5 days.
4. On the 6th day, switch to DMEM supplemented with penicillin (50 u/mL), streptomycin (50 μg/mL), L-glutamine (2 mM), and BDNF (20 ng/mL) but without FBS.
5. Change media every 2–3 days for an additional 7 days.
6. Differentiation can be monitored microscopically for neuron-like morphology.
7. To obtain sufficient mt yield, about 8 × 150 mm dishes of differentiated cells are needed.

### 6.3.5 Cell Harvesting, Mt Fractionation, and Protein Purification

***Note***: *Prior to harvesting, cells expressing VA-tagged target proteins can be cross-linked using the cell membrane-permeable reversible cross-linking reagent, dithiobis succinimidyl propionate (DSP). Our cross-linking protocol is ideal for attached cells growing in 150 mm diameter dishes.*

#### 6.3.5.1 Cell Preparation

1. Remove media from the tissue culture dish.
2. Wash each plate with 10 mL ice-cold PBS twice to remove traces of media.

#### 6.3.5.2 Cross-Linking with DSP

1. Prepare 0.25 M stock of DSP in DMSO.
2. Dilute 0.25 M DSP stock 1:500 in PBS, to a working concentration of 0.5 mM, immediately before use.
3. Add 4 mL of working solution to each plate.
4. Incubate plates at room temperature for 30 min.

5. Quench the cross-linking reaction by adding 400 μL quenching solution, mix the solutions well by pipetting or rocking each plate, and then incubate at room temperature for 10 min, and periodically tap the plate for cells to dissociate.
6. Gently pipette the liquid from the slanted plates to wash cells off the dish, and centrifuge the cell suspension at 1300 *xg* for 5 min. After discarding the supernatant, either flash-freeze the cell pellet with liquid nitrogen for storage or continue onto mt extraction.

#### 6.3.5.3 Mt Extraction

1. Mt isolation is based on methods previously described [21], with minor modifications.
2. Depending on the cell line, cross-linked cells should be harvested from 2 to 5 × 150 mm tissue culture dishes. After harvesting, decant supernatant and resuspend cells in 1.1 mL of ice-cold hypotonic buffer. Allow cells to swell by incubating for 10 min on ice.
3. Homogenize cells by passing through a syringe fitted with a 22-gauge needle for 30 strokes. ***Note:*** *alternatively, a pre-chilled Dounce glass tissue homogenizer can be used.*
4. Add the homogenate into a conical centrifuge tube containing 800 μL of ice-cold mt stabilization buffer and mix gently. Pellet unbroken cells, nuclei, and large debris by centrifuging at 1300 *x g* for 5 min at 4 °C.
5. Transfer the supernatant to another conical tube, and centrifuge at 17,000 *xg* for 15 min at 4 °C. The cytosolic (supernatant) and mt (pellet) fractions can be now used for affinity purification or frozen for future use.

#### 6.3.5.4 Tandem Affinity Purification Using VA-Tag

1. Resuspend the mt pellet in mt lysis buffer, and incubate on ice for 30 min. Centrifuge the sample mixture at 23,000 *xg* for 20 min at 4 °C. Transfer 100 μL of packed anti-Flag M2 agarose beads (Sigma-Aldrich) or 200 μL of bead slurry into a fresh Eppendorf tube, and wash the beads twice with TBS buffer.
2. Add the harvested supernatant to the tube containing pre-washed anti-FLAG M2 agarose beads, and incubate at 4 °C for 4 h with gentle rotation. Transfer the supernatant along with the beads to a Bio-spin disposable chromatography column.
3. Wash the column four times with 1 mL ice-cold TBS buffer, and wait for 10 min between washes.
4. Incubate the beads with 100 μL of TEV cleavage buffer, 5 μL TEV protease (2 mg/mL), and 2 μL 3×FLAG peptide (5 mg/mL), and rotate overnight at 4 °C.
5. The next day, in a fresh column, add 20 μL Strep-Tactin-Sepharose bead slurry and wash three times with 1 mL TEV cleavage buffer. Remove the top and bottom caps of the TEV cleavage column and drain the eluates into the column containing pre-washed Strep-Tactin-Sepharose beads. Incubate the column at 4 °C for 4 h, followed by washing the column four times with 1 mL TEV buffer, and 10 min wait between washes.
6. Elute with 50 mM ammonium bicarbonate and/or 2 mM D-biotin. Dry the eluted samples with a Speedvac prior to MS analysis, which is described below.

### 6.3.6 Sample Preparation for MS Analyses

#### 6.3.6.1 Digestion

1. Dilute the dried samples with 50 μL digestion buffer. Add 0.9 μL of 2 mM Tris (2-carboxyethyl) phosphine and incubate at room temperature for 45 min to reduce protein disulfide bonds.
2. Add 1 μL of alkylating solution and incubate the samples in the dark for 40 min. Digest the mixture with trypsin gold (Promega, cat # V5280). ***Note:*** *typically, we use 12 μg per digestion.*
3. Incubate the samples overnight at room temperature with gentle agitation by a rocking

shaker. The next day, stop the digestion by adding 1 μL of acetic acid.

#### 6.3.6.2 Desalting

1. Prepare a 96 well plate with 350 μL of wash solution and elution solution. Equilibrate the zip-tip by aspirating 20 μL of elution solution and discarding to waste three times. Wash by aspirating and dispensing three times the washing solution. After washing, draw 20 μL of digested sample into the tip, and then expel the liquid back into the sample, repeat this five to ten times. This will allow peptides to bind to the zip-tip.
2. Wash the zip-tip by drawing 20 μL of wash solution into the tip, then expelling into a waste beaker. Repeat this step three times. This step should remove all unwanted salts and detergents.
3. Add 5–10 μL of the fresh elution solution into a new, clean tube, and draw this solution through the zip-tip before expelling back into the same tube. Repeat this step two to three times in order to disassociate as much of the bound peptide as possible. ***Note:*** *Be careful not to draw air into the zip tip, which may make it difficult to expel all the solution.*
4. Repeat the equilibration, washing, binding and elution steps for two more times. Dry the samples by evaporation. This can be achieved either by air-drying for 2–3 h in a fume hood or under vacuum (e.g. Speedvac).
5. Resuspend the peptides in 20 μL of 0.1% formic acid, and then subject to MS. ***Note:*** At this point either proceed with MS or store peptides at −80 °C (or on dry ice) for future use.

### 6.3.7 MS for Identifying Interacting Proteins

#### 6.3.7.1 Preparation of Nanocolumn

1. Cut a 30–40 cm length of fused-silica capillary tubing. Place the length of tubing in the P-2000 laser puller (Sutter Instrument) so that the clear section is in the mirrored chamber of the puller. The laser is then focused on the center of the tubing and the fused silica can be melted. Select the program on the laser puller that will result in a 2–5 mm tip.

#### 6.3.7.2 Packing the Column

1. Place a small amount (5 mg) of C18 RP packing material (3 mm) in a microcentrifuge tube. Add 1 mL of methanol. Agitate the tube to create a slurry of the packing material. Place the open microcentrifuge tube into the pressure injection cell, Model PC1000, (Next Advance) and secure the lid by tightening the screws.
2. Insert the flat end of the pulled capillary column through the ferrule until it reaches the bottom of the microcentrifuge tube. Tighten the ferrule until the capillary does not move when gently tugged.
3. Adjust the pressure on the helium tank to 800–900 psi. Slowly pressurize the injection cell by opening the valve on the high-pressure line. If the injection cell is pressurized too rapidly, the microcentrifuge tube can rupture. Stop filling the capillary when the packing material reaches 10 cm of the capillary length.

#### 6.3.7.3 LC-MS/MS Analysis

1. Place the packed column in line with the LC-MS instrument. A Proxeon nanoLC pump is used to deliver a stable tip flow rate of 300 nL/min during the peptide separations.

   Resuspend the desalted peptide samples in 25 μL of 1% formic acid.
2. Inject 2 μL of each sample onto a Proxeon EASY nLC 1000 nano high-performance liquid chromatography connected to an Orbitrap Elite mass spectrometer.
3. Directly inject samples onto a nano column (C18 column, 10 cm × 75 μm ID, 3 μm, 100 Å) employing a water/acetonitrile/0.1% formic acid gradient as following. Over the course of 100 min, peptides can be separated using: 1% acetonitrile and increasing to 3% acetonitrile in the first 2 min; and a linear gradient is set up from 3% to 24% acetonitrile for 74 min, and then from 24% to 100% acetonitrile for 14 min, followed by 10 min wash at 100% acetonitrile.

4. Eluted peptides can then be directly sprayed into mass spectrometer using positive electrospray ionization (ESI) at an ion source temperature of 250 °C and an ion spray voltage of 2.1 kV.
5. Acquire full-scan MS spectra (m/z 350–2000) in the Orbitrap at 60,000 (m/z 400) resolution.
6. Using the Thermo Fisher Scientific XCalibur software, set the automatic gain control settings at 1e6 for full FTMS (Fourier Transform Mass Spectrometry) scans and 5e4 for MS/MS scans. Sequentially isolate the 15 most intense peptide ions with charge states ≥2 and fragment in the linear ion trap by low-energy collision-induced dissociation (CID).
7. Perform CID when ions intensity exceed 1500 counts, using a normalized collision energy set at 35%, activation Q at 0.250 and an activation time of 10 ms.

#### 6.3.7.4 Protein and Peptide Identification

1. The identified spectra can be obtained from the MS run as raw files. For database searching, convert the raw files into the standard mzXML format. Subject each identified spectrum to peptide spectrum matching against candidate peptides from human protein sequences using the SEQUEST ver. 2.7 rev. 9 search engine.
2. Run the putative matches identified by SEQUEST through the STATQUEST [22] filtering algorithm to assign statistical confidence.

### 6.3.8 Bioinformatics Pipeline

To detect high-confidence PPIs, we followed two major steps: (1) pre-processing, as well as (2) scoring, benchmarking, and clustering.

#### 6.3.8.1 Pre-processing

1. Filter the identified prey (or interacting) proteins from STATQUEST search engine at 90% probability cut-off.
2. Retain interacting proteins from AP-MS experiments in which targeted bait protein was captured.
3. Remove the most commonly reoccurring prey proteins (e.g., ribosomes, elongation or splicing factors) found in 80% or more in the AP-MS experiments. As well, prey proteins present in both the negative control and AP-MS experiments should be eliminated.

#### 6.3.8.2 Scoring, Benchmarking, and Clustering

1. Apply CompPASS [15, 23] or hypergeometric [13] algorithms to compute statistical scores for each interacting protein pairs.
2. The scored PPIs are then benchmarked against the reference set that is compiled either from the manually curated interactions from literatures or curated human protein complexes containing PPIs from CORUM (Comprehensive Resource of Mammalian Protein Complexes) database. To do so, we consider PPIs within complexes as true positive (TP), while PPIs between complexes as true negative (TN) dataset. Using the TP and TN data, the scored PPIs were then plotted using a Receiver Operate Characteristic (ROC). The area-under-the curve from the ROC plot is then computed and compared to evaluate the scoring performance, where 0.5 indicating completely random and 1 indicating perfect scoring.
3. After eliminating associations below a stringent threshold cut-off from the ROC analysis, the quality of scored PPIs can be assessed by cross-referencing to external sources such the experimentally derived interactions from literature sources or to those deposited in BioGRID database. As well, the newly-derived PPIs from the network can be compared to the computationally predicted interactions from STRING, GeneMANIA, and HumanNet. Interactions from the network can also be validated against the existing large-scale PPI studies by computing the functional coherence and similarity based on Gene Ontology annotations, and benchmarked against literature-curated PPIs (e.g. CORUM)

for the performance measures using fivefold cross validation.

4. The selected high-confidence PPIs were then used to generate macromolecular protein complexes using clustering algorithms, such as coreMethod [24] or MCL (Markov Clustering) [25] for downstream analysis.

## 6.4 Conclusion

In recent years, network biology has risen to the forefront in the exploration of the molecular basis of human health and disease [1, 7–9, 12, 13, 15]. Since macromolecular protein complexes are the direct mediators of most cellular functions [26], the systematic elucidation of PPIs and subunit composition of these complexes provides a powerful means for studying the pathophysiology of mt in the context of human health and disease [12, 27]. However, the construction of such a large-scale PPI networks requires reliable and efficient methods for elucidating proteome-wide mt interactions in a human cellular context, so in this chapter, we outlined a tagging-based AP-MS approach that can provide a viable, systematic workflow for discovering new mt links and providing future directions for understanding the role of mt in human physiology and pathology. To confirm the successful integration and expression of VA-tagged mt proteins, immunoblotting was performed in human embryonic kidney (HEK293) cells, using anti-FLAG antibody against the FLAG epitope, as shown in Fig. 6.2.

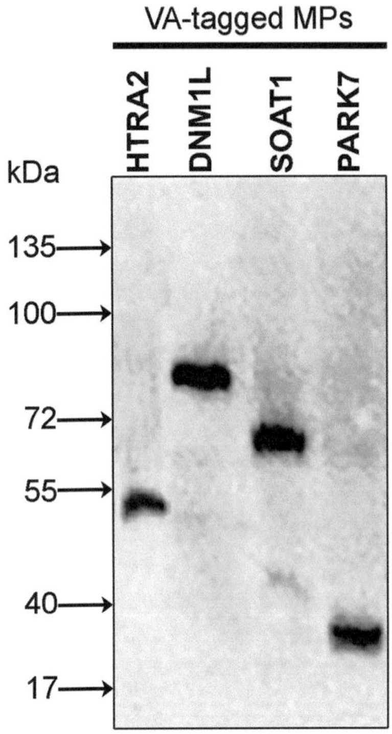

**Fig. 6.2** FLAG-epitope detection for VA-tagged mt proteins. Immunoblots showing the expression of a representative set of human mt proteins in lentivirus-transduced HEK293 cells. FLAG epitopes detected using anti-FLAG antibody. *MPs* mt proteins

A key component of the AP-MS workflow is the cross-linking step. Functionally relevant protein interactions in a biological system can be transient, and cross-linking chemicals provide means for capturing the transient interactions by covalently binding the interacting proteins together, facilitating the subsequent capture and identification of the interacting proteins [28]. Given this transient nature of most PPIs, the number of co-purified proteins, or "preys", of the tagged bait can depend on whether cross-linking was performed. In other words, chemical cross-linking is a critical step towards the detection of transient or weak interactions. As seen in Fig. 6.3, the mt protein, dynamin 1-like protein (DNM1L) was only recovered after cross-linking. This is consistent with past evidence suggesting that cross-linking can help to capture transient interactions in large-scale purifications [29, 30].

The AP-MS approach outlined in this chapter has been employed in a 2017 study from our group to purify a set of 27 nuclear-encoded human mt proteins associated with various neurodegenerative disorders. These mt bait proteins were selected broadly in terms of mt processes and were localized to the mt matrix, inner or outer membrane, and intermembrane space. In the assessment of the FLAG mt protein network generated by VA-tagging coupled with MS, 1964 high-confidence associations encompassing 772 prey proteins were identified using precision Orbitrap MS [8]. The overall reliability of the FLAG mt protein network was supported by the observation that nearly 75% of these interactions

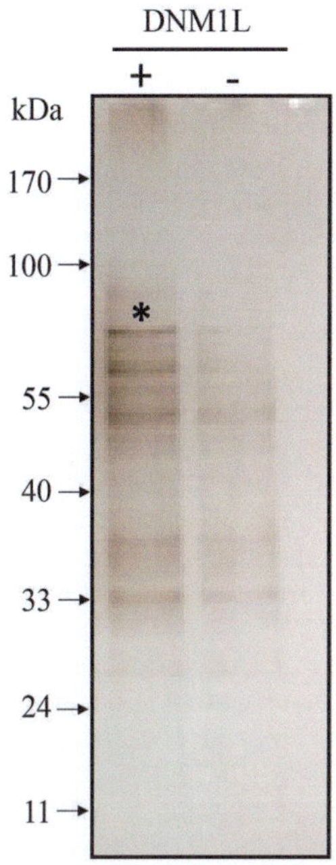


**Fig. 6.3** Detection of VA-tagged DNM1L in HEK293 cells with or without cross-linking. Silver stained SDS-PAGE gel showing affinity purified VA-tagged mt DNM1L in the presence (+) and absence (−) of chemical cross-linking step. Asterisk indicate the gel band corresponding to the recovery of DNM1L with cross-linking

were confirmed in human HEK293 and SH-SY5Y cell lines by primary antibody immunoprecipitation and MS, as well as the significant enrichment ($p < 2.2 \times 10^{-16}$) for literature-curated human mt PPIs in the FLAG MP network [8]. Notably, among the binding partners of the mt protein baits, an enrichment for 14-3-3 and canonical RNA-binding domains were observed. These domains are associated with mt biogenesis, apoptosis, and stress response, which provides support for the mt relevance of the network generated using the AP-MS workflow [8].

In summary, in this chapter we outlined a tagging-based AP-MS workflow that can be used to effectively identify mt PPIs and elucidate the subunit compositions of mt protein complexes. This system is compatible with Gateway cloning, and importantly, enables efficient and stable delivery of affinity-tagged ORFs into most mammalian cell types. Successful implementation of this work-flow will help to expand the human mt interaction landscape, thereby generating valuable insight into the molecular interactions underlining human mt health and disease.

**Acknowledgments** We thank Babu's group for their helpful comments. This work was funded by National Institutes of Health (R01GM106019) and Canadian Institutes of Health Research (MOP-125952; RSN-124512, 132191; FDN-154318) to MB. MB is a CIHR New Investigator (MSH-130178).

## References

1. Jiang Y, Wang X (2012) Comparative mitochondrial proteomics: perspective in human diseases. J Hematol Oncol 5:11. https://doi.org/10.1186/1756-8722-5-11
2. Cassarino DS, Bennett JP (1999) An evaluation of the role of mitochondria in neurodegenerative diseases: mitochondrial mutations and oxidative pathology, protective nuclear responses, and cell death in neurodegeneration. Brain Res Rev 29:1–25. https://doi.org/10.1016/S0165-0173(98)00046-0
3. Lin MT, Beal MF (2006) Mitochondrial dysfunction and oxidative stress in neurodegenerative diseases. Nat Lond 443:787–795. https://doi.org/10.1038/nature05292
4. Ren J, Pulakat L, Whaley-Connell A, Sowers JR (2010) Mitochondrial biogenesis in the metabolic syndrome and cardiovascular disease. J Mol Med 88:993–1001. https://doi.org/10.1007/s00109-010-0663-9
5. Ballinger SW (2005) Mitochondrial dysfunction in cardiovascular disease. Free Radic Biol Med 38:1278–1295. https://doi.org/10.1016/j.freeradbiomed.2005.02.014
6. Wallace DC (2005) A mitochondrial paradigm of metabolic and degenerative diseases, aging, and cancer: a dawn for evolutionary medicine. Annu Rev Genet 39:359–407. https://doi.org/10.1146/annurev.genet.39.110304.095751
7. Floyd BJ, Wilkerson EM, Veling MT et al (2016) Mitochondrial protein interaction mapping identifies regulators of respiratory chain function. Mol Cell 63:621–632. https://doi.org/10.1016/j.molcel.2016.06.033
8. Malty RH, Aoki H, Kumar A et al (2017) A map of human mitochondrial protein interactions linked to neurodegeneration reveals new mechanisms of redox homeostasis and NF-κB signaling. Cell Syst 5:564–577.e12. https://doi.org/10.1016/j.cels.2017.10.010
9. Basso M, Giraudo S, Corpillo D et al (2004) Proteome analysis of human substantia nigra in Parkinson's disease. Proteomics 4:3943–3952. https://doi.org/10.1002/pmic.200400848
10. Sokolina K, Kittanakom S, Snider J et al (2017) Systematic protein-protein interaction mapping for clinically relevant human GPCRs. Mol Syst Biol 13:918

11. Yao Z, Darowski K, St-Denis N et al (2017) A global analysis of the receptor tyrosine kinase-protein phosphatase interactome. Mol Cell 65:347–360. https://doi.org/10.1016/j.molcel.2016.12.004
12. Havugimana PC, Hart GT, Nepusz T et al (2012) A census of human soluble protein complexes. Cell 150:1068–1081. https://doi.org/10.1016/j.cell.2012.08.011
13. Wan C, Borgeson B, Phanse S et al (2015) Panorama of ancient metazoan macromolecular complexes. Nature 525:339–344. https://doi.org/10.1038/nature14877
14. Hein MY, Hubner NC, Poser I et al (2015) A human interactome in three quantitative dimensions organized by stoichiometries and abundances. Cell 163:712–723. https://doi.org/10.1016/j.cell.2015.09.053
15. Huttlin EL, Ting L, Bruckner RJ et al (2015) The BioPlex network: a systematic exploration of the human interactome. Cell 162:425–440. https://doi.org/10.1016/j.cell.2015.06.043
16. Krogan NJ, Cagney G, Yu H et al (2006) Global landscape of protein complexes in the yeast *Saccharomyces cerevisiae*. Nature 440:637–643. https://doi.org/10.1038/nature04670
17. Morgenstern M, Stiller SB, Lübbert P et al (2017) Definition of a high-confidence mitochondrial proteome at quantitative scale. Cell Rep 19:2836–2852. https://doi.org/10.1016/j.celrep.2017.06.014
18. ten Have S, Boulon S, Ahmad Y, Lamond AI (2011) Mass spectrometry-based immuno-precipitation proteomics – the user's guide. Proteomics 11:1153–1159. https://doi.org/10.1002/pmic.201000548
19. Mak AB, Ni Z, Hewel JA et al (2010) A lentiviral functional proteomics approach identifies chromatin remodeling complexes important for the induction of pluripotency. Mol Cell Proteomics MCP 9:811–823. https://doi.org/10.1074/mcp.M000002-MCP201
20. Encinas M, Iglesias M, Liu Y et al (2000) Sequential treatment of SH-SY5Y cells with retinoic acid and brain-derived neurotrophic factor gives rise to fully differentiated, neurotrophic factor-dependent, human neuron-like cells. J Neurochem 75:991–1003
21. Clayton DA, Shadel GS (2014) Purification of mitochondria by sucrose step density gradient centrifugation. Cold Spring Harb Protoc 2014:pdb.prot080028. https://doi.org/10.1101/pdb.prot080028
22. Kislinger T, Cox B, Kannan A et al (2006) Global survey of organ and organelle protein expression in mouse: combined proteomic and transcriptomic profiling. Cell 125:173–186. https://doi.org/10.1016/j.cell.2006.01.044
23. Sowa ME, Bennett EJ, Gygi SP, Harper JW (2009) Defining the human deubiquitinating enzyme interaction landscape. Cell 138:389–403. https://doi.org/10.1016/j.cell.2009.04.042
24. Leung HCM, Xiang Q, Yiu SM, Chin FYL (2009) Predicting protein complexes from PPI data: a core-attachment approach. J Comput Biol J Comput Mol Cell Biol 16:133–144. https://doi.org/10.1089/cmb.2008.01TT
25. van Dongen S, Abreu-Goodger C (2012) Using MCL to extract clusters from networks. Methods Mol Biol Clifton NJ 804:281–295. https://doi.org/10.1007/978-1-61779-361-5_15
26. Westermarck J, Ivaska J, Corthals GL (2013) Identification of protein interactions involved in cellular signaling. Mol Cell Proteomics MCP 12:1752–1763. https://doi.org/10.1074/mcp.R113.027771
27. Sahni N, Yi S, Taipale M et al (2015) Widespread macromolecular interaction perturbations in human genetic disorders. Cell 161:647–660. https://doi.org/10.1016/j.cell.2015.04.013
28. Yang B, Tang S, Ma C et al (2017) Spontaneous and specific chemical cross-linking in live cells to capture and identify protein interactions. Nat Commun 8:2240. https://doi.org/10.1038/s41467-017-02409-z
29. Vlasblom J, Jin K, Kassir S, Babu M (2014) Exploring mitochondrial system properties of neurodegenerative diseases through interactome mapping. J Proteome 100:8–24. https://doi.org/10.1016/j.jprot.2013.11.008
30. Hyung S-J, Ruotolo BT (2012) Integrating mass spectrometry of intact protein complexes into structural proteomics. Proteomics 12:1547–1564. https://doi.org/10.1002/pmic.201100520

# 7 Targeted Proteomics to Study Mitochondrial Biology

Justina C. Wolters, Hjalmar P. Permentier, Barbara M. Bakker, and Rainer Bischoff

**Abstract**

Targeted mass spectrometry in the selected or parallel reaction monitoring (SRM or PRM) mode is a widely used methodology to quantify proteins based on so-called signature or proteotypic peptides. SRM has the advantage of being able to quantify a range of proteins in a single analysis, for example, to measure the level of enzymes comprising a biochemical pathway. In this chapter, we will detail how to set up an SRM assay on the example of the mitochondrial protein succinate dehydrogenase [ubiquinone] flavoprotein subunit (mouse UniProt-code Q8K2B3). First, we will outline the in silico assay design including the choice of peptides based on a range of properties. We will further delineate different quantification strategies and introduce the reader to LC-MS assay development including the selection of the optimal peptide charge state and fragment ions as well as a discussion of the dynamic range of detection. The chapter will close with an application from the area of mitochondrial biology related to the quantification of a set of proteins isolated from mouse liver mitochondria in a study on mitochondrial respiratory flux decline in aging mouse muscle.

**Keywords**

Targeted proteomics · Mitochondria · Mass spectrometry · Protein quantification

## 7.1 The Role of Proteins in Mitochondria

Mitochondria are plastic organelles, which tend to change their protein composition and functionality depending on cell type, genetics, diet, age, lifestyle, and disease [1, 2]). In patients suffering from mitochondrial diseases, alterations in the proteome composition may compensate for malfunction of mutated or lowly expressed disease-related proteins [2, 3]. On the other hand, during

J. C. Wolters
Department of Analytical Biochemistry, Groningen Research Institute of Pharmacy, University of Groningen, Groningen, The Netherlands

Laboratory of Pediatrics, University of Groningen, University Medical Center Groningen, Groningen, The Netherlands

Systems Biology Centre for Energy Metabolism and Ageing, University of Groningen, University Medical Center Groningen, Groningen, The Netherlands

H. P. Permentier · R. Bischoff (✉)
Department of Analytical Biochemistry, Groningen Research Institute of Pharmacy, University of Groningen, Groningen, The Netherlands
e-mail: r.p.h.bischoff@rug.nl

B. M. Bakker
Laboratory of Pediatrics, University of Groningen, University Medical Center Groningen, Groningen, The Netherlands

Systems Biology Centre for Energy Metabolism and Ageing, University of Groningen, University Medical Center Groningen, Groningen, The Netherlands


A. Urbani, M. Babu (eds.), *Mitochondria in Health and in Sickness*, Advances in Experimental Medicine and Biology 1158, https://doi.org/10.1007/978-981-13-8367-0_7

biological aging the protein composition of mitochondria tends to change in a detrimental way. For instance, the strict subunit stoichiometry of respiratory complexes is lost in mitochondria from senescent cells [4]. Also the reduction of mitochondrial respiratory capacity in aging muscle is a well-described phenomenon [5–7].

Changes in the mitochondrial proteome can be regulated at different levels. Most of the mitochondrial proteins are encoded by nuclear genes. They are synthesized in the cytosol and subsequently targeted to the mitochondria [8]. The abundance of these proteins depends on the same mechanisms that regulate abundance of other nuclear encoded proteins: transcription, RNA processing, translation efficiency and protein stability. In addition, the efficiency of mitochondrial targeting plays a role [8]. A small subset of proteins, however, is encoded on the mitochondrial genome [9]. The expression regulation of the mitochondrial genes is less well understood. Recently, it has been shown that epigenetic regulation is not confined to the nuclear genome, but also plays a role in the mitochondrial genome [10].

To understand the functional consequences of mitochondrial proteome plasticity, it is important to measure alterations in protein concentrations with sensitive and accurate analytical techniques. Until now, mitochondrial proteomics studies have focused mostly on the qualitative characterization of the mitochondrial proteome, while quantitative analysis is in its infancy. The latter is the focus of this chapter.

## 7.2 Quantifying Proteins in Complex Biological Samples

Antibody-based methods (ELISA, immunoblotting and immunofluorescence) have long been the gold standard for determining protein concentrations [11–13]. However, results are highly dependent on the affinity, selectivity and sensitivity of the antibody used for each target protein. Unfortunately, information about the specificity, selectivity of antibodies and the location of epitopes is often lacking. Cross-reactivity with other proteins has been recognized as a major caveat of antibodies and recognition of the protein target is also dependent on the modification state of the protein (for example phosphorylation) or the interaction partners, which may mask the epitope under the chosen experimental conditions. All these factors create variability which is often not taken into account [14–16]. Targeted proteomics based on selected reaction monitoring (SRM) provides a powerful alternative to antibody-based protein quantification. In targeted proteomics, mass spectrometry (MS) is applied to selectively quantify the proteins of interest via peptides that are specific (proteotypic) for each of the target proteins (explained in more detail in the next paragraphs). The risk of targeting only sub-fractions of the protein species, like with antibodies, is reduced, due to the conversion of proteins into peptides and flexibility in the choice of suitable peptide(s) for the detection of a given target protein. Proteins can be targeted by multiple peptides per protein to improve selectivity and specificity. Modifications of targeted peptides, such as phosphorylation, will affect quantification by mass spectrometry. This may be turned into an advantage, however, since modifications will yield a mass difference, which can be targeted separately.

A typical targeted proteomics workflow offers the possibility to screen for many different proteins in a single run, with current techniques addressing 50–100 proteins, each targeted with multiple peptides. Targeted proteomics has advantages over classical immunoblotting in terms of accuracy and precision of the measurements. An important difference is that targeted proteomics quantifies at the peptide level whereas immunoblotting works at the protein level [14]. Although immunochemical techniques allow for multiplexing when using specific approaches (e.g. antibody arrays or FACS-based approaches), multiplexing is an inherent advantage of targeted proteomics, with faster assay development (considering the time it takes to raise specific antibodies against the target of interest and to validate them).

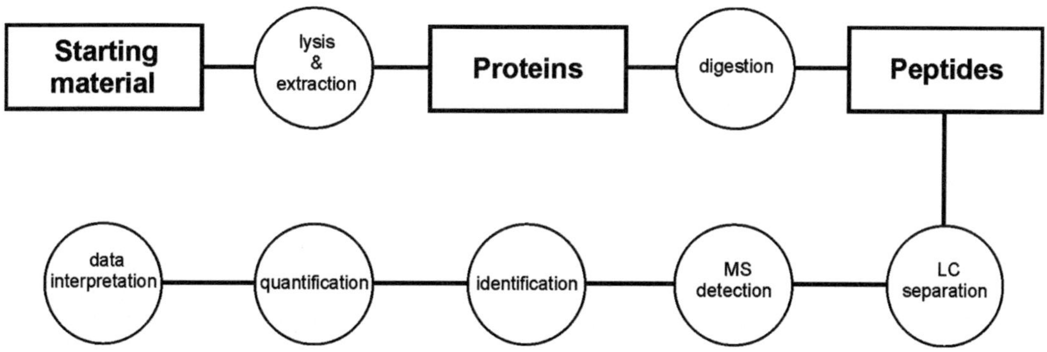


**Fig. 7.1** Proteomics workflow. The proteomics workflow starts with the lysis and protein extraction from the desired starting material. The proteins are digested prior to separation of the resulting peptides on a liquid chromatography (LC) column. The information of mass spectrometric (MS) analyses are used for the identification and quantification of the peptides and their related proteins by comparison with in silico generated MS/MS spectra from protein or nucleic acid sequence databases. These results can be used to interpret the biological model that is being studied

### 7.2.1 Introduction to Mass Spectrometry (MS)-Based Proteomics Workflows

The general workflow of mass spectrometry-based proteomics is summarized in Fig. 7.1. In the first step, proteins are isolated from the starting material of interest. This can be organ/tissue samples, cells harvested after culturing or (body) fluids such as plasma, urine or cerebrospinal fluid. In order to extract the proteins from cells or tissue samples a lysis or homogenization step [17] is required to disrupt the structure.

Since MS detection of large biomolecules, including intact proteins, has significant limitations, proteins are usually measured at the peptide level after digestion with a protease. The protease trypsin is commonly used, because it cleaves after the C-terminus of the positively charged amino acids lysine and arginine. The digestion step increases the complexity of the already complex protein mixture significantly. The human proteome map website (http://www.humanproteomemap.org) currently lists over 30.000 proteins that were detected based on a total of more than 290.000 non-redundant peptides. Even the proteome of a subcellular organelle like the mitochondrion is still predicted to contain at least 1158 or 1408 proteins, depending on the estimate [18, 19]. Reduction of complexity prior to MS detection is therefore essential, and is most commonly achieved by reverse phase liquid chromatography, which can be coupled directly to the mass spectrometer (LC-MS) for online detection of the separated peptides. Peptides are separated based on their hydrophobicity via a gradient of increasing organic solvent content in water at an acidic pH.

The mass spectrometer measures the mass-to-charge ratio (*m/z*) of peptide ions. Peptide ions are generated by positive electrospray ionization ($ESI^+$) after elution from the LC column. For this reason, the digestion with trypsin is beneficial, as it yields peptides with basic amino acids (lysine or arginine) at the C-terminus of each peptide, which are positively charged at low pH. In protein extracts from complex samples, containing peptides from many different proteins, the chemical composition and hence the mass-to-charge ratio will not be unique for a single peptide ion. It is therefore insufficient for identification of the corresponding protein. In an additional step, selected peptides are fragmented in the mass spectrometer via collisions with an inert gas. The generated fragments yield information about the sequence of the peptide via comparison of the generated mass fragments to in-silico generated fragment mass spectra for all peptides that can be derived from a given protein or nucleic acid sequence database [20].

The number of peptides that are fragmented and yield useful information throughout the LC-MS run is dependent on sample complexity and efficiency of the peptide separation as well as on the mass spectrometer (speed, sensitivity and resolution of the instrument). In a classical discovery-proteomics experiment (also called a shotgun proteomics experiment) a maximum of several thousand proteins can be identified in a single analysis [21]. More in-depth analyses require multiple fractionation steps, protein enrichment or the depletion of interfering (high-abundant) proteins.

### 7.2.2 Application of Proteomics to Mitochondria

Proteomics was first applied to the field of mitochondrial biology in 1998 [22], with the initial focus on the annotation of the mitochondrial proteome [23]. Multiple larger discovery proteomics studies appeared in the following years [24–27]. All screens were based on pre-fractionation of the proteins from isolated mitochondria in order to obtain a comprehensive view of the mitochondrial proteome. Various databases were created covering the mammalian mitochondrial proteome (an overview of the various databases can be found in the review by Chen [28]). Based on these databases 1200–1500 proteins are localized in mitochondria, even though the mitochondrial genome encodes for only 13 proteins (all components of the respiratory chain), while all other proteins are derived from proteins encoded by the nuclear genome [9]. Next to identification of mitochondrial proteins, relative quantification has been performed using comparative proteomics. These studies used the discovery proteomics approach to compare protein profiles under different conditions (for example healthy and disease states) either directly or via differential stable-isotope labelling of the proteins or peptides (see [29] for a review). Comparative mitochondrial proteomics has been used to study protein changes in mitochondria in various disease areas where mitochondria play a role, including metabolic disorders, neurodegenerative diseases, various myopathies, cancer, diabetes, obesity and cardiovascular disease as well as in aging. Several reviews list overviews of studies performed in these areas [2, 28, 30–32].

## 7.3 Targeted Proteomics

Disadvantages of discovery proteomics for the relative quantification of proteins are related to a bias for peptides with high signal intensities, the effect of sample complexity on coverage and the rather limited reproducibility and repeatability. In discovery proteomics, peptides are fragmented in order of decreasing intensity and coverage of sample complexity depends on the maximum number of peptides that can be selected and fragmented per time unit, which depends on the speed and sensitivity of the mass spectrometer and the complexity of the peptide mixture eluting from the LC column at a given point in time. In complex samples, the lower abundant peptides are therefore only detected when sample complexity is reduced prior to LC-MS, for example by fractionating proteins electrophoretically or chromatographically. This, however, means a significant increase in MS measurement time and usually requires a need for more starting material. Fractionation may also negatively affect reproducibility and repeatability of the measurements, as peptides may distribute over a number of fractions.

Targeted proteomics is a good alternative to overcome the limitations of discovery proteomics for protein quantification. Targeted proteomics differs from discovery proteomics in the sense that the entire MS measurement time is focused on the target(s) of interest, rather than screening as many peptides as possible. This improves the limit of detection, the linear range, repeatability and reproducibility and reduces the effect of sample complexity on the results as summarized in Fig. 7.2. While this method limits the number of protein targets studied per measurement from thousands to at most a hundred, it fits very well if the focus is already on a limited set, such as a part

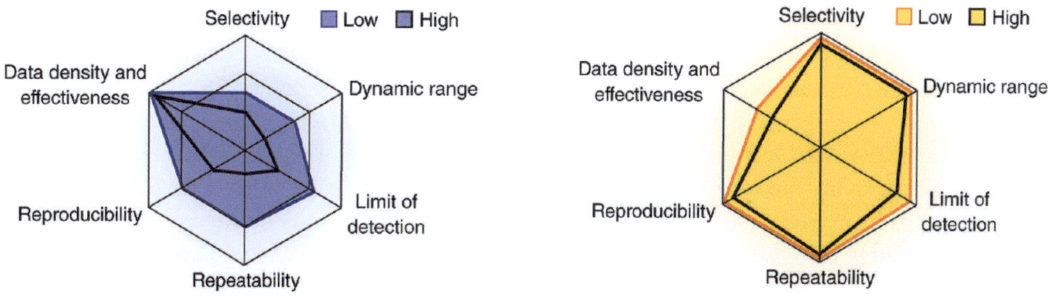


**Fig. 7.2** Comparison of discovery proteomics (left) and targeted proteomics (right) with respect to high and low complexity samples. (Figure adapted by permission from Macmillan Publishers Ltd.: Nature Biotechnology [33], copyright (2010))

of a specific subcellular fraction or specific pathways affected by disease.

We have successfully developed a targeted proteomics pipeline for a subset of mitochondrial proteins related to energy metabolism, including fatty acid beta-oxidation, the TCA cycle and oxidative phosphorylation [3]. The method has been shown to be quantitative, which opens the possibility of screening individual patients with aberrant mitochondrial function related to disease. In the next part, we will provide a step-by-step tutorial for the development of this type of assay, using this study as an illustration.

### 7.3.1 Getting Started with a Targeted Proteomics Workflow

The basic sample preparation steps of protein extraction and digestion are similar to the discovery proteomics workflow, but there are important changes in the LC-MS measurement protocol as well as data analysis and interpretation. In contrast to discovery proteomics, where the time investment is mainly in the interpretation of the generated data, the largest time investment in targeted proteomics is at the assay development stage prior to the LC-MS measurements, namely in selection of the proteins and peptides to be targeted.

The development of targeted proteomics assays can be divided into three main steps. The first step, the design of the assay, is most important as this will define the boundaries of the applied assay. Therefore, this steps requires a well-defined biological study hypothesis. The second step is the development of the LC-MS method for the selected targets including the choice of the quantification strategy (e.g. the kind of internal standard) and the third step is the application of the developed assays to the biological questions of interest.

### 7.3.2 In Silico Assay Design

The in silico assay design covers the selection of proteins to be targeted and the choice of peptides to be used to target these proteins.

#### 7.3.2.1 Choice of Protein Targets

Based on the research question (for example the study of specific biochemical pathways in mitochondria) a given set of proteins is selected to be targeted. This set depends also on the research focus. For example, for a systems biology or systems medicine approach it is important to quantify all proteins in a pathway to understand its regulation and adaptation to pathophysiological conditions (e.g., in one of our applications all proteins from the fatty acid beta-oxidation pathway were selected and quantified in order to use the data in a comprehensive computational model [34]). A complicating factor is that the mitochondrial respiratory complexes contain many subunits, complex I alone already contains 45 subunits [35]. With a typical assay size of 50–100 proteins in targeted proteomics, a full characterization of each of the mitochondrial proteins requires large investments. Since our focus was on the integration of all pathways involved in mitochondrial ATP generation, we initially selected only two protein subunits per respiratory complex. For a more specialized study of the respiratory chain, it would be advantageous to target all members of these protein complexes. This would allow to investigate if the stoichiometry in the complexes is altered under different conditions, as was suggested by untargeted proteomics in report on biological ageing [4]. Another rationale for selecting target proteins may be to combine panels of biomarkers related to a given disease to improve diagnostic or prognostic sensitivity and specificity [36, 37].

#### 7.3.2.2 Choice of Peptides

Amino acid sequences of protein targets can be found in the Uniprot-database (http://www.uniprot.org). One should keep in mind though, that the protein sequence represented here is a consensus sequence that does not take changes due to genetic variability or post-translational modifications into account. There is, however, increasing information about genetic variability that can, for example, be checked on the ExAC Browser (http://exac.broadinstitute.org/) and taken into consideration when selecting proteotypic peptides.

Expected tryptic peptides for each target protein can be found by screening the protein sequence for lysines and arginines, as trypsin cleaves the protein at the C-terminal side of these residues (except if they are followed by a proline). In silico digestion can also be done online via the ExPASy server (http://web.expasy.org/peptide_cutter) or with dedicated software for targeted proteomics (for example, the freely available targeted proteomics software package Skyline ([38]).

In general, a number of aspects need to be considered when selecting the best peptides for each protein target. The peptides should be unique for the protein target of interest, should have optimal properties for LC-MS detection (called proteotypic [39, 40]) and for quantification (called quantotypic [41]), which are affected by multiple characteristics of the peptides. This means that suitable peptides should bind to the reverse phase column material used for fractionation and elute within the range of the applied gradient. They should further ionize well in $ESI^+$ and provide intense fragment ions for maximal sensitivity. A detailed description of the selection criteria to be considered is given below.

The first criterion is whether the selected peptide is unique for the protein target, meaning that the same peptide must not be present in any other protein in the analyzed proteome. The uniqueness of a peptide can be checked via a sequence homology search (e.g. a BLAST search) against the consensus (sub)proteome of the organism (https://blast.ncbi.nlm.nih.gov). If none of the peptides suitable for LC-MS measurements is unique (this may happen when targeting a series of protein isoforms), one could consider quantifying all isoforms together as a group.

The second selection criterion is peptide length. The preferred peptide length is between 9 and 25 amino acids. Shorter peptides are often very hydrophilic and are thus not retained on $C_{18}$ reversed phase columns and they are also rarely unique. Peptides longer than 25 amino acids are problematic due to their often strongly hydrophobic nature, which may lead to losses during sample preparation (for example by non-specific binding to tubes and pipette tips) or LC separation. Fragmentation is also less efficient, decreasing the sensitivity of the measurements. A good indication that a peptide can be analyzed by mass spectrometry is its presence in databases from the NIST (http://chemdata.nist.gov) or PeptideAtlas (http://www.peptideatlas.org/speclib), which confirms that a peptide has been detected before in proteomics experiments. Alternatively, the bioinformatics tool CONSeQuence [42] may be used, which uses machine learning algorithms to predict peptides with the highest detectability by LC-MS.

The third selection criterion takes the risk for modifications in a given peptide into account. Modifications on amino acids will alter the molecular mass of a peptide and this will affect quantification (if one does not take the mass shift into consideration). These modifications can have a biological or chemical background. Known biological modifications of a protein like phosphorylation, acetylation or glycosylation are listed in the UniProt database (http://www.uniprot.org) and affected peptides can be avoided. Several amino acids are sensitive to chemical modification and should preferably be excluded due to possible variability at the quantification stage, even though these peptides might have a good signal in the MS. The amino acids or amino acid combinations are: methionine (M, oxidation), aspartic acid-proline (DP, hydrolysis of the peptide bond) and asparagine-glycine (NG, deamidation) [41]. Although cysteines are susceptible to chemical modifications, this reactivity may be blocked by reduction (for example by dithiothreitol) and subsequent alkylation (for example by iodoacetamide) [43], so that all peptides are completely modified and do not have to be excluded a priori.

The fourth selection criterion relates to the quantotypic nature of peptides, and assesses whether a given peptide will be generated reproducibly. If multiple lysines or arginines are close to each other (adjacent to or with one amino acid in between), trypsin sometimes misses one of the lysines/arginines, creating so-called miscleaved peptides. A second factor decreasing digestion efficiency of trypsin is the presence of an acidic residue near the cleavage site, at a distance of one or two amino acids [41, 44]. In practice, a mix of completely cleaved (tryptic) and miscleaved peptides is generated, which will affect reproducibility of the results significantly. The software program MC:Pred [36, 44–46] predicts the risk for miscleavages. If in silico analysis results only in peptides with a high propensity for miscleavages, one may consider using a different digestion method to achieve accurate quantification of the targeted protein. However, it must be kept in mind that trypsin is the most commonly used digestion enzyme in proteomics experiments and that therefore most prediction tools, knowledge and data is available for this enzyme. An overview of different digestion enzymes and their specificities can be found on the ExPASy server (http://web.expasy.org/peptide_cutter). Meyer et al. reviewed a number of alternative digestion methods that are suitable for reproducibly generating peptides [45]. In this way, peptides with different lengths and properties are created that may be more suitable for LC-MS detection.

### 7.3.3 How to Detect the Protein SDHA: An Example

The protein SDHA, the mitochondrial succinate dehydrogenase [ubiquinone] flavoprotein subunit (mouse UniProt-code Q8K2B3), is used as an example to illustrate the peptide selection workflow. The protein sequences were downloaded from UniProt and Table 7.1 contains all peptides for the mouse SDHA protein in the range between 9 and 25 amino acids. All these peptides are unique for the intended mouse protein (Table 7.1, column 1A). However, for this specific study it was the goal to create a single assay for the mouse and human orthologues of the same protein. This was done in order to easily switch between analyses in clinical samples and in a preclinical mouse model to study the disease mechanism in greater detail [3]. Therefore, all mouse peptides were also checked for their presence and uniqueness in the human SDHA orthologue (human UniProt-code P31040), and peptides with overlap were preferred for selection of the target peptides (Table 7.1, column 1B).

The Knowledgebase entry in UniProt shows that the first 42 amino acids form part of the signal peptide (transit peptide), which is used to translocate the protein to mitochondria. The signal peptide is subsequently removed by proteolytic cleavage and peptides from this region should therefore be excluded for quantification of the intra-mitochondrial mature form of SDHA (first 3 peptides in the Table 7.1, column 2A).

For detectability in LC-MS, peptides are selected with optimal CONSeQuence scores (Table 7.1, column 3A-C; SVM and ANN scores preferably above 0.5 and binary scores 3 or 4). The presence of a peptide in a spectral library is an alternative positive indication of good detectability (Table 7.1, column 2C, in this example the NIST spectral library was chosen), though absence in the NIST library is not a reason to exclude a peptide.

**Table 7.1** All tryptic peptides from mouse SDHA in the range between 9 and 25 amino acids with the properties used for selection of the optimal peptides for the targeted proteomics assay

| Protein target SDHA | 1. Uniqueness | | 2. Properties | | | 3. CONSeQuence analysis | | | 4. Possible modifications | | | 5. MC:Pred analysis | | 6. Possible digestion issues | |
|---|---|---|---|---|---|---|---|---|---|---|---|---|---|---|---|
| Peptides (9-25 AA) | 1A. Unique in mouse | 1B. Unique in human | 2A. Position | 2B. Length | 2C. Presence in NIST library | 3A. SVM | 3B. ANN | 3C. Binary | 4A. Containing M | 4B. Containing NG | 4C. Containing DP | 5A. SVM Nterm | 5B. SVM Cterm | 6A. R or K near cleavage | 6B. D or E near cleavage |
| MAGVGAVSR | Yes | No | 1 | 9 | No | 0.31 | 0.30 | 1 | 1 | 0 | 0 | -- | 0.47 | 0 | 0 |
| LALTGAWPGTLQK | Yes | No | 16 | 13 | No | 0.59 | 0.56 | 4 | 0 | 0 | 0 | 0.91 | 0.42 | 1 | 0 |
| QTCGFHFSVGENK | Yes | No | 29 | 13 | No | 0.65 | 0.62 | 3 | 0 | 0 | 0 | 0.42 | 0.84 | 1 | 1 |
| AAFGLSEAGFNTACLTK | Yes | No | 76 | 17 | Yes | 0.65 | 0.59 | 4 | 0 | 0 | 0 | 0.42 | 0.38 | 0 | 0 |
| SHTVAAQGGINAALGNMEEDNWR | Yes | Yes | 98 | 23 | No | 0.54 | 0.46 | 3 | 1 | 0 | 0 | 0.43 | 0.46 | 0 | 0 |
| TGHSLLHTLYGR | Yes | Yes | 196 | 12 | Yes | 0.54 | 0.56 | 3 | 0 | 0 | 0 | 0.48 | 0.47 | 0 | 1 |
| YDTSYFVEYFALDLLMENGECR | Yes | Yes | 211 | 22 | No | 0.37 | 0.35 | 2 | 1 | 1 | 0 | 0.56 | 0.47 | 0 | 1 |
| GVIALCIEDGSIHR | Yes | Yes | 233 | 14 | Yes | 0.65 | 0.63 | 3 | 0 | 0 | 0 | 0.47 | 0.61 | 1 | 1 |
| NTVIATGGYGR | Yes | No | 251 | 11 | Yes | 0.47 | 0.49 | 2 | 0 | 0 | 0 | 0.61 | 0.41 | 1 | 0 |
| TYFSCTSAHTSTGDGTAMVTR | Yes | No | 262 | 21 | Yes | 0.49 | 0.46 | 2 | 1 | 0 | 0 | 0.41 | 0.43 | 0 | 0 |
| GEGGILINSQGER | Yes | Yes | 313 | 13 | Yes | 0.61 | 0.60 | 3 | 0 | 0 | 0 | 0.44 | 0.5 | 0 | 1 |
| DHVYLQLHHLPPEQLATR | Yes | Yes | 362 | 18 | Yes | 0.39 | 0.38 | 2 | 0 | 0 | 0 | 0.63 | 0.46 | 0 | 1 |
| LPGISETAMIFAGVDVTK | Yes | Yes | 380 | 18 | Yes | 0.72 | 0.61 | 4 | 1 | 0 | 0 | 0.46 | 0.6 | 0 | 1 |
| EPIPVLPTVHYNMGGIPTNYK | Yes | Yes | 398 | 21 | Yes | 0.83 | 0.78 | 4 | 1 | 0 | 0 | 0.6 | 0.4 | 0 | 1 |
| LGANSLLDLVVFGR | Yes | Yes | 452 | 14 | Yes | 0.59 | 0.55 | 4 | 0 | 0 | 0 | 0.4 | 0.42 | 0 | 0 |
| ACALSIAESCRPGDK | Yes | No | 466 | 15 | Yes | 0.26 | 0.29 | 0 | 0 | 0 | 0 | 0.42 | 0.83 | 0 | 1 |
| ANAGEESVMNLDK | Yes | No | 486 | 13 | Yes | 0.45 | 0.43 | 2 | 1 | 0 | 0 | 0.41 | 0.64 | 1 | 1 |
| SMQNHAAVFR | Yes | Yes | 518 | 10 | Yes | 0.42 | 0.43 | 1 | 1 | 0 | 0 | 0.38 | 0.41 | 0 | 0 |
| VGSVLQEGCEK | Yes | No | 528 | 11 | Yes | 0.45 | 0.44 | 1 | 0 | 0 | 0 | 0.41 | 0.42 | 0 | 1 |
| ISQLYGDLK | Yes | No | 539 | 9 | Yes | 0.39 | 0.38 | 1 | 0 | 0 | 0 | 0.42 | 0.48 | 0 | 1 |
| GMVWNTDLVETLELQNLMLCALQTIYGAEAR | Yes | Yes | 555 | 31 | Yes | 0.28 | 0.27 | 2 | 2 | 0 | 0 | 0.79 | 0.89 | 1 | 1 |
| VDEYDYSKPIQGQQK | Yes | No | 601 | 15 | Yes | 0.32 | 0.35 | 1 | 0 | 0 | 0 | 0.77 | 0.6 | 1 | 1 |
| HTLSYVDIK | Yes | No | 625 | 9 | Yes | 0.49 | 0.48 | 1 | 0 | 0 | 0 | 0.67 | 0.45 | 1 | 1 |
| VTLEYRPVIDK | Yes | Yes | 637 | 11 | Yes | 0.33 | 0.34 | 1 | 0 | 0 | 0 | 0.63 | 0.47 | 0 | 1 |
| TLNEADCATVPPAIR | Yes | Yes | 648 | 15 | Yes | 0.80 | 0.73 | 4 | 0 | 0 | 0 | 0.47 | 0.62 | 0 | 1 |

Column 1, uniqueness of the peptide for the intended target in multiple orthologues. Column 2, properties of the peptides. Peptides in the signal sequence of the protein are indicated in red. Column 3, bio-informatics analyses to predict the proteotypic properties of the peptide using CONSeQuence [42]. The output from this software gives a score between 0 and 1 (1 being the highest scoring) based on linear Support Vector Machine (SVM) or Artificial Neural Network (ANN) models or a binary score between 1 and 4, which represents the number of algorithms that predict the peptide to be detectable by MS with 4 being the highest score. SVM and ANN scores above 0.5 and binary scores 3 or 4, were considered optimal. Column 4, presence of possible modification sites in the peptides. Column 5, bio-informatics analyses to predict the risk of miscleavages using MC:Pred [44]. The SVM output from this algorithm gives a score between 0 and 1, with high scores indicating a large chance of miscleavages, which is calculated for both the N- and C-terminus of each peptide. Scores below 0.5 were considered optimal. Column 6, presence of amino acids near the digestion side that might affect tryptic digestion. Factors that are a reason to exclude the peptide for selection are indicated in red and factors that could be avoided if possible are indicated in orange. Taking all factors into account resulted in three peptides with optimal properties for targeting SDHA, indicated by a peptide sequence in green text

With respect to quantotypic properties, peptides with possible modification sites like M, NG and DP were excluded (Table 7.1, column 4A-C). Prediction of the risk for miscleavages was based on MC:Pred analyses (Table 7.1, colums 5A-B, excluding SVM scores above 0.5). These scores are mainly based on the presence of R or K doublets or another R or K near the cleavage site (Table 7.1, column 6A). The presence of acidic amino acids (D or E) near the cleavage site was also considered detrimental to repeatable quantification of that particular peptide (Table 7.1, column 6B). Taking all of these criteria into account, a single peptide (LGANSLLDLVVFGR) remained. However, on closer inspection there are another two peptides (GEGGILINSQGER and TGHSLLHTLYGR), that have an acidic amino acid near the cleavage side, but for which the theoretical MC:Pred scores are within the limit. From these two peptides, the peptide with the highest CONSeQuence score was chosen (GEGGILINSQGER).

### 7.3.4 Quantification Strategies

Although MS quantification without internal standards (so-called label-free quantification) is widely used in proteomics [29], such analyses are susceptible to experimental variation and require rigorous standard procedures from sample preparation to final data analysis [29, 46]. The use of isotopically labeled standards is for that reason preferred for accurate quantification. Isotopically labeled standards have identical physicochemical properties except for an altered mass due to the introduced isotopic label (most often by introducing $^{13}C$- or $^{13}C^{15}N$-isotopes into lysines and arginines), which makes them the ideal standard to correct for variations within and between MS measurements.

The use of isotopically labeled standard proteins [47] is the best option for accurate protein quantification as it corrects for variation during sample preparation and notably during digestion. It should, however, be kept in mind that recombinant, isotopically labeled proteins may differ from their endogenous counterparts in terms of post-translational modifications, which may affect sample preparation. Moreover, this strategy is only possible if the isotopically labeled protein can be obtained at the required amount and in sufficient purity, which is often not possible or practical when targeting a large number of proteins. For this reason, commercially available isotopically labeled synthetic peptides are often used instead as internal standards for quantification [48]. A major disadvantage is that the isotopically labeled synthetic peptides do not correct for variation in the entire sample preparation workflow, notably trypsin digestion. For relative quantification, relatively cheap peptides can be used (not purified and without amino acid analysis to determine the accurate concentration). For absolute quantification high-grade synthetic peptides (isotopic purity >99%, peptide purity >97%, accurately quantified by amino acid analysis) need to be used, which are relatively costly.

The concatemer technology [49], (PolyQuant GmbH, Bad Abbach, Germany) offers a powerful alternative where all target peptides are concatenated into a synthetic protein with tryptic cleavage sites in between and expressed in *E. coli* in a medium containing $^{13}$C-labeled lysine and arginine. The stable isotope-labeled, synthetic protein is purified and used as internal standard for a range of target proteins. Since the synthetic protein is added to the sample immediately after protein extraction, the concatemers correct for losses during sample processing, such as during the digestion workflow. This is an advantage as compared to the use of synthetic peptides, which are in general added after the digestion workflow. We determined the loss during sample processing to be approximately 60%, demonstrating that this is a relevant correction. The examples shown in this chapter for a set of mitochondrial protein targets were all based on this technology, making use of three concatemers containing a total of 134 peptides belonging to 57 different protein targets [3].

### 7.3.5 LC-MS Assay Development

Targeted proteomics is typically done on a triple quadrupole MS instrument measuring in ESI$^+$ mode for positively charged peptide ions. A (nano)-liquid chromatography system with a reversed phase HPLC column is coupled on-line to the MS for separation of the peptide mixtures.

In a triple quadrupole MS, two selection steps are performed. At first, peptide ions are selected based on their mass-to-charge ratio ($m/z$) filtering out all other ions with different $m/z$. This is followed by fragmentation of the selected target peptide in a so-called collision cell and filtering of specific fragment ions based on their characteristic $m/z$ values. The combination of a peptide and a fragment ion $m/z$ value is called a transition. Using transitions improves the limit of detection by decreasing background signals and by significantly increasing the measurement time for each transition (compared to discovery proteomics, where the full $m/z$ range needs to be covered).

#### 7.3.5.1 Selection of the Optimal Peptide Charge State and Fragment Ions

In order to select transitions for each of the targeted peptides, it is necessary to determine the optimal charge state of the peptide ion for MS detection first. Note that the charge state of a peptide depends on the solvent pH, the peptide sequence, the instrument type and the operational settings and cannot be accurately predicted. The optimal MS detection range for peptides is between $m/z$ 300 and 1500 in most triple quadrupole mass spectrometers. For most tryptic peptides, the doubly- and/or triply-charged peptide ions fall within this $m/z$ range. For small peptides, the singly-charged form may be selected as well but this is rarely the case. The availability of synthetic (or concatemer-derived) peptides allows for determination of the optimal peptide charge state and optimization of the operational parameters of the mass spectrometer. Synthetic peptides can also be used for selection of the most intense fragment ions for each of the targeted peptides. Alternatively, peptide fragmentation (MS/MS) data from online libraries such as the PeptideAtlas/NIST-database (http://www.peptideatlas.org/speclib, containing multiple spectral libraries including a copy of the NIST peptide spectral library) can be used for pre-selection, but one should keep in mind that instrument type and settings affect the intensity of the fragment ions. The most important feature affecting fragment ion intensity is the collision energy used to fragment the peptide precursor ions. If the energy is too low, the peptide is not fragmented at all; if the energy is too high, peptides are broken into many small fragment ions that are not helpful for generating specific transitions. An example of the effect of the collision energy on signal intensities of an SDHA-derived peptide is shown in Fig. 7.3d. It can be seen that the highest signals are obtained in a window of around 5 eV and that the predicted optimal collision energy of 25 eV (based on prediction of the Skyline software [38]) is within this window. The selected transitions must be verified in complex samples, since other peptides may give rise to the same or very similar transitions at almost the same retention time thus leading to interfering signals that affect quantification. These interferences may occur for the endogenous peptide, the isotopically labeled internal standard peptide, or both, but in all cases the reliability of quantification is affected and these transitions should be excluded. An important quality check to reveal potential interferences is that the relative intensity ratios of all fragments between the isotopically labeled standard and the endogenous peptide should remain constant across all measurements. An example for one of the SDHA derived peptides, where no interferences are observed for the three transitions from the endogenous and the labeled internal standards, is shown in Fig. 7.3. In the spectral library the y3 fragment ion and y7 fragment ion have similar intensities (Fig. 7.3a), but the y7 fragment ion was selected in this case (Fig. 7.3b), because interferences are in general more likely to occur in the low *m/z*-range. The sum of the three transitions (y4, y7 and y8 fragments) is used to calculated the ratio between the endogenous peptide and the isotopically labeled standard peptide (Fig. 7.3c), which can be converted to the protein amount via the known amount of the added isotopically labeled standard.

The number of transitions that can be detected depends on the type of instrument and the settings. A detailed description of the different combinations of settings, like the dwell time (the time during which each transition is measured) and the cycle time (the time needed to measure all transitions once) and the consequences for sensitivity and multiplexing capacity is given by Lange et al. [50]. In the triple quadrupole mass spectrometer used for the mitochondrial assays [3], 134 peptides were addressed in a single run with 3 transitions for each endogenous and 3 transitions for each isotopically labeled peptide (>800 transitions) in a 100 min LC run. In order to measure all these peptides in a single run, we used the so-called 'scheduled SRM approach' where the transitions for each peptide were measured across a 4-min time window around the elution time of the respective peptide.

#### 7.3.5.2 Dynamic Range of Detection

The disadvantage of using concatemers is that all peptides in the same concatemer are added at the same amount without the possibility to adapt the added amount to the level of individual endogenous proteins. The optimal amount of added isotopically labeled standards for quantification purposes would be equal to the endogenous protein amount. Amounts of the isotopically labeled standard peptides that are too low amplify measurement variation, while too high amounts interfere with quantification of the endogenous peptide due to the fact that isotope labeling is never complete. For the concatemer-derived peptides of the mitochondrial study, labeling efficiency was 97% [3], meaning that 3% non-labeled peptide is added, which is identical to the peptide derived from the endogenous protein. Although the calculated protein amounts can be corrected for incomplete stable isotope labeling, quantifi-

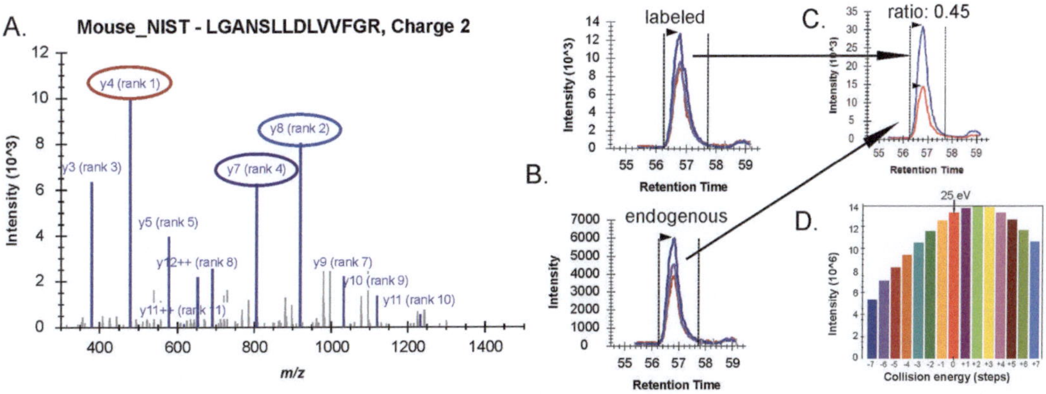


**Fig. 7.3** LC-MS assay development. (**a**) MS/MS spectrum derived from the database at the National Institute of Standards and Technology (NIST, http://www.peptideatlas.org/speclib). Three fragments (y4, y7 and y8) from this MS/MS spectrum were used to target the SDHA-peptide LGANSLLDLVVFGR. (**b**) LC-MS traces for the endogenous peptide and the isotopically labeled standard, combining signals for the three selected fragments. (**c**) Trace after integration of the summed peak areas for both peptides (panel **b**). The amount of the endogenous peptide can be calculated from the known amount of the added isotopically labeled standard. (**d**) Signal intensity of the isotopically labeled standard at different collision energies. The predicted collision energy (25 eV, based on prediction of the Skyline software [38])) is indicated and 14 steps of 1 eV difference were measured around this predicted optimum (The figure was created using Skyline software [38])

cation near the lower limit of detection is strongly affected by slightly varying labelling efficiency.

Next to these effects, it is important to check the linearity of the MS response around the concentration range of the endogenous proteins. To avoid having to add the internal concatemer standard at various amounts in separate experiments, one must show that the added amount is suitable for all protein targets in the assay and that the amounts of the concatemer-derived isotopically labeled peptides fall within the linear response range of the LC-MS measurements. In the mitochondrial study [3], this was done by adding the concatemers at amounts ranging from 0.15 to 15 ng (around 2.5–250 fmol) per microgram total protein and checking consistency of the calculated endogenous amounts at each isotopically labeled standard amount. The linear range will vary for the individual peptides within the concatemers, which means that some peptides can only be used at certain added amounts of isotopically labeled internal standard.

The signals of both SDHA-derived peptides are linear throughout the entire tested concentration range (Fig. 7.4, panel a, individual measurement results for both peptides; panel b, calculated amounts from three replicates). In order to summarize the results for each protein in the developed mitochondrial protein assay, the amount of each individual protein was normalized to the median of that protein amount determined at all five concatemer spiking levels. This allows to represent all proteins at the same scale, despite different absolute protein amounts (Fig. 7.4, panel c, the grey box plot shows the results from all targeted mitochondrial proteins together [3], and the red line indicates the individual results of SDHA within the complete series of targeted mitochondrial proteins). If the amount of the isotopically labeled standard peptide is affected by the endogenous protein amount, a deviation from the median is expected, which is mainly observed when adding the isotopically labeled standards at relatively high amounts (7.5 and 15 ng). The proteins that suddenly appear to have a much higher amount (median >100) are indeed the proteins that are present at relatively low endogenous levels. For these proteins, the incomplete labeling of the isotopically labeled standards most likely affects the determined amount, as discussed above. In summary, the highest amounts of isotopically labeled standards are not suitable for quantifying endogenous proteins at low amounts, but the lower amounts up to 1.5 ng concatemer are

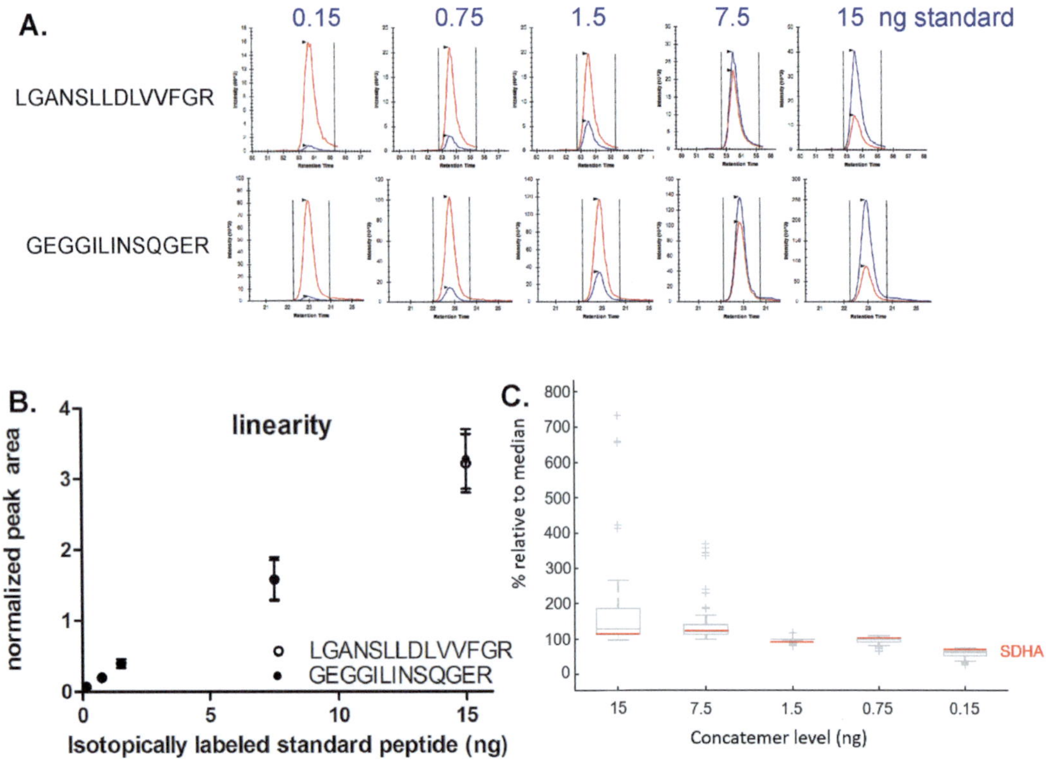


**Fig. 7.4** Detection range of the assay. (**a**) LC-MS traces for the two SDHA signature peptides at different amounts of added concatemers (red peak is the signal from the endogenous peptide, blue peak is the signal from the isotopically labeled internal standard peptide). (**b**) Linearity of the response of the two SDHA isotopically labeled concatemer-derived peptides (n = 3) normalized to the amount of the endogenous peptide. (**c**) Protein variability of all targeted mitochondrial proteins in the developed mitochondrial protein assay in response to different amounts of isotopically labeled concatemers. The amount of each individual protein was normalized to the median of that protein amount determined at all five concatemer spiking levels, to plot all proteins at the same scale, although they have different absolute protein amounts. The red line shows the individual result of SDHA, compared to the complete set of more than 50 targeted proteins. Panels (**b** and **c**) were reprinted with permission from [3]. (Copyright (2016) American Chemical Society)

suitable to quantify all targeted proteins with the lowest variation at 1.5 ng concatemer. In this way, we showed that a single-point addition was applicable for the quantification of the complete set of protein targets.

## 7.4 Applications of the Assay

With the targeted proteomics assays ready for application, one can start screening for the target proteins in different types of samples, though one should keep in mind that assay responses and interferences might vary per sample type. As example, selected results from the 134 peptides targeted in the mitochondrial protein assay to serve are presented in Fig. 7.5 [3].

The total ion current chromatogram (TIC; this is the ion current across all MS signals) of all 134 peptides is shown in Fig. 7.5a and the two peptides of SDHA are highlighted within the dedicated time windows of 4 min, where measurements were scheduled (similar windows were used for all targeted peptides). The amount of each peptide was calculated based on the known amount of the added, concatemer-derived, isotopically labeled standards, resulting in plots reporting the amounts of all targeted peptides/proteins (Fig. 7.5c). In this way, protein amounts were obtained for all targeted proteins based on mea-

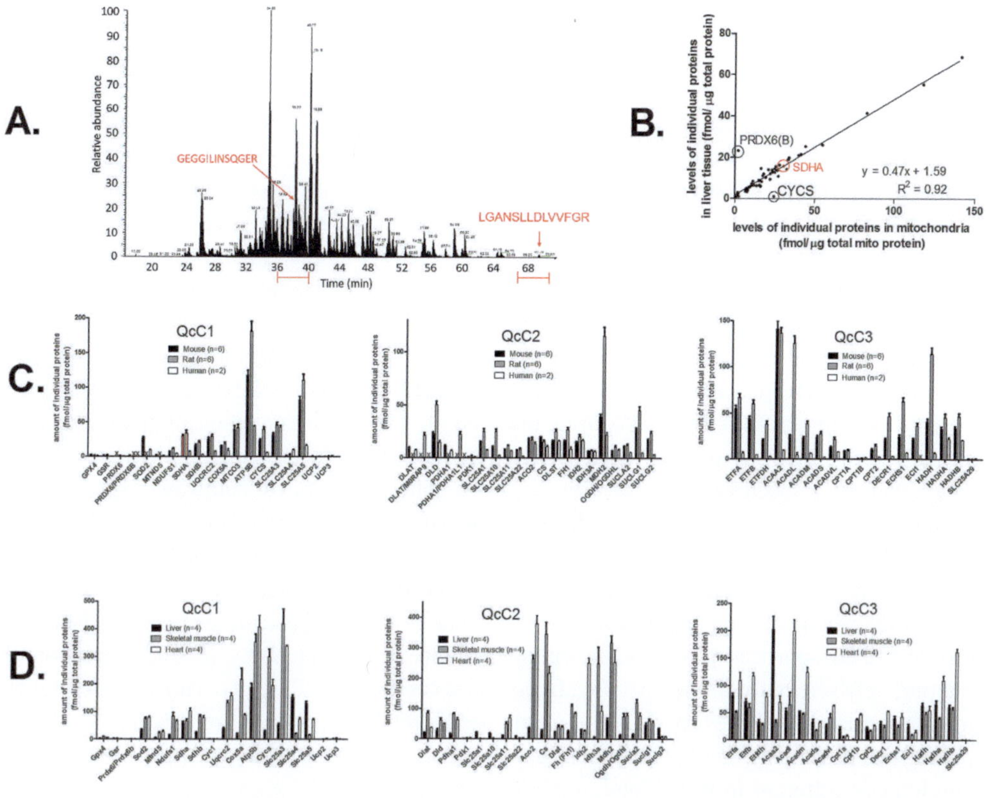


**Fig. 7.5** Application of a proteomics assay targeting a defined set of mitochondrial proteins [3]. (**a**) Total ion current (TIC; this is the ion current across all MS signals) of all targeted peptides. The two peptides from SDHA are indicated with red arrows and the scheduled time window of 4 min is shown below. (**b**). Protein quantification of the targeted mitochondrial proteins in isolated mouse liver mitochondria compared to quantification in mouse liver tissue from the same mice. (**c**) Amounts of all targeted proteins (in fmol/μg total protein) based on peak area ratios to isotopically labeled peptides derived from 3 concatemers (QcC1 to QcC3) in mitochondria isolated from mouse and rat liver and human fibroblasts. (**d**) Amounts of all targeted proteins (in fmol/μg total protein) based on peak area ratios to isotopically labeled peptides derived from 3 concatemers (QcC1 to QcC3) in mitochondria isolated from mouse liver, skeletal muscle and heart tissue. Panel (**b** and **c**) were reprinted with permission from [3]. (Copyright (2016) American Chemical Society)

suring 134 peptides in a single MS run injecting only 1 μg total mitochondrial protein plus 1.5 ng (around 25 fmol) of the concatemers, without the need for prior fractionation of the samples. The results shown in Fig. 7.5c are protein amounts determined in isolated mitochondria [51]. However, mitochondrial isolations can only be performed on fresh tissue, which is not always available or practical. Figure 7.5b shows a comparison of all protein amounts determined from mitochondria isolated from mouse liver tissues (mouse results from Fig. 7.5c) and amounts of the same proteins determined directly in liver tissue from the same mice. The protein amounts correlate well indicating that mitochondrial isolation is not required for this type of measurements. Figure 7.5b shows two exceptions that do not correlate between isolated mitochondria and liver tissue, namely peroxiredoxin-6 (PRDX6), which is present at much higher amounts in liver tissue compared to isolated mitochondria, and cytochrome c (CYCS), which is unexpectedly low in tissue, while it was easily detected in mitochondria. The PRDX6 observation fits with the reported preferential cytoplasmic localization of this protein under physiological conditions [52], but there is no obvious explanation for the apparent absence of cytochrome c in tissue. It might be due to different modification states that escape detection with the current targeted method but this requires further study.

A second application is the screening of these mitochondrial proteins in various organisms. The assay was designed in such way that it could be used not only for the mouse proteins, but also the human orthologues of the same proteins. Although the assay was not specifically designed for rat orthologues, additional comparison showed that the majority of the selected peptide targets could also be applied to the rat orthologues [3]. This provides the benefit that one can switch during the course of a study between patient material, for example related to a specific disease, and the corresponding animal model, for more detailed mechanistic studies. An example of this is given in Fig. 7.5c, where the proteins were quantified in mitochondria from mouse, rat and human.

The described approach provides the option to quantitatively screen for tissue-specific effects on mitochondrial protein levels. Mootha et al. [26] described previously that the amount of certain proteins is different in mitochondria originating from different tissues. Figure 7.5d shows the quantification of the targeted mitochondrial proteins originating from mouse liver, heart and skeletal muscle (part from the study from Stolle et al. [53] accepted in Aging Cell).

While developing a targeted proteomics assay represents a considerable time investment prior to the generation of results, there is a long-term benefit because the developed assay can be applied to any research question involving the protein targets in the future. New assays are thus a growing asset of bioanalytical tools in the lab. The described assay is useful for studying diseases related to mitochondrial energy metabolism. The targeted processes play important roles in many diseases like metabolic disorders, neurodegenerative diseases, various myopathies, cancer, diabetes, obesity and cardiovascular disease as well as in aging of cells and organisms [2, 28, 30–32]. The assays allow to study protein levels without the need for extensive sample enrichment and fractionation. At the same time it yields information about a large set of proteins in a single measurement. If properly designed, it can be used for protein orthologues from various species (e.g. mouse, rat and human) and may serve to delineate patient-specific as well as tissue-specific effects.

The above mentioned targeted mitochondrial protein assay has been applied to a variety of research questions ranging from the screening of patients with metabolic disorders (for example to MCAD patients and a patient with a newly identified modification [3]), to the effects of aging on skeletal muscle (Stolle et al. [53] accepted in Aging Cell), the screening of mouse models to evaluate the effect of single protein knockouts on an entire biochemical pathway [34], the effect of nutrition, for example of sweeteners, on liver lipid metabolism [54], and of malnutrition-induced hepatic dysfunction [55]. Developing such a targeted proteomics assay is thus recommended when planning to study questions that are central to biology in a wide range of contexts, such as the biology of mitochondria, which are involved in many biological mechanisms and disease pathophysiologies.

## 7.5 Additional Information

This chapter has been focused on the development of targeted proteomics experiments using concatemers for the accurate quantification of target proteins. For this reason we choose to exclude various other general aspects of the proteomics workflows. The reader can find more detailed descriptions of these topics in the literature. Details about sample preparation can be found in [56] and a more thorough introduction to software tools for developing targeted proteomics assays in [38]. Screenshots used in this chapter were prepared with the software program Skyline; this freely-available software provides support both for assay development (for example: selection of peptides within the defined criteria, checking uniqueness, predicting collision energies and retention times) and data-processing with clear on-line tutorials. Similar types of targeted proteomics experiments on high-resolution mass spectrometers (so-called parallel reaction monitoring (PRM)), where transitions do not have to be pre-selected, are described in [57, 58].

## 7.6 Summary

This chapter summarizes the role of targeted proteomics to study mitochondria by providing accurate quantification of multiple proteins in a single experiment. It provides a step-by-step tutorial on how to develop targeted proteomics experiments, from the design of the experiment, the choice of quantification strategy to the generation of results, illustrated with a mitochondrial protein target set.

Although the workflow has been illustrated with a mitochondrial protein target set, this type of assays can be applied to any research question related to protein targets, for example to studying complete protein pathways (like the fatty acid beta oxidation pathway), protein complexes (like the protein complexes involved in oxidative phosphorylation) or biomarker panels. Even post-translational modifications can be targeted, since such modifications will result in a mass shift that can be detected by MS if signature peptides can be found comprising the modification.

## References


1. Padrao AI, Ferreira R, Amado F, Vitorino R, Duarte JA (2016) Uncovering the exercise-related proteome signature in skeletal muscle. Proteomics 16:816–830
2. Gregersen N, Hansen J, Palmfeldt J (2012) Mitochondrial proteomics – a tool for the study of metabolic disorders. J Inherit Metab Dis 35:715–726
3. Wolters JC, Ciapaite J, van Eunen K, Niezen-Koning KE, Matton A, Porte RJ, Horvatovich P, Bakker BM, Bischoff R, Permentier HP (2016) Translational targeted proteomics profiling of mitochondrial energy metabolic pathways in mouse and human samples. J Proteome Res 15:3204–3213
4. Janssens GE, Meinema AC, Gonzalez J, Wolters JC, Schmidt A, Guryev V, Bischoff R, Wit EC, Veenhoff LM, Heinemann M (2015) Protein biogenesis machinery is a driver of replicative aging in yeast. elife 4:e08527. https://doi.org/10.7554/eLife.08527
5. Alves RM, Vitorino R, Figueiredo P, Duarte JA, Ferreira R, Amado F (2010) Lifelong physical activity modulation of the skeletal muscle mitochondrial proteome in mice. J Gerontol A Biol Sci Med Sci 65:832–842
6. Hwang CY, Kim K, Choi JY, Bahn YJ, Lee SM, Kim YK, Lee C, Kwon KS (2014) Quantitative proteome analysis of age-related changes in mouse gastrocnemius muscle using mTRAQ. Proteomics 14:121–132
7. Ibebunjo C, Chick JM, Kendall T, Eash JK, Li C, Zhang Y, Vickers C, Wu Z, Clarke BA, Shi J, Cruz J, Fournier B, Brachat S, Gutzwiller S, Ma Q, Markovits J, Broome M, Steinkrauss M, Skuba E, Galarneau JR, Gygi SP, Glass DJ (2013) Genomic and proteomic profiling reveals reduced mitochondrial function and disruption of the neuromuscular junction driving rat sarcopenia. Mol Cell Biol 33:194–212
8. Kang Y, Fielden LF, Stojanovski D (2017) Mitochondrial protein transport in health and disease. Semin Cell Dev Biol 76:142
9. Calvo SE, Mootha VK (2010) The mitochondrial proteome and human disease. Annu Rev Genomics Hum Genet 11:25–44
10. van der Wijst MG, van Tilburg AY, Ruiters MH, Rots MG (2017) Experimental mitochondria-targeted DNA methylation identifies GpC methylation, not CpG methylation, as potential regulator of mitochondrial gene expression. Sci Rep 7:177. 017-00263-z
11. McDonough AA, Veiras LC, Minas JN, Ralph DL (2015) Considerations when quantitating protein abundance by immunoblot. Am J Physiol Cell Physiol 308:C426–C433
12. Ivell R, Teerds K, Hoffman GE (2014) Proper application of antibodies for immunohistochemical detection: antibody crimes and how to prevent them. Endocrinology 155:676–687
13. Voller A, Bartlett A, Bidwell DE (1978) Enzyme immunoassays with special reference to ELISA techniques. J Clin Pathol 31:507–520
14. Aebersold R, Burlingame AL, Bradshaw RA (2013) Western blots versus selected reaction monitoring assays: time to turn the tables? Mol Cell Proteomics 12:2381–2382
15. Hoofnagle AN, Wener MH (2009) The fundamental flaws of immunoassays and potential solutions using tandem mass spectrometry. J Immunol Methods 347:3–11
16. Baker M (2015) Reproducibility crisis: blame it on the antibodies. Nature 521:274–276
17. Goldberg S (2015) Mechanical/physical methods of cell distribution and tissue homogenization. Methods Mol Biol 1295:1–20
18. Calvo SE, Clauser KR, Mootha VK (2016) MitoCarta2.0: an updated inventory of mammalian mitochondrial proteins. Nucleic Acids Res 44:D1251–D1257
19. Smith AC, Robinson AJ (2016) MitoMiner v3.1, an update on the mitochondrial proteomics database. Nucleic Acids Res 44:D1258–D1261
20. Nesvizhskii AI (2007) Protein identification by tandem mass spectrometry and sequence database searching. Methods Mol Biol 367:87–119

21. Thakur SS, Geiger T, Chatterjee B, Bandilla P, Frohlich F, Cox J, Mann M (2011) Deep and highly sensitive proteome coverage by LC-MS/MS without prefractionation. Mol Cell Proteomics 10:M110.003699
22. Rabilloud T, Kieffer S, Procaccio V, Louwagie M, Courchesne PL, Patterson SD, Martinez P, Garin J, Lunardi J (1998) Two-dimensional electrophoresis of human placental mitochondria and protein identification by mass spectrometry: toward a human mitochondrial proteome. Electrophoresis 19:1006–1014
23. Pagliarini DJ, Rutter J (2013) Hallmarks of a new era in mitochondrial biochemistry. Genes Dev 27:2615–2627
24. Taylor SW, Fahy E, Zhang B, Glenn GM, Warnock DE, Wiley S, Murphy AN, Gaucher SP, Capaldi RA, Gibson BW, Ghosh SS (2003) Characterization of the human heart mitochondrial proteome. Nat Biotechnol 21:281–286
25. Sickmann A, Reinders J, Wagner Y, Joppich C, Zahedi R, Meyer HE, Schonfisch B, Perschil I, Chacinska A, Guiard B, Rehling P, Pfanner N, Meisinger C (2003) The proteome of Saccharomyces cerevisiae mitochondria. Proc Natl Acad Sci U S A 100:13207–13212
26. Mootha VK, Bunkenborg J, Olsen JV, Hjerrild M, Wisniewski JR, Stahl E, Bolouri MS, Ray HN, Sihag S, Kamal M, Patterson N, Lander ES, Mann M (2003) Integrated analysis of protein composition, tissue diversity, and gene regulation in mouse mitochondria. Cell 115:629–640
27. Pagliarini DJ, Calvo SE, Chang B, Sheth SA, Vafai SB, Ong SE, Walford GA, Sugiana C, Boneh A, Chen WK, Hill DE, Vidal M, Evans JG, Thorburn DR, Carr SA, Mootha VK (2008) A mitochondrial protein compendium elucidates complex I disease biology. Cell 134:112–123
28. Chen X, Li J, Hou J, Xie Z, Yang F (2010) Mammalian mitochondrial proteomics: insights into mitochondrial functions and mitochondria-related diseases. Expert Rev Proteomics 7:333–345
29. Bantscheff M, Schirle M, Sweetman G, Rick J, Kuster B (2007) Quantitative mass spectrometry in proteomics: a critical review. Anal Bioanal Chem 389:1017–1031
30. Jiang Y, Wang X (2012) Comparative mitochondrial proteomics: perspective in human diseases. J Hematol Oncol 5:11. 8722-5-11
31. Chou HC, Chan HL (2012) Targeting proteomics to investigate metastasis-associated mitochondrial proteins. J Bioenerg Biomembr 44:629–634
32. Da Cruz S, Parone PA, Martinou JC (2005) Building the mitochondrial proteome. Expert Rev Proteomics 2:541–551
33. Domon B, Aebersold R (2010) Options and considerations when selecting a quantitative proteomics strategy. Nat Biotechnol 28:710–721
34. van Eunen K, Volker-Touw CM, Gerding A, Bleeker A, Wolters JC, van Rijt WJ, Martines AM, Niezen-Koning KE, Heiner RM, Permentier H, Groen AK, Reijngoud DJ, Derks TG, Bakker BM (2016) Living on the edge: substrate competition explains loss of robustness in mitochondrial fatty-acid oxidation disorders. BMC Biol 14:107
35. Zhu J, Vinothkumar KR, Hirst J (2016) Structure of mammalian respiratory complex I. Nature 536:354–358
36. Parker CE, Borchers CH (2014) Mass spectrometry based biomarker discovery, verification, and validation – quality assurance and control of protein biomarker assays. Mol Oncol 8:840–858
37. Boichenko AP, Govorukhina N, Klip HG, van der Zee AG, Guzel C, Luider TM, Bischoff R (2014) A panel of regulated proteins in serum from patients with cervical intraepithelial neoplasia and cervical cancer. J Proteome Res 13:4995–5007
38. MacLean B, Tomazela DM, Shulman N, Chambers M, Finney GL, Frewen B, Kern R, Tabb DL, Liebler DC, MacCoss MJ (2010) Skyline: an open source document editor for creating and analyzing targeted proteomics experiments. Bioinformatics 26:966–968
39. Craig R, Cortens JP, Beavis RC (2005) The use of proteotypic peptide libraries for protein identification. Rapid Commun Mass Spectrom 19:1844–1850
40. Kuster B, Schirle M, Mallick P, Aebersold R (2005) Scoring proteomes with proteotypic peptide probes. Nat Rev Mol Cell Biol 6:577–583
41. Brownridge P, Beynon RJ (2011) The importance of the digest: proteolysis and absolute quantification in proteomics. Methods 54:351–360
42. Eyers CE, Lawless C, Wedge DC, Lau KW, Gaskell SJ, Hubbard SJ (2011) CONSeQuence: prediction of reference peptides for absolute quantitative proteomics using consensus machine learning approaches. Mol Cell Proteomics 10:M110.003384
43. Sechi S, Chait BT (1998) Modification of cysteine residues by alkylation. A tool in peptide mapping and protein identification. Anal Chem 70:5150–5158
44. Lawless C, Hubbard SJ (2012) Prediction of missed proteolytic cleavages for the selection of surrogate peptides for quantitative proteomics. OMICS 16:449–456
45. Meyer B, Papasotiriou DG, Karas M (2011) 100% protein sequence coverage: a modern form of surrealism in proteomics. Amino Acids 41:291–310
46. Mitra V, Govorukhina N, Zwanenburg G, Hoefsloot H, Westra I, Smilde A, Reijmers T, van der Zee AG, Suits F, Bischoff R, Horvatovich P (2016) Identification of analytical factors affecting complex proteomics profiles acquired in a factorial design study with analysis of variance: simultaneous component analysis. Anal Chem 88:4229–4238
47. Lebert D, Dupuis A, Garin J, Bruley C, Brun V (2011) Production and use of stable isotope-labeled proteins for absolute quantitative proteomics. Methods Mol Biol 753:93–115
48. Kirkpatrick DS, Gerber SA, Gygi SP (2005) The absolute quantification strategy: a general procedure for the quantification of proteins and post-translational modifications. Methods 35:265–273
49. Beynon RJ, Doherty MK, Pratt JM, Gaskell SJ (2005) Multiplexed absolute quantification in proteomics

using artificial QCAT proteins of concatenated signature peptides. Nat Methods 2:587–589

50. Lange V, Picotti P, Domon B, Aebersold R (2008) Selected reaction monitoring for quantitative proteomics: a tutorial. Mol Syst Biol 4:222
51. Mildaziene V, Nauciene Z, Baniene R, Grigiene J (2002) Multiple effects of 2,2′,5,5′-tetrachlorobiphenyl on oxidative phosphorylation in rat liver mitochondria. Toxicol Sci 65:220–227
52. Eismann T, Huber N, Shin T, Kuboki S, Galloway E, Wyder M, Edwards MJ, Greis KD, Shertzer HG, Fisher AB, Lentsch AB (2009) Peroxiredoxin-6 protects against mitochondrial dysfunction and liver injury during ischemia-reperfusion in mice. Am J Physiol Gastrointest Liver Physiol 296:G266–G274
53. Stolle S, Ciapaite J, Reijne AC, Talarovicova A, Wolters JC, Aguirre-Gamboa R, van der Vlies P, de Lange K, Neerincx PB, van der Vries G, Deelen P, Swertz MA, Li Y, Bischoff R, Permentier HP, Horvatovitch PL, Groen AK, van Dijk G, Reijngoud DJ, Bakker BM (2018) Running-wheel activity delays mitochondrial respiratory flux decline in aging mouse muscle via a post-transcriptional mechanism. Aging Cell 17(1). https://doi.org/10.1111/acel.12700. Epub 2017 Nov 9
54. Janssens S, Ciapaite J, Wolters JC, van Riel NA, Nicolay K, Prompers JJ (2017) An in vivo magnetic resonance spectroscopy study of the effects of caloric and non-caloric sweeteners on liver lipid metabolism in rats. Nutrients 9:476. https://doi.org/10.3390/nu9050476
55. van Zutphen T, Ciapaite J, Bloks VW, Ackereley C, Gerding A, Jurdzinski A, de Moraes RA, Zhang L, Wolters JC, Bischoff R, Wanders RJ, Houten SM, Bronte-Tinkew D, Shatseva T, Lewis GF, Groen AK, Reijngoud DJ, Bakker BM, Jonker JW, Kim PK, Bandsma RH (2016) Malnutrition-associated liver steatosis and ATP depletion is caused by peroxisomal and mitochondrial dysfunction. J Hepatol 65:1198–1208
56. Feist P, Hummon AB (2015) Proteomic challenges: sample preparation techniques for microgram-quantity protein analysis from biological samples. Int J Mol Sci 16:3537–3563
57. Bourmaud A, Gallien S, Domon B (2016) Parallel reaction monitoring using quadrupole-Orbitrap mass spectrometer: principle and applications. Proteomics 16:2146–2159
58. Gallien S, Bourmaud A, Kim SY, Domon B (2014) Technical considerations for large-scale parallel reaction monitoring analysis. J Proteome 100:147–159

# Recent Advances in Targeting Human Mitochondrial AAA+ Proteases to Develop Novel Cancer Therapeutics

8

Keith S. Wong and Walid A. Houry

**Abstract**

The mitochondrion is a vital organelle that performs diverse cellular functions. In this regard, the cell has evolved various mechanisms dedicated to the maintenance of the mitochondrial proteome. Among them, AAA+ ATPase-associated proteases (AAA+ proteases) such as the Lon protease (LonP1), ClpXP complex, and the membrane-bound i-AAA, m-AAA and paraplegin facilitate the clearance of misfolded mitochondrial proteins to prevent the accumulation of cytotoxic protein aggregates. Furthermore, these proteases have additional regulatory functions in multiple biological processes that include amino acid metabolism, mitochondria DNA transcription, metabolite and cofactor biosynthesis, maturation and turnover of specific respiratory and metabolic proteins, and modulation of apoptosis, among others. In cancer cells, the increase in intracellular ROS levels promotes tumorigenic phenotypes and increases the frequency of protein oxidation and misfolding, which is compensated by the increased expression of specific AAA+ proteases as part of the adaptation mechanism. The targeting of AAA+ proteases has led to the discovery and development of novel anticancer compounds. Here, we provide an overview of the molecular characteristics and functions of the major mitochondrial AAA+ proteases and summarize recent research efforts in the development of compounds that target these proteases.

**Keywords**

Human mitochondria · Mitochondrial AAA+ proteases · Chemical modulators of AAA+ protease activity · Cancer therapeutics

## 8.1 Introduction

The mitochondrion is a vital organelle in the human cell with diverse biological functions such as energy metabolism, the biosynthesis of ATP and other important metabolites, signalling, cellular chemotaxis, development, and stress responses such as autophagy and apoptosis [73]. In this regard, the cell has evolved multiple mechanisms dedicated to maintaining mitochondrial health and function. These include the

K. S. Wong
Department of Biochemistry, University of Toronto, Toronto, ON, Canada

W. A. Houry (✉)
Department of Biochemistry, University of Toronto, Toronto, ON, Canada

Department of Chemistry, University of Toronto, Toronto, ON, Canada
e-mail: walid.houry@utoronto.ca





A. Urbani, M. Babu (eds.), *Mitochondria in Health and in Sickness*, Advances in Experimental Medicine and Biology 1158, https://doi.org/10.1007/978-981-13-8367-0_8

expression of various stress response proteins, the controlled biosynthesis and import of mitochondrial proteins, the assisted folding of these proteins and degradation of misfolded ones, the isolation of damaged mitochondria by modulating mitochondrial fission-fusion dynamics and their clearance from the cell via mitophagy [59].

The assisted folding of mitochondrial proteins and degradation of misfolded ones are mediated by specialized molecular chaperones and proteases that can be found in different mitochondrial compartments [59]. Among them, the ATPases Associated with diverse cellular Activities (AAA+) ATPase-coupled proteases (AAA+ proteases) mediate the active remodelling, unfolding and degradation of mitochondrial proteins using energy derived from ATP hydrolysis. Notable members of this class include the mitochondrial Lon protease (LonP1) and ClpXP complex that are found in the mitochondrial matrix, and the integral membrane proteases i-AAA, m-AAA and paraplegin that are localized to the inner mitochondrial membrane [59].

AAA+ proteases are important in facilitating the clearance of misfolded proteins from the mitochondria [59]. Accordingly, mutations in the AAA+ proteases that hinder or abolish their activities are associated with various developmental, neurosensory and metabolic disorders. These include CODAS syndrome (CODASS) due to mutations in LonP1 [27], Perrault syndrome 3 (PRLTS3) caused by mutations in ClpP [46], erythropoietic protoporphyria (EPP) that is caused by mutations in ClpX [101], optic atrophy 11 (OPA11) arising from mutations in i-AAA [43], spinocerebellar ataxia 28 (SCA28) [25] caused by mutations in m-AAA, and hereditary spastic paraplegia 7 (HSP7) due to mutations in paraplegin [75].

In addition to the aforementioned genetic diseases, AAA+ proteases are also important for the proliferation and metastasis of various human cancers. Notably, the increase in intracellular ROS level is commonly observed among cancer cells, as it promotes tumorigenesis and facilitates additional mutations leading to metastasis [81]. Consequently, a higher ROS level increases the frequency of mitochondrial protein oxidation and misfolding, which in turn induces an increase in the expression of AAA+ proteases as part of the cell's adaptation mechanism [81]. In this regard, the targeting of AAA+ proteases has been explored as a potential therapeutic strategy, which led to the discovery and development of various compounds with anti-cancer potential. This article provides an overview of the molecular characteristics and functions of the major mitochondrial AAA+ proteases, as well as, discusses recent research efforts aimed at the development of compounds targeting these proteases.

## 8.2 Mitochondrial Lon Protease (LonP1)

### 8.2.1 Molecular Characteristics of LonP1

LonP1 is a 959-residue long AAA+ serine protease found in the mitochondrial matrix and is encoded by the nuclear *LONP1* gene [79]. It has three distinct domains – the N-terminal domain, followed by the AAA+ ATPase domain and a C-terminal proteolytic domain (Fig. 8.1a). At the N-terminus of LonP1 is a 67-residue long mitochondrial targeting sequence (MTS) (Fig. 8.1a) that directs the localization of LonP1 to the mitochondrial matrix [99].

The atomic structure of human LonP1's proteolytic domain (Fig. 8.1b) has been solved to 2 Å resolution by X-ray crystallography (PDB: 2X36) [31]. The domain shares the same structural layout as the proteolytic domain of bacterial and archaeal Lon proteases with an active site that carries the conserved catalytic Ser-Lys dyad (S855-K898) (Fig. 8.1b; proteolytic site indicated with black dotted circle) that is non-canonical among serine proteases [31]. Structural analysis of the full-length LonP1 S855A proteolytic mutant by cryo-electron microscopy (cryo-EM) revealed the formation of a homo-hexamer when LonP1 is in complex with ADP (Fig. 8.1c, left) or AMP-PNP (Fig. 8.1c, right) [52]. In both cases, the AAA+ ATPase and proteolytic domains occupy the head region of the hexamer and provide the site for ATP hydrolysis and proteolysis of substrate proteins, while the N-terminal domain occupies the hexamer's legs region [52]. The hexameric form is the most common oligomeric state adopted by proteins carrying the AAA+ ATPase domain [85].

The proteolytic activity of LonP1 is driven by a cycle of conformational changes resulting from the binding and hydrolysis of ATP that is mediated by its AAA+ ATPase domain. Notably, the ATP-bound state of LonP1 (simulated by using AMP-PNP, the non-hydrolysable ATP analog) shows a shuttered entry pore to its proteolytic chamber that is gated by the axial pore loops (RTYVG; residues 563–567) [52]. In contrast, the ADP-bound state of LonP1 shows an unobstructed entry pore, suggesting the requirement of ATP hydrolysis in driving the entry of substrates into LonP1's proteolytic chamber [52]. Furthermore, the binding of ADP induces significant molecular rearrangements in LonP1, causing it to adopt a split-ring structure in the head region (Fig. 8.1c, left; the ring-splitting indicated by red dotted circles) that is not observed in the ATP-bound state (Fig. 8.1c, right). Nevertheless, it remains unclear if splitting the head of LonP1 has any functional significance with respect to its proteolytic cycle. Importantly, the N-terminal domain of LonP1 is shown to be essential for oligomerization and proteolysis, as deletion of the first 270 residues covering its MTS and part of the N-terminal domain prohibits the closing of the hexameric ring and abolishes its proteolytic activity *in vitro* [52].

### 8.2.2 Cellular Function of LonP1

The primary function of LonP1 is to maintain the mitochondrial proteome, mainly via the clearance of misfolded mitochondrial proteins that arise from oxidative damage [10, 13, 74]. The targeted degradation of oxidized mitochondrial aconitase (ACO2) is a well-characterized example of this process [10]. Notably, the clearance of misfolded proteins by LonP1 is an important step in the ATF5-mediated mitochondrial unfolded protein response ($UPR^{mt}$) during cellular stress induced by the formation of toxic protein aggregates in the mitochondria [8, 29].

In addition to clearing misfolded proteins from the mitochondria, LonP1 targets specific subunits of respiratory complexes for degradation as part of the cell's adaptation mechanism to stress or changes in respiratory condition. For example, LonP1 has been shown to directly interact and degrade specific peripheral arm subunits of Complex I (NDUFA9, NDUFS1, NDUFV1 and NDUFV2) to prevent production of reactive oxygen species (ROS) upon mitochondrial depolarization [77]. Similarly, LonP1 facilitates the turnover of SDHFA2, the flavination factor for subunit A of Complex II that is essential for its respiratory activity [9]. Furthermore, LonP1 has been shown to rapidly degrade isoform 1 of the cytochrome c oxidase subunit 4 (COX4-1) and is important for the cell's adaptation to hypoxia [30]. Outside of the respiratory chain, LonP1 has been shown to specifically mediate the rapid turnover of pyruvate dehydrogenase kinase 4 (PDK4), an inhibitory protein of the pyruvate dehydrogenase (PDH) complex, in response to changes in metabolic state of the mitochondria [22].

LonP1 has additional regulatory functions in multiple metabolic pathways. In renal glutamine catabolism, downregulation of the kidney isoform of mitochondrial glutaminase (glutaminase C) in the presence of diphenylarsinic acid (DPAA) occurs via its degradation by LonP1 [55]. In steroid biosynthesis, LonP1 mediates the turnover of the steroidogenic acute regulatory protein (StAR) [38], thereby suppressing the transfer of cholesterol from the outer mitochondrial membrane (OMM) to the inner mitochondrial membrane (IMM) [38]. In heme biogenesis, LonP1 selectively targets the mature form of the nonspecific 5′-aminolevulinic acid synthase (ALAS1) for degradation as part of a heme-induced negative feedback regulatory mechanism for heme biosynthesis [92]. LonP1 also participates in the oxygen-sensitive degradation of specific heme-carrying proteins, such as cystathionine β-synthase (CBS) that catalyzes hydrogen sulfide formation, and heme oxygenase 1 (HO-1) that converts heme into biliverdin during heme catabolism [90]. With respect

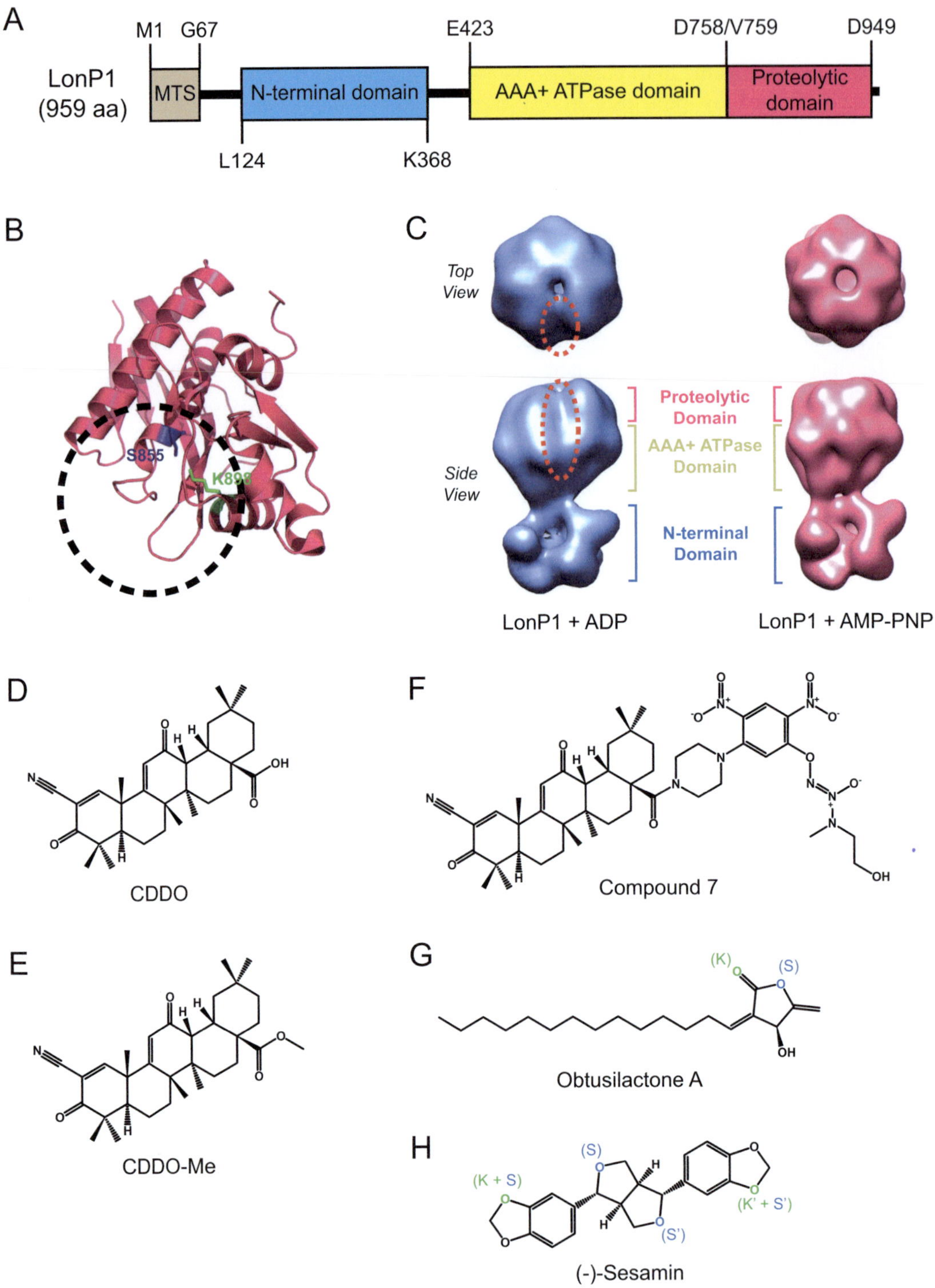


**Fig. 8.1** Structure of mitochondrial LonP1 protease and its inhibitors. (**a**) Domain architecture of LonP1 illustrated as a bar diagram. Domain boundaries were determined from curated information obtained from UniProt (www.uniprot.org) or other public bioinformatics databases. *MTS* mitochondrial targeting sequence, *AAA* ATPase associated with diverse cellular activities. (**b**) Atomic structure of LonP1 (PDB: 2X36) [31]. Residues

to mitochondrial DNA (mtDNA) transcription, LonP1 mediates the downregulation of mitochondrial transcription factor A (TFAM) by selectively degrading its phosphorylated form that has less affinity for mtDNA, [69]. This in turn promotes mtDNA transcriptional activities that are necessary for mitochondrial homeostasis.

### 8.2.3 Role of LonP1 in Cancer

LonP1 is expressed at high levels in various types of solid cancers of the skin, lung, colorectal, bladder, cervix and the central nervous system [15, 26, 68, 71, 79], and in non-solid cancers such as B-cell lymphoma and acute myeloid leukemia [7, 37]. Suppressing LonP1 expression halts the proliferation of cancer cells and sensitizes them to chemotherapeutic reagents or sub-optimal growth conditions. For example, siRNA knockdown (KD) of LonP1 suppresses the proliferation of human bladder cancer cells ScaBER and UM-UC3 and increases the sensitivity of UM-UC3 cells to doxorubicin leading to caspase-dependent apoptosis [68]. Similarly, siRNA KD of LonP1 decreases the viability of human malignant glioma cells D-54 and U-251 and compromises their ability to survive under hypoxic conditions [26]. In non-small-cell lung cancer (NSCLC) H1299, shRNA KD of LonP1 suppresses cell proliferation and increases the occurrence of caspase-dependent apoptosis [98].

Conversely, the overexpression of a functional LonP1 promotes cancer cell proliferation. In hypopharyngeal squamous cell carcinoma (HSCC) FADU cells and oral cavity squamous cell carcinoma (OCSCC) OEC-M1 cells, overexpressing LonP1 significantly increases cellular proliferation [15]. Furthermore, LonP1-overexpressing cells exhibit enhanced colony formation compared to control cells [15]. Importantly, overexpression of LonP1 increases cellular resistance to apoptosis-inducing reagents. For example, overexpressing LonP1 in colon carcinoma cells RKO confers cellular resistance to the triterpenoid 2-cyano-3, 12-dioxooleana-1,9(11)-dien-28-oic acid (CDDO) and its C-28 methyl ester derivative (CDDO-Me) [33]. Both CDDO and CDDO-Me (Fig. 8.1d, e) have been characterized in detail with regard to their anti-cancer activity and will be discussed in Sect. 8.2.4.1.

A notable cellular effect resulting from high LonP1 expression is the altered expression and remodelling of specific respiratory complexes. In FADU and OEC-M1 cells, high LonP1 level leads to increased expression of Complex I subunits NDUSF3 and NDUFS8 [15]. This in turn raises intracellular ROS levels and promotes cell proliferation via Ras-ERK signalling, and cell migration and metastatic activity via both ERK and p38 MAPK signalling pathways [15]. Conversely, KD of LonP1 decreases ROS production and suppresses cell proliferation through downregulation of JNK phosphorylation and subsequent inhibition of MAPK signalling downstream [68]. In murine melanoma cells B16F10, high LonP1 levels increase the expression of specific subunits of Complex I and Complex V while decreasing the expression of other subunits, leading to significant remodelling of both complexes [79]. Consequently, respiratory oxidative phosphoryla-

**Fig. 8.1** (continued) of the catalytic dyad are shown as sticks and coloured with Ser855 in blue and Lys898 in green. The active site of LonP1 in indicated with the black, dotted circle as shown. (**c**) Cryo-EM structures of LonP1 in complex with ADP (left; in blue) or AMP-PNP (right; in magenta). Both top views and side views of the structures are shown. Regions corresponding to the three functional domains are indicated. The red, dotted circles on LonP1 + ADP indicate the site of ring-splitting in LonP1's catalytic chamber as it adopts the split-ring conformation (see Sect. 8.2.1 for details). (**d**) Chemical structure of CDDO. (E) Chemical structure of CDDO-Me. (**f**) Chemical structure of Compound 7. (**g**) Chemical structure of Obtusilactone A. The oxygen atoms forming potential interactions with LonP1's catalytic dyad are coloured, with the oxygen interacting with Ser855 in blue and denoted with (S) and the oxygen interacting with Lys898 in green and denoted with (**k**). (**h**) Chemical structure of (−)-Sesamin. As in G, the oxygen atoms forming potential interactions with LonP1's catalytic dyad residues are coloured and denoted as before, with the exception that the piperonyl oxygen (shown in green) can likely interact with both Ser855 and Lys898, denoted by (K + S). The alternative LonP1-binding interface resulting from molecular symmetry is indicated by the coloured oxygen atoms denoted with (S′) or (K′ + S′).

tion (OXPHOS) is compromised, prompting the upregulation of the glycolytic pathway [79]. An increase in expression of the subunits of the spliceosome, the chaperone-containing TCP-1 (CCT) complex and the proteasome is also observed [79]. Collectively, these changes confer additional cellular protection against oncogene-induced senescence [79].

### 8.2.4 Chemical Inhibitors of LonP1 and Their Anti-cancer Properties

Given the central role of LonP1 in facilitating cancer cell proliferation and contribution to chemotherapeutic resistance, its inhibition using specific chemical inhibitors has been explored as a potential therapeutic strategy. Several classes of compounds have been identified to both inhibit the proteolytic activity of LonP1 and induce cytotoxicity in various types of cancer.

#### 8.2.4.1 The Triterpenoid CDDO and Its Derivatives

Triterpenoids are structurally diverse compounds with a wide range of biological effects [105]. One notable example is the synthetic oleanane 2-cyano-3,12-dioxooleana-1,9-dien-28-oic acid (CDDO) (Fig. 8.1d), also known as Bardoxolone (RTA-401) that specifically inhibits the proteolytic activity of purified LonP1 [7]. Inhibition of LonP1 by CDDO occurs via direct molecular interactions, although the exact binding site has not been determined. In vivo, CDDO induces caspase-dependent apoptosis in various human cancer cells with high LonP1 expression, including Granta MCL (B-cell lymphoma), RKO (colon carcinoma), HepG2 (liver hepatocellular carcinoma) and MCF7 (breast ductal carcinoma) [7, 33] (Table 8.1). Furthermore, CDDO-induced cytotoxicity is increased when LonP1 is overexpressed, thus providing further confirmation on the specific targeting of LonP1 by the drug in vivo [33]. Importantly, CDDO exerts no cytotoxic effects on normal primary human fibroblasts [33], which supports its potential use as an anti-cancer agent against cancer cells that are CDDO-sensitive due to high LonP1 levels. Phase I clinical trials have been conducted on CDDO in treating acute myeloid leukemia [95] and advanced solid tumors [86], but so far they have fell short of confirming the drug's anti-cancer effects in patients.

Derivatives of CDDO have been developed and the biochemical and physiological effects characterized. For example, CDDO-Me, also known as Bardoxolone methyl (RTA-402) (Fig. 8.1e), is the C-28 methyl ester of CDDO, and it inhibits the proteolytic activity of LonP1 in vitro with greater efficacy than CDDO [7]. In vivo, CDDO-Me induces greater cytotoxicity than CDDO in RKO, HepG2, MCF7 and the taxol-resistant A549 (A549/Taxol) cells, but not in normal primary human fibroblasts [33, 47] (Table 8.1). Furthermore, the cytotoxicity of CDDO-Me is reduced when LonP1 is overexpressed [33], thus providing additional proof for the specific targeting of LonP1 by CDDO-Me.

Nevertheless, CDDO-Me exhibits potential off-target effects, given a recent report on the interaction and inhibitory effect of CDDO-Me on Hsp90 in vivo [78]. Similarly, CDDO-Me has been shown to interact with Keap1, a central regulator of the nuclear factor (erythroid-derived 2)-like 2 (Nrf2) mediated stress responses against electrophilic and oxidative conditions [18]. The interaction inhibits Keap1 and in turn activates Nrf2-mediated cytoprotective responses [102].

With regard to its clinical application, a Phase I clinical trial conducted with CDDO-Me for treating advanced solid tumors and lymphomas indicates that the drug exhibits objective anti-cancer activity with minimal toxicity on the patients [44], prompting further development of CDDO derivatives, such as Omaveloxolone (RTA-408), which has also undergone a Phase I trial for assessment against advanced solid tumors [21]. Aside from its anti-cancer properties, a Phase III trial has also been conducted in evaluating CDDO-Me as a treatment for stage 4

**Table 8.1** Biological activities of small molecule compounds targeting different mitochondrial AAA+ proteases

| Compound | Protease targeted | Molecular mechanism | Cell line | Disease model | Comments and references |
|---|---|---|---|---|---|
| CDDO | LonP1 | Inhibition | Granta MCL | B-cell lymphoma | $IC_{50}$ = 1.3 μM [7] |
| | | | | | Apoptosis induced at 2.5 μM [7] |
| | | | LS174T | Colon carcinoma | No effect on cell viability [7] |
| | | | RKO | Colon carcinoma | Inhibits cell proliferation at ≥2.5 μM [33] |
| | | | | | 30–40% of cells in early apoptosis after 16-h exposure at 2.5 μM [33] |
| | | | HepG2 | Liver hepatocellular carcinoma | Inhibits cell proliferation at ≥5 μM [33] |
| | | | | | 20–30% of cells in late apoptosis after 16-h exposure at 5 μM [33] |
| | | | MCF7 | Breast ductal carcinoma | Inhibits cell proliferation at ≥5 μM [33] |
| | | | | | 30–40% of cells in late apoptosis after 24-h exposure at 5 μM [33] |
| | | | Human primary fibroblasts | Normal/ non-cancerous cells | No effect on cell proliferation [33] |
| CDDO-Me | LonP1 | Inhibition | RKO | Colon carcinoma | Inhibits cell proliferation at ≥1 μM [33] |
| | | | | | 50–60% of cells in early apoptosis after 16-h exposure at 2.5 μM [33] |
| | | | HepG2 | Liver hepatocellular carcinoma | Inhibits cell proliferation at ≥2.5 μM [33] |
| | | | | | 40–50% of cells in late apoptosis after 16-h exposure at 2.5 μM [33] |
| | | | MCF7 | Breast ductal carcinoma | Inhibits cell proliferation at ≥2.5 μM [33] |
| | | | | | Apoptosis induced in 60–70% of cells in late apoptosis after 24-h exposure at 2.5 μM [33] |
| | | | Human primary fibroblasts | Normal/ non-cancerous cells | 20–30% of cells in early apoptosis after 16-h exposure at 2.5 μM [33] |
| | | | A549/Taxol | Drug-resistant lung carcinoma | ~10% of cells in apoptosis after 24-h exposure at 4.8 μM [47] |
| Compound 7 (CDDO derivative) | LonP1 | Inhibition | A549/Taxol | Drug-resistant lung carcinoma | $IC_{50}$ = 0.35 μM [47] |
| | | | | | 30–40% of cells in apoptosis after 24-h exposure at 4.8 μM [47] |
| | | | A549 | Lung carcinoma | $IC_{50}$ = 1.07 μM [47] |

(continued)

**Table 8.1** (continued)

| Compound | Protease targeted | Molecular mechanism | Cell line | Disease model | Comments and references |
|---|---|---|---|---|---|
| Obtusilactone A | LonP1 | Inhibition | A549 | Lung carcinoma | $IC_{50}$ = 26.50 μM [98] |
| | | | H1299 | Lung carcinoma | $IC_{50}$ = 33.96 μM [98] |
| | | | | | Causes extensive DNA double-strand breaks [98] |
| | | | | | Induces JNK-mediated apoptosis after 12-h exposure at 40 μM [98] |
| | | | MRC-5 | Normal/non-cancerous cells | $IC_{50}$ = 49.43 μM [98] |
| (−)-Sesamin | LonP1 | Inhibition | H1299 | Lung carcinoma | Causes DNA double-strand breaks, but less extensive than Obtusilactone A [98] |
| | | | | | No additional data provided on cytotoxicity or apoptotic induction [98] |
| A2-32-01 | ClpP | Inhibition | TEX | Acute myeloid leukemia | $IC_{50}$ ~25 μM [19] |
| | | | OCI-AML2 | Acute myeloid leukemia | $IC_{50}$ ~25 μM [19] |
| | | | | | Impairs complex II activity and delays growth of cells xenografted in SCID mice; no observable liver, muscle or renal toxicity [19] |
| | | | K562 | Chronic myeloid leukemia | $IC_{50}$ ~25 μM [19] |
| | | | HL-60 | Acute promyelocytic leukemia | No effect on cell viability [19] |
| | | | 143B | Osteosarcoma | $IC_{50}$ ~25 μM [19] |
| | | | 143B Rho (0) | Mitochondria depletion control for 143B | No effect on cell viability [19] |
| | | | Human primary AML cells | Acute myeloid leukemia | $IC_{50}$ ranges from ~25 to ~125 μM, with cells expressing high ClpP levels being the most sensitive ($R^2$ = 0.9264) [19] |
| | | | | | Impairs proliferation of cells engrafted in NOD-SCID mice; no host toxicity observed [19] |
| | | | Human primary normal hematopoietic cells | Normal/non-cancerous cells | Minimal to no effect on cell viability [19] |
| ADEP-28 | ClpP | Dysregulated activation | HEK293 T-REx | N/A | $IC_{50}$ = 0.36 μM [100] |
| | | | HEK293 T-REx *CLPP*$^{-/-}$ | ClpP-deletion control for HEK293 T-REx | No effect on cell viability [100] |

(continued)

**Table 8.1** (continued)

| Compound | Protease targeted | Molecular mechanism | Cell line | Disease model | Comments and references |
|---|---|---|---|---|---|
| ADEP-41 | ClpP | Dysregulated activation | HEK293 T-REx | N/A | $IC_{50}$ = 0.49 μM [100] |
| | | | | | Mcl-1-mediated apoptosis induced in ~10% of cells after 72-h exposure at 10 μM [100] |
| | | | HEK293 T-REx *CLPP*$^{-/-}$ | ClpP-deletion control for HEK293 T-REx | No effect on cell viability [100] |
| | | | | | No apoptosis observed [100] |
| | | | HeLa | Endocervical carcinoma | $IC_{50}$ = 0.54 μM [100] |
| | | | HeLa T-REx | Endocervical carcinoma | $IC_{50}$ = 0.48 μM [100] |
| | | | U2OS | Osteosarcoma | $IC_{50}$ = 0.58 μM [100] |
| | | | SH-SY5Y | Neuroblastoma | $IC_{50}$ = 0.86 μM [100] |

chronic kidney disease and type 2 diabetes mellitus, although the study has revealed serious side effects of the drug such as heart failure and fluid retention in susceptible patient subgroups [17].

Chemical hybrids of CDDO with different apoptotic inducers have also been explored. For example, Compound 7 (Fig. 8.1f) is a hybrid of CDDO and an $O^2$-(2,4-dinitrophenyl) diazeniumdiolate moiety [47]. $O^2$-(2,4-dinitrophenyl) diazeniumdiolates (e.g. PABA/NO, JS-K) generate intracellular nitric oxide (NO) upon activation by glutathione S-transferase π (GSTπ) that is highly expressed in many cancers [94]. The increase in NO level in turn creates nitrosative stress leading to cytotoxicity [45, 93]. Thus, a CDDO-diazeniumdiolate hybrid can target and inhibit LonP1 while simultaneously inducing nitrosative stress upon activation by GSTπ. Indeed, the hybrid Compound 7 is cytotoxic and induces apoptosis in lung carcinoma cells A549 and the Taxol-resistant equivalent A549/Taxol (Table 8.1) with greater efficacy than CDDO-Me (Table 8.1) and JS-K [47]. Furthermore, Compound 7 generates the highest levels of intracellular NO and is the most cytotoxic to A549/Taxol cells expressing a much higher level of GSTπ compared to A549 and the non-cancerous MRC-5 cells [47]. This highlights the potential of CDDO-diazeniumdiolate hybrids in targeting cancer cells that have developed resistance to traditional chemotherapeutic agents such as Taxol.

#### 8.2.4.2 Obtusilactone A

In addition to CDDO and its derivatives, new types of LonP1-inhibiting compounds have also been discovered and characterized. For example, the γ-lactone Obtusilactone A (OA) (Fig. 8.1g) is a natural product isolated from the aromatic evergreen tree *Cinnamomum kotoense* Kanehira and has shown potential antioxidant and anti-cancer properties [16]. Further investigation has determined that OA inhibits the proteolytic activity of LonP1 both in vitro and in vivo [98]. Simulated docking analysis suggests that OA may bind to the active site of LonP1, with each of the two carbonyl oxygen atoms in its γ-lactone ring forming hydrogen bonds with Ser855 and Lys898 (Fig. 8.1b) of LonP1's catalytic dyad, respectively (Fig. 8.1g; interacting residues for each oxygen atom denoted with S for Ser855 and K for Lys898). At the cellular level, OA induces cytotoxicity in lung carcinoma cells A549 and H1299, resulting in extensive DNA double-strand breaks and subsequent apoptotic induction that is mediated through the c-Jun N-terminal kinase (JNK) pathway (Table 8.1) [98]. Treating the non-cancerous MRC-5 cells with OA also induces cytotoxicity, albeit it is less severe compared to A549 and H1299 (Table 8.1), which suggests higher drug sensitivity of cancer cells that express higher levels of LonP1 [98].

#### 8.2.4.3 (−)-Sesamin

(−)-Seasamin (Fig. 8.1h) is another natural product isolated from *C. kotoense* that shows potential antioxidant and anti-cancer properties [16]. Like OA, (−)-Sesamin also inhibits LonP1 both in vitro and in vivo [98]. Docking analysis suggests that (−)-Sesamin may bind to LonP1's active site, with the oxygen atoms of its tetracyclic ether ring (Fig. 8.1h; shown in blue) and piperonyl group (Fig. 8.1h; shown in green) forming hydrogen bonds with Ser855 and Lys898 (Fig. 8.1h; S denotes the tetracyclic ether oxygen atom interacting with Ser855 and K + S denotes the piperonyl oxygen atom interacting with both Ser855 and Lys898) [98]. Notably, the structural symmetry of (−)-Sesamin suggests that the molecule has two putative binding interfaces for LonP1 (Fig. 8.1H; the alternative binding interface denoted with S′ for the tetracyclic ether oxygen and K′ + S′ for the piperonyl oxygen). Similar to OA, (−)-Sesamin has been shown to induce DNA double-strand breaks in H1299 cells (Table 8.1) albeit at lesser severity than OA [98], but its cytotoxicity in cancer cells has not been reported.

## 8.3 Mitochondrial ClpXP Protease Complex

### 8.3.1 Molecular Characteristics of Mitochondrial ClpXP

The mitochondrial ClpXP complex is another major AAA+ protease found in the mitochondrial matrix. ClpXP is constituted of two proteins: the 277-residue long serine protease ClpP and the 633-residue long AAA+ ATPase ClpX (Fig. 8.2a). ClpP and ClpX are encoded by the nuclear *CLPP* and *CLPX* genes, respectively. ClpP consists almost entirely of the serine proteolytic domain, with the addition of a 56-residue long MTS at its N-terminus that directs its localization to the mitochondrial matrix, which is cleaved off upon protein maturation [23]. It also has a 26-residue long C-terminal tail that modulates the ClpP-ClpX interaction and the proteolytic activity of the complex in vitro [49] (Fig. 8.2a).

ClpX consists of two domains: the N-terminal zinc-binding domain (ZBD) that facilitates the recognition and binding of substrate proteins [6, 91], followed by the AAA+ ATPase domain that mediates the oligomerization, ATP hydrolysis and protein unfoldase activity of ClpX [70, 88]. Like ClpP, a 56-residue long MTS is present at the N-terminus of ClpX.

The atomic structure of human mitochondrial ClpP has been solved at 2.1 Å resolution by X-ray crystallography (PDB: 1TG6) (Fig. 8.2b, c) [48]. The active site of ClpP carries the Ser153-His178-Asp227 catalytic triad (Fig. 8.2b) [28]. Like its bacterial counterparts, the functional state of mitochondrial ClpP is a cylindrical tetradecamer of two identical heptameric rings (Fig. 8.2c) [48]. In this configuration, axial entry pores are formed at the top and bottom ends of the ClpP tetradecamer (Fig. 8.2c, top view) that allows entry of substrates for degradation, while the generated peptidyl fragments likely exit via transient side pores (Fig. 8.2c, side view) [67]. Importantly, the 14 active sites are shielded and are exposed only within ClpP's proteolytic chamber, thereby effectively prohibiting unspecific proteolysis.

The functional form of mitochondrial ClpX is a hexamer [49], which is the most common among AAA+ ATPases [79]. Notably, the hexamerization of ClpX leads to the formation of six complete ATP-binding pockets with all the necessary conserved functional motifs (i.e. Walker A, Walker B, Sensor I, Sensor II and the Arginine finger) for ATP hydrolysis. The energy derived from ATP hydrolysis drives the allosteric movements and structural rearrangements in ClpX that are necessary for its substrate-unfolding activity [79].

To form the complete, functional ClpXP complex, one ClpX hexamer docks on each end of the ClpP tetradecamer such that the axial pores of both proteins are aligned on the same axis [49]. Importantly, interaction between ClpX and ClpP is facilitated by the docking of the essential IGF loops of ClpX at the hydrophobic binding sites on ClpP apical surface that are formed between every two ClpP subunits at the ClpX-ClpP interface. Furthermore, the docking of IGF loops at these hydrophobic pockets is a highly dynamic process that contributes significantly to stabilizing the ClpX-ClpP interaction [4].

Protein degradation by ClpXP occurs via multiple steps. The first step involves the recognition and binding of the substrate protein to ClpX. Subsequently, the bound substrate is then unfolded and threaded through the axial pore of ClpX by the coordinated power strokes of specialized pore loops, utilizing energy derived from ATP hydrolysis [70]. Notably, this is a highly processive process that translocates the unfolded substrate into ClpP's proteolytic chamber, where it is degraded into small peptidyl fragments [67]. Ultimately, these fragments are expelled from the ClpP lumen through side pores that are formed via the dynamic, allosteric rearrangements of ClpP subunits [53]. The degradation cycle can then be repeated upon the binding of a new substrate molecule.

### 8.3.2 Cellular Function of Mitochondrial ClpXP

ClpXP contributes to mitochondrial proteome maintenance by facilitating the turnover of specific substrate proteins. Loss of ClpXP function thus results in the accumulation of misfolded proteins, impairing various cellular processes. For example, ClpP KD has been shown to induce the accumulation of misfolded subunit B of Complex II (SDHB), leading to the loss of Complex II function, impaired OXPHOS and ATP production, and resulting in oxidative stress [83]. Furthermore, the loss of ClpP expression has been shown to hinder the expression of mitochondrial unfolded protein response ($UPR^{mt}$) marker proteins, increase mitochondrial fragmentation, and attenuate the normal activities of respiratory complexes in muscle cells [24]. Notably, ClpXP expression has been shown to increase in response to the accumulation of misfolded mitochondrial proteins during $UPR^{mt}$ [104], mitochondrial frataxin deficiency [40], and respiratory deficiency induced by mitochondrial DNA (mtDNA) mutagenesis [20].

In addition to maintaining the mitochondrial proteome, ClpXP provides additional modulatory functions in multiple biological processes. For example, ClpXP degrades both ALAS1 and the erythroid-specific ALAS2 as part of a negative feedback mechanism that regulates heme biosynthesis [61, 101]. ClpXP is also involved in the turnover of phosphatase and tensin homologue-induced kinase 1 (PINK1), which in turn affects the recruitment of the E3 ubiquitin ligase Parkin to the mitochondrial surface, potentially modulating the dynamics of mitophagy [39]. Furthermore, ClpXP has been shown to facilitate the turnover of the dynamin-related protein 1 (Drp1) and to modulate mitochondrial fission-fusion dynamics [24]. ClpXP has also been shown to recognize and degrade the nitric oxide-associated protein 1 (NOA1) that regulates mitochondrial translation [2]. Notably, ClpXP alters OXPHOS in response to stress or other external stimuli via the turnover of specific respiratory proteins, such as the peripheral arm subunits NDUFV1 and NDUFV2 of Complex I, to quench ROS production in depolarized mitochondria [77].

Unlike bacterial ClpXP, human mitochondrial ClpX has been shown to perform dedicated biochemical functions that are independent of ClpP. For example, ClpX functions as a molecular chaperone in the maturation of ALAS2 during erythropoiesis [50]. ClpX may also stabilize the mitochondrial transcription factor A (TFAM) to maintain proper mtDNA segregation [51]. In addition, ClpX physically interacts with the mitochondrial E3 ubiquitin-protein ligase XIAP (X-linked inhibitor of apoptosis proteins) and may potentially promote apoptosis by antagonizing XIAP from caspase inhibition in a similar manner as other known IAP-binding proteins [96]. Furthermore, ClpX alone has been shown to induce the upregulation of the transcriptional regulator C/EBP homologous protein (CHOP) during myogenesis and potentially stimulates $UPR^{mt}$ [1].

### 8.3.3 Role of Mitochondrial ClpXP in Cancer

Expression of mitochondrial ClpXP has been characterized in various types of cancer. Notably, ClpP is upregulated to various degrees in solid

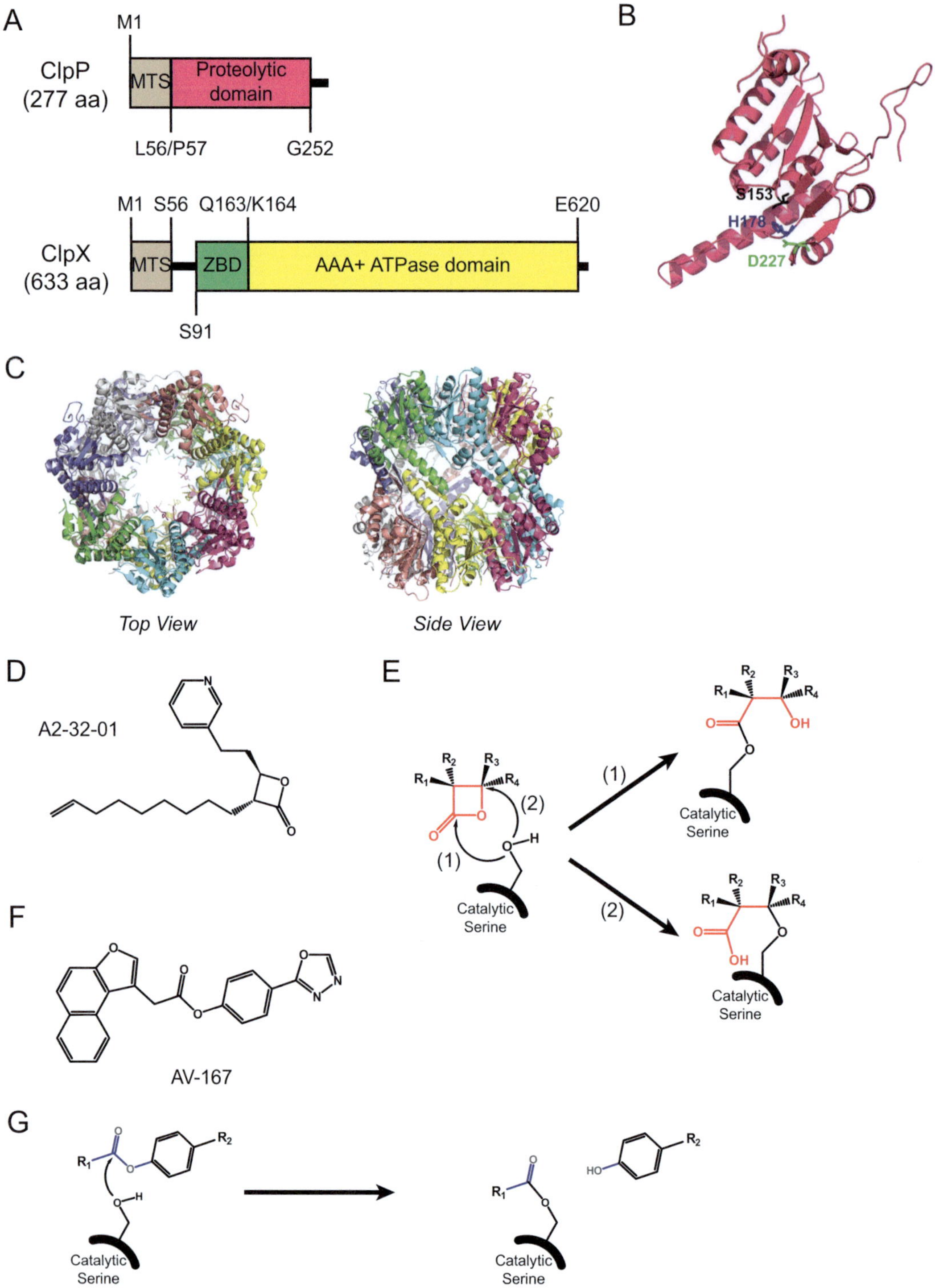


**Fig. 8.2** Structure of mitochondrial ClpXP and its inhibitors. (**a**) Domain architecture of ClpP (top) and ClpX (bottom) illustrated as bar diagrams. Boundaries were determined from curated information in UniProt or other public bioinformatics databases. *ZBD* zinc-binding domain. (**b**) Atomic structure of the ClpP subunit in the apo tetradecamer (ZBD: 1TG6) [48]. The catalytic triad is shown as sticks and coloured with Ser153 in black,

cancers of almost all major organs or tissues, including breast, ovary, prostate, uterus, bladder, lung, stomach, liver, thyroid, prostate, testis and central nervous system [20, 72, 83], as well as in non-solid cancers such as acute myeloid leukemia [19].

Despite the fact that the functional roles of ClpXP in cancer pathology are not as well characterized as that of LonP1, the currently available evidence indicates the importance of ClpXP in the proliferation and metastasis of specific cancers. For example, depletion of ClpX or ClpP suppresses the proliferation and hinders colony formation of prostate cancer cells PC3 [83]. Furthermore, PC3 cells depleted of ClpX or ClpP are compromised in their ability to invade normal tissues and metastasize when xenografted into immunocompromised mice [83]. Nevertheless, the importance of ClpXP in cancer pathology depends heavily on cell type, as illustrated in the observation that KD of ClpX or ClpP has minimal effects on the proliferation of breast cancer cells MCF7 [83].

For non-solid cancer, ClpP depletion reduces the growth and viability of multiple leukemic cell lines expressing elevated levels of the protease [19]. The loss of ClpP expression also compromises the integrity and activity of Complex II, leading to reduction in OXPHOS and increase in ROS production [19]. Interestingly, while an increase in $UPR^{mt}$ is consistently observed in leukemic cells with elevated ClpP expression, reduction in their viability as a result of ClpP KD has no observable effect on either the expression of $UPR^{mt}$ markers or the mass and morphology of the mitochondria [19], which contradicts the findings in immortalized muscle cells [24] and fibroblasts [104], highlighting a potential variation in ClpP's role in $UPR^{mt}$ among different cell types.

### 8.3.4 Chemical Inhibitors of Mitochondrial ClpXP as Potential Cancer Therapeutics

Relative to LonP1, the utilization of ClpXP as a potential drug target in cancer therapy has been explored to a lesser degree. Nevertheless, specific chemical inhibitors of ClpXP have been identified and their potential therapeutic effects against specific types of cancer have been characterized.

#### 8.3.4.1 β-Lactones

β-Lactones have previously been developed as potential antibiotics to inhibit ClpP in *Staphylococcus aureus* and suppress its pathogenesis [60, 103]. Among the analogs developed, A2-32-01 (Fig. 8.2d) has been investigated further with regard to its inhibition of human mitochondrial ClpP and its therapeutic potential against acute myeloid leukemia [19]. Although the inhibitory mechanism of A2-32-01 on mitochondrial ClpP has not been investigated in detail, the drug might covalently modify ClpP's catalytic Ser residue via one of two possible reactions (Fig. 8.2e) that have been proposed for β-lactones in general [11].

In vivo, A2-32-01 induces cytotoxicity in OCI-AML2, TEX and the chronic myeloid leukemic K562 cells with $IC_{50}$ at ~25 μM but has no effect on the acute promyelocytic leukemic HL-60 cells (Table 8.1) [19]. Notably, OCI-AML2, TEX and K562 cells all show highly elevated ClpP expression compared to HL-60 [19]. Furthermore, A2-32-01 induces cytotoxicity in osteosarcoma cells 143B ($IC_{50}$ at ~25 μM) but not in the mitochondria-depleted 143B Rho (0) cells (Table 8.1) [19]. The combined data suggests that cellular sensitivity to A2-32-01 originates from the mitochondria.

**Fig. 8.2** (continued) His178 in blue and Asp227 in green. (**c**) Top view (left) and side view (right) of the human apo ClpP tetradecamer [100]. Individual ClpP subunits are distinguished by different colours. (**d**) Chemical structure of A2-32-01. (**e**) Proposed general chemical mechanism in the covalent modification of ClpP's catalytic Ser residue by a β-lactone. The essential β-lactone ring is highlighted in red. $R_1$, $R_2$, $R_3$ and $R_4$ represent structural elements that are unique to individual compounds and do not participate in the reaction. (**f**) Chemical structure of AV-167. (**g**) Proposed general chemical mechanism in the covalent modification of ClpP's catalytic Ser residue by a phenyl ester. The essential ester group is highlighted in blue. $R_1$ and $R_2$ represent structural elements that are unique to individual compounds and that do not participate in the reaction

To further establish the therapeutic potential of A2-32-01, treating human primary leukemic cells expressing high levels of ClpP with the drug induces cytotoxicity, with $IC_{50}$ ranging from ~25 to ~125 μM (Table 8.1) [19]. The specific targeting of A2-32-01 against leukemic cells is confirmed as the drug has minimal to no effect on the viability of normal primary hematopoietic cells. Furthermore, A2-32-01 has been shown to suppress the proliferation of OCI-AML2 and primary human leukemic xenografts in immunodeficient mice with no observable toxic effects on the animals (Table 8.1).

#### 8.3.4.2 Phenyl Esters

Phenyl esters are a new class of ClpP inhibitors that are more potent and chemically stable than β-lactones [42]. They were originally developed as potential antibiotics against *S. aureus*, but, in a counter screen, AV-167 (Fig. 8.2f) was found to inhibit human ClpP from degrading fluorogenic peptidyl substrates [42]. Subsequent mass spectrometric analysis on the covalent modification of the catalytic Ser residue of SaClpP confirmed that AV-167 shares a common reaction mechanism with other active phenyl ester analogs, in which the catalytic serine's hydroxyl oxygen performs a nucleophilic attack on the carboxyl carbon of the drug and displaces the phenyl moiety that gets converted to an equivalent phenol moiety in the process (Fig. 8.2g) [42]. Data on the cytotoxicity of AV-167 on human cells is currently unavailable.

### 8.3.5 Activators of ClpP

The architecture of the ClpXP complex that segregates its ATP-dependent protein unfolding activity to ClpX and its proteolytic activity to ClpP provides a unique window for chemical interference via disabling the regulatory function of ClpX. Indeed, small molecules such as acyldepsipeptides (ADEPs) [54], activators of ClpP protease (ACPs) [64] and related compounds [14] have been identified that activate and dysregulate the protease activity of ClpP via physical displacement of ClpX, while keeping ClpP in its activated form. Importantly, dysregulation of ClpP by ADEP has been shown to be a promising venue for antibiotics development, with these molecules showing potent bactericidal activities against various pathogenic bacteria, including *S. aureus*, *Streptococcus pneumoniae*, *Neisseria meningitidis*, *Neisseria gonorrhoeae*, *Bacillus subtilis*, *Enterococcus faecalis* and *Listeria innocua* [35, 36].

Recently, ADEP analogs that can specifically target the human mitochondrial ClpP have been shown to induce apoptotic cell death in immortalized human cancer cell lines with high potency [100], thus paving the way to the exploration of using ADEPs as novel anti-cancer drugs.

#### 8.3.5.1 General Molecular Mechanism of ADEP-Based ClpP Dysregulation

The molecular mechanism for ClpP's activation by ADEP has been characterized in detail. Binding of ADEP occurs at the same hydrophobic pocket between ClpP subunits that normally accommodate the docking of ClpX's IGF loops [62]. The higher affinity of ADEP allows it to effectively out-compete the IGF loops for ClpP-binding [4]. Importantly, the docking and occupancy of ClpP's hydrophobic pockets by ADEP disrupts the dynamic docking of IGF loops that is essential in maintaining ClpX-ClpP interaction, resulting in the rapid dissociation of ClpX from ClpP [4]. Furthermore, ADEP-binding simulates the structural effects of ClpX-ClpP interaction, which forces ClpP to remain in its active state that is characterized by a widened axial pore [62] and structuring of its pore loops [66]. This in turn enables the unregulated access of peptides, molten globules and even folded proteins into ClpP's proteolytic chamber, where they are degraded in an unspecific manner [82].

#### 8.3.5.2 Dysregulation of ClpP by ADEP Causes Cell Death via Intrinsic Apoptosis

ADEP analogs with high specificity for human mitochondrial ClpP have been identified and characterized. Among them, ADEP-28 (Fig. 8.3a) and ADEP-41 (Fig. 8.3b) were found to be potent activators of human ClpP's peptidase and protease activities against fluorogenic model substrates [100]. The presence of ADEP at low concentrations is sufficient in dissociating ClpX from ClpP. Notably, both ADEP-28 and ADEP-41 induce cytotoxicity in ClpP-expressing HEK293 T-REx cells with $IC_{50}$ measured at 0.36 μM and 0.49 μM, respectively (Table 8.1) [100]. Importantly, ADEP-induced cytotoxicity is dependent on ClpP expression, as the drug has minimal to no observable effect on HEK293 T-REx *CLPP*$^{-/-}$ cells even with high doses of the drug (Table 8.1) [100]. Furthermore, the severity of ADEP-induced cytotoxicity increases with respect to an increase in intracellular ClpP level.

In support of ADEP's anti-cancer potential, ADEP-41 was shown to induce cytotoxicity in various cancer-derived cell lines, including HeLa (endocervical carcinoma), U2OS (osteosarcoma) and undifferentiated SH-SY5Y (neuroblastoma), with $IC_{50}$ ranging from 0.48 to 0.86 μM (Table 8.1) [100]. Analysis of the underlying cell death mechanism revealed that treatment of cells with ADEP activates the intrinsic, caspase-dependent apoptotic pathway, with the ADEP-treated cells showing characteristic phenotypes such as increase in DNA strand breaks, mitochondrial fragmentation, loss of OXPHOS and the proteolytic activation of caspase-9 and caspase-3 [100].

#### 8.3.5.3 Structural Evidence for a Highly Dynamic ADEP-ClpP Complex

To further understand the molecular mechanism underlying ClpP's activation by ADEP, atomic structure of the mitochondrial ClpP in complex with ADEP-28 was solved at 2.8 Å resolution [100]. ADEP-28 binds to a highly complementary hydrophobic pocket that forms between two adjacent ClpP subunits (Fig. 8.3c) in the same manner as observed in other ADEP-bound ClpP. Notably, binding of ADEP-28 induces the formation of structured axial loops at the entrance pore (Fig. 8.3c). Furthermore, as with other ADEP-ClpP interaction, the binding of ADEP-28 to human ClpP widens the axial pore (Fig. 8.3d) so to enable the unregulated access of protein substrates, resulting in unspecific proteolysis [100].

Surprisingly, the ADEP-28-bound human ClpP oligomer adopts a compact conformation (Fig. 8.3e) that can be visualized in the rotation of one of its heptameric rings relative to the apposing one (Fig. 8.3e; indicated with tilt in the purple dotted line), resulting in the compaction of ClpP as indicated by the distance between the two rings of catalytic Ser residues being reduced by ~7.8 Å (Fig. 8.3e; upper and lower Ser rings coloured in red and blue, respectively). This provides the first structural evidence that ADEP-bound ClpP is in fact a dynamic structure that adopts multiple conformations over its proteolytic cycle, contrary to a previously proposed model in which ADEP locks ClpP into its active state with the extended conformation [32].

## 8.4 i-AAA and m-AAA Metalloproteases

The compartmentalization of mitochondria and the differential localization of mitochondrial proteins to these compartments give rise to the need for additional specialized AAA+ proteases for the clearance of proteins that are misfolded locally and that are not accessible by the matrix-localized LonP1 and ClpXP. In this regard, the membrane-bound i-AAA and m-AAA metalloproteases and paraplegin are responsible of maintaining the protein homeostasis of mitochondrial membranes and the inter-membrane space (IMS) in addition to the mitochondrial matrix. Other important functions of these proteases have also been identified.

### 8.4.1 i-AAA Metalloprotease

#### 8.4.1.1 Molecular Characteristics of i-AAA

i-AAA is a 733-residue long integral membrane metalloprotease encoded by the nuclear *YME1L1* gene and is found in the inner membrane of mitochondria (IMM) [87]. It consists of a N-terminal domain carrying a cleavable MTS for localization to the IMM, followed by a single trans-

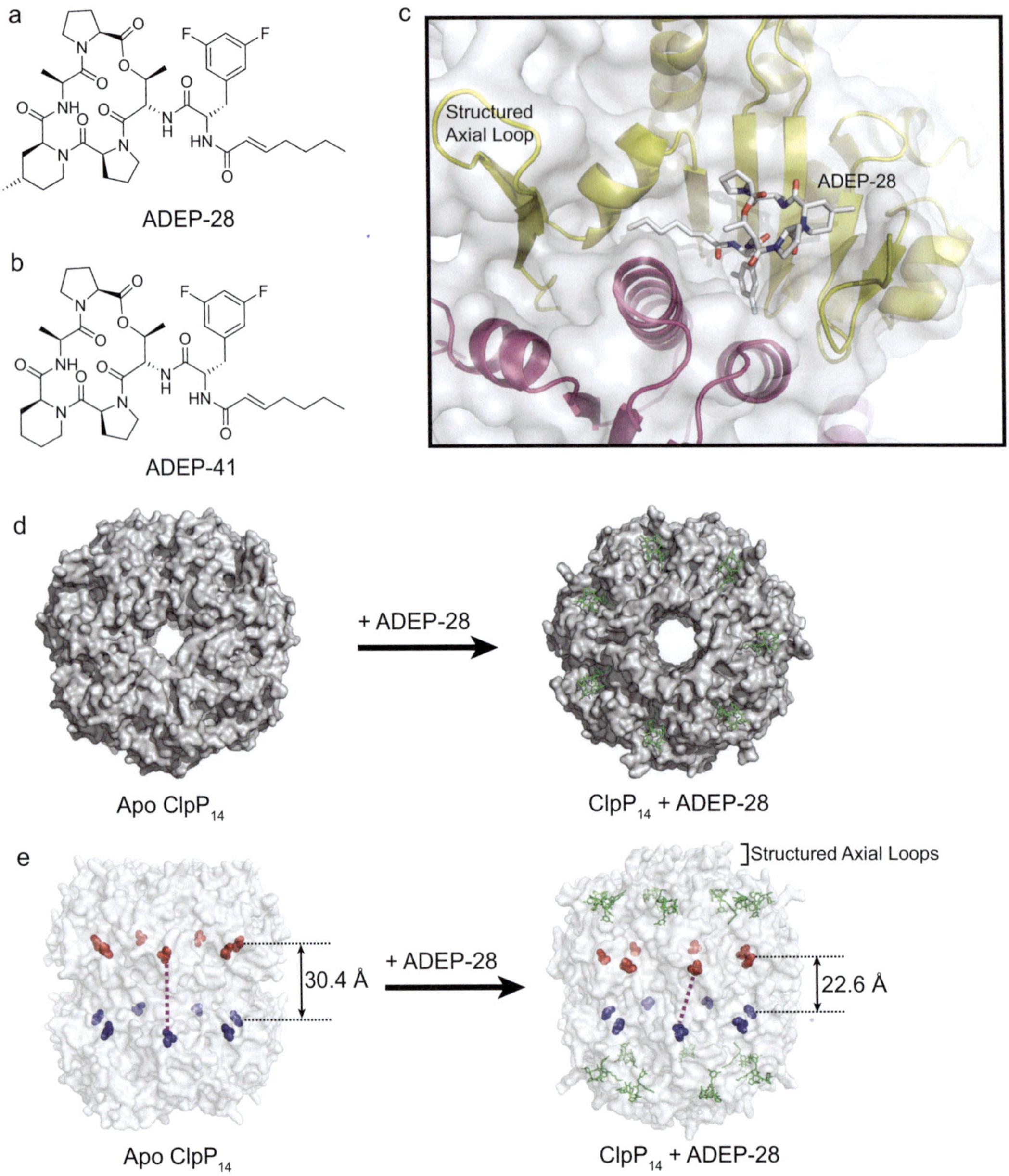


**Fig. 8.3** Structural information of the activation of mitochondrial ClpP by ADEP. (**a**) Chemical structure of ADEP-28. (**b**) Chemical structure of ADEP-41. (**c**) ADEP-binding pocket in the atomic structure of human mitochondrial ClpP in complex with ADEP-28 (PDB: 6BBA) [100]. The bound ADEP-28 is shown as sticks and coloured by elements. The two ClpP subunits forming the ADEP-binding pocket are illustrated as ribbons with the structured axial loop of one subunit denoted. The molecular surface of the ClpP oligomer is used to highlight the contours of the ADEP-binding pocket. (**d**) Atomic structures of apo tetradecameric ClpP and ADEP-28-bound tetradecameric ClpP, showing the enlargement of ClpP's axial pore caused by ADEP-28-binding. ADEP-28 is shown as sticks and coloured in green. (**e**) Compaction of tetradecameric ClpP upon ADEP-28 binding is illustrated by the distance between two catalytic Ser residues from apposing rings (shown as spheres; top ring coloured in red and bottom ring coloured in blue). Rotation of the top ring relative to the bottom ring is illustrated as the tilting of the purple, dotted line linking the two foremost Ser residues. ADEP-28 is shown as sticks and coloured in green. Structured axial pores of ClpP that are present in ADEP-28-bound form are as indicated

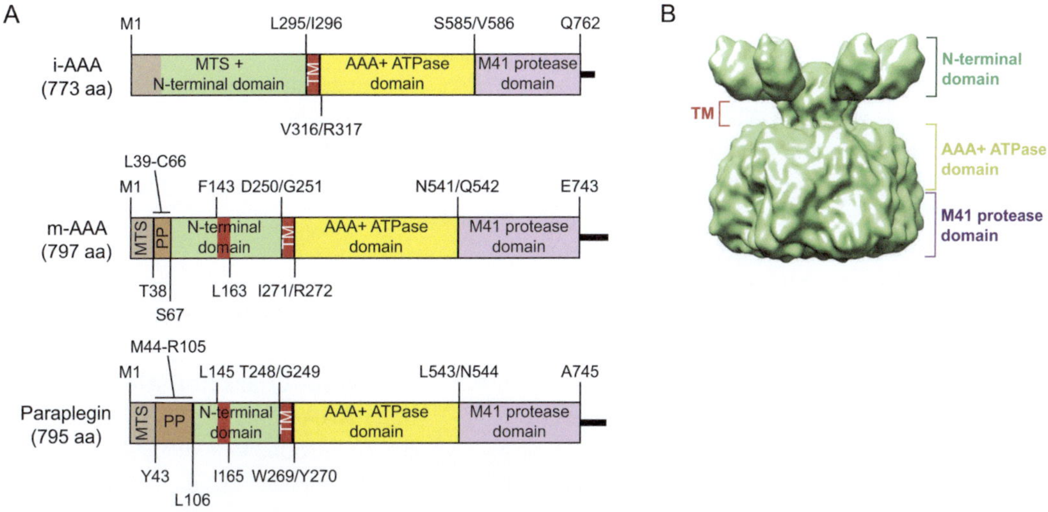


**Fig. 8.4** Domain arrangement and structures of i-AAA, m-AAA and paraplegin. (**a**) Domain arrangement of i-AAA (top), m-AAA (middle) and paraplegin (bottom) are illustrated as bar diagrams. All domain boundaries were determined using curated information from UniProt or other public bioinformatics databases. For i-AAA, the boundary between the MTS and N-terminal domain is not well defined and thus is not explicitly shown. For m-AAA and paraplegin, the trans-membrane domain located within the N-terminal domain is indicated in the same red colour as the trans-membrane domain preceding the AAA+ ATPase domain. *PP* propeptide, *TM* trans-membrane domain. (**b**) Cryo-EM structure of the yeast Yta10/12 hetero-oligomer. The location of each functional domain is as indicated

membrane (TM) domain, an AAA+ ATPase domain, and a C-terminal proteolytic domain (Fig. 8.4a), which bears resemblance to LonP1's domain architecture (Fig. 8.1a). The N-terminal domain of i-AAA faces the mitochondrial matrix; its sequence is poorly conserved but is likely to mediate protein-protein interactions [65]. The singular TM domain forms the amphiphilic α-helical barrel upon oligomerization of i-AAA and anchor the protease to the IMM [65]. The AAA+ ATPase domain facilitates the hexamerization of i-AAA and provides the sites of ATP binding and hydrolysis [65]. The proteolytic domain of i-AAA belongs to the M41 metalloprotease family and has an active site with the sequence TAYHESGHAI (residues 596–615; essential residues are underlined) that coordinately binds a single divalent cation (usually $Zn^{2+}$), unlike LonP1 and ClpP that are serine proteases. Notably, HEXXH is a highly conserved sequence motif among M41 metalloproteases, in which the two His residues bind $Zn^{2+}$ and the Glu residue is the proteolytic residue [65]. Both the AAA+ ATPase and proteolytic domains of i-AAA face the IMS [65].

#### 8.4.1.2 Cellular Function of i-AAA

i-AAA generally functions in the proteolytic clearance of specific proteins that are misfolded in the IMM and IMS. Recognition by i-AAA is mediated by N-terminal or C-terminal degrons on the substrate that become exposed as the substrate protein misfolds [84]. This in turn allows i-AAA to engage in the active unfolding and degradation of these substrates in a processive manner [84]. Notably, in the degradation of integral membrane substrates, the energy required for extracting their TM domains is believed to be supplied by ATP hydrolysis, although the exact mechanism involved remains unclear [34].

In addition, i-AAA performs important regulatory functions via complete or partial proteolysis of specific protein targets in response to environmental stimuli. Notably, i-AAA participates in reciprocal proteolysis with the integral IMM metalloprotease OMA1 under stress conditions causing mitochondrial depolarization. Specifically, i-AAA degrades OMA1 when the cellular ATP supply is ample, whereas OMA1 degrades i-AAA when the ATP pool is depleted [80]. Importantly, the mutual proteolytic regulation of i-AAA and OMA1 in turn modulates the differential processing of the integral IMM dynamin-like GTPase OPA1. The processing of OPA1 by i-AAA results in the GTPase's activation that promotes maintenance of normal mitochondrial morphology and protein homeostasis, while processing by OMA1 yields to an inactive form of OPA1 that promotes mitochondrial fragmentation instead [80].

Other cellular functions of i-AAA have also been identified. For example, i-AAA was shown to have a direct impact on the lamellar morphology of mitochondrial cristae and the turnover rates of the subunits of Complex I (NDUFB6 and ND1) and Complex IV (COX4) [87]. KD of i-AAA thus results in accumulation of these respiratory subunits that remain unassembled, loss of Complex I activity, increase in oxidative stress protein carbonylation and increased mitochondrial fragmentation, leading to reduced cellular proliferation and sensitization to apoptotic inducing reagents [87]. Furthermore, i-AAA modulates apoptosis by mediating the turnover of PRELID1 (protein of relevant evolutionary and lymphoid interest 1) [43], which facilitates the accumulation of cardiolipin in mitochondrial membranes and suppresses apoptotic onset [76].

#### 8.4.1.3 i-AAA Expression Is Linked to Cancer Progression

Compared to LonP1 and ClpP, the link between i-AAA and cancer is not as well established. The limited available data nevertheless suggests that it may suppress cancer cell growth. For example, the expression of i-AAA has been shown to be downregulated by the proto-oncogene product c-Myc in rat pheochromocytoma cells PC12 [41]. On the other hand, i-AAA overexpression has been shown to suppress the growth of human hepatocellular carcinoma cells SMMC7721 [97]. Furthermore, *YME1L1* is among the genes showing consistent genetic alterations in primary human glioma cells that are associated with poor prognosis [12].

Presently, there are no published data on the development of chemical reagents that can specifically target and modulate i-AAA, or on utilizing i-AAA as a drug target in cancer therapeutics.

### 8.4.2 m-AAA Metalloprotease

#### 8.4.2.1 Molecular Characteristics of m-AAA

m-AAA is a 797-residue long integral membrane metalloprotease encoded by the *AFG3L2* gene that preferentially localizes to the inner boundary membrane (versus the cristae membrane) of IMM [89]. It shares the same domain architecture as i-AAA (Fig. 8.4a), but with the addition of a propeptide (PP) located between the MTS and the N-terminal domain that is removed upon maturation of the protease, and a second TM region that is located within the N-terminal domain (Fig. 8.4a; TM represented as red bars). Notably, the presence of two TM segments contributes to m-AAA having the opposite topology compared to i-AAA, with the N-terminal domain of m-AAA facing the IMS and its AAA+ and proteolytic domains facing the mitochondrial matrix. As with i-AAA, the formation of m-AAA hexamers is mediated by its AAA+ ATPase domain, and its proteolytic activity is contributed by its M41 metalloprotease domain with the active site sequence VAY<u>H</u><u>E</u>AG<u>H</u>AV (residues 571–580; essential residues are underlined) for $Zn^{2+}$-binding and proteolytic catalysis, as discussed previously in Sect. 8.4.1.1.

m-AAA shows normal proteolytic activity as a homo-oligomer or as a hetero-oligomer when in complex with the architecturally similar paraplegin (Fig. 8.4a), a homologue of m-AAA encoded by the nuclear *SPG7* gene that is co-localized to IMM and shares the same topology [25]. Notably, the interaction of m-AAA with paraplegin has been proposed to be important for modulating its substrate specificity [58]. This is further supported by the observation that in hereditary spastic paraplegia (HSP7), paraplegin deficiency causes axonal degradation in neurons, despite m-AAA homo-oligomers being a functional substitute for m-AAA-paraplegin hetero-oligomers in enzymatic activities [58].

Structural analyses on full-length or individual domains of m-AAA and paraplegin homologues from various species have revealed collectively that these proteins share a highly similar structure, as represented by the yeast m-AAA hetero-oligomer Yta10/12 [63] (Fig. 8.4b). Notably, docking analysis using the atomic structures of domains of m-AAA and paraplegin homologues from human, yeast and *Thermatoga maritima* has shown highly complementary spatial fittings of these domains to Yta10/12 [34].

#### 8.4.2.2 Cellular Function of m-AAA

Both m-AAA and paraplegin are important in maintaining normal cellular respiration by mediating the processing and maturation of various respiratory proteins. Depletion of these proteases leads to various respiratory defects and gives rise to deleterious cellular phenotypes. For example, loss of m-AAA and paraplegin has been shown to impair the proper assembly of Complex I and to increase cellular sensitivity to oxidative stress in human fibroblasts [5]. Similarly, expression of catalytically inactive mutants of m-AAA in respiratory-deficient yeast cells results in impaired OXPHOS due to the reduced activity and expression of cytochrome *c* oxidase subunits, highlighting the role of m-AAA in Complex IV maturation [25].

Notably, m-AAA performs vital functions that are indispensable in neuronal cells. For example, m-AAA is essential in maintaining normal mitochondrial structure and in facilitating proper mitochondrial ribosome assembly in Purkinje cells (large neurons found in the cortex of the cerebellum) [3]. Furthermore, m-AAA also suppresses the hyperphosphorylation of the tau protein and facilitates normal anterograde mitochondrial transport in mouse cortical neurons [56]. Importantly, m-AAA mediates the turnover of the essential subunit of the mitochondrial $Ca^{2+}$ uniporter (MCU), EMRE [57]. This process counters the activity of the MAIP1 complex that promotes EMRE biogenesis and ensures the efficient assembly of MCU with gate-keeping subunits, thereby preventing mitochondrial $Ca^{2+}$ overload and the accelerated opening of the mitochondrial permeability transition pore (MPTP), preserving neuronal cell viability [57].

While mutations in m-AAA and paraplegin that give rise to spinocerebellar ataxia (SCA28) and hereditary spastic paraplegia (HSP7), respectively, have been well-documented and the underlying pathological mechanisms have been characterized [25, 75], there are no current reports that link either m-AAA or paraplegin to cancer.

## 8.5 Concluding Remarks

The studies discussed here have provided new insights for deciphering the molecular mechanisms by which mitochondrial AAA+ proteases contribute to cancer pathology. These insights are invaluable for the continuous development of new cancer therapeutic strategies. Importantly, the structural characterization of AAA+ proteases in complex with their respective activity-modulating compounds has revealed key allosteric events to provide the essential structural basis for improving the potency and specificity of these compounds and for developing new compounds that will produce the same biomolecular effects in the AAA+ proteases via alternative mechanisms. Furthermore, the characterization of cellular events induced by these compounds leading up to apoptosis has revealed distinct pharmacological vulnerabilities in cancer cells that can be exploited via the synergistic use of multiple drugs, or by creating hybrid molecules that combine the pharmacological characteristics of the originals. Hence, continued efforts at understanding the structure and function of these AAA+ proteases are critical to the development of novel anticancer compounds.

**Acknowledgments** We thank Mr. Vaibhav Bhandari and Mr. Kamran Rizzolo of the Houry group for their helpful suggestions in the writing of this article. This was funded by Natural Sciences and Engineering Research Council of Canada grant (RGPIN 2014-05393) to WAH.

## References


1. Al-Furoukh N, Ianni A, Nolte H, Holper S, Kruger M, Wanrooij S, Braun T (2015) ClpX stimulates the mitochondrial unfolded protein response (UPRmt) in mammalian cells. Biochim Biophys Acta 1853:2580–2591. doi:S0167-4889(15)00206-2 [pii]. https://doi.org/10.1016/j.bbamcr.2015.06.016
2. Al-Furoukh N, Kardon JR, Kruger M, Szibor M, Baker TA, Braun T (2014) NOA1, a novel ClpXP substrate, takes an unexpected nuclear detour prior to mitochondrial import. PLoS One 9:e103141. https://doi.org/10.1371/journal. pone.0103141 [doi] PONE-D-14-12184 [pii]
3. Almajan ER et al (2012) AFG3L2 supports mitochondrial protein synthesis and Purkinje cell survival. J Clin Invest 122:4048–4058. :64604 [pii]. https://doi.org/10.1172/JCI64604
4. Amor AJ, Schmitz KR, Sello JK, Baker TA, Sauer RT (2016) Highly dynamic interactions maintain kinetic stability of the ClpXP protease during the ATP-fueled mechanical cycle. ACS Chem Biol 11:1552–1560. https://doi.org/10.1021/acschembio.6b00083
5. Atorino L et al (2003) Loss of m-AAA protease in mitochondria causes complex I deficiency and increased sensitivity to oxidative stress in hereditary spastic paraplegia. J Cell Biol 163:777–787. https://doi.org/10.1083/jcb.200304112. [doi] jcb.200304112 [pii]
6. Banecki B, Wawrzynow A, Puzewicz J, Georgopoulos C, Zylicz M (2001) Structure-function analysis of the zinc-binding region of the Clpx molecular chaperone. J Biol Chem 276:18843–18848. https://doi.org/10.1074/jbc.M007507200. doi] M007507200 [pii
7. Bernstein SH et al (2012) The mitochondrial ATP-dependent Lon protease: a novel target in lymphoma death mediated by the synthetic triterpenoid CDDO and its derivatives. Blood 119:3321–3329. blood-2011-02-340075 [pii] 10.1182/blood-2011-02-340075
8. Bezawork-Geleta A, Brodie EJ, Dougan DA, Truscott KN (2015) LON is the master protease that protects against protein aggregation in human mitochondria through direct degradation of misfolded proteins. Sci Rep 5:17397. srep17397 [pii]. https://doi.org/10.1038/srep17397
9. Bezawork-Geleta A, Saiyed T, Dougan DA, Truscott KN (2014) Mitochondrial matrix proteostasis is linked to hereditary paraganglioma: LON-mediated turnover of the human flavinylation factor SDH5 is regulated by its interaction with SDHA. FASEB J 28:1794–1804. :fj.13-242420 [pii]. https://doi.org/10.1096/fj.13-242420
10. Bota DA, Davies KJ (2002) Lon protease preferentially degrades oxidized mitochondrial aconitase by an ATP-stimulated mechanism. Nat Cell Biol 4:674–680. https://doi.org/10.1038/ncb836. doi] ncb836 [pii
11. Bottcher T, Sieber SA (2012) β-Lactams and β-lactones as activity-based probes in chemical biology. MedChemComm 3:408–417. https://doi.org/10.1039/c2md00275b
12. Bredel M et al (2009) A network model of a cooperative genetic landscape in brain tumors. JAMA 302:261–275. :302/3/261 [pii]. https://doi.org/10.1001/jama.2009.997
13. Bulteau AL et al (2017) Dysfunction of mitochondrial Lon protease and identification of oxidized protein in mouse brain following exposure to MPTP: implications for Parkinson disease free Radic. Biol Med 108:236–246. S0891-5849(17)30188-0 [pii]. https://doi.org/10.1016/j.freeradbiomed.2017.03.036
14. Carney DW, Compton CL, Schmitz KR, Stevens JP, Sauer RT, Sello JK (2014) A simple fragment of cyclic acyldepsipeptides is necessary and sufficient for ClpP activation and antibacterial activity. Chembiochem 15:2216–2220. https://doi.org/10.1002/cbic.201402358
15. Cheng CW et al (2013) Overexpression of Lon contributes to survival and aggressive phenotype of cancer cells through mitochondrial complex I-mediated generation of reactive oxygen species. Cell Death Dis 4:e681. cddis2013204 [pii]. https://doi.org/10.1038/cddis.2013.204
16. Cheng KC, Hsueh MC, Chang HC, Lee AY, Wang HM, Chen CY (2010) Antioxidants from the leaves of Cinnamomum kotoense. Nat Prod Commun 5:911–912
17. Chin MP et al (2014) Risk factors for heart failure in patients with type 2 diabetes mellitus and stage 4 chronic kidney disease treated with bardoxolone methyl. J Card Fail 20:953–958. S1071-9164(14)01224-X [pii]. https://doi.org/10.1016/j.cardfail.2014.10.001
18. Cleasby A et al (2014) Structure of the BTB domain of Keap1 and its interaction with the triterpenoid antagonist CDDO. PLoS One 9:e98896. https://doi.org/10.1371/journal.pone.0098896. PONE-D-14-06787 [pii]
19. Cole A et al (2015) Inhibition of the mitochondrial protease ClpP as a therapeutic strategy for human acute myeloid Leukemia. Cancer Cell 27:864–876. S1535-6108(15)00180-4 [pii]. https://doi.org/10.1016/j.ccell.2015.05.004
20. Cormio A, Musicco C, Gasparre G, Cormio G, Pesce V, Sardanelli AM, Gadaleta MN (2017) Increase in proteins involved in mitochondrial fission, mitophagy, proteolysis and antioxidant response in type I endometrial cancer as an adaptive response to respiratory complex I deficiency. Biochem Biophys Res Commun 491:85–90. S0006-291X(17)31389-X [pii]. https://doi.org/10.1016/j.bbrc.2017.07.047
21. Creelan BC et al (2017) Safety, pharmacokinetics, and pharmacodynamics of oral omaveloxolone (RTA 408), a synthetic triterpenoid, in a first-in-human trial of patients with advanced solid tumors. Onco Targets Ther 10:4239–4250. https://doi.org/10.2147/OTT.S136992. ott-10-4239 [pii]

22. Crewe C, Schafer C, Lee I, Kinter M, Szweda LI (2017) Regulation of pyruvate dehydrogenase kinase 4 in the heart through degradation by the Lon protease in response to mitochondrial substrate availability. J Biol Chem 292:305–312. M116.754127 [pii]. https://doi.org/10.1074/jbc.M116.754127
23. de Sagarra MR, Mayo I, Marco S, Rodriguez-Vilarino S, Oliva J, Carrascosa JL, Casta n JG (1999) Mitochondrial localization and oligomeric structure of HClpP, the human homologue of E. coli ClpP. J Mol Biol 292:819–825. https://doi.org/10.1006/jmbi.1999.3121. S0022283699931212 [pii
24. Deepa SS et al (2016) Down-regulation of the mitochondrial matrix peptidase ClpP in muscle cells causes mitochondrial dysfunction and decreases cell proliferation free Radic. Biol Med 91:281–292. S0891-5849(15)01174-0 [pii]. https://doi.org/10.1016/j.freeradbiomed.2015.12.021
25. Di Bella D et al (2010) Mutations in the mitochondrial protease gene AFG3L2 cause dominant hereditary ataxia SCA28. Nat Genet 42:313
26. Di K, Lomeli N, Wood SD, Vanderwal CD, Bota DA (2016) Mitochondrial Lon is over-expressed in high-grade gliomas, and mediates hypoxic adaptation: potential role of Lon as a therapeutic target in glioma. Oncotarget 7:77457–77467. 12681 [pii]. https://doi.org/10.18632/oncotarget.12681
27. Dikoglu E et al (2015) Mutations in LONP1, a mitochondrial matrix protease, cause CODAS syndrome. Am J Med Genet A 167:1501–1509. https://doi.org/10.1002/ajmg.a.37029
28. Ekici OD, Paetzel M, Dalbey RE (2008) Unconventional serine proteases: variations on the catalytic Ser/His/Asp triad configuration. Protein Sci 17:2023–2037. ps.035436.108 [pii]. https://doi.org/10.1110/ps.035436.108
29. Fiorese CJ, Schulz AM, Lin YF, Rosin N, Pellegrino MW, Haynes CM (2016) The transcription factor ATF5 mediates a mammalian mitochondrial UPR. Curr Biol 26:2037–2043. S0960-9822(16)30614-5 [pii]. https://doi.org/10.1016/j.cub.2016.06.002
30. Fukuda R, Zhang H, Kim JW, Shimoda L, Dang CV, Semenza GL (2007) HIF-1 regulates cytochrome oxidase subunits to optimize efficiency of respiration in hypoxic cells. Cell 129:111–122. S0092-8674(07)00307-8 [pii]. https://doi.org/10.1016/j.cell.2007.01.047
31. Garcia-Nafria J et al (2010) Structure of the catalytic domain of the human mitochondrial Lon protease: proposed relation of oligomer formation and activity. Protein Sci 19:987–999. https://doi.org/10.1002/pro.376
32. Gersch M et al (2015) AAA+ chaperones and acyldepsipeptides activate the ClpP protease via conformational control. Nat Commun 6:6320. doi:ncomms7320 [pii]. https://doi.org/10.1038/ncomms7320
33. Gibellini L et al (2015) Inhibition of Lon protease by triterpenoids alters mitochondria and is associated to cell death in human cancer cells. Oncotarget 6:25466–25483. 4510 [pii]. https://doi.org/10.18632/oncotarget.4510
34. Glynn SE (2017) Multifunctional mitochondrial AAA proteases. Front Mol Biosci 4:34. https://doi.org/10.3389/fmolb.2017.00034
35. Goodreid JD et al (2016) Development and characterization of potent cyclic Acyldepsipeptide analogues with increased antimicrobial activity. J Med Chem 59:624–646. https://doi.org/10.1021/acs.jmedchem.5b01451
36. Goodreid JD et al (2014) Total synthesis and antibacterial testing of the A54556 cyclic acyldepsipeptides isolated from Streptomyces hawaiiensis. J Nat Prod 77:2170–2181. https://doi.org/10.1021/np500158q
37. Goto M et al (2014) Adaptation of leukemia cells to hypoxic condition through switching the energy metabolism or avoiding the oxidative stress. BMC Cancer 14:76. 1471-2407-14-76 [pii]. https://doi.org/10.1186/1471-2407-14-76
38. Granot Z et al (2007) Turnover of mitochondrial steroidogenic acute regulatory (StAR) protein by Lon protease: the unexpected effect of proteasome inhibitors. Mol Endocrinol 21:2164–2177. me.2005-0458 [pii]. https://doi.org/10.1210/me.2005-0458
39. Greene AW et al (2012) Mitochondrial processing peptidase regulates PINK1 processing, import and Parkin recruitment. EMBO Rep 13:378–385embor201214 [pii]. https://doi.org/10.1038/embor.2012.14
40. Guillon B et al (2009) Frataxin deficiency causes upregulation of mitochondrial Lon and ClpP proteases and severe loss of mitochondrial Fe-S proteins. FEBS J 276:1036–1047. EJB6847 [pii]. https://doi.org/10.1111/j.1742-4658.2008.06847.x
41. Guo QM et al (2000) Identification of c-myc responsive genes using rat cDNA microarray. Cancer Res 60:5922–5928
42. Hackl MW et al (2015) Phenyl esters are potent inhibitors of Caseinolytic protease P and reveal a Stereogenic switch for Deoligomerization. J Am Chem Soc 137:8475–8483. https://doi.org/10.1021/jacs.5b03084
43. Hartmann B et al (2016) Homozygous YME1L1 mutation causes mitochondriopathy with optic atrophy and mitochondrial network fragmentation. Elife:5. https://doi.org/10.7554/eLife.16078
44. Hong DS et al (2012) A phase I first-in-human trial of bardoxolone methyl in patients with advanced solid tumors and lymphomas. Clin Cancer Res 18:3396–3406. 1078-0432.CCR-11-2703 [pii]. https://doi.org/10.1158/1078-0432.CCR-11-2703
45. Huang Z et al (2018) Glutathione S-transferase pi-Activatable O(2)-(Sulfonylethyl derived) Diazeniumdiolates potently suppress melanoma in vitro and in vivo. J Med Chem 61:1833–1844. https://doi.org/10.1021/acs.jmedchem.7b01178
46. Jenkinson EM et al (2013) Perrault syndrome is caused by recessive mutations in CLPP, encoding a mitochondrial ATP-dependent chambered protease. Am J Hum Genet 92:605–613. S0002-9297(13)00108-0 [pii]. https://doi.org/10.1016/j.ajhg.2013.02.013

47. Kang F, Ai Y, Zhang Y, Huang Z (2018) Design and synthesis of new hybrids from 2-cyano-3,12-dioxooleana- 9-dien-28-oic acid and O(2)-(2,4-dinitrophenyl) diazeniumdiolate for intervention of drug-resistant lung cancer. Eur J Med Chem 149:269–280. S0223-5234(18)30199-5 [pii]. https://doi.org/10.1016/j.ejmech.2018.02.062
48. Kang SG, Maurizi MR, Thompson M, Mueser T, Ahvazi B (2004) Crystallography and mutagenesis point to an essential role for the N-terminus of human mitochondrial ClpP. J Struct Biol 148:338–352. S1047-8477(04)00146-7 [pii]. https://doi.org/10.1016/j.jsb.2004.07.004
49. Kang SG, Ortega J, Singh SK, Wang N, Huang NN, Steven AC, Maurizi MR (2002) Functional proteolytic complexes of the human mitochondrial ATP-dependent protease, hClpXP. J Biol Chem 277:21095–21102. https://doi.org/10.1074/jbc.M201642200. doi] M201642200 [pii
50. Kardon JR et al (2015) Mitochondrial ClpX activates a key enzyme for heme biosynthesis and erythropoiesis. Cell 161:858–867. S0092-8674(15)00434-1 [pii]. https://doi.org/10.1016/j.cell.2015.04.017
51. Kasashima K, Sumitani M, Endo H (2012) Maintenance of mitochondrial genome distribution by mitochondrial AAA+ protein ClpX. Exp Cell Res 318:2335–2343. S0014-4827(12)00343-6 [pii]. https://doi.org/10.1016/j.yexcr.2012.07.012
52. Kereiche S et al (2016) The N-terminal domain plays a crucial role in the structure of a full-length human mitochondrial Lon protease. Sci Rep 6:33631. srep33631 [pii]. https://doi.org/10.1038/srep33631
53. Kimber MS, Yu AY, Borg M, Leung E, Chan HS, Houry WA (2010) Structural and theoretical studies indicate that the cylindrical protease ClpP samples extended and compact conformations. Structure 18:798–808. S0969-2126(10)00189-9 [pii]. https://doi.org/10.1016/j.str.2010.04.008
54. Kirstein J et al (2009) The antibiotic ADEP reprogrammes ClpP, switching it from a regulated to an uncontrolled protease. EMBO Mol Med 1:37–49. https://doi.org/10.1002/emmm.200900002
55. Kita K, Suzuki T, Ochi T (2012) Diphenylarsinic acid promotes degradation of glutaminase C by mitochondrial Lon protease. J Biol Chem 287:18163–18172. M112.362699 [pii]. https://doi.org/10.1074/jbc.M112.362699
56. Kondadi AK et al (2014) Loss of the m-AAA protease subunit AFG(3)L(2) causes mitochondrial transport defects and tau hyperphosphorylation. EMBO J 33:1011–1026. embj.201387009 [pii]. https://doi.org/10.1002/embj.201387009
57. Konig T et al (2016) The m-AAA protease associated with Neurodegeneration Limits MCU Activity in mitochondria. Mol Cell 64:148–162. S1097-2765(16)30462-2 [pii]. https://doi.org/10.1016/j.molcel.2016.08.020
58. Koppen M, Metodiev MD, Casari G, Rugarli EI, Langer T (2007) Variable and tissue-specific subunit composition of mitochondrial m-AAA protease complexes linked to hereditary spastic paraplegia. Mol Cell Biol 27:758–767. MCB.01470-06 [pii]. https://doi.org/10.1128/MCB.01470-06
59. Kotiadis VN, Duchen MR, Osellame LD (2014) Mitochondrial quality control and communications with the nucleus are important in maintaining mitochondrial function and cell health. Biochim Biophys Acta 1840:1254–1265. S0304-4165(13)00481-9 [pii]. https://doi.org/10.1016/j.bbagen.2013.10.041
60. Krysiak J, Stahl M, Vomacka J, Fetzer C, Lakemeyer M, Fux A, Sieber SA (2017) Quantitative map of beta-lactone-induced virulence regulation. J Proteome Res 16:1180–1192. https://doi.org/10.1021/acs.jproteome.6b00705
61. Kubota Y, Nomura K, Katoh Y, Yamashita R, Kaneko K, Furuyama K (2016) Novel mechanisms for Heme-dependent degradation of ALAS1 protein as a component of negative feedback regulation of Heme biosynthesis. J Biol Chem 291:20516–20529. M116.719161 [pii]. https://doi.org/10.1074/jbc.M116.719161
62. Lee BG et al (2010) Structures of ClpP in complex with acyldepsipeptide antibiotics reveal its activation mechanism. Nat Struct Mol Biol 17:471–478. nsmb.1787 [pii]. https://doi.org/10.1038/nsmb.1787
63. Lee S, Augustin S, Tatsuta T, Gerdes F, Langer T, Tsai FT (2011) Electron cryomicroscopy structure of a membrane-anchored mitochondrial AAA protease. J Biol Chem 286:4404–4411. M110.158741 [pii]. https://doi.org/10.1074/jbc.M110.158741
64. Leung E et al (2011) Activators of cylindrical proteases as antimicrobials: identification and development of small molecule activators of ClpP protease. Chem Biol 18:1167–1178. S1074-5521(11)00306-1 [pii]. https://doi.org/10.1016/j.chembiol.2011.07.023
65. Levytskyy RM, Bohovych I, Khalimonchuk O (2017) Metalloproteases of the inner mitochondrial membrane. Biochemistry 56:4737–4746. https://doi.org/10.1021/acs.biochem.7b00663
66. Li DH et al (2010) Acyldepsipeptide antibiotics induce the formation of a structured axial channel in ClpP: a model for the ClpX/ClpA-bound state of ClpP. Chem Biol 17:959–969. S1074-5521(10)00286-3 [pii]. https://doi.org/10.1016/j.chembiol.2010.07.008
67. Liu K, Ologbenla A, Houry WA (2014a) Dynamics of the ClpP serine protease: a model for self-compartmentalized proteases. Crit Rev Biochem Mol Biol 49:400–412. https://doi.org/10.3109/10409238.2014.925421
68. Liu Y et al (2014b) Inhibition of Lon blocks cell proliferation, enhances chemosensitivity by promoting apoptosis and decreases cellular bioenergetics of bladder cancer: potential roles of Lon as a prognostic marker and therapeutic target in baldder cancer. Oncotarget 5:11209–11224. 2026 [pii]. https://doi.org/10.18632/oncotarget.2026
69. Lu B et al (2013) Phosphorylation of human TFAM in mitochondria impairs DNA binding and promotes degradation by the AAA+ Lon protease. Mol Cell

49:121–132. S1097-2765(12)00904-5 [pii]. https://doi.org/10.1016/j.molcel.2012.10.023
70. Martin A, Baker TA, Sauer RT (2008) Pore loops of the AAA+ ClpX machine grip substrates to drive translocation and unfolding. Nat Struct Mol Biol 15:1147–1151. nsmb.1503 [pii]. https://doi.org/10.1038/nsmb.1503
71. Nie X, Li M, Lu B, Zhang Y, Lan L, Chen L, Lu J (2013) Down-regulating overexpressed human Lon in cervical cancer suppresses cell proliferation and bioenergetics. PLoS One 8:e81084. https://doi.org/10.1371/journal. pone.0081084 [doi] PONE-D-13-29172 [pii]
72. Nishigaki R et al (2005) Proteomic identification of differentially-expressed genes in human gastric carcinomas. Proteomics 5:3205–3213. https://doi.org/10.1002/pmic.200401307
73. Nunnari J, Suomalainen A (2012) Mitochondria: in sickness and in health. Cell 148:1145–1159. S0092-8674(12)00235-8 [pii]. https://doi.org/10.1016/j.cell.2012.02.035
74. Padrao AI et al (2012) Impaired protein quality control system underlies mitochondrial dysfunction in skeletal muscle of streptozotocin-induced diabetic rats. Biochim Biophys Acta 1822:1189–1197. S0925-4439(12)00096-8 [pii]. https://doi.org/10.1016/j.bbadis.2012.04.009
75. Pierson TM et al (2011) Whole-exome sequencing identifies homozygous AFG3L2 mutations in a spastic ataxia-neuropathy syndrome linked to mitochondrial m-AAA proteases. PLoS Genet 7:e1002325. https://doi.org/10.1371/journal.pgen.1002325. [doi] PGENETICS-D-11-00928 [pii]
76. Potting C, Tatsuta T, Konig T, Haag M, Wai T, Aaltonen MJ, Langer T (2013) TRIAP1/PRELI complexes prevent apoptosis by mediating intramitochondrial transport of phosphatidic acid. Cell Metab 18:287–295. S1550-4131(13)00297-0 [pii]. https://doi.org/10.1016/j.cmet.2013.07.008
77. Pryde KR, Taanman JW, Schapira AH (2016) A LON-ClpP proteolytic Axis degrades complex I to extinguish ROS production in depolarized mitochondria. Cell Rep 17:2522–2531. S2211-1247(16)31585-6 [pii]. https://doi.org/10.1016/j.celrep.2016.11.027
78. Qin DJ et al (2015) Hsp90 is a novel target molecule of CDDO-me in inhibiting proliferation of ovarian Cancer cells. PLoS One 10:e0132337. https://doi.org/10.1371/journal. pone.0132337 [doi] PONE-D-15-20212 [pii]
79. Quiros PM et al (2014) ATP-dependent Lon protease controls tumor bioenergetics by reprogramming mitochondrial activity. Cell Rep 8:542–556. S2211-1247(14)00486-0 [pii]. https://doi.org/10.1016/j.celrep.2014.06.018
80. Rainbolt TK, Lebeau J, Puchades C, Wiseman RL (2016) Reciprocal degradation of YME1L and OMA1 adapts mitochondrial proteolytic activity during stress. Cell Rep 14:2041–2049. S2211-1247(16)30095-X [pii]. https://doi.org/10.1016/j.celrep.2016.02.011
81. Sabharwal SS, Schumacker PT (2014) Mitochondrial ROS in cancer: initiators, amplifiers or an Achilles' heel? Nat Rev Cancer 14:709–721. :nrc3803 [pii]. https://doi.org/10.1038/nrc3803
82. Sass P, Josten M, Famulla K, Schiffer G, Sahl HG, Hamoen L, Brotz-Oesterhelt H (2011) Antibiotic acyldepsipeptides activate ClpP peptidase to degrade the cell division protein FtsZ. Proc Natl Acad Sci U S A 108:17474–17479. 1110385108 [pii]. https://doi.org/10.1073/pnas.1110385108
83. Seo JH et al (2016) The mitochondrial Unfoldase-peptidase complex ClpXP controls bioenergetics stress and metastasis. PLoS Biol 14:e1002507. https://doi.org/10.1371/journal.pbio.1002507. [doi] PBIOLOGY-D-16-00117 [pii]
84. Shi H, Rampello AJ, Glynn SE (2016) Engineered AAA+ proteases reveal principles of proteolysis at the mitochondrial inner membrane. Nat Commun 7:13301. ncomms13301 [pii]. https://doi.org/10.1038/ncomms13301
85. Snider J, Thibault G, Houry WA (2008) The AAA+ superfamily of functionally diverse proteins. Genome Biol 9:216. gb-2008-9-4-216 [pii]. https://doi.org/10.1186/gb-2008-9-4-216
86. Speranza G et al (2012) Phase I study of the synthetic triterpenoid, 2-cyano-3, 12-dioxoolean-1, 9-dien-28-oic acid (CDDO), in advanced solid tumors. Cancer Chemother Pharmacol 69:431–438. https://doi.org/10.1007/s00280-011-1712-y
87. Stiburek L et al (2012) YME1L controls the accumulation of respiratory chain subunits and is required for apoptotic resistance, cristae morphogenesis, and cell proliferation. Mol Biol Cell 23:1010–1023. mbc.E11-08-0674 [pii]. https://doi.org/10.1091/mbc.E11-08-0674
88. Stinson BM, Nager AR, Glynn SE, Schmitz KR, Baker TA, Sauer RT (2013) Nucleotide binding and conformational switching in the hexameric ring of a AAA+ machine. Cell 153:628–639. S0092-8674(13)00351-6 [pii]. https://doi.org/10.1016/j.cell.2013.03.029
89. Suppanz IE, Wurm CA, Wenzel D, Jakobs S (2009) The m-AAA protease processes cytochrome c peroxidase preferentially at the inner boundary membrane of mitochondria. Mol Biol Cell 20:572–580. E07-11-1112 [pii]. https://doi.org/10.1091/mbc.E07-11-1112
90. Teng H, Wu B, Zhao K, Yang G, Wu L, Wang R (2013) Oxygen-sensitive mitochondrial accumulation of cystathionine beta-synthase mediated by Lon protease. Proc Natl Acad Sci U S A 110:12679–12684. 1308487110 [pii]. https://doi.org/10.1073/pnas.1308487110
91. Thibault G, Houry WA (2012) Role of the N-terminal domain of the chaperone ClpX in the recognition and degradation of lambda phage protein O. J Phys Chem B 116:6717–6724. https://doi.org/10.1021/jp212024b
92. Tian Q, Li T, Hou W, Zheng J, Schrum LW, Bonkovsky HL (2011) Lon peptidase 1 (LONP1)-dependent

breakdown of mitochondrial 5-aminolevulinic acid synthase protein by heme in human liver cells. J Biol Chem 286:26424–26430. M110.215772 [pii]. https://doi.org/10.1074/jbc.M110.215772
93. Townsend DM et al (2006) A glutathione S-transferase pi-activated prodrug causes kinase activation concurrent with S-glutathionylation of proteins. Mol Pharmacol 69:501–508. mol.105.018523 [pii]. https://doi.org/10.1124/mol.105.018523
94. Townsend DM, Findlay VL, Tew KD (2005) Glutathione S-transferases as regulators of kinase pathways and anticancer drug targets. Methods Enzymol 401:287–307. S0076-6879(05)01019-0 [pii]. https://doi.org/10.1016/S0076-6879(05)01019-0
95. Tsao T et al (2010) Role of peroxisome proliferator-activated receptor-gamma and its coactivator DRIP205 in cellular responses to CDDO (RTA-401) in acute myelogenous leukemia. Cancer Res 70:4949–4960. 0008-5472.CAN-09-1962 [pii]. https://doi.org/10.1158/0008-5472.CAN-09-1962
96. Verhagen AM, Kratina TK, Hawkins CJ, Silke J, Ekert PG, Vaux DL (2007) Identification of mammalian mitochondrial proteins that interact with IAPs via N-terminal IAP binding motifs. Cell Death Differ 14:348–357. 4402001 [pii]. https://doi.org/10.1038/sj.cdd.4402001
97. Wan D et al (2004) Large-scale cDNA transfection screening for genes related to cancer development and progression. Proc Natl Acad Sci U S A 101:15724–15729. 0404089101 [pii]. https://doi.org/10.1073/pnas.0404089101
98. Wang HM et al (2010) Obtusilactone A and (-)-sesamin induce apoptosis in human lung cancer cells by inhibiting mitochondrial Lon protease and activating DNA damage checkpoints. Cancer Sci 101:2612–2620. https://doi.org/10.1111/j.1349-7006.2010.01701.x
99. Wang N, Maurizi MR, Emmert-Buck L, Gottesman MM (1994) Synthesis, processing, and localization of human Lon protease. J Biol Chem 269:29308–29313
100. Wong KS et al (2018) Acyldepsipeptide analogs dysregulate human mitochondrial ClpP protease activity and cause apoptotic cell death. Cell Chem Biol 25:1–14. https://doi.org/10.1016/j.chembiol.2018.05.014
101. Yien YY et al (2017) Mutation in human CLPX elevates levels of delta-aminolevulinate synthase and protoporphyrin IX to promote erythropoietic protoporphyria. Proc Natl Acad Sci U S A 114:E8045–E8052. 1700632114 [pii]. https://doi.org/10.1073/pnas.1700632114
102. Yonchuk JG et al (2017) Characterization of the potent, selective Nrf2 activator, 3-(Pyridin-3-Ylsulfonyl)-5-(Trifluoromethyl)-2H-Chromen-2-one, in cellular and in vivo models of pulmonary oxidative stress. J Pharmacol Exp Ther 363:114–125. jpet.117.241794 [pii]. https://doi.org/10.1124/jpet.117.241794
103. Zeiler E, Korotkov VS, Lorenz-Baath K, Bottcher T, Sieber SA (2012) Development and characterization of improved beta-lactone-based anti-virulence drugs targeting ClpP. Bioorg Med Chem 20:583–591. S0968-0896(11)00596-7 [pii]. https://doi.org/10.1016/j.bmc.2011.07.047
104. Zhao Q, Wang J, Levichkin IV, Stasinopoulos S, Ryan MT, Hoogenraad NJ (2002) A mitochondrial specific stress response in mammalian cells. EMBO J 21:4411–4419
105. Zhou M, Zhang RH, Wang M, Xu GB, Liao SG (2017) Prodrugs of triterpenoids and their derivatives. Eur J Med Chem 131:222–236. S0223-5234(17)30154-X [pii]. https://doi.org/10.1016/j.ejmech.2017.03.005

# Investigating the Role of Mitochondria in Type 2 Diabetes – Lessons from Lipidomics and Proteomics Studies of Skeletal Muscle and Liver

9

Lisa Kappler, Laxmikanth Kollipara, Rainer Lehmann, and Albert Sickmann

**Abstract**

Mitochondrial dysfunction is discussed as a key player in the pathogenesis of type 2 diabetes mellitus (T2Dm), a highly prevalent disease rapidly developing as one of the greatest global health challenges of this century. Data however about the involvement of mitochondria, central hubs in bioenergetic processes, in the disease development are still controversial. Lipid and protein homeostasis are under intense discussion to be crucial for proper mitochondrial function. Consequently proteomics and lipidomics analyses might help to understand how molecular changes in mitochondria translate to alterations in energy transduction as observed in the healthy and metabolic diseases such as T2Dm and other related disorders. Mitochondrial lipids integrated in a tool covering proteomic and functional analyses were up to now rarely investigated, although mitochondrial lipids might provide a possible lynchpin in the understanding of type 2 diabetes development and thereby prevention. In this chapter state-of-the-art analytical strategies, pre-analytical aspects, potential pitfalls as well as current proteomics and lipidomics-based knowledge about the pathophysiological role of mitochondria in the pathogenesis of type 2 diabetes will be discussed.

**Keywords**

Type 2 diabetes · Mitochondria · Lipidomics · Proteomics · Mass spectrometry · Function

L. Kappler
Institute for Clinical Chemistry and Pathobiochemistry, Department for Diagnostic Laboratory Medicine, University Hospital Tuebingen, Tuebingen, Germany

L. Kollipara
Leibniz-Institut für Analytische Wissenschaften – ISAS – e.V., Dortmund, Germany

R. Lehmann
Institute for Clinical Chemistry and Pathobiochemistry, Department for Diagnostic Laboratory Medicine, University Hospital Tuebingen, Tuebingen, Germany

Institute for Diabetes Research and Metabolic Diseases of the Helmholtz Centre Munich at the University of Tuebingen, Tuebingen, Germany

German Center for Diabetes Research (DZD e.V.), Tuebingen, Germany

A. Sickmann (✉)
Leibniz-Institut für Analytische Wissenschaften – ISAS – e.V., Dortmund, Germany

Medical Proteome Centre, Ruhr Universität Bochum, Bochum, Germany

Department of Chemistry, College of Physical Sciences, University of Aberdeen, Aberdeen, UK
e-mail: sickmann@isas.de





A. Urbani, M. Babu (eds.), *Mitochondria in Health and in Sickness*, Advances in Experimental Medicine and Biology 1158, https://doi.org/10.1007/978-981-13-8367-0_9

## Abbreviations

| | |
|---|---|
| 1D | One-dimensional |
| 2D | Two-dimensional |
| ADP | Adenosine diphosphate |
| ATP | Adenosine-5′-triphosphate |
| CER | Ceramide |
| CL | Cardiolipin |
| DAG | Diacylglycerol |
| DC | Differential centrifugation |
| DHA | Docosahexaenoic acid; 22:6 |
| ESI | Electrospray ionization |
| ETC | Electron transport chain |
| FA | Fatty acid |
| FASP | Filter-aided sample preparation |
| FFA | Free fatty acid |
| GC | Gas chromatography |
| HPLC | High-performance liquid chromatography |
| HRR | High resolution respirometry |
| IMM | Inner mitochondrial membrane |
| IR | Insulin resistance |
| LC | Liquid chromatography |
| MACS | Magnetic cell isolation and separation |
| MLCL | Monolyso-CL |
| MS | Mass spectrometry |
| MS/MS | Tandem mass spectrometry |
| MTBE | Methyl tert-butyl ether |
| NAD | Nicotinamide adenine dinucleotide |
| NAFL | Non-alcoholic fatty liver |
| OMM | Outer mitochondrial membrane |
| OXPHOS | Oxidative phosphorylation |
| PA | Phosphatidic acid |
| PAGE | Polyacrylamide gel electrophoresis |
| PC | Phosphatidylcholine |
| PE | Phosphatidylethanolamine |
| PG | Phosphatidylglycerol |
| PI | Phosphatidylinositol |
| PL | Phospholipid |
| PS | Phosphatidylserine |
| PTMs | Post-translational modifications |
| SIMPLEX | Simultaneous metabolite, protein, lipid extraction |
| SM | Sphingomyelin |
| STZ | Streptozotocin |
| T1Dm | Type 1 Diabetes mellitus |
| T2Dm | Type 2 Diabetes mellitus |
| TAG | Triacylglycerol |
| TCA cycle | Tricarboxylic acid cycle |
| TLC | Thin layer chromatography |
| UC | Ultracentrifugation |
| UPLC | Ultra-performance liquid chromatography |
| WB | Western blotting |
| WHO | World health organization |
| wt | Wild type |

## 9.1 Introduction

### 9.1.1 Mitochondrial Dysfunction, Insulin Resistance and Type 2 Diabetes Mellitus

Type 2 diabetes mellitus (T2Dm) is rapidly developing as one of the greatest global health challenges of this century. In 2014, 422 million people over 18 years suffered from diabetes according to the World Health Organization (WHO). The worldwide prevalence of diabetes rose from 4.7% in 1980 to 8.5% in 2014. Despite the report of rare monogenetic forms of diabetes, [263] T2Dm is a polygenic disease and therefore often covered in GWAS (genome wide association studies) [14]. Insulin resistance (IR) plays a major role in the development of T2Dm, the so called pre-diabetic phase [151, 172, 315]. However, the underlying mechanisms in the pathogenesis of T2Dm are still not fully understood. In IR the insulin target organs such as liver, white adipose tissue and skeletal muscle, show a reduced response/sensitivity to insulin. Despite marked hyperinsulinemia the body fails to overcome this resistance to insulin. Tremendous changes in lifestyle (over- and malnutrition; physical inactivity) during the past decades around the world are the leading causes for the epidemic increase in IR and T2Dm [92].

The skeletal muscle and the liver are among the major insulin target organs and play essential roles in glucose homeostasis with skeletal muscle being responsible for over 80% of the glucose disposal [58, 59]. Decades before the development of β-cell failure and hyperglycemia, skeletal muscle IR is considered to be one of the initiating or primary defects in the pathogenesis [315]. Another currently intensely discussed and investigated key factor in the pathogenic scenario on the way to the manifestation of T2Dm is the non-alcoholic fatty liver (NAFL) disease [6]. Different theories about mechanisms underlying peripheral IR development are discussed, which are the keys to elucidate the etiology of T2Dm such as the accumulation of fatty acid metabolites causing lipotoxicity after elevated secretion of fatty acids by the liver in a state of obesity [38], or e.g. triacylglycerol (TAG) accumulation in skeletal muscle tissue [179, 212] and liver [6].

However, in recent years the link of mitochondrial dysfunction to IR came into focus, due to the central metabolic function of mitochondria, their essential role in the regulation of bioenergetic processes of cell stress response and epigenetic modulations [1, 208, 311]. A link between mitochondrial dysfunction and IR was first reported nearly more than 40 years ago [324]. Mitochondrial dysfunctions are discussed as the cause, the consequence and as an innocent bystander in IR [24, 122, 144, 145, 230]. An impaired substrate flux in IR could lower mitochondrial oxidative capacity [250]. Alternatively, lower rates of substrate oxidation could be caused by a deficiency in the electron transport chain (ETC) and/or a lower mitochondrial mass could lead to inhibited metabolic fluxes leading to accumulation of toxic metabolites from incomplete oxidation of glucose and lipids such as diacylglycerols (DAG) [193], in general acyl-CoAs and their respective acylcarnitines, [158] TAGs [37] and ceramides [279] thereby causing IR [158, 193].

The research focus since many years was mainly on skeletal muscle mitochondria. In human studies performed 20 years ago it was hypothesised that the mitochondrial dysfunction decreases lipid oxidation in muscle in obesity and diabetes (measured across the leg) [144, 150]. Kelley et al. eventually described smaller mitochondria and a decreased ETC activity (measured as NADH:$O_2$ oxidoreductase) in mitochondrial preparations of T2Dm and obese patients [145]. Lately, various features of mitochondrial dysfunctions were uncovered in muscles of insulin resistant and/or type 2 diabetic compared to healthy subjects such as lowered oxidative capacity in muscle biopsies from T2Dm patients [24, 145, 234, 246], which was confirmed in isolated mitochondria [188]. After normalisation to mitochondrial mass these findings revealed to be robust, except in one study [24]. The reduced oxidative capacity was often concomitant with a reduced mitochondrial content [44, 145], mostly investigated using electron microscopy [44, 145, 245] or citrate synthase activity [24, 117, 145]. Decreased mitochondrial enzyme activities [24, 117, 145, 188] or lowered protein abundance [117, 188], in particular of those involved in the ETC [145, 246], were also associated with IR and T2Dm. The hypothesis that mitochondrial alterations are the cause and not the consequence of IR was supported by the detection of a lower mitochondrial content and oxidative capacity in non-diabetic individuals with a family history of type 2 diabetes [230]. Mitochondrial dysfunction and its role in skeletal muscle IR however remains still conflicting, since some research groups did not observe alterations such as the reduced oxidative capacity [24, 121], or the reduced mitochondrial content in IR [246]. In animal studies, high-fat feeding-induced IR, but even an opposite relationship for oxidative capacity was observed [57, 87, 110, 123, 302, 303]. This was interpreted as a compensatory effect to overcome excess substrate flows at early stages of the disease development due to increased fatty acid oxidation capacity, mitochondrial content and increased oxidative protein activities in skeletal muscle [57, 87, 110, 123, 302, 303]. High-fat diet-induced IR in rats was observed albeit an increased mitochondrial amount [110]. An ETC deficiency caused by an iron-deficient diet (by decreasing the iron-containing ETC proteins) even protected rats from IR induced by high-fat feeding [109]. Due to this inconsistency of data, the role of mitochondrial dysfunction in the etiology of muscle IR remains still unclear.

Liver as the major site for endogenous glucose production and one of the key organs in lipid metabolism is also in the focus of IR and diabetes research [6, 292]. However, human studies investigating mitochondrial alterations of the liver in T2Dm were rarely conducted [281]. Mainly non-invasive assessments of hepatic mitochondrial metabolism were applied [228]. A higher oxidative capacity of mitochondria from liver biopsies was observed in insulin resistant obese compared to lean patients despite a non-alcoholic fatty liver (NAFL) [156], also seen in non-invasive MRS methods assessing hepatic mitochondria [154, 280]. Contrarily, high resolution respirometry (HRR) revealed evidence for a normal oxidative phosphorylation in T2Dm compared to lean and obese patients [177]. Even an impaired hepatic mitochondrial function in T2Dm was described [260, 282], also reported in diabetic rats after high-caloric feeding [23]. Overall, the role of alterations of hepatic mitochondria in context of IR is still up for debate and lipidomics and proteomics investigations are so far limited.

Changes in lipid and protein homeostasis are very likely to affect mitochondrial molecular profile which, in turn, regulates mitochondrial structure and function [126, 217]. Therefore, proteomics and lipidomics analyses (see Sect. 9.3) might help to understand how molecular changes in mitochondria translate to alterations in energy transduction [143] (respiratory analyses; Sect. 9.1.7) as observed in the healthy and metabolic diseases such as T2Dm and other related disorders.

### 9.1.2 Mitochondrial Lipids and Their Physiological Role

Lipids are structurally distinct and diverse biomolecules, which are hydrophobic or amphiphilic. More than 40,000 lipid structures are listed in the LIPID MAPS database (http://www.lipidmaps.org) as of January 2018. In 2005, lipids were grouped based on chemical and biochemical properties from polar via neutral to nonpolar into eight categories by the LIPID MAPS consortium, the National Nomenclature and the International Lipid Classification Committee: fatty acyls, glycerolipids, glycerophospholipids, sphingolipids, sterol lipids, prenol lipids, saccharolipids, and polyketides [67, 68]. Lipids play a fundamental role in pathophysiological processes since they are fundamental components of the cellular membranes, function as energy storage units and signalling molecules, but are also key players in membrane transport and anchoring [235, 252, 306]. Many lipid groups are involved in membrane structure and scaffolding of its proteins [29, 252]. Phospholipids including phosphatidylcholines, phosphatidylethanolamines, phosphatidylserines, phosphatidylinositols, phosphatidic acids, sterols, and sphingolipids including sphingomyelins and glycosphingolipids form together with proteins the lipid bilayers of cell membranes [255]. Therefore, metabolic homeostasis of the lipidome as part of the metabolome is fundamental for health.

Changes in lipid homeostasis are very likely to affect the mitochondrial membrane composition and, in turn, mitochondrial structure and function [126, 217]. Mitochondria are the site for β-oxidation, the main pathway for fatty acid oxidation [160]. Moreover, mitochondria are also sites of lipid synthesis, remodelling and inter-organellar lipid trafficking [176, 287]. Approximately 45% of their own phospholipids are synthesised within the mitochondria [307], mostly phosphatidylethanolamines (PE), phosphatidylglycerols (PG), phosphatidic acids (PA) and cardiolipins (CL) [53, 176]. Mitochondrial lipids are involved in diverse physiological processes including mitochondrial fusion and fission, membrane structure and fluidity, ETC assemblage, protein biogenesis, and apoptosis [46, 47, 83, 126, 147, 207], making them to more than simply membrane constituents. Exemplarily for the role of lipids in mitochondrial function, cardiolipin – the mitochondrial signature phospholipid, has been shown to be a key player in the organization of the general mitochondrial membrane structure, but also in the organization of the essential ETC components into higher order assemblies called supercomplexes [46, 207, 256, 304]. It contributes typically to 13–14% of the lipid composition of mitochondria depending on the cell type. The other major components are phosphatidylcholines (PC; 40–44%), phosphatidylethanolamines (27–34%), phosphatidylserines

(PS; 1–3%) and phosphatidylinositols (PI; 5–15%) [53, 307, 333]. The inner mitochondrial membrane (IMM) and the outer mitochondrial membrane (OMM) however play rather different roles in the function of mitochondria, related also to their differing lipid composition. The IMM is a protein-enriched membrane with only 20% lipids of total mass, whereas the OMM is rich in lipids [53, 56, 333]. While IMM's signature is a high CL content (13–14%), whereas the OMM has a high cholesterol content up to 10% and is known to have rather low or even depleted CL levels [56, 75]. Since most lipids besides the mitochondrial signature lipid CL are not organelle-specific, a sophisticated mitochondrial isolation and purification procedure is a prerequisite for an accurate and comprehensive lipid profiling strategy to investigate the contribution of individual lipids to mitochondrial (dys-)function.

### 9.1.3 Diversification of Mitochondrial Lipids Between Species and Tissues

Tissue-specific differences in mitochondrial equipment and function could play a major role in disease pathologies and form a basis for drug targeting. As demonstrated by morphology, but also enzymatic equipment, mitochondrial DNA (mtDNA) copy numbers, protein abundances and oxidative capacity between cell types [12, 137, 138, 140, 161, 190], mitochondria were shown to be tailored to the specific needs and demands of different tissues. Despite this apparent physiological diversity, little data exists combining several approaches to investigate overall molecular mechanisms from mitochondrial lipid and protein composition up to mitochondrial function for these tissue-specific differences. However, given such widely varying morphologies, metabolic pathways and tissue-specific demands, it is reasonable to expect that the oxidative function as well as the lipid and the protein composition of these organelles may vary significantly from tissue to tissue. Proteomics approaches showed distinct tissue-specific differences between mitochondria from liver and muscle tissues [77], and since the mitochondrial phospholipid content and composition have been shown to affect ETC activity and respiratory function [182, 288, 305], a lipidomics and proteomics approach combined with functional studies might help understanding mitochondrial organisation by identifying common and unique tissue specificities.

Between different species a great variation of the acyl chain compositions of CLs was detected by Minkler et al. (2010) [186]. In cardiac mitochondria from dogs 77% were tetra-linoleoyl-CL, as published previously for humans (78–79%) [258]. In mouse heart however, only 22% of the CL was tetra-linoleoyl-CL, whereas 53% CL species contained 22:6 acyl chains. Rat liver was about 57% tetra-linoleoyl-CL [186]. Tetra-linoleoyl-CLs however are the most abundant PUFA-CL species in liver and in muscle mitochondria as already published previously in rats [69, 257].

Cheng et al. [42] hypothesised the CL composition to be tailored to the metabolic needs of the surrounding tissue. Tissues highly relying on mitochondrial activity, such as muscle and liver, are equipped with CLs suited for supporting mitochondrial function. Other tissues such as brain that mostly rely on glucose have no need for the specifically adapted tetralinoleyl-CL and lipid profiles can be expected to be much more diverse. The lipid profile of murine brain mitochondria and its alteration with age was shown to be distinctly different from skeletal muscle mitochondria [236]. Another reason for the tissue-specific CL composition could be the responsiveness of tissue to certain fatty acids. A significantly higher proportion of radiolabelled LDL ($^{3}$H-cholesteryl oleate tracer) enriched in docosahexaenoic acid (DHA; 22:6) was found in heart, brown adipose and brain tissues relative to liver LDL [237]. Heart might be similar to skeletal muscle, due to their myogenic precursor [263, 293]. A more efficient uptake of these fatty acids could increase the supply of 22:6 acyl chains for remodelling and synthesis of CLs and PEs. Additionally, as the major site of *de novo* lipogenesis, liver synthesises the majority of lipids [220]. Investigations on mitochondrial specificities of liver and skeletal muscle as two insulin target organs could help to elucidate organ cross-talks and might lead to more target-specific treatments of mitochondrial dysfunctions associated with e.g. IR and T2Dm.

### 9.1.4 Brief Introduction of State-of-the-Art Lipidomic Analysis of Mitochondria

With advances in soft ionisation technologies and mass analysers, lipidomics approaches have greatly been driven by state-of-the-art mass spectrometry (MS) technologies during the last decade (Fig. 9.5 and Table 9.1). Soft ionisation technologies frequently used in lipidomics include electrospray ionisation (ESI), atmospheric pressure chemical ionisation (APCI), as well as matrix-assisted laser desorption/ionisation (MALDI). Mass analysers include ion trap (IT), time-of-flight (TOF), quadrupole (Q), Fourier transform ion cyclotron resonance (FT-ICR), Orbitrap, and hybrid MS (e.g., triple quadrupole (QqQ), IT-TOF, QTOF, and IT-FT-ICR). MS coupled with liquid chromatography (LC-MS) provides a tool for detection and separation of specific lipid classes providing analytical specificity for each lipid species within a lipid class [178]. LC-MS based lipidomics approaches reduce ion suppression in analysing complex biological samples by adding the retention of the compounds of interest by the stationary phase in

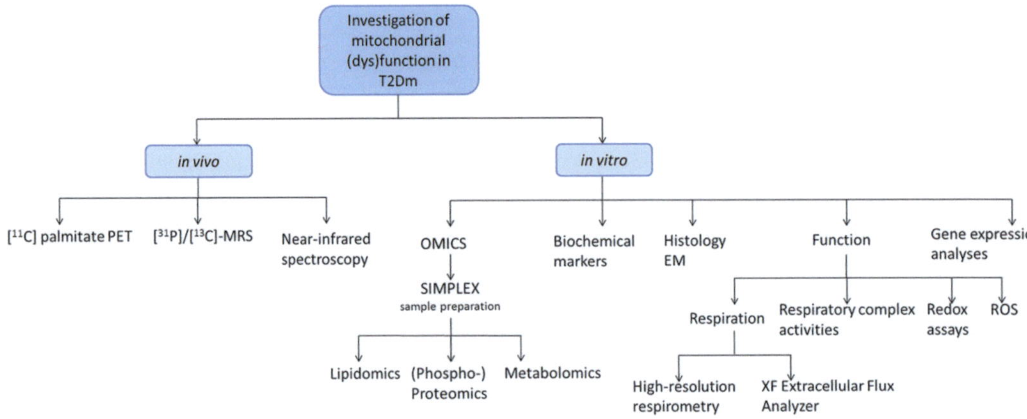


**Fig. 9.5** Selected methods assessing features of mitochondrial (dys)function in T2Dm research investigating liver and muscle. *EM* electron microscopy, *PET* positron-emission tomography, *MRS* magnetic resonance spectroscopy, *ROS* reactive oxygen species production

**Table 9.1** Methods applied in liver and muscle assessing features of mitochondrial (dys)function in T2Dm research investigating liver and muscle

| In vivo | | In vitro | |
|---|---|---|---|
| Technique | References | Technique | References |
| [$^{11}$C] palmitate PET | [132] | Microscopy (histology, TEM) | [44, 145] |
| [$^{31}$P]/[$^{13}$C]-MRS | [82, 196, 230, 260, 282] | Omics: lipidomics, proteomics, (metabolomics) | [167, 289] |
| Near-infrared spectroscopy | [320] (reviewed) | Respiratory complex activities | [245] |
| | | High-resolution respirometry | [156, 188, 234] |
| | | Biochemical markers | [157] |
| | | Redox assays | [17] (reviewed) |
| | | XF extracellular flux analyser | [201, 326] |
| | | Gene expression analyses | [187, 223, 286] |

Selected references applying the listed techniques are given; *TEM* trans emission microscopy, *PET* positron-emission tomography, *MRS* magnetic resonance spectroscopy the column as another factor to selectivity and allow the identification of isobaric compounds. In contrast, shotgun lipidomics employs direct infusion of ionised lipids (via suitable interface e.g. ESI) into the MS that detects the lipids solely on basis of mass to charge (m/z) ratio allowing high-throughput untargeted lipid profiling [105]. Untargeted approaches cover broad lipid profiles, whereas targeted approaches are applied for specific classes or pathways.

The lipidomics approach is defined as "top-down" or "bottom-up" based on the strategies that are applied for lipid screening. The specificity of both depends on the resolving power or resolution (R) of a MS, which is usually the ratio of the ion peak mass to the peak width at 50% of its height to the mass of the peak i.e. full width at half maximum (FWHM) [101]. The shift between the measured and the theoretical mass value is defined as the mass accuracy and described relatively e.g. parts per million (ppm).

The "top-down" lipidomics approach takes the unique elemental composition of lipid species of specific classes into account using exact mass measurement by high-resolution MS, e.g. the Orbitrap and FT-ICR [30, 194]. For example, the LTQ-Orbitrap (Thermo Scientific) enables a high mass resolution exceeding 100,000, a mass error less than 2 ppm using internal standards and 5 ppm with external calibration [180, 264]. Data acquisition is performed at high resolution and accuracy with the Orbitrap (e.g. $R = 60{,}000$, $\leq 5$ ppm) whereas MS/MS is recorded in a linear ion trap mass analyser at high speed/sensitivity with low resolution (e.g. $R = \sim 4000$, ~100 ppm).

The so called "bottom-up", approach in tandem-MS utilizes the "building block" feature of lipids. Specific fragment ions are used to identify and quantify precursor ions. Different chemical properties of lipid classes such as polarity and charge can be used to determine individual molecular species of a class by scanning precursor ions or neutral losses that are indicative for head groups, fatty acyls or long-chain sphingoid bases, etc. [31].

LC-MS based mitochondrial lipidomics found 8814 lipid species in mitochondria enriched fractions from mouse brain and skeletal muscle altering with age [236]. Three hundred and ninety seven lipid species were detected in liver-derived cell culture samples from only 100 μg mitochondrial protein [142]. LC-MS was also applied for the comparison of CL species in mitochondria from mouse heart, dog heart and rat liver [186], the latter also linked with the feeding of different diets [15]. Andreyev et al. [4] detected around 220 lipid species in mitochondria from macrophages using an LC-MS approach. As an example for shotgun analyses, 183 lipid species were revealed in brain mitochondria from mice using QqQ [147, 148]. Another shotgun approach covered 217 lipid species in liver mitochondria for investigation of temporal and spatial lipid organisation to uncover daily oscillations in mitochondria [8]. In human skeletal muscle Herbst et al. detected 28 lipid species investigating effects of omega-3 supplementation using gas chromatography [118], which was also used by Barzanti et al. to determine the fatty acid composition in brain, liver and heart mitochondria from rats at a different age [11] and by Timmons et al. [295] for the comparison of the fatty acid profiles of brain mitochondria to those from heart, kidney and liver isolated from Sprague Dawley rats. Fourier transformation (FT)-MS was applied for mitochondrial lipid profiling on aging yeast [168] and also MALDI-TOF-MS techniques were applied to investigate mitochondria from tissue and yeast [5]. For lipidomics analyses covering the mitochondrial signature lipid cardiolipin, 100 μg [142] to 1.5 mg [15, 186] mitochondrial protein was required revealing 32 cardiolipin species [142] or less [4, 5, 148, 186], but even the detection of up to 100 species was reported [147].

### 9.1.5 Mitochondrial Proteins and Their Physiological Role

Evolutionarily the mitochondria still continue as the powerhouse of living cell by generating adenosine triphosphate (ATP) – the major source of energy for performing cellular and metabolic functions [81]. The origin, genetic composition and physiological role of the mitochondria in eukaryotes are well-described in the literature [22, 65, 81, 99, 162, 209]. Briefly, the human mitochondrial genome or mtDNA is a circular molecule of 16,569 base pairs and comprises 37 genes coding for two ribosomal RNAs, 22 transfer RNAs and 13 proteins [284]. However, the majority (~99%) of the mitochondrial proteome constituents are synthesized on the ribosomes in the cytosol (encoded via nuclear genome) as precursor proteins and later imported either post- or co-translationally into this sub-cellular organelle [62, 319]. This transport process is regulated by means of highly specialized and dynamic translocase machineries/pathways located mainly in the (i) OMM; translocase of the outer membrane or TOM complex, sorting and assembly machinery or SAM pathway, (ii) IMS (intermembrane space); mitochondrial IMS assembly or MIA pathway and (iii) IMM; translocase of the inner membrane or TIM complex of the mitochondria [36, 319]. Owing to the advancements in various omics-based technologies, microscopy studies and prediction algorithms the current estimate of eukaryotic mitochondrial proteins is 1000–1500 [32, 36, 269] and it is believed that the complexity of the mitochondrial proteome can be attributed to tissue-specificity, evolutionary heredity and biochemical variability [32, 210], apparently identical to the mitochondrial lipidome. Moreover, the diversity of mitochondrial proteins is exemplified by so-called dual-localization i.e. their presence on either side of membrane (OMM and IMM) and dynamic nature of the mitochondria *per se* that enables them to communicate with other sub-cellular organelles/compartments e.g. the endoplasmic reticulum and the nucleus [272]. Previous reports suggest that the mitochondrial proteins have a wide dynamic range spanning roughly five or six orders of magnitude of abundance [32]. According to the latest MitoCarta2.0 [33] – an evidence based inventory, which consists of 1158 genes (in man and mouse) that encode mitochondrial proteins, whereas MitoMiner v3.1 curated database [272] consists of information from 12 different organisms and currently reports localization of 1839 human mitochondrial genes. Interestingly, of these ~1100 candidates (MitoCarta), there is a similar share of proteins i.e. ~26% (i) whose function is not yet known and (ii) have only domain annotations based on sequence homology [32].

Despite their inherent functional versatility, most of the mitochondrial proteins are involved in important cellular processes such as (i) energy metabolism, (ii) metabolism of amino acids, lipids and heme, (iii) protein transport, folding and their degradation, (iv) signalling pathways (especially $Ca^{2+}$) and (v) programmed cell death or apoptosis [36, 198, 285]. Besides energy production, the mitochondria also hosts proteins that are involved in maintaining $Ca^{2+}$ homeostasis, reactive oxygen species (ROS) and reactive nitrogen species (RNS); all deemed as signalling molecules and their imbalance i.e. altered cellular concentrations is often associated with a myriad of human disorders including cancer and neurodegenerative diseases e.g. Alzheimer's [97, 267, 325]. Lastly, the role of mitochondria in triggering (i.e. BCL-2 family members) and mediating apoptosis by a cascade of enzymatic steps (i.e. caspases) has been regarded as pivotal in determining the fate of a living cell in health and disease states [312].

#### 9.1.5.1 Mitochondrial Proteins and Post-translational Modifications (PTMs)

In response to their internal or external environmental stimuli, proteins localized in different cellular compartments are subjected to a vast variety of reversible or irreversible post-translational modifications (PTMs) in their lifetime and thus impacting their biological function and also further imparting dynamic nature to these biomolecules. Mitochondrial proteins, identical to other cellular proteins undergo several PTMs and of recent interest include: acetylation, proteolytic cleavage, oxidation, ADP-ribosylation, nitration, *O*-linked glycosylation, phosphorylation, succinylation, SUMOlyation, S-nitrosylation and ubiquitination [153, 274, 277]. Recent review articles highlight some of these PTMs on the mitochondrial proteins and their role in regulating important biological functions such as ATP production, mitochondrial fission – fusion, autophagy and apoptosis [74, 274, 277]. Evidently, perturbations in the levels of PTMs occurring mainly due to the different stimuli or genetic variations can cause mitochondrial dysfunction leading to tissue damage and disease progression. Among others, transient reversible phosphorylation/dephosphorylation (*O*-linked on Ser/Thr/Tyr amino acid residues) is one of the most widely studied PTMs in the mitochondria due to its occurrence and significance in signalling pathways that are involved in various cellular processes and in maintaining metabolic homeostasis [131]. It has been known that the phosphorylation/dephosphorylation events – primarily catalyzed by protein kinases/phosphatases, dictate mitochondrial function and a distinct set of these counteracting enzymes are either localized in different compartments (OMM, IMS, IMM and matrix) or transported into the mitochondria [173]. Besides the mitochondrial OXPHOS in the ETC complex 5 of respiratory activity [113], phosphorylation plays a critical role in the (i) protein import machinery e.g. TIM and TOM complexes [111], (ii) mitophagy – a selective degradation of the damaged mitochondria by autophagy [64] and (iii) apoptosis [199]. In general, the challenge however, lies in the detection of PTMs as most of them are usually present in low-abundance/low stoichiometry and oftentimes rely on biochemical techniques and PTM-specific enrichment strategies preferably in conjunction with LC-MS analysis [331].

### 9.1.6 Brief Introduction of State-of-the-Art Proteomics Profiling Approaches

In the past decade, research on mitochondria using proteomics has gradually increased (Fig. 9.1) with mass spectrometry (MS) playing an integral role. Conventionally, proteome analysis is performed using either gel-based (e.g. 2D-PAGE) or gel-free (high-performance liquid chromatography, HPLC) approaches. However, due to the practical limitations associated with the gel-based techniques especially in detection of PTMs, low abundant and hydrophobic proteins there has been a shift to utilize gel-free methods more frequently. Nevertheless, despite the debate on which is the most suitable way for proteome analysis, it is believed that both strategies can complement each other [88]. In general, likewise to lipidomics, LC-MS based proteomics can be broadly categorized as "bottom-up" proteomics i.e. analysis of peptide mixtures resulting from proteolytic digests and "top-down proteomics" i.e. analysis of intact proteins. Although bottom-up is the most commonly used method in majority of proteomics laboratories worldwide, recently the top-down strategy is slowly garnering popularity [35].

Bottom-up proteomics typically involves MS analysis of peptides that are generated by the

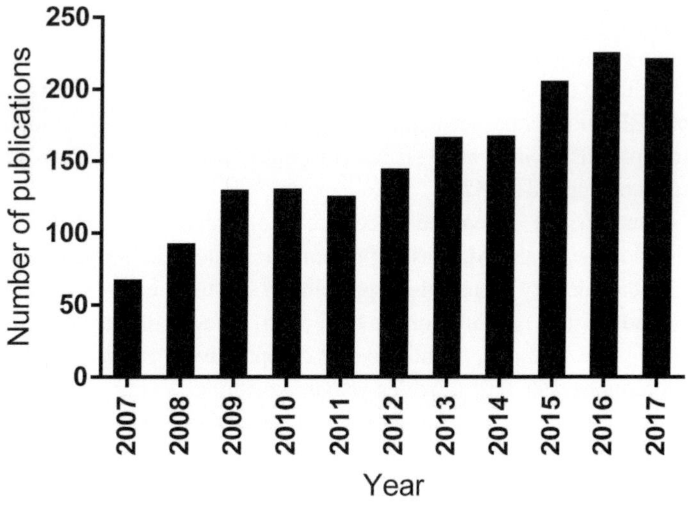


**Fig. 9.1** Total number of summed research articles that contain key words "mitochondria proteomics" in their abstracts indicate a steady increase in application of proteomics approaches for studying mitochondria since January-2007 till December-2017 (http://www.ncbi.nlm.nih.gov/pubmed)

enzymatic (mostly trypsin) hydrolysis of proteins. To minimize the complexity of the several thousands of peptide mixtures, a fractionation step such as reversed phase (RP) HPLC is performed prior to MS analysis, which could be operated in either online or offline mode. Despite routinely employed, 1D RP chromatography has been deemed insufficient to effectively resolve highly complex peptide mixtures (https://doi.org/10.17877/DE290R-17852). In order to enhance the peptides separation performance, multi-dimensional techniques that rely on the physicochemical properties of proteins/peptides have emerged [330]. Currently, 2D liquid chromatography (2D-LC) methods are commonly used to ameliorate proteome coverage, which utilize at least two different LC separation mechanisms to achieve desired orthogonality. Yates and co-workers introduced the multidimensional protein identification technology (MudPIT) by combining strong cation exchange (SCX) chromatography (peptides net charge-based separation) for the 1D separation followed by RP (2D) fractionation before MS analysis [174, 316]. Recently, employing RP-HPLC fractionation for both 1D and 2D is gaining popularity. Here, the separation of peptides is carried out at two different pH values, which is also known as high–/low-pH RP-RP fractionation [91]. Briefly, the peptide mixtures are fractionated at high-pH (8.0–10.0) RP-HPLC (1D) followed by 2D separation on low pH ($\leq$2.0) RP-HPLC of each individual fraction and subsequent MS analysis. Moreover, it was showed that by selectively combining fractions (known as concatenation) that are eluting at different retention times from high-pH RP HPLC and subsequent downstream 2D low pH RP LC-MS analysis leads to better protein coverage [313, 327].

Due to the recent advanced technological developments, the MS-based proteomics strategies can identify thousands of proteins in shorter periods of time ranging from an hour [116] to few days [195]. Notably, quantitative or comparative proteomics (Fig. 9.2) provides valuable information about the relative proteome differences in different biological conditions such as healthy vs. disease [2]. In practice, MS analysis cannot be directly used for quantification purposes due to its inherent nature and peptides in particular, possess different types of physicochemical properties depending on their amino acid compositions, which could modulate their behavior during LC-MS analysis and thereby affecting protein quantification [10, 114, 200]. Therefore, stable isotopes i.e. $^{13}C$, $^{15}N$, $^{2}H$, $^{18}O$ have been incorporated in protein quantification workflows, which rely on the basis that the various physicochemical properties of the unlabeled (or natural) and stable isotope-labeled (SIL) peptides are nearly similar [90]. Furthermore, it is also assumed that the behaviour of both natural and SIL peptides is almost identical during sample preparation processes and downstream LC-MS analysis [98]. The current MS-based quantification technologies introduce these stable isotopes by several ways such as (a) metabolic labelling e.g. stable isotope labelling by amino acids in cell culture (SILAC) [204] (Fig. 9.2A) or (b) chemical labelling e.g. isobaric tags for relative and absolute quantification (iTRAQ) [249] (Fig. 9.2B) and tandem mass tags (TMT) [291]. Once labelled, the differentially labeled samples can be multiplexed either on protein (i.e. SILAC) or peptide level (i.e. iTRAQ and TMT) assuming ~100% labelling efficiency and subsequently, quantification of peptides is performed on MS or MS/MS level, respectively. The rationale behind this type of quantification is the absolute mass shift arising due to the incorporation of stable isotope(s) into the peptide or protein and based on the presumption that the relative signal intensities of differentially labeled peptides are directly proportional to their relative concentrations between the multiplexed samples. Lastly, by using their peak areas information peptide/protein quantification is performed in a relative manner. (https://doi.org/10.17877/DE290R-17852). Label-free protein quantification or LFQ, which is a relatively economical approach compared to labelling strategies can be performed by comparing precursor ion intensities or by spectral counting (Fig. 9.2C) and in both methods due to the absence of stable heavy isotopes, each sample is analyzed separately. In precursor ion intensity-based LFQ, the peak areas or peptide abundances

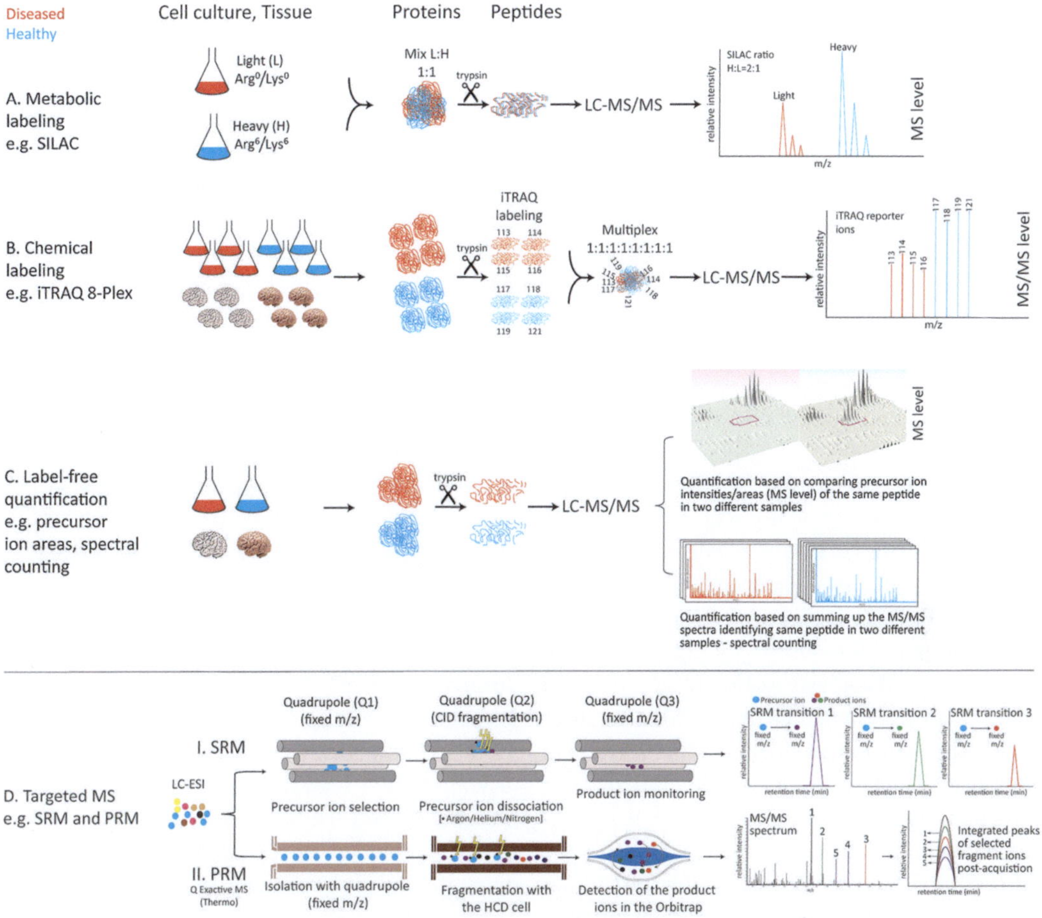


**Fig. 9.2** Schematic representation of the current quantitative proteomics strategies employed for analyzing biological samples. Quantification methods A, B, C (metabolic, chemical labelling and label-free) are mainly used for relative or comparative proteomics studies, whereas targeted-MS technologies such as SRM and PRM (D. I, II) can be used for both relative and absolute protein quantification in combination with SIL peptides. (Figure adapted from https://doi.org/10.17877/DE290R-17852)

are obtained from successive LC-MS runs whereas, in spectral counting based LFQ the number of MS/MS spectra of an identified peptide is summed up and they are then compared across all individual sample measurements (https://doi.org/10.17877/DE290R-17852). There are several studies about the applications of the aforementioned technologies and a recent review by Wang et al. highlighted the utilization of bottom-up quantitative proteomics to analyze mitochondrial proteome with a focus on the wide-range of biological functions and human diseases in which the mitochondria are involved [314]. On the other hand, top-down proteomics is an evolving strategy that promises direct identification of the so-called *proteoforms* – a term used to describe various forms of protein arising from a specific gene [297]. Recently, Kelleher and coworkers employed subcellular fractionation of organelles in combination with top-down LC-MS analysis and concluded the feasibility of this approach to measure relatively large biomolecules (<60 kDa) such as integral membrane proteins and complexes in human mitochondria [34].

Besides, chemical cross-linking and MS (*XL-MS*) is a viable approach to study large protein complexes and protein-protein interactions. *XL-MS* is a process of covalently joining two reactive groups within a single or between several protein molecules using a suitable cross-linking reagent followed by either bottom-up or top-down MS-based proteomics analysis. Presently, both *in vitro* and *in vivo XL-MS* experiments are performed to elucidate the 3D/native confirmation of a protein or to identify the interacting sites between the subunits and large protein assemblies [271]. This combined technique (*XL-MS*) has been recently employed to gain new insights into the enzyme organization in the Krebs cycle [322] and the mitochondrial protein complexes involved in the respiratory chain [175, 262].

#### 9.1.6.1 MS Data Acquisition Strategies

MS raw data can be typically acquired in two different modes i.e. (i) data dependent acquisition (DDA) and (ii) data independent acquisition (DIA) [170]. DDA is a commonly used strategy in bottom-up proteomics for qualitative or discovery-based analysis as highly selective MS and MS/MS scans and high quality MS/MS spectra are obtained for confident protein identifications using database search algorithms. However, one of the major drawbacks of DDA is its stochastic nature of picking precursor ions for the MS/MS analysis and for this reason, oftentimes there is a bias toward the most abundant peptides in a complex mixture (e.g. human). This can cause undersampling of the low abundant species and thereby hinders the dynamic range of the MS measurements (https://doi.org/10.17877/DE290R-17852). In DIA mode the selection criterion for fragmentation is not based on the precursor ions intensities in the full MS scan. Instead, MS/MS analysis is sequentially performed of the entire set of precursor ions in the visible m/z range of a mass spectrometer [296]. Due its unbiased nature of precursor ion selection, DIA is not limited by the dynamic range problem and provides less missing values compared to DDA and therefore it could be a promising alternative for deeper proteome coverages [61]. However, in DIA mode there is a reduction of precursor ion selectivity due to wider segments of the mass range and typically it generates chimeric and complex MS/MS spectra that might be problematic during data analysis and interpretation using typical proteomics-based software tools [296]. Nevertheless, owing to the collaborative efforts of proteomics researchers, instrument and software developers robust analysis platforms are made available for evaluating DIA data [197]. In the recent past, Villeneuve LM et al. applied DIA based LFQ approach to relatively quantify differentially regulated mitochondrial proteins in the brain of PTEN-induced kinase 1 (PINK1) knock-out rat model. Their DIA results in conjunction with magnetic resonance spectroscopy findings indicated altered mitochondrial functions in the absence of PINK1 and provided hints toward early detection of potential biomarkers associated to neurodegeneration [309].

#### 9.1.6.2 Targeted-MS Based Approaches

The targeted-MS technologies include selected reaction monitoring (SRM)/multiple reaction monitoring (MRM) and parallel reaction monitoring (PRM) [232] (Fig. 9.2D) – comparable to immunoblot assays e.g. Western blotting (WB) [248]. Unlike DDA and DIA, these techniques usually do not scan the complete m/z range, but instead rely on the list of specific precursor ion species provided by the user. This enables to increase the (i) sensitivity of the analytes, (ii) achieve linear response over a wide dynamic range, and detection of very low-abundant peptides/proteins in highly complex matrices [164]. Typically, SRM (Fig. 9.2D.I) can be performed on a triple quadrupole instrument where the first (Q1) and the third (Q3) quadrupoles specifically and selectively filter predefined m/z values that are related to a given a precursor and product ions, respectively. The fragmentation process known as beam type collision induced dissociation of the precursor ions is carried out by the second quadrupole (Q2). The main criterion for signal detection therefore, relies on the exact pair of precursor and product ion m/z values, which is usually known as a "transition". However, in the PRM based targeted-MS (Fig. 9.2D.II), the last quadrupole is exchanged with a high resolution, accurate mass accuracy (HRAM) mass analyzer (e.g. the Orbitrap). The high quality i.e. HRAM full MS/MS spectrum consisting of product ions of the corresponding targeted peptide is used to integrate and subsequently generate respective peak areas that facilitate peptide/protein quantification [171, 232]. Moreover, SRM and PRM can be used for absolute quantification (AQUA) of desired proteins [89]. By spiking-in SIL peptides i.e. adding a known amount of SIL counterpart of the respective endogenous peptide as internal standards/surrogates to the sample digests prior to LC-MS analysis enables absolute peptide/protein quantification. Recently, Lam MPY et al. developed a MRM-based workflow to quantify some of the key phosphorylated mitochondrial membrane proteins in the cardiac tissue of both man and mouse [163]. Furthermore, Ordureau A et al. used PRM technology to study PINK1-dependent phosphorylation of parkin and ubiquitin, which play an important role in mitochondrial ubiquitination and mitophagy [205].

### 9.1.7 Brief Introduction of State-of-the-Art Respiratory Analysis of Mitochondria

Although changes in lipid and protein homeostasis are very likely to affect the mitochondrial molecular profile which, in turn, regulates mitochondrial structure and function [126, 217], little is known on the linkage between the mitochondrial molecular fingerprints between different tissues, but also different species, and their impact on mitochondrial function. Since this could provide an important tool to disentangle tissue-specific mitochondrial alterations especially upon external stimuli such as diet-induced excess substrate flux causing IR and diabetes, mitochondrial function is also covered in the following.

Mitochondria are highly dynamic double-membraned organelles producing energy by oxidative phosphorylation (OXPHOS). Four respiratory complexes NADH-ubiquinone (coenzyme Q; CoQ) reductase (complex I), succinate-CoQ reductase (complex II), CoQ-cytochrome c reductase (complex III) and cytochrome c oxidase (complex IV) form the ETC. Complexes I, III, and IV pump reduction- and oxidation-driven protons out of the matrix building up an electrochemical proton gradient. This gradient is used by the ATP synthase (complex V) to convert ADP to ATP. Freely diffusible CoQ and cytochrome c transfer electrons between complexes (Fig. 9.3). *In vitro* investigations of isolated mitochondria or tissue, permeabilised applying detergents such as saponin, can be used to study electron flows through specific sites/complexes allowing the detection of rather subtle changes in the respiration rate of mitochondria such as consequences of gene mutations, diet, and environmental insults (Fig. 9.5 and Table 9.1). Electron transfer to specific sites/complexes results from metabolisation of different substrates. For example fatty acyl-CoAs added to carnitine (such as octanoylcarnitine) to enable transport into mitochondria via carnitine-palmitoyl-transferase 1 (CPT1) and CPT2 can be used as substrates for β-oxidation. It provides electrons to complexes I and II by the metabolisation of acetyl-CoA through the TCA cycle, but also through electron transfer flavoprotein (ETF) via $FADH_2$ as a product from β-oxidation. Pyruvate however after crossing the membrane via the mitochondrial pyruvate carrier (MPC) is decarboxylated by the pyruvate dehydrogenase complex to acetyl-CoA, $CO_2$ and NADH($H^+$) and therefore solely metabolised in the TCA cycle, transfers electrons to complex 1 [219, 226, 241]. The definition of mitochondrial respiratory states was originally introduced by Chance and Williams [40], with state 1 being defined as mitochondria in respiration buffer without substrates and adenylates, state 2 being supplemented with ADP before endogenous substrate exhaustion, state 3 being defined as respiration with excess substrate and ADP to ATP conversion, state 4 being defined as respiration with excess substrate but absence of ADP and state 5 after exhaustion of oxygen. This nomenclature however was redefined amongst others by Gnaiger [94], defining the intrinsically uncoupled non-phosphorylating respiration (without ADP present) or inhibition of the ATP synthesis as LEAK state and the oxidative phosphorylation, when ADP is added in kinetically saturated concentrations additionally to the substrate as OXPHOS state. The maximal uncoupled respiration in presence of an external uncoupler is defined as ET (electron transfer) pathway.

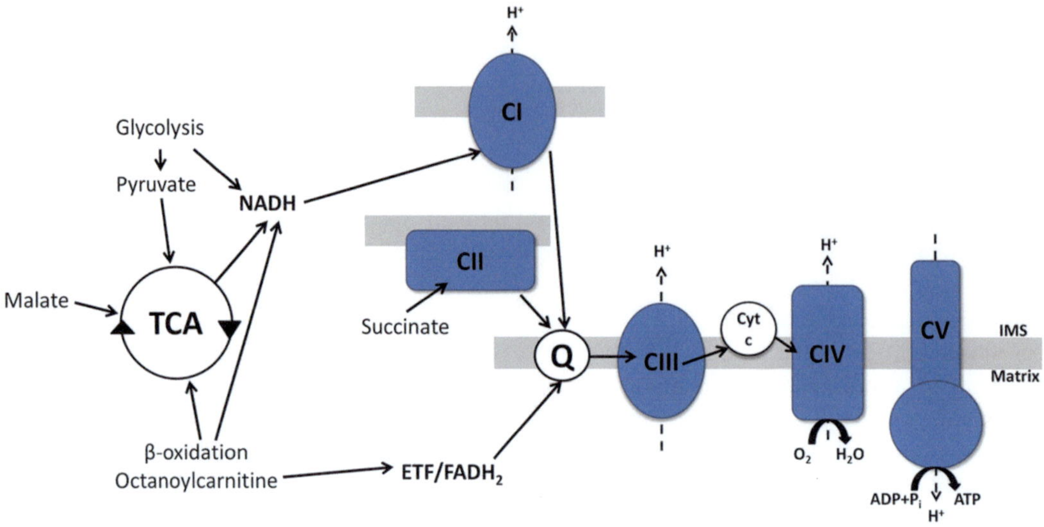


**Fig. 9.3** Scheme of the mitochondrial electron transport chain and frequently used substrates in respiratory analyses. Pyruvate is used as complex I-associated substrate. Malate is added to replenish the TCA cycle. Succinate is a complex II-linked substrate, while fatty acyl carnitines such as octanoylcarnitine are broken down via β-oxidation to acetyl-CoA (further metabolised in the TCA cycle), NADH (electron transfer to complex 1) and $FADH_2$ provides electrons via ETF to CoQ. *TCA* tricarboxylic acid cycle. *CI* NADH-ubiquinone reductase, *CII* succinate-CoQ reductase, *CIII* CoQ-cytochrome c reductase, *CIV* cytochrome c oxidase, *CV* ATP synthase, *Cyt c* cytochrome c, *Q/CoQ* ubiquinone/coenzyme Q, *IMS* intermembrane space, *ETF* electron transfer flavoprotein complex

This state reveals the limitation of OXPHOS capacity by the electron transport chain. After full inhibition of ET pathways or absence of fuel substrates the state is called residual oxygen consumption (ROX).

Over the past 60 years, analyses of mitochondrial (dys) function were mostly performed as mitochondrial respiration analyses with classical oxygen electrodes [39]. They were however limited for example in regard to throughput, background noise and resolution [93]. Two new systems have been lately developed: the high-resolution Oxygraph-2k (O2k, Oroboros Instruments, Austria) and the high-throughput Seahorse XF Extracellular Flux Analyzer (Seahorse XF, Seahorse Bioscience Inc.) for functional mitochondria investigations. Both analyse the mitochondrial respiratory function in real time in *in vitro* experiments on isolated mitochondria, but also in cultured cells and tissues with high resolution and high sensitivity. The O2k was developed in mid 1990s by Gnaiger et al. [95]. By performing the injections into the O2k chambers manually, each complex of the electron transport system can be studied separately and sequentially. However, this technique is rather labour intense, since constant monitoring and adjustment are needed and the Oxygraph is not capable of high-throughput performance since only two samples can be analysed at a time. The Seahorse XF was introduced by Wu et al. [323] to address the need for high-throughput systems in mitochondrial analyses. Oxygen consumption rate and glycolysis (extracellular acidification rate) [72] are measured by a phosphorescent probe-containing bio cartridge as a lid. It is based on 24 or 96 microplate assays allowing the analysis of multiple samples at the same time. Optimisation of the concentrations before the experiment and running costs are rather high, since the special plates are single use products.

## 9.2 Mitochondrial Sample Preparation and Purity Control

A causative role for alterations in mitochondrial lipid and protein composition in mediating mitochondrial dysfunction is very likely, but in comprehensive studies, that combine functional and systems biological approaches, it is rather challenging to isolate adequate amounts of sufficiently pure mitochondria. This is often hampered by a small sample amount such as human biopsies. Diverse mitochondrial isolation procedures exist with different requirements when it comes to time and equipment that is needed. Additionally, different principles of separation are applied, why choosing the most suitable method is quite challenging for a scientist planning analyses of mitochondrial fractions. Contaminations by other organelles may result in misleading analytical findings and impure samples may also hamper the comparability of data between different groups [102, 207]. Since most lipids and many proteins are not specific for mitochondria and therefore quantification in a mitochondria-specific fashion is challenging, adequate mitochondrial isolation and purification protocols are a prerequisite for valid lipid and protein profiling strategies.

### 9.2.1 Purification of Mitochondria from Tissues or Cell Cultures

The most commonly used and simple method to isolate mitochondria from tissues or cell cultures is by differential centrifugation (DC) [11, 15, 186, 236, 295]. This might be due to the fact that it is relatively fast and provides intact and vital mitochondria and is therefore well suited for investigations of mitochondrial physiology [80, 261]. Briefly, tissues or cells are minced and homogenised, commonly using a Dounce homogeniser [5, 16, 54]. Cell disruption supported by a detergent such as digitonin has also been reported [129, 152, 192]. A slow centrifugation step sediments unbroken cells and cell debris (800–1000 g), whereas a subsequent fast centrifugation step of the supernatant pellets the mitochondria (8000–10,000 g). To achieve a higher purity than acquired by DC, a subsequent high-speed centrifugation on a density gradient (for example percoll or sucrose) is added (Fig. 9.4) [4, 148, 168, 186, 294]. This step might however lead to losses of mitochondria and requires an ultracentrifuge, which might not be accessible for every lab/scientist. However, methods based on centrifugation tend to be more time-consuming if a large number of samples are simultaneously extracted. Commercialised more rapid approaches to isolate mitochondria exist based on antibody-coupled bead separation as for example the so-called MACS (magnetic cell isolation and separation) procedure, which was first described by Hornig-Do et al. [125]. Mitochondria from a homogenised sample are coupled to anti-TOM22-antibodies, which are conjugated to paramagnetic beads and can then be isolated in a magnetic field. The MACS procedure achieves higher yields of mitochondria [125]. Liver mitochondria isolated by MACS showed higher oxygen consumption rates than mitochondria isolated by DC and less contaminations by ER, microsomes and peroxisomes as observed in proteomics analyses [78]. In an approach combining WB analyses with lipidomics analyses, however, mitochondria isolation by UC was revealed as superior for enriching mitochondria and depleting contaminations by other organelles, followed by DC and then MACS [142]. Mitochondrial fractions purified by UC were shown to contain higher amounts of the only mitochondria-specific lipid, cardiolipin, and higher abundance of mitochondrial markers for ATP-synthase [142].

In addition to different analytical approaches, lipidomics data published so far originated from various mitochondria isolation methods: either DC [11, 15, 186, 295] or UC on density gradients like iodixanol [4], ficoll and sucrose [147, 148], or percoll [143, 186, 294, 295].

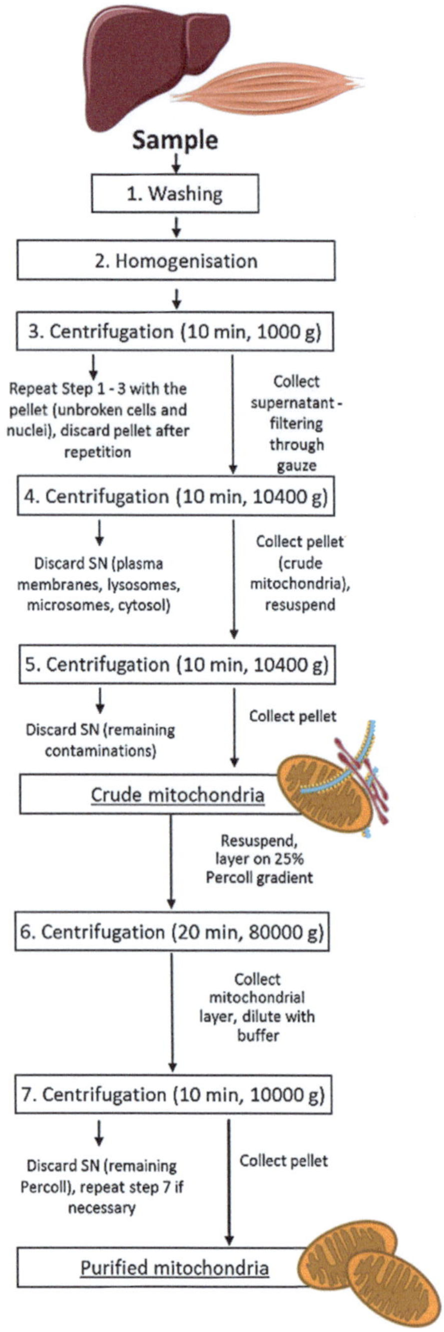


**Fig. 9.4** Scheme of a suggested mitochondrial isolation strategy applying differential centrifugation followed by ultracentrifugation (according to Kappler et al. [142]). Figure created using illustrations provided by Servier medical art. *SN* supernatant

### 9.2.2 Sample Preparation for Lipidomics

Prior to detection, mitochondrial lipids are extracted using organic solvents (see below). A determination of the protein content (e.g. by Bradford or bicinchoninic acid assay) in an aliquot of the mitochondrial fraction, which is usually suspended in buffer solution (e.g. sucrose, Tris and EGTA; [142]), before the extraction allows the quantification of lipids per amount mitochondrial protein. To avoid oxidation of polyunsaturated fatty acids (PUFA) freeze-thaw cycles of the mitochondria should be avoided, which can be achieved by aliquoting the mitochondrial fraction prior to storage at −80 °C. Before the lipid extraction procedure, preferably on ice or at 4 °C, internal standards can be spiked in. 80% methanol can universally be applied to extract amphiphilic metabolites with mixed polarity [266]. The extraction of neutral lipids like for example triglycerides however requires the addition of a lipophilic organic solvent. Most lipids can be extracted using the classic Folch method (chloroform:methanol = 2:1) [76] and the modified Bligh/Dyer method (chloroform:methanol = 1:2) [18], developed in the late 1950s, using a mixture of chloroform, methanol and water, resulting in a separation of polar and non-polar compounds in two phases [4, 5, 11, 15, 148, 168, 186, 236]. Proteins are simultaneously precipitated using this mixture. However, the precipitated proteins are located in between the upper polar phase and the apolar, lipid containing phase, which makes the contamination-free access to the lipid fraction quite challenging. Besides with the addition of methanol or isopropanol, proteins can be removed by an acid. This however might lead to hydrolysis of some lipids [266]. Several alternative solvents such as isopropanol, ethyl acetate, hexane and methyl tert-butyl ether (MTBE) can be used for comparable lipid extraction efficiency for major lipids, but with a varying recovery of low abundant lipids [242]. Matyash et al. [184] developed a sample extraction method for lipidomics replacing chloroform by the less hazardous MTBE. Non-extractable constituents are pelleted at the bottom and the upper non-polar phase and the lower polar phase are easy accessible. This method was modified for the complementary analyses of lipophilic and amphiphilic metabolites from small tissue samples [41] and was successfully applied for lipid extraction from cell culture mitochondria [142] and tissue mitochondria [143]. Briefly, water was added to 100 μg (by protein) of mitochondria to reach a total volume of 100 μl. After this, 350 μl of ice-cold methanol (MeOH) including internal standards were added. Samples were briefly vortexed, 1 ml of MTBE was added, and the samples were shaken for 30 min at room temperature. After adding 250 μl of water and incubating the sample at room temperature for 10 min, samples were centrifuged for 20 min at 1000 g and 4 °C to induce phase separation. With the relative amounts of MTBE:methanol:$H_2O$ (20:6:7) a 2:1 ratio of upper (non-polar) to lower (polar) phase was yielded. However, the extraction is dependent on the following applied analytical method. Determination of fatty acids via gas chromatography e.g. needs a further methylation step [11, 118]. The extracted lipids are often separated by chromatography such as thin layer chromatography [118], and HPLC [236] or are directly infused into a mass spectrometer [146].

Recently, Coman et al. introduced a sample preparation strategy that facilitates concurrent identification of different classes of biomolecules from the same starting material [50]. This elegant, comprehensive sample preparation strategy allows the implementation of systems biology oriented research even in projects with very low sample amounts, like human tissue samples. In their so-called simultaneous metabolite, protein, lipid extraction (SIMPLEX) approach, which is performed in one sample tube, apolar lipids are separated from more polar metabolites and the precipitated proteins are used for proteomics analysis to simultaneously study the interaction between key players/pathways in a biological system (see also flowchart, Fig. 9.5). Coman et al. have employed well-established protocols including: (i) MTBE, methanol, water based method for separating lipids and more polar metabolites [184], (ii) filter aided sample preparation (FASP) [321] for proteolytic digestion (see Sect. 9.2.3) and (iii) titanium dioxide chromatography for enrichment of phosphopeptides [290]. This multi-omics workflow allowed them to relatively quantify 3762 biomolecules (i.e. 360 lipids, 75 metabolites, and 3327 proteins) in parallel in a untreated vs. a peroxisomal proliferator-activated receptor gamma activated OP9 cell line and notably, by using only ~one million cells per condition.

### 9.2.3 Sample Preparation for Proteomics

In bottom-up proteomics workflow, the isolated mitochondria irrespective of the tissue of origin should be lysed to release proteins for the downstream proteolysis followed by LC-MS analysis. Typically, the lysis buffer should consist of a detergent e.g. sodium dodecyl sulfate (SDS) to rupture the double-membrane structure of the mitochondria. Inclusion of protease inhibitors (e.g. complete mini, Sigma-Aldrich) in the lysis buffer is beneficial to prevent unspecific proteolytic activity. For studies directed towards the analysis of PTMs such as phosphorylation, it is necessary to add phosphatase inhibitors (e.g. PhosSTOP, Sigma-Aldrich) to all the buffers that are used during the mitochondria preparations. Furthermore, it is recommended to perform freezing and thawing cycles of the lysates that might enable to release membrane-bound proteins into the surrounding environment. After lysis, protein concentration of lysates is determined (e.g. BCA assay) and carbamidomethylation step is performed i.e. reduction of the disulfide bonds (Cys) followed by subsequent alkylation of free thiol (sulfhydryl) groups. Next, the sample clean-up and proteolysis can be done in two ways: (i) organic-solvent (e.g. acetone, ethanol or trichloroacetic acid) based protein precipitation followed by in solution enzymatic digestion and (ii) using molecular weight cut off (MWCO) membrane spin filters. The second one is relatively more popular approach that was introduced by Manza et al. [181] and later adapted as filter aided sample preparation (FASP) by Wiśniewski et al. [321]. The FASP protocol showed promising results in eliminating majority of the cell/tissue lysis buffer ingredients that are incompatible with downstream LC-MS analysis particularly detergents such as SDS. Briefly, sample lysate is first diluted with highly concentrated (~8.0 M) urea containing buffer and then placed on a centrifugal device that has s a membrane filtration unit with nominal MWCO of 10, 30 or 50 kDa. The usage of high concentrated urea buffer was demonstrated to be highly effective in removal of SDS. Upon centrifugation, proteins are captured on the MWCO filter unit and most of the lysis buffer components including salts and detergents are passed into the flow through. Next, the device is treated with an alkaline buffer (e.g. ammonium bicarbonate, pH 7.8) to eliminate residual urea from the membrane filter. Subsequently, proteolysis is carried out on the membrane unit by addition of an enzyme (e.g. trypsin) containing buffer and incubation under suitable conditions for the type of protease used (https://doi.org/10.17877/DE290R-17852). After FASP, the clean-up of peptides can be done by solid phase extraction (SPE) to remove undesirable digestion buffer components prior to LC-MS analysis. For phosphoproteome analysis, enrichment of the phosphopeptides is a prerequisite owing to their low stoichiometry in complex sample digests. Hence dedicated, sensitive and well-established protocols (e.g. immobilized metal ion and metal oxide affinity chromatography based enrichment) are established, which usually complement one another and facilitate detection of the phosphopeptides by LC-MS [63, 332].

### 9.2.4 Purity Control of the Mitochondrial Components

Tests for contaminations or quality controls of mitochondrial enrichment are not always reported in analytical experiments. However, analytical studies performed on poorly purified mitochondria may result in erroneous conclusions if the samples contain relevant contaminations by the ER and other organelles, as discussed by Kappler et al. [142] for mitochondrial lipids. Investigations in skeletal muscle tissue revealed cardiolipin (CL) as a superior marker of mitochondrial content [165]. Lipids other than the mitochondrial signature lipid CL cannot exclusively be assigned to individual subcellular components. Published CL contents however range between 10 to 20% [48, 53] of total phospholipids reported for mitochondria from liver, yeast and cauliflower. This variation was suggested to be based on not solely different experimental conditions, but also by a variation in the degrees of purity of the mitochondrial samples [207] and a tissue- or species-specific cardiolipin contribution [126, 143, 186, 258]. The activity of citrate synthase (CS), located in the mitochondrial matrix, is also commonly used as a quantitative marker for mitochondrial content in tissue and cells [24, 70, 165]. However, CS is discussed to be regulated for example acutely by exercise in humans questioning its role as a marker in exercise studies [166, 300]. This leads to the assumption that citrate synthase protein might be a better measure. Organelle-specific protein markers were reported as reliable markers such as ATP5 for mitochondria [165]. This marker as well as markers for potential contaminations by the ER (e.g. GRP78, calnexin), the nucleus (e.g. lamin B, fibrillarin), cytoplasm (e.g. glutamine fructose-6-phosphate amidotransferase), lysosomes (e.g. lysosomal membrane glycoprotein 1) and the Golgi apparatus (e.g. giantin) as well as lipid droplets (e.g. perilipin 2) should be investigated as a purity examination by WB or comparable assays. This is recommended in the early phase of an experiment to ensure the quality and reliability of subsequent publications. Additionally, LC-MS data acquired from the isolated mitochondria digests can be used to determine the purity of this organelle. This is done by comparing the list containing identifiers i.e. gene-IDs/UniProt IDs of all identified proteins (preferably with $\geq$2 unique peptides and 1% FDR settings) obtained in a particular bottom-up proteomics LC-MS measurement with the entries in mitochondrial database e.g. MitoCarta 2.0 to calculate the share of proteins that are localized in the mitochondria [310].

## 9.3 The Mitochondrial Lipidome of Liver and Muscle in Type 2 Diabetes Mellitus

### 9.3.1 Type 2 Diabetes-Associated Alterations of Mitochondrial Lipids

Obesity, a major risk factor for T2Dm, is characterised by lipid accumulation (e.g. TAGs) in lipid droplets, adipocytes, but also in ectopic tissues such as liver and skeletal muscle [254]. Since decades the lipids stored ectopically are suggested as key players in IR development [133, 159, 229]. A current working model of lipid-mediated IR is a release of lipids as FAs from adipose tissue into the circulation when the storage capacity of adipocytes is exceeded [247, 273], but also by a concomitantly occurrent decrease of insulin-mediated suppression of lipolysis [19, 268]. These lipids are proposed to accumulate in skeletal muscle and liver as lipid droplets potentially triggering IR and T2Dm [247, 273]. A variety of lipids such as diacylglycerols (DAGs), ceramides (CERs) and associated metabolites like acylcarnitines have been implicated as mediators of IR [253]. The molecular mechanism, however, linking obesity with IR and T2Dm is poorly understood. A link between IR and mitochondrial (dys)function came in recent years more and more into the focus of research due to the central metabolic function of mitochondria and their essential role in regulation of bioenergetic processes, of cell response and epigenetic modulation [1, 208, 311]. In IR and T2Dm, mitochondrial dysfunctions were reported extensively but quite controversially for skeletal muscle and liver, including studies showing increased respiratory capacities, unchanged respiratory capacities and decreased capacities (see in Sect. 9.1.1). The IR- and T2Dm-associated mitochondrial (dys)function might be based on alterations in lipid and protein homeostasis affecting the mitochondrial molecular profile and in turn mitochondrial structure and function (Table 9.2).

Exemplarily, CERs amongst other lipids and proposed mechanisms were discussed to activate the intrinsic apoptosis pathway by increasing the membrane permeability of mitochondria, leading to a release of cytochrome c and by that trigger apoptosis of β-cells via the caspase cascade [49, 86, 243]. These channels are assumed to be BAX (proapoptotic protein) CER oligomers [86, 243] or self-assembled CER channels [49]. Palmitate stimulation of a murine skeletal muscle cell line (C2C12) leading to an inhibition of insulin signalling caused an accumulation of sphingomyelin (SM) precursors such as sphinganine, dihydroceramide, and CER [218]. Accumulated CERs but not SMs were also seen in skeletal muscle mouse tissue from obese mice [218]. Bajpeyi et al. [9] observed an inverse relationship of CER and DAG levels with insulin sensitivity in muscle biopsies from T2Dm patients and healthy donors. Insulin-deprived streptozocin (STZ)-diabetic C57BL/J6 mice also showed increased CER content in quadriceps muscle [328]. In these mice, content and composition of sarcoplasmic fraction sphingolipids were also affected, but in mitochondria only sphingosine was increased [328]. Lipidomics brought attention to individual CER species like increased C16:0 CERs inhibiting fatty acid oxidation and negatively regulating insulin signalling and energy expenditure in obese humans and mice [84]. Induction of anabolic processes might promote lipid storage to avoid this lipotoxicity. Indeed, ablation of the dihydroceramide desaturase 1, an enzyme that produces CERs, in mouse embryonic fibroblasts (MEFs) enhanced starvation responses, dissociated from a strong activation of the antiapoptotic and anabolic signalling pathway regulated by Akt/protein kinase B and showed a replacement of the most common sphingolipids by dihydro forms [270].

Another hypothesis linking mitochondria to lipid-mediated IR concerns altered β-oxidation in obesity and T2Dm. This leads to incomplete fat oxidation concomitant with an impaired switch to carbohydrate oxidation and partially TCA cycle substrate depletion associated with an accumulation of acyl-CoAs, their respective acylcarnitines and IR [135, 158]. Increased DAG levels were reported in total membrane fractions (after removal of the cytosolic fraction) from

**Table 9.2** Survey of selected reports on total tissue and mitochondrial lipids from muscle and liver in insulin resistance and diabetes

| Tissue | Method | Findings | Material | References |
|---|---|---|---|---|
| Human skeletal muscle | 52 male volunteers (Pima Indians), euglycemic clamp, GC | Unsaturation of PLs correlated with insulin action | Muscle biopsy | [211] |
| Mouse heart | Diabetes wt mice (4 mo) induced by single i.v. injection of STZ/ob/ob mice; shotgun lipidomics (ESI-MS) | CL content decreased, tetralinoeyl-CL depleted and replaced by longer and more unsaturated FAs | Tissue | [108] |
| Rat heart | Male Sprague Dawley rats; diabetes induced by STZ, ESI-MS | TAG remodelling (fivefold increase in tripalmitin mass; 60% decrease in polyunsaturated TAG molecular species mass); 46% increase in PI mass; 44% increase plasmenylethanolamine; 22% decrease in 1-stearoyl-2-arachidonoyl PE | Tissue | [106] |
| Mouse heart | Diabetes wt mice (4mo) induced by single i.v. injection of STZ; ESI-MS | Decreased CL; metabolic precursor of CL (=PG) also substantially depleted; glycerol 3-phosphate, necessary for the penultimate step in PG production, decreased by 58% in diabetic myocardium | Tissue | [107] |
| Rat heart | Male Sprague Dawley rats; diabetes induced by STZ; ESI-MS | Increase of both calcium-independent phospholipase A2 (iPLA2) mRNA and iPLA2 activity in rat myocardium; dramatic increase in acylcarnitine molecular species | Tissue | [278] |
| Rat heart | Alloxan-diabetic male Wistar rats; TLC | PC: decreased 20:4(n−6) and 16:0, increased 18:2(n−6) and 18:0 | Tissue | [103] |
| | | PE: 20:4(n−6) significantly reduced | | |
| | | CL, PS, PI: unchanged | | |
| | | TGs: increased by 90% and contained significantly higher levels of 18:2(n−6) | | |
| Rat heart | 4 days or 28 days after STZ injection; TLC | No difference in the amount of radioactivity incorporated into CL, PG or other PLs; decrease in PG; did not affect the activity of the enzymes of PG and CL biosynthesis in mitochondrial fractions | Tissue/mitochondria | [112] |
| Rat heart | STZ-induced diabetes or hyperinsulinemia in rats; TLC | Unchanged CL content and MLCL acyl transferase and CL synthase activities | Mitochondria | [289] |
| Mouse skeletal muscle cell line (C2C12), skeletal muscle | BSA-conjugated palmitate to increase synthesis of endogenous sphingolipids and to inhibit insulin signalling and oxidative phosphorylation in cell culture; quadriceps muscles from obese mice with impaired glucose tolerance; LC/MS-MS | Cells: accumulation of SM precursors such as sphinganine, dihydro-CER, and CER; inhibition of insulin stimulation of a central modulator of anabolic metabolism, Akt/PKB; inhibition of insulin-stimulated glycogen synthesis; decrease of oxygen consumption and ATP synthesis | Cells, tissue | [218] |
| | | Tissue: accumulated CERs but not SMs | | |

(continued)

**Table 9.2** (continued)

| Tissue | Method | Findings | Material | References |
|---|---|---|---|---|
| Mouse liver | Mice (C57/BL6) were fed a high-trans-fat high-fructose diet (TFD) for 8 wk. to induce simple steatosis and NASH by 24 wk.; LC-MS/MS | Steatosis: triglyceride accumulation and IR, TCA cycle fluxes remained normal. | Tissue | [221] |
| | | NASH: twofold induction of mitochondrial fluxes through the TCA cycle, anaplerosis and pyruvate cycling; blunted ketogenesis; accumulation of hepatic DAGs, CERs and long-chain acylcarnitines, suggesting inefficient oxidation and disposal of excess FFAs. | | |
| Human muscle | Lean individuals, endurance-trained athletes, obese men and women with and without T2Dm; LC-MS/MS, insulin sensitivity was measured using hyperinsulinemic-euglycemic clamp | Sarcolemma: 1,2-DAGs were not significantly related to insulin sensitivity, CERs were inversely related to insulin sensitivity (significant for the C18:0), SMs also inversely related to insulin sensitivity (strongest relationships for the C18:1, C18:0, and C18:2 species) | Muscle biopsies; fractionated into sarcolemmal, cytosolic, mitochondrial/ER, and nuclear compartments | [227] |
| | | Mitochondrial/ER and nuclear fractions: 1,2-DAGs were positively related to, while ceramides were inversely related to, insulin sensitivity. All sphingolipids but only specific DAGs administered to isolated mitochondria decreased mitochondrial state 3 respiration. | | |
| | | Cytosol: lipids not related to insulin sensitivity | | |
| Mouse heart and liver tissue | Type 2 diabetes in obese (NZO/NON)F1 male mice; TLC-GC | Liver: rosiglitazone-mediated substantial accumulation of TAGs within the liver | Tissue | [318] |
| | | Heart: treatment of diabetic mice with rosiglitazone, a drug increasing insulin sensitivity, increased CL levels and led to a remodelling towards higher tetralinoleyl-CL and reduced C22:6 fatty acyl chains in heart | | |
| Mouse heart | STZ-induced diabetes in 6-wk-old mice; LC-MS/MS | IFM: CL content decreased, CRLS activity decreased, ATP synthase activity decreased | Mitochondrial subpopulations (IFM, SSM) | [52] |
| | | SSM: no changes | | |
| Rat liver | STZ-induced diabetes in 48-h-old male Wistar rats; GC | Mitochondria: when lipoperoxidation still was not significant, the membrane fluidity increased because of the increment in the unsaturated to saturated fatty acids ratio (U/S) | Mitochondria and tissue | [225] |
| | | Tissue: diabetes induced a decrease in the U/S fatty acids ratio of liver total lipids | | |

(continued)

**Table 9.2** (continued)

| Tissue | Method | Findings | Material | References |
|---|---|---|---|---|
| Mouse liver | Fed either a regular normal chow diet or a HFD for a total of 8 wk., 4 wk. exercise; LC-MS/MS | DAG and cholesterol esters accumulated in HFD, decreased with training; the PC/PE ratio (associated with membrane integrity and linked to hepatic disease progression) increased by training | Tissue | [139] |
| Human muscle | Healthy sedentary obese controls, individuals with type 2 diabetes and lean endurance-trained athletes; fractionated using ultracentrifugation, LC-MS/MS | Higher total DAG levels in obesity and T2Dm, mostly in membranes<br>Membranes: C18:0/C20:4, Di-C16:0 and Di-C18:0 more in T2Dm, total membrane DAG and Di-C18:0 inversely correlated with insulin sensitivity | Muscle biopsy, fractionation into membrane and cytosolic fractions | [13] |
| Human muscle | Serial muscle biopsies in healthy, lean subjects before and during a lipid infusion, fractionated using ultracentrifugation, LC-MS/MS | Acute induction of muscle IR was associated with a transient increase in total and cytosolic DAG content | Serial muscle biopsies, fractionation into membrane and cytosolic fractions | [283] |
| Human muscle | Lean insulin-sensitive volunteers; i.v. fat, oral fat, i.v. endotoxin, and i.v. glycerol as control, LC-MS/MS | After 6 h, whole-body insulin sensitivity was reduced by both fats and LPS, while hepatic insulin sensitivity was unaffected.<br>Membrane $(C18:2)_2$ DAG species doubled after i.v. fat and correlated with PKCθ activation after oral fat, whereas CERs were unchanged | Muscle biopsy, fractionation into membrane and cytosolic fractions | [202] |
| Human muscle | Trained and sedentary participants; 6-h infusion of lipid or glycerol; serial muscle biopsies, (UP) LC-MS/MS | Lipid infusion reduced insulin sensitivity.<br>Increased linoleic and linolenic acid content of TAGs without changing total TAGs.<br>In the sedentary group, lipid infusion increased total, oleic, and linoleic and linolenic acid content of DAG.<br>Regardless of training status, lipid infusion did not alter total CERs, saturated CERs, palmitoyl-carnitine, or oleoyl-carnitine. | Muscle biopsy | [45] |
| Mouse muscle | C57BL/6 J mice were fed an 8-wk or 20-wk low fat diet or HFD (insulin sensitivity reduced); skeletal muscle mitochondria were isolated and FA composition of skeletal muscle mitochondrial PLs was analysed by TLC-GC; HRR | 8 weeks of HFD: mono-unsaturated FAs (=MUFAs) (16:1n7, 18:1n7 and 18:1n9) were decreased, whereas saturated FA (16:0) were increased in phospholipids<br>20 weeks of HFD: decreased MUFA while n-6 PUFA (18:2n6, 20:4n6, 22:5n6) showed a pronounced increase<br>HRR: after 20 weeks, enhanced maximal capacity of the electron transport chain and a tendency for increased ADP-stimulated respiration, but only when fuelled by a lipid-derived substrate | Mitochondria | [120] |

(continued)

**Table 9.2** (continued)

| Tissue | Method | Findings | Material | References |
|---|---|---|---|---|
| Rat heart | STZ-induced diabetes in 6–8-wk-old male Wistar rat; PLs by TLC-MS | Mitochondria from T1Dm heart presented lower OXPHOS activity and PL remodeling characterized by higher PC levels, lower PG, PI and SM content, higher amounts of long fatty acyl side chains and increased lipid peroxidation, particularly of CL | Mitochondria | [71] |
| Human muscle | Six physically active donors were compared to six sedentary lean donors and six donors with T2Dm; LC-MS/MS and GC-FID | DAGs and ceramides in myotubes were inversely associated with insulin sensitivity | Biopsy, biopsy-derived cells | [9] |
| Mouse muscle | STZ-induced diabetes in 13-wk-old male C57BL/6 J mice, LC-MS/MS | Insulin deprivation in STZ diabetic C57BL/6 J mice increases quadriceps muscle CER content; relationships between the features of the diabetic phenotype, the content of long-chain FA-CoAs and CERs containing C18-fatty acids in sarcoplasm, but not in mitochondria. | Homogenates and sarcoplasmic and mitochondrial fractions | [328] |
| | | Sarcoplasmic fraction: both the content and composition of sphingolipids were most affected by insulin deprivation. | | |
| | | Mitochondrial fraction: Sphingosine was the sole sphingolipid affected by insulin deprivation (level nearly 50% higher). | | |
| Mouse liver | HFD mice were fed with HFD for 5 mo. | HFD led to a marked accumulation of DAGs and a fivefold increase of TAGs, but not FFAs. CL and PE were decreased with HFD. | Tissue | [79] |
| | Second group: STZ was injected intraperitoneally and mice were analysed 2 mo later. | STZ led to severe depletion of hepatic DAGs, TAGs and also FFAs. Unchanged CL and PE levels. | | |
| | TLC | | | |

*CL* cardiolipin, *MLCL* monolyso-CL, *PG* phosphatidylglycerol, *PI* phosphatidylinositol, *CER* ceramide, *PL* phospholipid, *PE* phosphatidylethanolamine, *PC* phosphatidylcholine, *PS* phosphatidylserine, *TAG* triacylglycerol, *DAG* diacylglycerol, *FA* fatty acid, *FFA* free fatty acids, *SM* sphingomyelin, *IR* insulin resistance, *wt* wild type, *i.v.* intravenous, *STZ* streptozotocin, *mo* months, *wk.* week/s, *TLC* thin layer chromatography, *ESI* electron spray ionisation, *MS* mass spectrometry, *GC* gas chromatography, *IFM* interfibrillar mitochondria, *SSM* subsarcolemmal mitochondria, *LC* liquid chromatography, *UPLC* ultra-performance liquid chromatography

diabetic human muscle, which were positively correlated with T2Dm [13]. These increased levels of DAGs in IR were also found additionally (to total levels) in cytosolic fractions [283]. In line with that was an increase in $(C18:2)_2$-DAG in muscle membrane fractions after a 6-h lipid infusion in healthy humans [202]. Serial muscle biopsies during a 6-h lipid infusion contained an increased content of total, oleic, linoleic acid as well as linolenic acid in DAGs (in sedentary participants), while levels of CERs and acylcarnitines remained unchanged (sedentary and trained participants) and activation of insulin signalling transducing molecules was reduced [45]. In mitochondrial/ER fractions from human muscle biopsies, however, 1,2-DAGs were positively related to insulin sensitivity, while CERs were inversely related [227]. Given this finding of alterations in tissue and some studies about mitochondrial lipidome levels, a closer investigation of the mitochondrial lipidome itself in combination with proteomics and subcellular fractionations might be essential to elucidate the spatio-temporal dynamics of lipid species and possible effector proteins underlying IR development and the role of mitochondrial (dys)function hereby.

Besides cardiolipin (CL), the mitochondrial signature lipid, most lipids are not specific for individual organelles and therefore challenging to quantify in tissue lysates in a mitochondria-specific fashion. CL is therefore, because of its organelle specificity, quite often investigated in whole tissue. For the other lipid species an accurate, comprehensive lipid profiling strategy, which enriches mitochondria and minimizes contaminations by membranes from other organelles, is a prerequisite (see Sect. 9.2.1) to investigate the contribution of individual lipids to mitochondrial composition and in turn function.

#### 9.3.1.1 Cardiolipin, a Key Lipid in Insulin Resistance (IR) and Type 2 Diabetes (T2Dm)

Most investigations about cardiolipin, the signature lipid of mitochondria, were performed in animal heart muscle [107, 108, 115]. Its isolation from heart for the first time led to its name designation [213]. CLs are known to be a key player in the organisation of mitochondrial membrane structure and in the organisation of the ETC components into higher order assemblies called supercomplexes [46, 176, 207, 233, 256, 304, 329]. CL was therefore discussed to be required for OXPHOS and to have a broad impact on mitochondrial physiology often investigated in yeast [46, 176, 207, 233, 256, 304, 329]. In heart, high levels of especially tetra-linoleoyl-CL were proposed to be essential for mitochondrial function, whereas decreased levels of this specific CL are associated with cardiac malfunctions including ischemia and reperfusion, but also heart failure, Barth syndrome and diabetes [275]. A defective CL remodelling is discussed in both type 1 (T1Dm) and type 2 diabetes (T2Dm) [107, 115, 318]. "Shotgun" lipidomics analysis of type 1 (streptozocin (STZ)-induced) and type 2 (ob/ob) diabetic mice showed similar alterations in the CL profile [107, 108]. A defective remodelling of CL was shown at a very early stage of pathological development in the heart: The CL content was decreased and the usually dominating tetra-linoleoyl-CL depleted and replaced by CLs with longer and more unsaturated fatty acids (mostly C22:6) [107, 108]. Decreased levels of CL were also found in liver homogenates of HFD-fed mice, but not of STZ-induced diabetic mice (T1Dm) [79]. PG, the precursor of CL, showed no increased 22:6 levels in myocardium [107], but total PG levels were also shown to be decreased in mitochondrial and microsomal membrane fractions of diabetic rat heart [112]. In these mitochondrial fractions from diabetic rat heart, the enzyme activity of CL and PG biosynthesis was however unchanged, as well as an unchanged amount of radioactivity incorporated into CL, PG or other phospholipids, leading to the assumption of the intrinsic PG pool being used for the newly synthesised CL [112]. In interfibrillar, but not subsarcolemmal, heart muscle mitochondria from type 1 diabetic mice the CL biosynthetic pathway was reported as impaired, contributing to decreased CL content, as well as impaired ATP synthase activity presumably by a decreased interaction of ATP synthase F0 complex subunits [52]. In line with the observed higher unsaturation of CL species in diabetic heart, a higher unsaturation of phospholipids (% C20-22 polyunsaturated fatty acids) inversely correlated with insulin action in human skeletal muscle, observed in whole tissue [211]. Accordingly, 20 weeks of high-fat diet in mice decreased the MUFA content and increased the PUFA content in muscle mitochondria [120]. Treatment of diabetic mice with rosiglitazone, a drug increasing insulin sensitivity, increased CL levels and led to a remodelling towards more tetra-linoleoyl-CL and less C22:6 fatty acyl chains in heart [318], a shift associated with amplified ROS production by modulating electron transport efficiency [317]. In an alloxan- or STZ-induced diabetes rat model however unchanged CL levels were found in rat heart tissue [103], but other alterations in lipid composition like decreased amounts of C20:4 in PCs and PEs or higher amounts of TAGs [103], remodelled TAGs [106] and increased acylcarnitine levels [278] were observed in rat heart tissue. Lower levels of CL were already suggested to be the cause for deficits in mitochondrial OXPHOS seen in the Barth syndrome patients having low CL levels due to mutated tafazzin [185]. Indeed, in skeletal muscle, research findings point to either a decreased mitochondrial function in diabetes at least in humans [24, 44, 144, 188, 191, 223, 234, 245], or to unchanged function reported in the diabetes context [55, 73, 167, 301, 308].

In contrast to muscle, mainly increased mitochondrial function was reported in liver of insulin resistant, diabetic or non-alcoholic fatty liver disease (NAFLD) patients [43, 82, 156, 187, 231, 286]. Much less data reporting a decreased mitochondrial function in liver in diabetes exist [281], being more strongly associated with a progression towards non-alcoholic steatohepatitis (NASH) [156, 224]. This inefficient oxidation and disposal of free fatty acids was suggested to cause the accumulation of hepatic DAGs, CERs and long-chain acylcarnitines observed in high-trans-fat high-fructose diet-induced NASH in mice [221]. Hepatic mitochondria seem to cope with chronic excess substrate supply in an early stage of NAFLD by increasing the maximal respiration rate, which seems to fail in the ongoing pathogenesis [156]. Mitochondria and the ER play an important role in NAFLD, since membrane phospholipids accumulate in the ER triggering an ER-stress response and a lipoinflammatory process [21], a hallmark for metabolic disease [127]. Although the role of especially altered subcellular structures on molecular level in disease progression is not entirely understood [155], it has been recently proposed that an increased ER-mitochondrial linkage with exchange of calcium, lipids and metabolites at the contact site of mitochondria associated-membranes (MAM) might be an essential component in the dysfunction of both organelles in obesity-related diseases [7].

#### 9.3.1.2 Mitochondrial Lipids Affected by Diet and Exercise

The availability of certain fatty acids are proposed to be the limiting factor determining the mitochondrial lipid composition due to the observation that the acyl chain composition of cardiolipins reflects the dominant acyl chains of the surrounding tissue [259]. The composition of mitochondrial membranes from murine liver and skeletal muscle is shown to be adaptable to the provided plant oil diet [239]. Supplementation with 22:6 and 20:5 acyl chains increased tetralinoleoyl-CLs and CLs containing 22:6 acyl chains in isolated heart mitochondria of rats [201], indicating an impact of dietary long-chain PUFAs on mitochondrial phospholipid composition [276]. In cardiac mitochondrial phospholipids, supplementation with 22:6 acyl chains was reported to increase both 22:6 and 20:5 acyl chains and decrease 20:4 acyl chains. This is discussed as a cardioprotective effect of these fatty acids [146]. Liver mitochondria from rats maintained on a diet containing trans-fat as the major constituent, which is associated with diabetes development [130], showed the greatest relative percentage of mitochondrial monolyso-CL (MLCL) $(18:2)_3$ compared to diets enriched with SFA, MUFA and ω3/ω6-PUFA [15]. These high levels of MLCL $(18:2)_3$ were discussed to reflect impaired CL maturation or increased steady-state oxidative stress [15]. Hence, changes in lipid homeostasis are likely to regulate mitochondrial structure and thereby function [126, 217]. Influencing or targeted manipulation of the mitochondrial molecular composition might be a way to address mitochondrial (dys)function in diabetes.

Physical exercise has been frequently postulated to combat obesity-related complications such as insulin resistance, T2Dm and NAFLD with proposed mechanisms likely due to its impact on cellular metabolism and mitochondrial alterations. Physical activity was shown to increase mitochondrial mass and oxidative capacity in skeletal muscle [134, 251, 298, 299]. In rat liver mitochondria higher amounts of saturated fatty acids, lower amounts of MUFAs and higher amounts of PUFAs were reported due to exercise [239]. In mouse muscle mitochondria less PUFAs were detected after exercise [239]. The lower amount of MUFAs solely seen for liver mitochondria, was discussed to be based on the higher metabolic rate of liver [239]. Another study from this group revealed, that rats fed with a diet enriched in sun flower oil showed increased SFA, MUFA and PUFA (n-6) percentages in liver but not in muscle mitochondrial lipids after 8 weeks of submaximal training [240]. Training decreased the high-fat feeding-induced accumulation of DAGs and cholesterol esters in murine livers, the PC/PE ratio, associated with membrane integrity, was increased by training and reduced fatty acid transporters CD36 and FATP4 suggested a reduced expression of fatty acid entry into hepatocytes [139]. The different response to exercise of liver and muscle on lipid level, might also play a role in context of exercise-caused oxidative stress. Eight weeks of regular treadmill training increased peroxidative stress (hydroperoxide amount) in mitochondrial membranes of rat livers after a sunflower and olive oil enriched diet and for rat muscles upon the latter [183]. In this study, almost sixfold lower amounts of hydroperoxide levels were reported in muscle compared to liver [183]. This led to the hypothesis that muscle might be more resistant to diet-induced oxidative stress (diet with high degree of unsaturated fatty acids) due to the capacity to increase mitochondrial oxidative defence (measured via SOD activity) with training [119].

#### 9.3.1.3 Mitochondrial Lipids and Oxidative Stress

Diabetes and obesity are often associated with oxidative stress proposed to cause insulin resistance [3, 20, 66, 85, 128]. Cardiolipins are sensitive to oxidation due to their unsaturation levels and their close proximity to the major sites for ROS production within the mitochondria [25, 26]. An accumulation of PUFA-CLs especially 22:6, as also seen at a very early stage of pathological development in the diabetic heart [107, 108], was positively correlated with ROS production and was shown to affect mitochondrial membrane potential in cultured mammalian cells [124, 317]. Heart mitochondria from STZ-induced diabetic rats had more lipid peroxidation, particularly of CL [71]. Peroxidised CL is associated with decreased activity of respiratory chain complexes (I, III, and IV) as well as decreased supercomplex assemblage [214–216]. Selective CL oxidation of all phospholipids was even discussed to trigger the release of proapoptotic factors from mitochondria [141]. mRNA and protein of lyso-CL acyltransferase 1 (ALCAT1) were shown to be induced by ROS in oxidative stress-associated diseases like obesity and diabetes [169]. ALCAT1 catalyses the remodelling of CL dominated by long-chain highly unsaturated fatty acids such as 22:6. ALCAT1 amongst others is therefore discussed as a major regulator of abnormal remodelling of CL in diet-induced obesity, which leads to oxidative stress, mitochondrial dysfunction and thereby insulin resistance, since ALCAT1$^{-/-}$ mice are fully protected from the onset of high-fat diet-induced obesity and T2Dm [169].

Unsaturated membrane lipids increase the membrane fluidity, probably helping mitochondria to cope with ROS over short term. But they also become more prone to peroxidation depending on the number of bisallylic methylene groups present in PUFA [51, 122]. The degree of polyunsaturation of the mitochondrial membrane lipids is positively correlated with the mitochondrial proton leak as observed in liver mitochondria from mammals and birds [27, 28, 238]. In diabetic context, the mitochondrial membrane fluidity of rat liver was reported to be increased in STZ-induced diabetes due to an increment in the unsaturated to saturated fatty acids ratio (U/S). This was the opposite to the observations in rat liver tissue, where diabetes induced a decrease in the U/S fatty acids ratio of liver total lipids [225], indicating an independent fatty acid composition regulation in mitochondria compared to tissue.

#### 9.3.1.4 Substitutes for Cardiolipins (CL)?

Since IR and T2Dm are associated with a depletion of CL as often reported in heart [107, 108], a possible substitution of CL might serve as rescuing factor. Phosphatidylethanolamine (PE) and CL, both non-bilayer-forming phospholipids [149], are functionally converging in mitochondria [96]. Mutant prokaryotic cells lacking the ability to synthesise phosphatidylserine, the precursor of PE, and therefore devoid of PE, showed compensatory increased CL, PG and to a lesser extent PA levels [244]. This convergent correlation was also seen in yeast under depletion of PE or CL [205]. A depletion of mitochondrial PE in CHO-K1 cells, a hamster ovarian cell line, led to a decrease in respiratory capacity and defective ETC complexes [288]. In line with this, a higher electron transport chain activity and respiration was observed in rat hepatoma cells accumulating PE due to a lack of PE N-methyltransferase (PEMT). This lack of PEMT protected mice against diet-induced obesity and IR and led to a shift in pyruvate metabolism in hepatic mitochondria routing towards decarboxylation and energy production instead of carboxylation and glucose production [305]. Absence of CL was also shown to be partially compensated in respiratory chain activity by its precursor phosphatidylglycerol (PG) in yeast mitochondria [136]. Accordingly, a decreased pool of PGs was observed in STZ-induced diabetic rats hearts probably being shuffled into synthesis of new CL [71, 112]. However elevated PG levels in a *crd1*Δ strain of yeast (null in expression of CL synthase) were shown to fail the substitution of CL in supercomplex formation of complex III and IV, indicating a specific role of CL in the critical interaction between electron transport chain complexes for supercomplex assembly [329]. The amount of such supercomplexes was reported to be dependent on the CL amount [329]. Hence, investigations of the lipid molecular composition and the linkage to its impact on mitochondrial function might provide further insights to elucidate mechanisms in diabetes development and possible targeted treatment.

## 9.4 The Mitochondrial Proteome of Liver and Muscle in in Type 2 Diabetes Mellitus

Mitochondrial dysfunction has been closely associated with the progressive pathological condition of T2Dm in conjunction with IR [189] although the exact mechanism/etiology underlying this relationship is still elusive and debatable [104]. Besides adipose tissue, pancreatic β-cells; liver and skeletal muscle tissues are predominantly prone to manifest impaired mitochondrial functions [222]. In this context, the number of proteomics and phosphoproteomics studies focusing on the mitochondrial dysfunction and T2Dm have been limited in the past. However, in the last decade, comparative proteomics studies performed using isolated mitochondria from liver and muscle tissues obtained from human biopsies or mouse models belonging to different biological states (e.g. lean and obese) revealed significant alterations in the mitochondrial proteome and phosphoproteome [100]. Some of the recent proteomics-based investigations related to T2Dm are briefly described below. Deng W-J et al. performed a comprehensive proteomics study using the mitochondria derived from the liver of a diabetic rat model (Goto-Kakizaki, GK) to monitor the progression of T2Dm in the GK rat i.e. early, prediabetic and diabetic. By employing 2D LC-MS methodology, they could identify 1091 mitochondrial proteins including several phosphoproteins, and hydroxyproteins [60]. Label-free quantification approach led them to reveal differential regulation of (i) mitochondrial metabolic enzymes, (ii) proteins involved in the ATP production i.e. TCA cycle, β-oxidation of fatty acids and OXPHOS, and (iii) phospholipid metabolism. In phosphoproteome data they found downregulation of *Pdp2*, a phosphatase and increased phosphorylation of *Pdha1*, its corresponding substrate. *Pdha1* is a component of the pyruvate dehydrogenase complex, which is involved in the energy production pathway and it is deactivated upon phosphorylation. Consequently, decreased level of *Pdp2* was attributed to the development of T2Dm. In another study, Guo and colleagues did quantitative proteomics profiling on liver mitochondria obtained from high fat diet (HFD) and normal diet (ND) fed mouse model. By combining iTRAQ technology with orthogonal LC fractionation and MS analysis, they could see 92 differentially regulated proteins between HFD and ND fed mice. Furthermore, pathway analysis

(KEGG, Cytoscape), functional analysis (oxygen consumption rate and ATP generation), and biochemical analysis (WB) were also performed to support their findings. These results were in agreement with earlier reports i.e. under obesity and HFD states, increased abundances of proteins involved in the bioenergetic pathways were observed. Additionally, they found some previously not known mechanisms such as upregulation of the mitochondrial intermembrane space bridging or mitochondrial contact site complex, which is required for maintaining the integrity of the mitochondria in HFD fed mice [104]. Besides animal models, Zhao and coworkers utilized mitochondria from resting human skeletal muscle biopsy to study phosphorylation status of IMM protein complexes. By applying a myriad of phosphopeptides enrichment protocols followed by LC-MS analysis, they could identify several site-specific phosphorylated proteins i.e. on Ser, Thr and Tyr residues. Of particular interest was the identification of relatively high number of Tyr-phosphorylation sites, which is considered to play a vital role in mitochondrial signalling pathways. As phosphorylation plays an important role in majority of the metabolic pathways, this study could serve as a map for current/future functional and comparative phosphoproteomics research on the mitochondrial dysfunction associated IR and T2Dm [332].

## References


1. Adam-Vizi V, Chinopoulos C (2006) Bioenergetics and the formation of mitochondrial reactive oxygen species. Trends Pharmacol Sci 27:639–645. https://doi.org/10.1016/j.tips.2006.10.005
2. Aebersold R (2003) Quantitative proteome analysis: methods and applications. J Infect Dis 187:S315–S320
3. Anderson EJ et al (2009) Mitochondrial $H_2O_2$ emission and cellular redox state link excess fat intake to insulin resistance in both rodents and humans. J Clin Investig 119:573–581. https://doi.org/10.1172/jci37048
4. Andreyev AY et al (2010) Subcellular organelle lipidomics in TLR-4-activated macrophages. J Lipid Res 51:2785–2797. https://doi.org/10.1194/jlr.M008748
5. Angelini R, Vitale R, Patil VA, Cocco T, Ludwig B, Greenberg ML, Corcelli A (2012) Lipidomics of intact mitochondria by MALDI-TOF/MS. J Lipid Res 53:1417–1425. https://doi.org/10.1194/jlr.D026203
6. Arab JP, Arrese M, Trauner M (2018) Recent insights into the pathogenesis of nonalcoholic fatty liver disease. Annu Rev Pathol 13:321–350. https://doi.org/10.1146/annurev-pathol-020117-043617
7. Arruda AP, Pers BM, Parlakgül G, Güney E, Inouye K, Hotamisligil GS (2014) Chronic enrichment of hepatic ER-mitochondria contact sites leads to calcium dependent mitochondrial dysfunction in obesity. Nat Med 20:1427–1435. https://doi.org/10.1038/nm.3735
8. Aviram R, Manella G, Kopelman N, Neufeld-Cohen A, Zwighaft Z, Elimelech M (2016) Lipidomics analyses reveal temporal and spatial lipid organization and uncover daily oscillations in intracellular organelles. Mol Cell 62. https://doi.org/10.1016/j.molcel.2016.04.002
9. Bajpeyi S et al (2014) Lipid in skeletal muscle myotubes is associated to the donors' insulin sensitivity and physical activity phenotypes. Obesity (Silver Spring, MD) 22:426–434. https://doi.org/10.1002/oby.20556
10. Bantscheff M, Schirle M, Sweetman G, Rick J, Kuster B (2007) Quantitative mass spectrometry in proteomics: a critical review. Anal Bioanal Chem 389:1017–1031. https://doi.org/10.1007/s00216-007-1486-6
11. Barzanti V et al (1994) The effect of dietary lipid changes on the fatty acid composition and function of liver, heart and brain mitochondria in the rat at different ages. Br J Nutr 71:193–202
12. Benard G et al (2006) Physiological diversity of mitochondrial oxidative phosphorylation. Am J Physiol Cell Physiol 291:C1172–C1182. https://doi.org/10.1152/ajpcell.00195.2006
13. Bergman BC, Hunerdosse DM, Kerege A, Playdon MC, Perreault L (2012) Localisation and composition of skeletal muscle diacylglycerol predicts insulin resistance in humans. Diabetologia 55:1140–1150. https://doi.org/10.1007/s00125-011-2419-7
14. Billings LK, Florez JC (2010) The genetics of type 2 diabetes: what have we learned from GWAS? Ann N Y Acad Sci 1212:59–77. https://doi.org/10.1111/j.1749-6632.2010.05838.x
15. Bird SS, Marur VR, Sniatynski MJ, Greenberg HK, Kristal BS (2011) Lipidomics profiling by high-resolution LC-MS and high-energy collisional dissociation fragmentation: focus on characterization of mitochondrial cardiolipins and monolysocardiolipins. Anal Chem 83:940–949. https://doi.org/10.1021/ac102598u
16. Bird SS, Stavrovskaya IG, Gathungu RM, Tousi F, Kristal BS (2015) Qualitative characterization of the rat liver mitochondrial lipidome using all ion fragmentation on an Exactive benchtop Orbitrap

MS. Methods Mol Biol 1264:441–452. https://doi.org/10.1007/978-1-4939-2257-4_36
17. Blacker TS, Duchen MR (2016) Investigating mitochondrial redox state using NADH and NADPH autofluorescence. Free Radic Biol Med 100:53–65. https://doi.org/10.1016/j.freeradbiomed.2016.08.010
18. Bligh EG, Dyer WJ (1959) A rapid method of total lipid extraction and purification. Can J Biochem Physiol 37:911–917. https://doi.org/10.1139/o59-099
19. Boden G (1997) Role of fatty acids in the pathogenesis of insulin resistance and NIDDM. Diabetes 46:3–10
20. Bonnard C et al (2008) Mitochondrial dysfunction results from oxidative stress in the skeletal muscle of diet-induced insulin-resistant mice. J Clin Invest 118:789–800. https://doi.org/10.1172/JCI32601
21. Borradaile NM, Han X, Harp JD, Gale SE, Ory DS, Schaffer JE (2006) Disruption of endoplasmic reticulum structure and integrity in lipotoxic cell death. J Lipid Res 47:2726–2737. https://doi.org/10.1194/jlr.M600299-JLR200
22. Borst P (1977) Structure and function of mitochondrial DNA. Trends Biochem Sci 2:31–34. https://doi.org/10.1016/0968-0004(77)90252-3
23. Bouderba S, Sanz MN, Sánchez-Martín C, El-Mir MY, Villanueva GR, Detaille D, Koceïr EA (2012) Hepatic mitochondrial alterations and increased oxidative stress in nutritional diabetes-prone psammomys obesus model. Exp Diabetes Res 2012. https://doi.org/10.1155/2012/430176
24. Boushel R, Gnaiger E, Schjerling P, Skovbro M, Kraunsøe R, Dela F (2007) Patients with type 2 diabetes have normal mitochondrial function in skeletal muscle. Diabetologia 50:790–796. https://doi.org/10.1007/s00125-007-0594-3
25. Boveris A, Chance B (1973) The mitochondrial generation of hydrogen peroxide. General properties and effect of hyperbaric oxygen. Biochem J 134:707–716
26. Brand MD (2016) Mitochondrial generation of superoxide and hydrogen peroxide as the source of mitochondrial redox signaling. Free Radic Biol Med 100:14–31. https://doi.org/10.1016/j.freeradbiomed.2016.04.001
27. Brand MD, Turner N, Ocloo A, Else PL, Hulbert AJ (2003) Proton conductance and fatty acyl composition of liver mitochondria correlates with body mass in birds. Biochem J 376:741–748. https://doi.org/10.1042/bj20030984
28. Brookes P, A Buckingham J, Tenreiro A, Hulbert A, Brand M (1998) The proton permeability of the inner membrane of liver mitochondria from ectothermic and endothermic vertebrates and from obese rats: correlations with standard metabolic rate and phospholipid fatty acid composition comparative biochemistry and physiology Part B, Biochem Mol Biol 119:325–334 doi:https://doi.org/10.1016/S0305-0491(97)00357-X
29. Brown HA, Murphy RC (2009) Working towards an exegesis for lipids in biology. Nat Chem Biol 5:602–606. https://doi.org/10.1038/nchembio0909-602
30. Brügger B (2014) Lipidomics: analysis of the lipid composition of cells and subcellular organelles by electrospray ionization mass spectrometry. Annu Rev Biochem 83:79–98. https://doi.org/10.1146/annurev-biochem-060713-035324
31. Brügger B, Erben G, Sandhoff R, Wieland FT, Lehmann WD (1997) Quantitative analysis of biological membrane lipids at the low picomole level by nano-electrospray ionization tandem mass spectrometry. Proc Natl Acad Sci U S A 94:2339–2344
32. Calvo SE, Mootha VK (2010) The mitochondrial proteome and human disease. Annu Rev Genomics Hum Genet 11:25–44. https://doi.org/10.1146/annurev-genom-082509-141720
33. Calvo SE, Clauser KR, Mootha VK (2016) MitoCarta2.0: an updated inventory of mammalian mitochondrial proteins. Nucleic Acids Res 44:D1251–D1257. https://doi.org/10.1093/nar/gkv1003
34. Catherman AD et al (2013) Top down proteomics of human membrane proteins from enriched mitochondrial fractions. Anal Chem 85:1880–1888. https://doi.org/10.1021/ac3031527
35. Catherman AD, Skinner OS, Kelleher NL (2014) Top down proteomics: facts and perspectives. Biochem Biophys Res Commun 445:683–693. https://doi.org/10.1016/j.bbrc.2014.02.041
36. Chacinska A, Koehler CM, Milenkovic D, Lithgow T, Pfanner N (2009) Importing mitochondrial proteins: machineries and mechanisms. Cell 138:628–644. https://doi.org/10.1016/j.cell.2009.08.005
37. Chalkley SM, Hettiarachchi M, Chisholm DJ, Kraegen EW (1998) Five-hour fatty acid elevation increases muscle lipids and impairs glycogen synthesis in the rat. Metabolism 47:1121–1126. https://doi.org/10.1016/S0026-0495(98)90287-6
38. Chan DC, Gan SK, Wong AT, Barrett PH, Watts GF (2014) Association between skeletal muscle fat content and very-low-density lipoprotein-apolipoprotein B-100 transport in obesity: effect of weight loss. Diabetes Obes Metab 16:994–1000. https://doi.org/10.1111/dom.12311
39. Chance B, Williams GR (1955a) Respiratory enzymes in oxidative phosphorylation. I. Kinetics of oxygen utilization. J Biol Chem 217:383–393
40. Chance B, Williams GR (1955b) Respiratory enzymes in oxidative phosphorylation. III. The steady state. J Biol Chem 217:409–427
41. Chen S et al (2013) Simultaneous extraction of metabolome and lipidome with methyl tert-butyl ether from a single small tissue sample for ultra-high performance liquid chromatography/mass spectrometry. J Chromatogr A 1298:9–16. https://doi.org/10.1016/j.chroma.2013.05.019
42. Cheng H et al (2008) Shotgun lipidomics reveals the temporally dependent, highly diversified cardiolipin profile in the mammalian brain: temporally coordi-

nated postnatal diversification of cardiolipin molecular species with neuronal remodeling. Biochemistry 47:5869–5880. https://doi.org/10.1021/bi7023282
43. Chiappini F et al (2006) Exploration of global gene expression in human liver steatosis by high-density oligonucleotide microarray. Lab Investig 86:154–165. https://doi.org/10.1038/labinvest.3700374
44. Chomentowski P, Coen PM, Radikova Z, Goodpaster BH, Toledo FG (2011) Skeletal muscle mitochondria in insulin resistance: differences in intermyofibrillar versus subsarcolemmal subpopulations and relationship to metabolic flexibility. J Clin Endocrinol Metabol 96:494–503. https://doi.org/10.1210/jc.2010-0822
45. Chow LS et al (2014) Training status diverges muscle diacylglycerol accumulation during free fatty acid elevation. Am J Physiol Endocrinol Metab 307:E124–E131. https://doi.org/10.1152/ajpendo.00166.2014
46. Claypool SM (2009) Cardiolipin, a critical determinant of mitochondrial carrier protein assembly and function. Biochim Biophys Acta 1788:2059–2068. https://doi.org/10.1016/j.bbamem.2009.04.020
47. Claypool SM, Koehler CM (2012) The complexity of cardiolipin in health and disease. Trends Biochem Sci 37:32–41. https://doi.org/10.1016/j.tibs.2011.09.003
48. Colbeau A, Nachbaur J, Vignais PM (1971) Enzymac characterization and lipid composition of rat liver subcellular membranes. Biochim Biophys Acta Biomembr 249:462–492. https://doi.org/10.1016/0005-2736(71)90123-4
49. Colombini M (2017) Ceramide channels and mitochondrial outer membrane permeability. J Bioenerg Biomembr 49:57–64. https://doi.org/10.1007/s10863-016-9646-z
50. Coman C, Solari FA, Hentschel A, Sickmann A, Zahedi RP, Ahrends R (2016) Simultaneous metabolite, protein, lipid extraction (SIMPLEX): a combinatorial multimolecular Omics approach for systems biology. Mol Cell Proteomics 15:1453–1466
51. Cosgrove JP, Church DF, Pryor WA (1987) The kinetics of the autoxidation of polyunsaturated fatty acids. Lipids 22:299–304. https://doi.org/10.1007/bf02533996
52. Croston TL et al (2013) Evaluation of the cardiolipin biosynthetic pathway and its interactions in the diabetic heart. Life Sci 93:313–322. https://doi.org/10.1016/j.lfs.2013.07.005
53. Daum G, Vance JE (1997) Import of lipids into mitochondria. Prog Lipid Res 36:103–130. https://doi.org/10.1016/S0163-7827(97)00006-4
54. Daum G, Böhni P, Schatz G (1982) Import of proteins into mitochondria. Cytochrome b2 and cytochrome c peroxidase are located in the intermembrane space of yeast mitochondria. J Biol Chem 257:13028–13033
55. De Feyter HM, van den Broek NM, Praet SF, Nicolay K, van Loon LJ, Prompers JJ (2008) Early or advanced stage type 2 diabetes is not accompanied by in vivo skeletal muscle mitochondrial dysfunction. Eur J Endocrinol 158:643–653. https://doi.org/10.1530/eje-07-0756
56. de Kroon AIPM, Dolis D, Mayer A, Lill R, de Kruijff B (1997) Phospholipid composition of highly purified mitochondrial outer membranes of rat liver and Neurospora crassa. Is cardiolipin present in the mitochondrial outer membrane? Biochim Biophys Acta Biomembr 1325:108–116. https://doi.org/10.1016/S0005-2736(96)00240-4
57. de Wilde J et al (2008) Short-term high fat-feeding results in morphological and metabolic adaptations in the skeletal muscle of C57BL/6J mice. Physiol Genomics 32:360–369. https://doi.org/10.1152/physiolgenomics.00219.2007
58. DeFronzo RA, Alvestrand A, Smith D, Hendler R, Hendler E, Wahren J (1981a) Insulin resistance in uremia. J Clin Investig 67:563
59. DeFronzo RA, Jacot E, Jequier E, Maeder E, Wahren J, Felber JP (1981b) The effect of insulin on the disposal of intravenous glucose. Results from indirect calorimetry and hepatic and femoral venous catheterization. Diabetes 30:1000–1007
60. Deng W-J, Nie S, Dai J, Wu J-R, Zeng R (2010) Proteome, phosphoproteome, and hydroxyproteome of liver mitochondria in diabetic rats at early pathogenic stages. Mol Cell Proteomics 9:100–116
61. Doerr A (2015) DIA mass spectrometry. Nat Method 12:35–35. https://doi.org/10.1038/nmeth.3234
62. Dudek J, Rehling P, van der Laan M (2013) Mitochondrial protein import: common principles and physiological networks. Biochimica et Biophysica Acta (BBA) Mol Cell Res 1833:274–285. https://doi.org/10.1016/j.bbamcr.2012.05.028
63. Dunn JD, Reid GE, Bruening ML (2010) Techniques for phosphopeptide enrichment prior to analysis by mass spectrometry. Mass Spectrom Rev 29:29–54. https://doi.org/10.1002/mas.20219
64. Durcan TM, Fon EA (2015) The three 'P's of mitophagy: PARKIN, PINK1, and post-translational modifications. Genes Dev 29:989–999. https://doi.org/10.1101/gad.262758.115
65. Ernster L, Schatz G (1981) Mitochondria: a historical review. J Cell Biol 91:227s
66. Evans JL, Goldfine ID, Maddux BA, Grodsky GM (2002) Oxidative stress and stress-activated signaling pathways: a unifying hypothesis of type 2 diabetes. Endocr Rev 23:599–622. https://doi.org/10.1210/er.2001-0039
67. Fahy E, Subramaniam S, Brown HA, Glass CK, Merrill AH, Murphy RC (2005) A comprehensive classification system for lipids. J Lipid Res 46. https://doi.org/10.1194/jlr.E400004-JLR200
68. Fahy E, Subramaniam S, Murphy RC, Nishijima M, Raetz CR, Shimizu T (2009) Update of the LIPID MAPS comprehensive classification system for lipids. J Lipid Res 50. https://doi.org/10.1194/jlr.R800095-JLR200
69. Fajardo VA, McMeekin L, Saint C, LeBlanc PJ (2015) Cardiolipin linoleic acid content and

mitochondrial cytochrome c oxidase activity are associated in rat skeletal muscle. Chem Phys Lipids 187:50–55. https://doi.org/10.1016/j.chemphyslip.2015.02.004
70. Fernstrom M et al (2007) Reduced efficiency, but increased fat oxidation, in mitochondria from human skeletal muscle after 24-h ultraendurance exercise. J Appl Physiol (Bethesda, MD: 1985) 102:1844–1849. https://doi.org/10.1152/japplphysiol.01173.2006
71. Ferreira R et al (2013) Lipidomic characterization of streptozotocin-induced heart mitochondrial dysfunction. Mitochondrion 13:762–771. https://doi.org/10.1016/j.mito.2013.05.001
72. Ferrick DA, Neilson A, Beeson C (2008) Advances in measuring cellular bioenergetics using extracellular flux. Drug Discov Today 13:268–274. https://doi.org/10.1016/j.drudis.2007.12.008
73. Fisher-Wellman KH et al (2014) Mitochondrial respiratory capacity and content are normal in young insulin-resistant obese humans. Diabetes 63:132–141. https://doi.org/10.2337/db13-0940
74. Flippo KH, Strack S (2017) Mitochondrial dynamics in neuronal injury, development and plasticity. J Cell Sci 130:671
75. Flis VV, Daum G (2013) Lipid transport between the endoplasmic reticulum and mitochondria. Cold Spring Harb Perspect Biol 5. https://doi.org/10.1101/cshperspect.a013235
76. Folch J, Lees M, Sloane Stanley GH (1957) A simple method for the isolation and purification of total lipides from animal tissues. J Biol Chem 226:497–509
77. Forner F, Foster LJ, Campanaro S, Valle G, Mann M (2006) Quantitative proteomic comparison of rat mitochondria from muscle, heart, and liver. Mol Cell Proteomics 5:608–619. https://doi.org/10.1074/mcp.M500298-MCP200
78. Franko A et al (2013) Efficient isolation of pure and functional mitochondria from mouse tissues using automated tissue disruption and enrichment with anti-TOM22 magnetic beads. PLoS One 8:e82392. https://doi.org/10.1371/journal.pone.0082392
79. Franko A et al (2014) Liver adapts mitochondrial function to insulin resistant and diabetic states in mice. J Hepatol 60:816–823. https://doi.org/10.1016/j.jhep.2013.11.020
80. Frezza C, Cipolat S, Scorrano L (2007) Organelle isolation: functional mitochondria from mouse liver, muscle and cultured filroblasts. Nat Protoc 2:287–295
81. Friedman JR, Nunnari J (2014) Mitochondrial form and function. Nature 505:335. https://doi.org/10.1038/nature12985
82. Fritsch M et al (2015) Time course of postprandial hepatic phosphorus metabolites in lean, obese, and type 2 diabetes patients. Am J Clin Nutr 102:1051–1058. https://doi.org/10.3945/ajcn.115.107599
83. Frohman M (2015) Role of mitochondrial lipids in guiding fission and fusion. J Mol Med 93:263–269. https://doi.org/10.1007/s00109-014-1237-z
84. Fucho R, Casals N, Serra D, Herrero L (2017) Ceramides and mitochondrial fatty acid oxidation in obesity. FASEB J 31:1263–1272. https://doi.org/10.1096/fj.201601156R
85. Furukawa S et al (2004) Increased oxidative stress in obesity and its impact on metabolic syndrome. J Clin Invest 114:1752–1761. https://doi.org/10.1172/jci21625
86. Galadari S, Rahman A, Pallichankandy S, Galadari A, Thayyullathil F (2013) Role of ceramide in diabetes mellitus: evidence and mechanisms. Lipids Health Dis 12:98. https://doi.org/10.1186/1476-511x-12-98
87. Garcia-Roves P, Huss JM, Han DH, Hancock CR, Iglesias-Gutierrez E, Chen M, Holloszy JO (2007) Raising plasma fatty acid concentration induces increased biogenesis of mitochondria in skeletal muscle. Proc Natl Acad Sci U S A 104:10709–10713. https://doi.org/10.1073/pnas.0704024104
88. Geert B, Evy V, Arnold De L, Liliane S (2005) Gel-based versus gel-free proteomics: a review. Comb Chem High Throughput Screen 8:669–677. https://doi.org/10.2174/138620705774962490
89. Gerber SA, Rush J, Stemman O, Kirschner MW, Gygi SP (2003) Absolute quantification of proteins and phosphoproteins from cell lysates by tandem MS. Proc Natl Acad Sci U S A 100:6940–6945. https://doi.org/10.1073/pnas.0832254100
90. Gevaert K, Impens F, Ghesquiere B, Van Damme P, Lambrechts A, Vandekerckhove J (2008) Stable isotopic labeling in proteomics. Proteomics 8:4873–4885. https://doi.org/10.1002/pmic.200800421
91. Gilar M, Olivova P, Daly AE, Gebler JC (2005) Two-dimensional separation of peptides using RP-RP-HPLC system with different pH in first and second separation dimensions. J Sep Sci 28:1694–1703. https://doi.org/10.1002/jssc.200500116
92. GLOBAL STATUS REPORT on noncommunicable diseases (2014)
93. Gnaiger E (2008) Polarographic oxygen sensors, the oxygraph, and high-resolution respiroemtry to assess mitochondrial function. In: Dykens J, Will Y (eds) Drug-induced mitochondrial dysfunction. Wiley, Hoboken, pp 327–352
94. Gnaiger E (2009) Capacity of oxidative phosphorylation in human skeletal muscle: new perspectives of mitochondrial physiology. Int J Biochem Cell Biol 41:1837–1845. https://doi.org/10.1016/j.biocel.2009.03.013
95. Gnaiger E, Steinlechner-Maran R, Mendez G, Eberl T, Margreiter R (1995) Control of mitochondrial and cellular respiration by oxygen. J Bioenerg Biomembr 27:583–596
96. Gohil VM, Thompson MN, Greenberg ML (2005) Synthetic lethal interaction of the mitochondrial phosphatidylethanolamine and cardiolipin biosynthetic pathways in Saccharomyces cerevisiae. J Biol Chem 280:35410–35416. https://doi.org/10.1074/jbc.M505478200
97. Görlach A, Bertram K, Hudecova S, Krizanova O (2015) Calcium and ROS: a mutual interplay.

Redox Biol 6:260–271. https://doi.org/10.1016/j.redox.2015.08.010

98. Gouw JW, Krijgsveld J, Heck AJR (2010) Quantitative proteomics by metabolic labeling of model organisms. Mol Cell Proteomics: MCP 9:11–24. https://doi.org/10.1074/mcp.R900001-MCP200
99. Gray MW, Burger G, Lang BF (1999) Mitochondrial evolution. Science 283:1476
100. Grimsrud PA et al (2012) A quantitative map of the liver mitochondrial phosphoproteome reveals posttranslational control of ketogenesis. Cell Metab 16:672–683. https://doi.org/10.1016/j.cmet.2012.10.004
101. Gross JH (2006) Mass spectrometry: a textbook. Springer, Berlin/Heidelberg
102. Gross VS, Greenberg HK, Baranov SV, Carlson GM, Stavrovskaya IG, Lazarev AV, Kristal BS (2011) Isolation of functional mitochondria from rat kidney and skeletal muscle without manual homogenization. Anal Biochem 418:213–223. https://doi.org/10.1016/j.ab.2011.07.017
103. Gudbjarnason S, el-Hage AN, Whitehurst VE, Simental F, Balazs T (1987) Reduced arachidonic acid levels in major phospholipids of heart muscle in the diabetic rat. J Mol Cell Cardiol 19:1141–1146
104. Guo Y et al (2013) Quantitative proteomic and functional analysis of liver mitochondria from high fat diet (HFD) diabetic mice. Mol Cell Proteomics 12:3744–3758
105. Han X, Gross RW (2003) Global analyses of cellular lipidomes directly from crude extracts of biological samples by ESI mass spectrometry: a bridge to lipidomics. J Lipid Res 44. https://doi.org/10.1194/jlr.R300004-JLR200
106. Han X, Abendschein DR, Kelley JG, Gross RW (2000) Diabetes-induced changes in specific lipid molecular species in rat myocardium. Biochem J 352:79–89
107. Han X, Yang J, Cheng H, Yang K, Abendschein DR, Gross RW (2005) Shotgun lipidomics identifies cardiolipin depletion in diabetic myocardium linking altered substrate utilization with mitochondrial dysfunction. Biochemistry 44:16684–16694. https://doi.org/10.1021/bi051908a
108. Han X, Yang J, Yang K, Zhao Z, Abendschein DR, Gross RW (2007) Alterations in myocardial cardiolipin content and composition occur at the very earliest stages of diabetes: a shotgun lipidomics study. Biochemistry 46:6417–6428. https://doi.org/10.1021/bi7004015
109. Han D-H, Hancock CR, Jung SR, Higashida K, Kim SH, Holloszy JO (2011) Deficiency of the mitochondrial electron transport chain in muscle does not cause insulin resistance. PLoS One 6:e19739. https://doi.org/10.1371/journal.pone.0019739
110. Hancock CR, Han D-H, Chen M, Terada S, Yasuda T, Wright DC, Holloszy JO (2008) High-fat diets cause insulin resistance despite an increase in muscle mitochondria. Proc Natl Acad Sci U S A 105:7815–7820. https://doi.org/10.1073/pnas.0802057105
111. Harbauer AB, Zahedi René P, Sickmann A, Pfanner N, Meisinger C (2014) The protein import machinery of mitochondria – a regulatory hub in metabolism, stress, and disease. Cell Metab 19:357–372. https://doi.org/10.1016/j.cmet.2014.01.010
112. Hatch GM, Cao SG, Angel A (1995) Decrease in cardiac phosphatidylglycerol in streptozotocin-induced diabetic rats does not affect cardiolipin biosynthesis: evidence for distinct pools of phosphatidylglycerol in the heart. Biochem J 306:759–764. https://doi.org/10.1042/bj3060759
113. Hatefi Y (1985) The mitochondrial Electron transport and oxidative phosphorylation system. Annu Rev Biochem 54:1015–1069. https://doi.org/10.1146/annurev.bi.54.070185.005055
114. Hawkridge AM (2014) CHAPTER 1 practical considerations and current limitations in quantitative mass spectrometry-based proteomics. In: Quantitative proteomics. The Royal Society of Chemistry, Cambridge, pp 1–25. https://doi.org/10.1039/9781782626985-00001
115. He Q, Han X (2014) Cardiolipin remodeling in diabetic heart. Chem Phys Lipids 179:75–81. https://doi.org/10.1016/j.chemphyslip.2013.10.007
116. Hebert AS, Richards AL, Bailey DJ, Ulbrich A, Coughlin EE, Westphall MS, Coon JJ (2014) The one hour yeast proteome. Mol Cell Proteomics 13:339–347
117. Heilbronn LK, Gan SK, Turner N, Campbell LV, Chisholm DJ (2007) Markers of mitochondrial biogenesis and metabolism are lower in overweight and obese insulin-resistant subjects. J Clin Endocrinol Metab 92:1467–1473. https://doi.org/10.1210/jc.2006-2210
118. Herbst EA et al (2014) Omega-3 supplementation alters mitochondrial membrane composition and respiration kinetics in human skeletal muscle. J Physiol 592:1341–1352. https://doi.org/10.1113/jphysiol.2013.267336
119. Higuchi M, Cartier LJ, Chen M, Holloszy JO (1985) Superoxide dismutase and catalase in skeletal muscle: adaptive response to exercise. J Gerontol 40:281–286
120. Hoeks J et al (2011) High fat diet-induced changes in mouse muscle mitochondrial phospholipids do not impair mitochondrial respiration despite insulin resistance. PLoS One 6:e27274. https://doi.org/10.1371/journal.pone.0027274
121. Holloway GP, Thrush AB, Heigenhauser GJ, Tandon NN, Dyck DJ, Bonen A, Spriet LL (2007) Skeletal muscle mitochondrial FAT/CD36 content and palmitate oxidation are not decreased in obese women. Am J Phys Endocrinol Metab 292:E1782–E1789. https://doi.org/10.1152/ajpendo.00639.2006
122. Holman RT (1954) Autoxidation of fats and related substances. Prog Chem Fats Other Lipids 2:51–98
123. Holmstrom MH, Iglesias-Gutierrez E, Zierath JR, Garcia-Roves PM (2012) Tissue-specific control of mitochondrial respiration in obesity-related insulin resistance and diabetes. Am J Phys Endocrinol

Metab 302:E731–E739. https://doi.org/10.1152/ajpendo.00159.2011
124. Hong MY et al (2002) Fish oil increases mitochondrial phospholipid unsaturation, upregulating reactive oxygen species and apoptosis in rat colonocytes. Carcinogenesis 23:1919–1925
125. Hornig-Do HT, Gunther G, Bust M, Lehnartz P, Bosio A, Wiesner RJ (2009) Isolation of functional pure mitochondria by superparamagnetic microbeads. Anal Biochem 389:1–5. https://doi.org/10.1016/j.ab.2009.02.040
126. Horvath SE, Daum G (2013) Lipids of mitochondria. Prog Lipid Res 52:590–614. https://doi.org/10.1016/j.plipres.2013.07.002
127. Hotamisligil GS (2010) Endoplasmic reticulum stress and the inflammatory basis of metabolic disease. Cell 140:900–917. https://doi.org/10.1016/j.cell.2010.02.034
128. Houstis N, Rosen ED, Lander ES (2006) Reactive oxygen species have a causal role in multiple forms of insulin resistance. Nature 440:944–948. https://doi.org/10.1038/nature04634
129. Hovius R, Lambrechts H, Nicolay K, de Kruijff B (1990) Improved methods to isolate and subfractionate rat liver mitochondria. Lipid composition of the inner and outer membrane. Biochim Biophys Acta Biomembr 1021:217–226. https://doi.org/10.1016/0005-2736(90)90036-N
130. Hu FB, van Dam RM, Liu S (2001) Diet and risk of type II diabetes: the role of types of fat and carbohydrate. Diabetologia 44:805–817. https://doi.org/10.1007/s001250100547
131. Humphrey SJ, James DE, Mann M (2015) Protein phosphorylation: a major switch mechanism for metabolic regulation. Trends Endocrinol Metab 26:676–687. https://doi.org/10.1016/j.tem.2015.09.013
132. Iozzo P et al (2010) Fatty acid metabolism in the liver, measured by positron emission tomography, is increased in obese individuals. Gastroenterology 139:846–856. , 856.e841–846. https://doi.org/10.1053/j.gastro.2010.05.039
133. Jacob S et al (1999) Association of increased intramyocellular lipid content with insulin resistance in lean nondiabetic offspring of type 2 diabetic subjects. Diabetes 48:1113–1119
134. Jacobs RA, Fluck D, Bonne TC, Burgi S, Christensen PM, Toigo M, Lundby C (2013) Improvements in exercise performance with high-intensity interval training coincide with an increase in skeletal muscle mitochondrial content and function. J Appl Physiol (Bethesda, MD: 1985) 115:785–793. https://doi.org/10.1152/japplphysiol.00445.2013
135. Jelenik T, Roden M (2013) Mitochondrial plasticity in obesity and diabetes mellitus. Antioxid Redox Signal 19:258–268. https://doi.org/10.1089/ars.2012.4910
136. Jiang F et al (2000) Absence of cardiolipin in the crd1 null mutant results in decreased mitochondrial membrane potential and reduced mitochondrial function. J Biol Chem 275:22387–22394. https://doi.org/10.1074/jbc.M909868199
137. Johnson DT, Harris RA, Blair PV, Balaban RS (2007a) Functional consequences of mitochondrial proteome heterogeneity. Am J Physiol Cell Physiol 292:C698–C707. https://doi.org/10.1152/ajpcell.00109.2006
138. Johnson DT et al (2007b) Tissue heterogeneity of the mammalian mitochondrial proteome. Am J Physiol Cell Physiol 292:C689–C697. https://doi.org/10.1152/ajpcell.00108.2006
139. Jordy AB et al (2015) Analysis of the liver lipidome reveals insights into the protective effect of exercise on high-fat diet-induced hepatosteatosis in mice. Am J Phys Endocrinol Metab 308:E778–E791. https://doi.org/10.1152/ajpendo.00547.2014
140. Jorgensen W, Jelnes P, Rud KA, Hansen LL, Grunnet N, Quistorff B (2012) Progression of type 2 diabetes in GK rats affects muscle and liver mitochondria differently: pronounced reduction of complex II flux is observed in liver only. Am J Phys Endocrinol Metab 303:E515–E523. https://doi.org/10.1152/ajpendo.00103.2012
141. Kagan VE et al (2005) Cytochrome c acts as a cardiolipin oxygenase required for release of proapoptotic factors. Nat Chem Biol 1:223–232. http://www.nature.com/nchembio/journal/v1/n4/suppinfo/nchembio727_S1.html
142. Kappler L, Li J, Häring H-U, Weigert C, Lehmann R, Xu G, Hoene M (2016) Purity matters: a workflow for the valid high-resolution lipid profiling of mitochondria from cell culture samples. Sci Rep 6:21107
143. Kappler L, Hoene M, Hu C, von Toerne C, Li J, Bleher D, Hoffmann C, Böhm A, Kollipara L, Zischka H, Königsrainer A, Häring H-U, Peter A, Xu G, Sickmann A, Hauck SM, Weigert C, Lehmann R (2019) Linking bioenergetic function of mitochondriato tissue-specific molecular fingerprints. Am J Physiol Endocrinol Metab
144. Kelley DE, Goodpaster B, Wing RR, Simoneau JA (1999) Skeletal muscle fatty acid metabolism in association with insulin resistance, obesity, and weight loss. Am J Phys 277:E1130–E1141
145. Kelley DE, He J, Menshikova EV, Ritov VB (2002) Dysfunction of mitochondria in human skeletal muscle in type 2 diabetes. Diabetes 51:2944–2950. https://doi.org/10.2337/diabetes.51.10.2944
146. Khairallah RJ et al (2010) Dietary supplementation with docosahexaenoic acid, but not eicosapentaenoic acid, dramatically alters cardiac mitochondrial phospholipid fatty acid composition and prevents permeability transition. Biochim Biophys Acta 1797:1555–1562. https://doi.org/10.1016/j.bbabio.2010.05.007
147. Kiebish MA et al (2008) Lipidomic analysis and electron transport chain activities in C57BL/6J mouse brain mitochondria. J Neurochem 106:299–312. https://doi.org/10.1111/j.1471-4159.2008.05383.x

148. Kiebish MA, Han X, Seyfried TN (2009) Examination of the brain mitochondrial lipidome using shotgun lipidomics. Methods Mol Biol 579:3–18. https://doi.org/10.1007/978-1-60761-322-0_1
149. Killian JA, de Kruijff B (2004) Nonbilayer lipids affect peripheral and integral membrane proteins via changes in the lateral pressure profile. Biochimica et Biophysica Acta (BBA)-Biomembranes 1666:275–288
150. Kim JY, Hickner RC, Cortright RL, Dohm GL, Houmard JA (2000) Lipid oxidation is reduced in obese human skeletal muscle. Am J Phys Endocrinol Metab 279:E1039–E1044
151. Kitabchi AE et al (2005) Role of insulin secretion and sensitivity in the evolution of type 2 diabetes in the diabetes prevention program: effects of lifestyle intervention and metformin. Diabetes 54:2404–2414
152. Klement P, Nijtmans LGJ, Vandenbogert C, Houstek J (1995) Analysis of oxidative phosphorylation complexes in cultured human fibroblasts and amniocytes by blue-native-electrophoresis using mitoplasts isolated with the help of digitonin. Anal Biochem 231:218–224. https://doi.org/10.1006/abio.1995.1523
153. Klimova N, Long A, Kristian T (2017) Significance of mitochondrial protein post-translational modifications in pathophysiology of brain injury. Transl Stroke Res. https://doi.org/10.1007/s12975-017-0569-8
154. Koliaki C, Roden M (2013) Hepatic energy metabolism in human diabetes mellitus, obesity and non-alcoholic fatty liver disease. Mol Cell Endocrinol 379:35–42. https://doi.org/10.1016/j.mce.2013.06.002
155. Koliaki C, Roden M (2016) Alterations of mitochondrial function and insulin sensitivity in human obesity and diabetes mellitus. Annu Rev Nutr 36:337–367. https://doi.org/10.1146/annurev-nutr-071715-050656
156. Koliaki C et al (2015) Adaptation of hepatic mitochondrial function in humans with non-alcoholic fatty liver is lost in steatohepatitis. Cell Metab 21:739–746. https://doi.org/10.1016/j.cmet.2015.04.004
157. Kotronen A, Seppala-Lindroos A, Vehkavaara S, Bergholm R, Frayn KN, Fielding BA, Yki-Jarvinen H (2009) Liver fat and lipid oxidation in humans. Liver Int 29:1439–1446. https://doi.org/10.1111/j.1478-3231.2009.02076.x
158. Koves TR et al (2008) Mitochondrial overload and incomplete fatty acid oxidation contribute to skeletal muscle insulin resistance. Cell Metab 7:45–56. https://doi.org/10.1016/j.cmet.2007.10.013
159. Krssak M et al (1999) Intramyocellular lipid concentrations are correlated with insulin sensitivity in humans: a 1H NMR spectroscopy study. Diabetologia 42:113–116. https://doi.org/10.1007/s001250051123
160. Kunau W-H, Dommes V, Schulz H (1995) β-Oxidation of fatty acids in mitochondria, peroxisomes, and bacteria: a century of continued progress. Prog Lipid Res 34:267–342. https://doi.org/10.1016/0163-7827(95)00011-9
161. Kunz WS (2003) Different metabolic properties of mitochondrial oxidative phosphorylation in different cell types – important implications for mitochondrial cytopathies. Exp Physiol 88:149–154. https://doi.org/10.1113/eph8802512
162. Kurland CG, Andersson SGE (2000) Origin and evolution of the mitochondrial proteome. Microbiol Mol Biol Rev 64:786–820
163. Lam MPY et al (2012) An MRM-based workflow for quantifying cardiac mitochondrial protein phosphorylation in murine and human tissue. J Proteome 75:4602–4609. https://doi.org/10.1016/j.jprot.2012.02.014
164. Lange V, Picotti P, Domon B, Aebersold R (2008) Selected reaction monitoring for quantitative proteomics: a tutorial. Mol Syst Biol 4:222–222. https://doi.org/10.1038/msb.2008.61
165. Larsen S et al (2012) Biomarkers of mitochondrial content in skeletal muscle of healthy young human subjects. J Physiol 590:3349–3360
166. Leek BT, Mudaliar SR, Henry R, Mathieu-Costello O, Richardson RS (2001) Effect of acute exercise on citrate synthase activity in untrained and trained human skeletal muscle. Am J Phys Regul Integr Comp Phys 280:R441–R447
167. Lefort N et al (2010) Increased reactive oxygen species production and lower abundance of complex I subunits and carnitine palmitoyltransferase 1B protein despite normal mitochondrial respiration in insulin-resistant human skeletal muscle. Diabetes 59:2444–2452. https://doi.org/10.2337/db10-0174
168. Leonov A et al (2017) Specific changes in mitochondrial lipidome alter mitochondrial proteome and increase the geroprotective efficiency of lithocholic acid in chronologically aging yeast. Oncotarget 8:30672–30691. https://doi.org/10.18632/oncotarget.16766
169. Li J et al (2010) Cardiolipin remodeling by ALCAT1 links oxidative stress and mitochondrial dysfunction to obesity. Cell Metab 12:154–165. https://doi.org/10.1016/j.cmet.2010.07.003
170. Liebler DC (2004) Shotgun mass spec goes independent. Nat Method 1:16. https://doi.org/10.1038/nmeth1004-16
171. Liebler DC, Zimmerman LJ (2013) Targeted quantitation of proteins by mass spectrometry. Biochemistry 52:3797–3806. https://doi.org/10.1021/bi400110b
172. Lillioja S et al (1993) Insulin resistance and insulin secretory dysfunction as precursors of non-insulin-dependent diabetes mellitus. Prospective studies of Pima Indians. New Engl J Med 329:1988–1992. https://doi.org/10.1056/nejm199312303292703
173. Lim S, Smith KR, Lim S-TS, Tian R, Lu J, Tan M (2016) Regulation of mitochondrial functions by protein phosphorylation and dephosphorylation. Cell Biosci 6:25. https://doi.org/10.1186/s13578-016-0089-3

174. Link AJ et al (1999) Direct analysis of protein complexes using mass spectrometry. Nat Biotechnol 17:676–682. https://doi.org/10.1038/10890
175. Liu F, Lössl P, Rabbitts BM, Balaban RS, Heck AJR (2018) The interactome of intact mitochondria by cross-linking mass spectrometry provides evidence for coexisting respiratory supercomplexes. Mol Cell Proteomics 17:216–232
176. Lu Y-W, Claypool SM (2015) Disorders of phospholipid metabolism: an emerging class of mitochondrial disease due to defects in nuclear genes. Front Genet 6:1–27. https://doi.org/10.3389/fgene.2015.00003
177. Lund MT et al (2016) Hepatic mitochondrial oxidative phosphorylation is normal in obese patients with and without type 2 diabetes. J Physiol 594:4351–4358. https://doi.org/10.1113/JP272105
178. Lydic TA, Busik JV, Reid GE (2014) A monophasic extraction strategy for the simultaneous lipidome analysis of polar and nonpolar retina lipids. J Lipid Res 55:1797–1809. https://doi.org/10.1194/jlr.D050302
179. Machann J, Häring H, Schick F, Stumvoll M (2004) Intramyocellular lipids and insulin resistance. Diabetes Obes Metab 6:239–248. https://doi.org/10.1111/j.1462-8902.2004.00339.x
180. Makarov A, Denisov E, Kholomeev A, Balschun W, Lange O, Strupat K, Horning S (2006) Performance evaluation of a hybrid linear ion trap/orbitrap mass spectrometer. Anal Chem 78:2113–2120. https://doi.org/10.1021/ac0518811
181. Manza LL, Stamer SL, Ham A-JL, Codreanu SG, Liebler DC (2005) Sample preparation and digestion for proteomic analyses using spin filters. Proteomics 5:1742–1745. https://doi.org/10.1002/pmic.200401063
182. Mårtensson CU, Doan KN, Becker T (2016) Effects of lipids on mitochondrial functions. Biochimica et Biophysica Acta (BBA) – Mol Cell Biol Lipid 1862:102–113. https://doi.org/10.1016/j.bbalip.2016.06.015
183. Mataix J, Quiles JL, Huertas JR, Battino M, Manas M (1998) Tissue specific interactions of exercise, dietary fatty acids, and vitamin E in lipid peroxidation. Free Radic Biol Med 24:511–521
184. Matyash V, Liebisch G, Kurzchalia TV, Shevchenko A, Schwudke D (2008) Lipid extraction by methyl-tert-butyl ether for high-throughput lipidomics. J Lipid Res 49:1137–1146. https://doi.org/10.1194/jlr.D700041-JLR200
185. McKenzie M, Lazarou M, Thorburn DR, Ryan MT (2006) Mitochondrial respiratory chain supercomplexes are destabilized in Barth syndrome patients. J Mol Biol 361:462–469. https://doi.org/10.1016/j.jmb.2006.06.057
186. Minkler PE, Hoppel CL (2010) Separation and characterization of cardiolipin molecular species by reverse-phase ion pair high-performance liquid chromatography-mass spectrometry. J Lipid Res 51:856–865. https://doi.org/10.1194/jlr.D002857
187. Misu H et al (2007) Genes involved in oxidative phosphorylation are coordinately upregulated with fasting hyperglycaemia in livers of patients with type 2 diabetes. Diabetologia 50:268–277. https://doi.org/10.1007/s00125-006-0489-8
188. Mogensen M, Sahlin K, Fernstrom M, Glintborg D, Vind BF, Beck-Nielsen H, Hojlund K (2007) Mitochondrial respiration is decreased in skeletal muscle of patients with type 2 diabetes. Diabetes 56:1592–1599. https://doi.org/10.2337/db06-0981
189. Montgomery MK, Turner N (2015) Mitochondrial dysfunction and insulin resistance: an update. Endocr Connect 4:R1–R15. https://doi.org/10.1530/EC-14-0092
190. Mootha VK et al (2003a) Integrated analysis of protein composition, tissue diversity, and gene regulation in mouse mitochondria. Cell 115:629–640. https://doi.org/10.1016/S0092-8674(03)00926-7
191. Mootha VK et al (2003b) PGC-1alpha-responsive genes involved in oxidative phosphorylation are coordinately downregulated in human diabetes. Nat Genet 34:267–273. https://doi.org/10.1038/ng1180
192. Moreadith RW, Fiskum G (1984) Isolation of mitochondria from ascites tumor cells permeabilized with digitonin. Anal Biochem 137:360–367. https://doi.org/10.1016/0003-2697(84)90098-8
193. Morino K, Petersen KF, Shulman GI (2006) Molecular mechanisms of insulin resistance in humans and their potential links with mitochondrial dysfunction. Diabetes 55
194. Murphy RC, Axelsen PH (2011) Mass spectrometric analysis of long-chain lipids. Mass Spectrom Rev 30:579–599. https://doi.org/10.1002/mas.20284
195. Nagaraj N et al (2011) Deep proteome and transcriptome mapping of a human cancer cell line. Mol Syst Biol 7
196. Nair S, V PC, Arnold C, Diehl AM (2003) Hepatic ATP reserve and efficiency of replenishing: comparison between obese and nonobese normal individuals. Am J Gastroenterol 98:466–470. https://doi.org/10.1111/j.1572-0241.2003.07221.x
197. Navarro P et al (2016) A multicenter study benchmarks software tools for label-free proteome quantification. Nat Biotechnol 34:1130. https://doi.org/10.1038/nbt.3685. https://www.nature.com/articles/nbt.3685#supplementary-information
198. Newmeyer DD, Ferguson-Miller S (2003) Mitochondria: releasing power for life and unleashing the machineries of death. Cell 112:481–490. https://doi.org/10.1016/S0092-8674(03)00116-8
199. Niemi NM, MacKeigan JP (2013) Mitochondrial phosphorylation in apoptosis: flipping the death switch. Antioxid Redox Signal 19:572–582. https://doi.org/10.1089/ars.2012.4982
200. Nikolov M, Schmidt C, Urlaub H (2012) Quantitative mass spectrometry-based proteomics: an overview. Method Mol Biol (Clifton, NJ) 893:85–100. https://doi.org/10.1007/978-1-61779-885-6_7
201. Nisr RB, Affourtit C (2014) Insulin acutely improves mitochondrial function of rat and human skeletal

muscle by increasing coupling efficiency of oxidative phosphorylation. Biochim Biophys Acta 1837:270–276. https://doi.org/10.1016/j.bbabio.2013.10.012
202. Nowotny B et al (2013) Mechanisms underlying the onset of oral lipid-induced skeletal muscle insulin resistance in humans. Diabetes 62:2240–2248. https://doi.org/10.2337/db12-1179
203. O'Shea KM et al (2009) Dietary omega-3 fatty acids alter cardiac mitochondrial phospholipid composition and delay Ca2+–induced permeability transition. J Mol Cell Cardiol 47:819–827. https://doi.org/10.1016/j.yjmcc.2009.08.014
204. Ong S-E, Blagoev B, Kratchmarova I, Kristensen DB, Steen H, Pandey A, Mann M (2002) Stable isotope labeling by amino acids in cell culture, SILAC, as a simple and accurate approach to expression proteomics. Mol Cell Proteomics 1:376–386
205. Ordureau A et al (2015) Defining roles of PARKIN and ubiquitin phosphorylation by PINK1 in mitochondrial quality control using a ubiquitin replacement strategy. Proc Natl Acad Sci 112:6637
206. Osman C et al (2009) The genetic interactome of prohibitins: coordinated control of cardiolipin and phosphatidylethanolamine by conserved regulators in mitochondria. J Cell Biol 184:583–596. https://doi.org/10.1083/jcb.200810189
207. Osman C, Voelker DR, Langer T (2011) Making heads or tails of phospholipids in mitochondria. J Cell Biol 192:7–16. https://doi.org/10.1083/jcb.201006159
208. Ott M, Gogvadze V, Orrenius S, Zhivotovsky B (2007) Mitochondria, oxidative stress and cell death. Apoptosis 12:913–922
209. Pagliarini DJ, Rutter J (2013) Hallmarks of a new era in mitochondrial biochemistry. Genes Dev 27:2615–2627
210. Palmfeldt J, Bross P (2017) Proteomics of human mitochondria. Mitochondrion 33:2–14. https://doi.org/10.1016/j.mito.2016.07.006
211. Pan DA, Lillioja S, Milner MR, Kriketos AD, Baur LA, Bogardus C, Storlien LH (1995) Skeletal muscle membrane lipid composition is related to adiposity and insulin action. J Clin Investig 96:2802–2808
212. Pan DA et al (1997) Skeletal muscle triglyceride levels are inversely related to insulin action. Diabetes 46:983–988. https://doi.org/10.2337/diab.46.6.983
213. Pangborn MC (1942) Isolation and purification of a serologically active phospholipid from beef heart. J Biol Chem 143:247–256
214. Paradies G, Petrosillo G, Pistolese M, Ruggiero FM (2000) The effect of reactive oxygen species generated from the mitochondrial electron transport chain on the cytochrome c oxidase activity and on the cardiolipin content in bovine heart submitochondrial particles. FEBS Lett 466:323–326
215. Paradies G, Petrosillo G, Pistolese M, Ruggiero FM (2001) Reactive oxygen species generated by the mitochondrial respiratory chain affect the complex III activity via cardiolipin peroxidation in beef-heart submitochondrial particles. Mitochondrion 1:151–159. https://doi.org/10.1016/S1567-7249(01)00011-3
216. Paradies G, Petrosillo G, Pistolese M, Ruggiero FM (2002) Reactive oxygen species affect mitochondrial electron transport complex I activity through oxidative cardiolipin damage. Gene 286:135–141. https://doi.org/10.1016/S0378-1119(01)00814-9
217. Paradies G, Paradies V, De Benedictis V, Ruggiero FM, Petrosillo G (2014) Functional role of cardiolipin in mitochondrial bioenergetics. Biochimica et Biophysica Acta (BBA) – Bioenergetics 1837:408–417. https://doi.org/10.1016/j.bbabio.2013.10.006
218. Park M, Kaddai V, Ching J, Fridianto KT, Sieli RJ, Sugii S, Summers SA (2016) A role for ceramides, but NOT sphingomyelins, as antagonists of insulin signaling and mitochondrial metabolism in C2C12 myotubes. J Biol Chem. https://doi.org/10.1074/jbc.M116.737684
219. Patel MS, Roche TE (1990) Molecular biology and biochemistry of pyruvate dehydrogenase complexes. FASEB J 4:3224–3233
220. Patel MS, Owen OE, Goldman LI, Hanson RW (1975) Fatty acid synthesis by human adipose tissue. Metabolism 24:161–173
221. Patterson RE et al (2016) Lipotoxicity in steatohepatitis occurs despite an increase in tricarboxylic acid cycle activity. Am J Physiol Endocrinol Metab 310:E484–E494. https://doi.org/10.1152/ajpendo.00492.2015
222. Patti M-E, Corvera S (2010) The role of mitochondria in the pathogenesis of type 2 diabetes. Endocr Rev 31:364–395. https://doi.org/10.1210/er.2009-0027
223. Patti ME et al (2003) Coordinated reduction of genes of oxidative metabolism in humans with insulin resistance and diabetes: potential role of PGC1 and NRF1. Proc Natl Acad Sci U S A 100:8466–8471. https://doi.org/10.1073/pnas.1032913100
224. Perez-Carreras M et al (2003) Defective hepatic mitochondrial respiratory chain in patients with nonalcoholic steatohepatitis. Hepatology 38:999–1007. https://doi.org/10.1053/jhep.2003.50398
225. Pérez-Hernández IH, Domínguez-Fuentes JM, Palomar-Morales M, Zazueta-Mendizabal AC, Baiza-Gutman A, Mejía-Zepeda R (2017) Liver mitochondrial membrane fluidity at early development of diabetes and its correlation with the respiration. J Bioenerg Biomembr 49(3):231–239. https://doi.org/10.1007/s10863-017-9700-5
226. Perham RN (1991) Domains, motifs, and linkers in 2-oxo acid dehydrogenase multienzyme complexes: a paradigm in the design of a multifunctional protein. Biochemistry 30:8501–8512
227. Perreault L et al (2018) Intracellular localization of diacylglycerols and sphingolipids influences insulin sensitivity and mitochondrial function in human skeletal muscle. JCI Insight 3. https://doi.org/10.1172/jci.insight.96805
228. Perry RJ et al (2017) Non-invasive assessment of hepatic mitochondrial metabolism by positional iso-

topomer NMR tracer analysis (PINTA). Nat Commun 8:798. https://doi.org/10.1038/s41467-017-01143-w
229. Perseghin G et al (1999) Intramyocellular triglyceride content is a determinant of in vivo insulin resistance in humans: a 1H-13C nuclear magnetic resonance spectroscopy assessment in offspring of type 2 diabetic parents. Diabetes 48:1600–1606
230. Petersen KF, Dufour S, Befroy D, Garcia R, Shulman GI (2004) Impaired mitochondrial activity in the insulin-resistant offspring of patients with type 2 diabetes. N Engl J Med 350:664–671. https://doi.org/10.1056/NEJMoa031314
231. Petersen KF, Befroy DE, Dufour S, Rothman DL, Shulman GI (2016) Direct assessment of hepatic mitochondrial oxidation and pyruvate cycling in non alcoholic fatty liver disease by (13)C magnetic resonance spectroscopy. Cell Metab 24:167–171. https://doi.org/10.1016/j.cmet.2016.06.005
232. Peterson AC, Russell JD, Bailey DJ, Westphall MS, Coon JJ (2012) Parallel reaction monitoring for high resolution and high mass accuracy quantitative, targeted proteomics. Mol Cell Proteomics 11:1475–1488
233. Pfeiffer K, Gohil V, Stuart RA, Hunte C, Brandt U, Greenberg ML, Schagger H (2003) Cardiolipin stabilizes respiratory chain supercomplexes. J Biol Chem 278:52873–52880. https://doi.org/10.1074/jbc.M308366200
234. Phielix E et al (2008) Lower intrinsic ADP-stimulated mitochondrial respiration underlies in vivo mitochondrial dysfunction in muscle of male type 2 diabetic patients. Diabetes 57:2943–2949. https://doi.org/10.2337/db08-0391
235. Plakkal Ayyappan J, Paul A, Goo YH (2016) Lipid droplet-associated proteins in atherosclerosis. Mol Med Rep 13:4527–4534
236. Pollard AK, Ortori CA, Stoger R, Barrett DA, Chakrabarti L (2017) Mouse mitochondrial lipid composition is defined by age in brain and muscle. Aging 9:986–998. https://doi.org/10.18632/aging.101204
237. Polozova A, Gionfriddo E, Salem N Jr (2006) Effect of docosahexaenoic acid on tissue targeting and metabolism of plasma lipoproteins. Prostaglandins Leukot Essent Fat Acids 75:183–190. https://doi.org/10.1016/j.plefa.2006.05.009
238. Porter RK, Hulbert AJ, Brand MD (1996) Allometry of mitochondrial proton leak: influence of membrane surface area and fatty acid composition. Am J Phys 271:R1550–R1560
239. Quiles JL, Huertas JR, Manas M, Battino M, Mataix J (1999) Physical exercise affects the lipid profile of mitochondrial membranes in rats fed with virgin olive oil or sunflower oil. Br J Nutr 81:21–24
240. Quiles JL, Huertas JR, Manas M, Ochoa JJ, Battino M, Mataix J (2001) Dietary fat type and regular exercise affect mitochondrial composition and function depending on specific tissue in the rat. J Bioenerg Biomembr 33:127–134
241. Reed LJ (2001) A trail of research from lipoic acid to alpha-keto acid dehydrogenase complexes. J Biol Chem 276:38329–38336. https://doi.org/10.1074/jbc.R100026200
242. Reis A, Rudnitskaya A, Blackburn GJ, Mohd Fauzi N, Pitt AR, Spickett CM (2013) A comparison of five lipid extraction solvent systems for lipidomic studies of human LDL. J Lipid Res 54:1812–1824. https://doi.org/10.1194/jlr.M034330
243. Riedl SJ, Shi Y (2004) Molecular mechanisms of caspase regulation during apoptosis. Nat Rev Mol Cell Biol 5:897–907. https://doi.org/10.1038/nrm1496
244. Rietveld AG, Killian JA, Dowhan W, de Kruijff B (1993) Polymorphic regulation of membrane phospholipid composition in Escherichia coli. J Biol Chem 268:12427–12433
245. Ritov VB, Menshikova EV, He J, Ferrell RE, Goodpaster BH, Kelley DE (2005) Deficiency of subsarcolemmal mitochondria in obesity and type 2 diabetes. Diabetes 54:8–14
246. Ritov VB et al (2010) Deficiency of electron transport chain in human skeletal muscle mitochondria in type 2 diabetes mellitus and obesity. Am J Phys Endocrinol Metab 298:E49–E58. https://doi.org/10.1152/ajpendo.00317.2009
247. Ritter O, Jelenik T, Roden M (2015) Lipid-mediated muscle insulin resistance: different fat, different pathways? J Mol Med (Berlin, Germany) 93:831–843. https://doi.org/10.1007/s00109-015-1310-2
248. Ronsein GE et al (2015) Parallel reaction monitoring (PRM) and selected reaction monitoring (SRM) exhibit comparable linearity, dynamic range and precision for targeted quantitative HDL proteomics. J Proteome 113:388–399. https://doi.org/10.1016/j.jprot.2014.10.017
249. Ross PL et al (2004) Multiplexed protein quantitation in Saccharomyces cerevisiae using amine-reactive isobaric tagging reagents. Mol Cell Proteomics 3:1154–1169
250. Ruderman NB, Saha AK, Vavvas D, Witters LA (1999) Malonyl-CoA, fuel sensing, and insulin resistance. Am J Phys 276:E1–e18
251. Russell AP, Foletta VC, Snow RJ, Wadley GD (2014) Skeletal muscle mitochondria: a major player in exercise, health and disease. Biochimica et Biophysica Acta (BBA) – Gen Subj 1840:1276–1284
252. Saliba A-E, Vonkova I, Gavin A-C (2015) The systematic analysis of protein–lipid interactions comes of age. Nat Rev Mol Cell Biol 16:753
253. Samuel VT, Shulman GI (2012) Mechanisms for insulin resistance: common threads and missing links. Cell 148:852–871. https://doi.org/10.1016/j.cell.2012.02.017
254. Samuel VT, Petersen KF, Shulman GI (2010) Lipid-induced insulin resistance: unravelling the mechanism. Lancet (London, England) 375:2267–2277. https://doi.org/10.1016/s0140-6736(10)60408-4

255. Santos CR, Schulze A (2012) Lipid metabolism in cancer. FEBS J 279:2610–2623. https://doi.org/10.1111/j.1742-4658.2012.08644.x
256. Schlame M, Ren M (2009) The role of cardiolipin in the structural organization of mitochondrial membranes. Biochim Biophys Acta Biomembr 1788:2080–2083. https://doi.org/10.1016/j.bbamem.2009.04.019
257. Schlame M, Brody S, Hostetler KY (1993) Mitochondrial cardiolipin in diverse eukaryotes. Eur J Biochem 212:727–733. https://doi.org/10.1111/j.1432-1033.1993.tb17711.x
258. Schlame M et al (2003) Phospholipid abnormalities in children with Barth syndrome. J Am Coll Cardiol 42:1994–1999
259. Schlame M, Ren M, Xu Y, Greenberg ML, Haller I (2005) Molecular symmetry in mitochondrial cardiolipins. Chem Phys Lipids 138:38–49. https://doi.org/10.1016/j.chemphyslip.2005.08.002
260. Schmid AI, Szendroedi J, Chmelik M, Krssak M, Moser E, Roden M (2011) Liver ATP synthesis is lower and relates to insulin sensitivity in patients with type 2 diabetes. Diabetes Care 34:448–453. https://doi.org/10.2337/dc10-1076
261. Schöttl T, Kappler L, Braun K, Fromme T, Klingenspor M (2014) Limited mitochondrial capacity of visceral versus subcutaneous white adipocytes in male C57BL/6N mice. Endocrinology 156:923–933. https://doi.org/10.1210/en.2014-1689
262. Schweppe DK et al (2017) Mitochondrial protein interactome elucidated by chemical cross-linking mass spectrometry. Proc Natl Acad Sci 114:1732
263. Schwitzgebel VM (2014) Many faces of monogenic diabetes. J Diabetes Investig 5:121–133. https://doi.org/10.1111/jdi.12197
264. Schwudke D et al (2007) Top-down lipidomic screens by multivariate analysis of high-resolution survey mass spectra. Anal Chem 79:4083–4093. https://doi.org/10.1021/ac062455y
265. Seale P et al (2008) PRDM16 controls a brown fat/skeletal muscle switch. Nature 454:961–967. http://www.nature.com/nature/journal/v454/n7207/suppinfo/nature07182_S1.html
266. Seppanen-Laakso T, Oresic M (2009) How to study lipidomes. J Mol Endocrinol 42:185–190. https://doi.org/10.1677/JME-08-0150
267. Sheehan JP, Swerdlow RH, Miller SW, Davis RE, Parks JK, Parker WD, Tuttle JB (1997) Calcium homeostasis and reactive oxygen species production in cells transformed by mitochondria from individuals with sporadic Alzheimer's disease. J Neurosci 17:4612
268. Shulman GI (2000) Cellular mechanisms of insulin resistance. J Clin Invest 106:171–176. https://doi.org/10.1172/jci10583
269. Sickmann A et al (2003) The proteome of Saccharomyces cerevisiae mitochondria. Proc Natl Acad Sci 100:13207
270. Siddique MM et al (2013) Ablation of dihydroceramide desaturase 1, a therapeutic target for the treatment of metabolic diseases, simultaneously stimulates anabolic and catabolic signaling. Mol Cell Biol 33:2353–2369. https://doi.org/10.1128/mcb.00226-13
271. Sinz A (2006) Chemical cross-linking and mass spectrometry to map three-dimensional protein structures and protein–protein interactions. Mass Spectrom Rev 25:663–682. https://doi.org/10.1002/mas.20082
272. Smith AC, Robinson AJ (2016) MitoMiner v3.1, an update on the mitochondrial proteomics database. Nucleic Acids Res 44:D1258–D1261. https://doi.org/10.1093/nar/gkv1001
273. Snel M et al (2012) Ectopic fat and insulin resistance: pathophysiology and effect of diet and lifestyle interventions. Int J Endocrinol 2012:983814. https://doi.org/10.1155/2012/983814
274. Song B-J, Akbar M, Abdelmegeed MA, Byun K, Lee B, Yoon SK, Hardwick JP (2014) Mitochondrial dysfunction and tissue injury by alcohol, high fat, nonalcoholic substances and pathological conditions through post-translational protein modifications. Redox Biol 3:109–123. https://doi.org/10.1016/j.redox.2014.10.004
275. Sparagna GC, Lesnefsky EJ (2009) Cardiolipin remodeling in the heart. J Cardiovasc Pharmacol 53:290–301. https://doi.org/10.1097/FJC.0b013e31819b5461
276. Stanley WC, Khairallah RJ, Dabkowski ER (2012) Update on lipids and mitochondrial function: impact of dietary n-3 polyunsaturated fatty acids. Curr Opin Clin Nutr Metab Care 15:122–126. https://doi.org/10.1097/MCO.0b013e32834fdaf7
277. Stram AR, Payne RM (2016) Posttranslational modifications in mitochondria: protein signaling in the powerhouse cellular and molecular life sciences. CMLS 73:4063–4073. https://doi.org/10.1007/s00018-016-2280-4
278. Su X, Han X, Mancuso DJ, Abendschein DR, Gross RW (2005) Accumulation of long-chain acylcarnitine and 3-hydroxy acylcarnitine molecular species in diabetic myocardium: identification of alterations in mitochondrial fatty acid processing in diabetic myocardium by shotgun lipidomics. Biochemistry 44:5234–5245. https://doi.org/10.1021/bi047773a
279. Summers SA, Garza LA, Zhou H, Birnbaum MJ (1998) Regulation of insulin-stimulated glucose transporter GLUT4 translocation and Akt kinase activity by ceramide. Mol Cell Biol 18:5457–5464. https://doi.org/10.1128/mcb.18.9.5457
280. Sunny NE, Parks EJ, Browning JD, Burgess SC (2011) Excessive hepatic mitochondrial TCA cycle and gluconeogenesis in humans with nonalcoholic fatty liver disease. Cell Metab 14:804–810. https://doi.org/10.1016/j.cmet.2011.11.004
281. Sunny NE, Bril F, Cusi K (2017) Mitochondrial adaptation in nonalcoholic fatty liver disease: novel mechanisms and treatment strategies trends in endocrinology and metabolism. TEM 28:250–260. https://doi.org/10.1016/j.tem.2016.11.006

282. Szendroedi J et al (2009) Abnormal hepatic energy homeostasis in type 2 diabetes. Hepatology 50:1079–1086. https://doi.org/10.1002/hep.23093
283. Szendroedi J et al (2014) Role of diacylglycerol activation of PKCtheta in lipid-induced muscle insulin resistance in humans. Proc Natl Acad Sci U S A 111:9597–9602. https://doi.org/10.1073/pnas.1409229111
284. Taanman J-W (1999) The mitochondrial genome: structure, transcription, translation and replication. Biochimica et Biophysica Acta (BBA) – Bioenergetics 1410:103–123. https://doi.org/10.1016/S0005-2728(98)00161-3
285. Tait SWG, Green DR (2012) Mitochondria and cell signalling. J Cell Sci 125:807–815. https://doi.org/10.1242/jcs.099234
286. Takamura T et al (2008) Obesity upregulates genes involved in oxidative phosphorylation in livers of diabetic patients. Obesity (Silver Spring, MD) 16:2601–2609. https://doi.org/10.1038/oby.2008.419
287. Tamura Y, Sesaki H, Endo T (2014) Phospholipid transport via mitochondria. Traffic 15:933–945. https://doi.org/10.1111/tra.12188
288. Tasseva G, Bai HD, Davidescu M, Haromy A, Michelakis E, Vance JE (2013) Phosphatidylethanolamine deficiency in mammalian mitochondria impairs oxidative phosphorylation and alters mitochondrial morphology. J Biol Chem 288:4158–4173. https://doi.org/10.1074/jbc.M112.434183
289. Taylor WA, Xu FY, Ma BJ, Mutter TC, Dolinsky VW, Hatch GM (2002) Expression of monolysocardiolipin acyltransferase activity is regulated in concert with the level of cardiolipin and cardiolipin biosynthesis in the mammalian heart. BMC Biochem 3:9
290. Thingholm TE, Jørgensen TJD, Jensen ON, Larsen MR (2006) Highly selective enrichment of phosphorylated peptides using titanium dioxide. Nat Protoc 1:1929. https://doi.org/10.1038/nprot.2006.185
291. Thompson A et al (2003) Tandem mass tags: a novel quantification strategy for comparative analysis of complex protein mixtures by MS/MS. Anal Chem 75:1895–1904. https://doi.org/10.1021/ac0262560
292. Tilg H, Moschen AR, Roden M (2017) NAFLD and diabetes mellitus. Nat Rev Gastroenterol Hepatol 14:32–42. https://doi.org/10.1038/nrgastro.2016.147
293. Timmons JA et al (2007) Myogenic gene expression signature establishes that brown and white adipocytes originate from distinct cell lineages. Proc Natl Acad Sci U S A 104:4401–4406
294. Timmons MD, Bradley MA, Lovell MA, Lynn BC (2011) Procedure for the isolation of mitochondria, cytosolic and nuclear material from a single piece of neurological tissue for high-throughput mass spectral analysis. J Neurosci Methods 197:279–282. https://doi.org/10.1016/j.jneumeth.2011.02.027
295. Timmons MD, Bradley M, Lovell MA, Lynn BC (2012) Comparison of Sprague Dawley rat mitochondrial fatty acid profiles isolated from various tissues. Neurobiol Lipid 11
296. Ting YS et al (2015) Peptide-centric proteome analysis: an alternative strategy for the analysis of tandem mass spectrometry data. Mol Cell Proteomics 14:2301–2307
297. Toby TK, Fornelli L, Kelleher NL (2016) Progress in top-down proteomics and the analysis of Proteoforms. Annu Rev Anal Chem 9:499–519. https://doi.org/10.1146/annurev-anchem-071015-041550
298. Toledo FG, Goodpaster BH (2013) The role of weight loss and exercise in correcting skeletal muscle mitochondrial abnormalities in obesity, diabetes and aging. Mol Cell Endocrinol 379:30–34
299. Toledo FGS, Menshikova EV, Ritov VB, Azuma K, Radikova Z, DeLany J, Kelley DE (2007) Effects of physical activity and weight loss on skeletal muscle mitochondria and relationship with glucose control in type 2 diabetes. Diabetes 56:2142–2147. https://doi.org/10.2337/db07-0141
300. Tonkonogi M, Harris B, Sahlin K (1997) Increased activity of citrate synthase in human skeletal muscle after a single bout of prolonged exercise. Acta Physiol Scand 161:435–436
301. Trenell MI, Hollingsworth KG, Lim EL, Taylor R (2008) Increased daily walking improves lipid oxidation without changes in mitochondrial function in type 2 diabetes. Diabetes Care 31:1644–1649. https://doi.org/10.2337/dc08-0303
302. Turner N, Bruce CR, Beale SM, Hoehn KL, So T, Rolph MS, Cooney GJ (2007) Excess lipid availability increases mitochondrial fatty acid oxidative capacity in muscle: evidence against a role for reduced fatty acid oxidation in lipid-induced insulin resistance in rodents. Diabetes 56:2085–2092. https://doi.org/10.2337/db07-0093
303. Turner N et al (2009) Enhancement of muscle mitochondrial oxidative capacity and alterations in insulin action are lipid species dependent- potent tissue-specific effects of medium-chain fatty acids. Diabetes 58:2547–2554. https://doi.org/10.2337/db09-0784
304. Unsay JD, Cosentino K, Subburaj Y, García-Sáez AJ (2013) Cardiolipin effects on membrane structure and dynamics. Langmuir 29:15878–15887. https://doi.org/10.1021/la402669z
305. van der Veen JN, Lingrell S, da Silva RP, Jacobs RL, Vance DE (2014) The concentration of phosphatidylethanolamine in mitochondria can modulate ATP production and glucose metabolism in mice. Diabetes 63:2620–2630. https://doi.org/10.2337/db13-0993
306. van Meer G (2005) Cellular lipidomics. EMBO J 24:3159–3165. https://doi.org/10.1038/sj.emboj.7600798
307. van Meer G, Voelker DR, Feigenson GW (2008) Membrane lipids: where they are and how they behave. Nat Rev Mol Cell Biol 9:112–124
308. van Tienen FH et al (2012) Physical activity is the key determinant of skeletal muscle mitochondrial function in type 2 diabetes. J Clin Endocrinol Metab 97:3261–3269. https://doi.org/10.1210/jc.2011-3454

309. Villeneuve LM, Purnell PR, Boska MD, Fox HS (2016) Early expression of Parkinson's disease-related mitochondrial abnormalities in PINK1 knockout rats. Mol Neurobiol 53:171–186. https://doi.org/10.1007/s12035-014-8927-y
310. Walheim E, Wiśniewski JR, Jastroch M (2018) Respiromics – an integrative analysis linking mitochondrial bioenergetics to molecular signatures. Mol Metab 9:4–14. https://doi.org/10.1016/j.molmet.2018.01.002
311. Wallace DC, Fan W (2010) Energetics, epigenetics, mitochondrial genetics. Mitochondrion 10:12–31. https://doi.org/10.1016/j.mito.2009.09.006
312. Wang C, Youle RJ (2009) The role of mitochondria in apoptosis. Annu Rev Genet 43:95–118. https://doi.org/10.1146/annurev-genet-102108-134850
313. Wang Y et al (2011) Reversed-phase chromatography with multiple fraction concatenation strategy for proteome profiling of human MCF10A cells. Proteomics 11:2019–2026. https://doi.org/10.1002/pmic.201000722
314. Wang Y, Zhang J, Li B, He Q-Y (2017) Proteomic analysis of mitochondria: biological and clinical progresses in cancer. Expert Rev Proteomics 14:891–903. https://doi.org/10.1080/14789450.2017.1374180
315. Warram JH, Martin BC, Krolewski AS, Soeldner JS, Kahn CR (1990) Slow glucose removal rate and hyperinsulinemia precede the development of type II diabetes in the offspring of diabetic parents. Ann Intern Med 113:909–915
316. Washburn MP, Wolters D, Yates JR (2001) Large-scale analysis of the yeast proteome by multidimensional protein identification technology. Nat Biotechnol 19:242–247. http://www.nature.com/nbt/journal/v19/n3/suppinfo/nbt0301_242_S1.html
317. Watkins SM, Carter LC, German JB (1998) Docosahexaenoic acid accumulates in cardiolipin and enhances HT-29 cell oxidant production. J Lipid Res 39:1583–1588
318. Watkins SM, Reifsnyder PR, Pan HJ, German JB, Leiter EH (2002) Lipid metabolome-wide effects of the PPARgamma agonist rosiglitazone. J Lipid Res 43:1809–1817
319. Wiedemann N, Frazier AE, Pfanner N (2004) The protein import machinery of mitochondria. J Biol Chem 279:14473–14476
320. Willingham TB, McCully KK (2017) In vivo assessment of mitochondrial dysfunction in clinical populations using near-infrared spectroscopy. Front Physiol 8:689. https://doi.org/10.3389/fphys.2017.00689
321. Wisniewski JR, Zougman A, Nagaraj N, Mann M (2009) Universal sample preparation method for proteome analysis. Nat Method 6:359–362. http://www.nature.com/nmeth/journal/v6/n5/suppinfo/nmeth.1322_S1.html
322. Wu F, Minteer S (2015) Krebs cycle metabolon: structural evidence of substrate channeling revealed by cross-linking and mass spectrometry. Angew Chem Int Ed 54:1851–1854. https://doi.org/10.1002/anie.201409336
323. Wu M et al (2007) Multiparameter metabolic analysis reveals a close link between attenuated mitochondrial bioenergetic function and enhanced glycolysis dependency in human tumor cells. Am J Physiol Cell Physiol 292:C125–C136. https://doi.org/10.1152/ajpcell.00247.2006
324. Yamada T, Ida T, Yamaoka Y, Ozawa K, Takasan H, Honjo I (1975) Two distinct patterns of glucose intolerance in icteric rats and rabbits. Relationship to impaired liver mitochondria function. J Lab Clin Med 86:38–45
325. Yan Y, C-L W, W-R Z, H-P C, Liu J (2006) Cross-talk between calcium and reactive oxygen species signaling. Acta Pharmacol Sin 27:821. https://doi.org/10.1111/j.1745-7254.2006.00390.x
326. Yang C, Aye Cho C, Li X, Diaz Ramos A, Zorzano A, Mora S (2012a) Mitochondrial dysfunction in insulin resistance: differential contributions of chronic insulin and saturated fatty acid exposure in muscle cells. Biosci Rep 32:465–478. https://doi.org/10.1042/bsr20120034
327. Yang F, Shen Y, Camp DG, Smith RD (2012b) High pH reversed-phase chromatography with fraction concatenation as an alternative to strong-cation exchange chromatography for two-dimensional proteomic analysis. Expert Rev Proteomics 9:129–134. https://doi.org/10.1586/epr.12.15
328. Zabielski P et al (2014) Impact of insulin deprivation and treatment on sphingolipid distribution in different muscle subcellular compartments of streptozotocin-diabetic C57Bl/6 mice. Am J Phys Endocrinol Metab 306:E529–E542. https://doi.org/10.1152/ajpendo.00610.2012
329. Zhang M, Mileykovskaya E, Dowhan W (2002) Gluing the respiratory chain together. Cardiolipin is required for supercomplex formation in the inner mitochondrial membrane. J Biol Chem 277:43553–43556. https://doi.org/10.1074/jbc.C200551200
330. Zhang X, Fang A, Riley CP, Wang M, Regnier FE, Buck C (2010) Multi-dimensional liquid chromatography. Proteomics Analytica Chimica Acta 664:101–113. https://doi.org/10.1016/j.aca.2010.02.001
331. Zhao Y, Jensen ON (2009) Modification-specific proteomics: strategies for characterization of post-translational modifications using enrichment techniques. Proteomics 9:4632–4641. https://doi.org/10.1002/pmic.200900398
332. Zhao X, León IR, Bak S, Mogensen M, Wrzesinski K, Højlund K, Jensen ON (2011) Phosphoproteome analysis of functional mitochondria isolated from resting human muscle reveals extensive phosphorylation of inner membrane protein complexes and enzymes. Mol Cell Proteomics: MCP 10:M110.000299. https://doi.org/10.1074/mcp.M110.000299
333. Zinser E, Daum G (1995) Isolation and biochemical characterization of organelles from the yeast, Saccharomyces cerevisiae. Yeast 11:493–536. https://doi.org/10.1002/yea.320110602

# Quantification of Mitochondrial Network Characteristics in Health and Disease

# 10

Andrew J. Valente, Joao Fonseca, Fereshteh Moradi, Gregory Foran, Alexander Necakov, and Jeffrey A. Stuart

**Abstract**

The term 'mitochondrial dynamics' is commonly used to refer to ongoing fusion and fission of mitochondrial structures within a living cell. A growing number of diseases, from Charcot Marie Tooth Type 2a neuropathies to cancer, is known to be associated with the dysregulation of mitochondrial dynamics, leading to irregularities of mitochondrial network morphology that are associated with aberrant metabolism and cellular dysfunction. Studying these phenomena, and potential pharmacological interventions to correct them, in cultured cells is a powerful approach to developing treatments or cures. Appropriately designed experiments and quantitative approaches for characterizing mitochondrial morphology and function are essential for furthering our understanding. In this chapter, we discuss the importance of cell incubation conditions, choices around imaging modalities, and data analysis tools with respect to experimental outcomes and the interpretation of results from studies of mitochondrial dynamics. We focus primarily on the quantitative analysis of mitochondrial morphology, providing an overview of the available tools and approaches currently being used and discussing some of the strengths and weaknesses associated with each. Finally, we discuss how the ongoing development of imaging and analysis tools continues to improve our ability to study normal and aberrant mitochondrial physiology in vitro and in vivo.

**Keywords**

MiNA · Mitochondrial networks · Mitochondrial dynamics · Fusion · Fission · Cell physiology · Live cell imaging

## 10.1 Introduction

In all nucleated animal cells, even post-mitotic terminally differentiated cells like neurons and myofibres, mitochondria are in a dynamic state of ongoing biogenesis, involving replication of mitochondrial DNA (mtDNA) and gain of mitochondrial mass, and mitophagy involving mtDNA degradation and loss of mitochondrial mass. Although these are clearly dynamic processes, the term 'mitochondrial dynamics' is more often applied to the more rapidly changing aspects of mitochondrial shape and organization within a cell. Within all cells, mitochondria exist as a population of organelles undergoing continual fusion into highly branched networks and fission into

A. J. Valente · J. Fonseca · F. Moradi · G. Foran · A. Necakov · J. A. Stuart (✉)
Department of Biological Sciences, Brock University, St. Catharines, ON, Canada
e-mail: jstuart@brocku.ca





A. Urbani, M. Babu (eds.), *Mitochondria in Health and in Sickness*, Advances in Experimental Medicine and Biology 1158, https://doi.org/10.1007/978-981-13-8367-0_10

smaller punctate and rod-like structures. The processes of fusion and fission, and their balance or imbalance to bring about remodeling and/or relocalization of mitochondria, are of biological and physiological significance.

The dynamic nature of mitochondria has been appreciated since, over a century ago, Lewis and Lewis [1] noted not only the heterogeneous morphology of mitochondria, but also how that morphology would change over time, morphing from long threadlike structures, to rings, to granules, and back. It was not until 1980, however, that mitochondrial labeling in live cells was demonstrated using rhodamine 123, a dye previously used for lasers [2]. Only a year later it was noted that rhodamine 123 could be used as an indicator of membrane potential as it was lost from the mitochondria of cells treated with uncouplers such as carbonyl cyanide-*4*-(trifluoromethoxy) phenylhydrazone (FCCP) [3]. This provided the means for probing not only the morphology but also the functional characteristics of mitochondria in live cells. Fluorescence microscopy studies have since benefited from the development of a wide variety of fluorescent small molecule probes and, more recently, fluorescent proteins and genetically encoded sensors. An extensive overview of many of the available mitochondrial labels and probes can be found in [4]. More recently, improvements in the computing technology for working with the large datasets that can be acquired using these methods have enabled significant leaps forward in the analysis of mitochondrial dynamics.

In many human mitochondrial diseases, relative rates of fusion and fission are affected directly via the inheritance of mutations in the genes involved in regulating these processes. In other diseases, such as cancer or diabetes, the connection is indirect but there are reasons to believe that treatment options could be productively targeted to the machinery of mitochondrial dynamics. It is, in general, important to understand how disease processes negatively affect mitochondrial dynamics and how pharmacological interventions might be targeted to correct this. A necessary step toward this goal is the development and refinement of methods for visualizing and analyzing mitochondria in live cells. In this chapter we focus primarily on fluorescence microscopy approaches for studying the steady-state characteristics of mitochondrial structure and function in live mammalian cells in the contexts of normal and diseased states.

### 10.1.1 Mitochondrial Dynamics in Cell Processes

Within an individual mammalian cell, mitochondria undergo continual structural and positional changes that lead to the highly heterogeneous morphologies we observe under the microscope. Rafelski and colleagues have provided a detailed description of the size, shape, and position of mitochondrial structures that occur [5, 6]. The machinery and molecular mechanisms of mitochondrial fusion/fission have been extensively studied, and readers are directed to several recent reviews for details of these [7, 8].

The overall architecture of mitochondrial networks, including the relative occurrence of punctate or tubular ‘individual’ mitochondrial structures versus highly interconnected and branched structures, represents the balance between fusion and fission processes at a given moment. Fusion and fission processes are in turn mediated by several molecular motors that use the energy in GTP to drive conformational changes. In the outer membrane of mammalian mitochondria, fusion is mediated by the mitofusins (Mfn1, Mfn2), and fission by Drp1. In the inner membrane, Opa1 participates in both fusion and fission processes. A number of accessory proteins interact with this core machinery to facilitate and direct the functions of fusion and fission proteins. Several post-translational modifications have been identified that further refine the functions of Mfn1/2 and Drp1 through changes in protein localization and specific activity [8, 9]. In addition, mitochondria are tethered to cytoskeletal elements and other organelles (e.g. endoplasmic reticulum) [10], and these interactions can influence the distribution and organization of mitochondria within the cell.

Since mitochondria are dynamic organelles, capable of fusion, fragmentation, biogenesis, mitophagy, and intra- or inter-cellular transport, the overall state of the mitochondrial network and its distribution within a cell is fluid. In studies performed in live cells in vitro, mitochondrial movement is variable, but can be sufficiently rapid as to interfere with imaging by introducing artifacts where movement rates exceed image acquisition rates. Alternatively, stable features of mitochondrial networks arise that are associated with specific cellular activities. For example, the relative distribution of mitochondria between branched networks and more fragmented individual structures changes over the course of the cell cycle [11]. In addition, metabolic switching from greater reliance on oxidative phosphorylation to glucose fermentation is associated with stable increases in fragmentation and a reduction of overall mitochondrial mass. Mitochondrial network characteristics are also related to cell stress and disease. Relatively rapid (in live cell culture, on the order of minutes) fragmentation of networks can occur when oxidative phosphorylation is uncoupled by protonophores like FCCP, or respiration is inhibited by, for example, hypoxia/anoxia. There are important direct and indirect links between mitochondrial network characteristics and cellular stress and disease. Pharmacological or hormonal targeting of the machinery of mitochondrial dynamics could represent a promising clinical approach to treating symptoms.

### 10.1.2 Mitochondrial Dynamics in Disease

The dynamic morphological characteristics of mitochondrial networks are affected directly and indirectly in many diseases. Peripheral neuronal pathologies are associated with mutations in Mfn2, Opa1, and various accessory proteins that coordinate the activities of these proteins, consequently affecting mitochondrial dynamics and the steady state morphology of mitochondrial networks. Several recent reviews discuss details of these diseases [12–14]. Barth syndrome, caused by mutations in the cardiolipin synthase enzyme Tafazzin, is characterized by mitochondrial fragmentation, perhaps related to the requirement of critical fusion/fission proteins for cardiolipin [15]. In addition, more indirect links between cellular dysfunctions and mitochondrial network fragmentation have been identified in a growing list of diseases. In cancer cells, where metabolism has shifted toward increased dependence on glycolysis and pentose phosphate pathway activity, mitochondrial networks are typically more fragmented. Pharmacological inhibition of Drp1 to reduce fission rates can improve the fusion state of mitochondrial networks and slow cancer cell growth [16]. Increased network fragmentation is also observed in neurodegenerative diseases and diabetes. The physiological significance of these observations has not always been established, but in some instances there is evidence that recovery of a highly fused mitochondrial state might improve cell function. There is therefore growing interest in mitochondrial dynamics as a pharmacological target in various diseases, and concomitantly an increasing need to determine the efficacy of putative molecular effectors by quantifying their effects on mitochondrial dynamics and/or steady state network morphology.

## 10.2 Fluorescence Microscopy for Studying Mitochondrial Dynamics

### 10.2.1 Fluorescence Microscopy Techniques

A wide range of fluorescence microscopy techniques has been employed for the study of mitochondrial networks. Resolution, acquisition speed, signal, and phototoxicity represent key experimental considerations and no single microscopy system can simultaneously maximize all of these. Therefore, the selection of a fluorescence microscopy modality should be made based on sample characteristics and the goals of the experiment. Here, we will briefly describe typical benefits and drawbacks of various fluorescence microscopy modalities in the context of different experimental needs.

Adherent mammalian cells are commonly used to study mitochondrial physiology. These cells exhibit a relatively flat morphology, only a few microns thick. For such thin samples, and many experimental contexts, conventional brightfield fluorescence microscopy is sufficient. Due to the contributions of out of focus light, conventional brightfield microscopy does not lend itself to three-dimensional imaging without further processing, but this can be accomplished by computational means. Deconvolution is used to redistribute photon signals to their correct origins, typically by using some information about the microscope's intrinsic characteristics that determine how the image is distorted.

All optical systems introduce some distortion and degradation of the real image. This leads to the notion of *diffraction limited imaging*, a topic we will return to later. The degradation of an image from acquisition may be described by the convolution of the original image and noise associated with the fluorescence, $I_d + n_f$, with the instrument response function or *point spread function*, $P$, and the addition of detection noise, $n_d$, as follows:

$$I_d = \left(I_o + n_f\right) \otimes P + n_d$$

The goal of deconvolution is to restore the true image, $I_o$, from the acquired dataset. As the point spread function is a three-dimensional entity, best results are obtained from deconvolving three-dimensional stacks with a three-dimensional empirical or theoretical point spread function. There are numerous reviews and book chapters devoted to describing the deconvolution approaches used in practice [17–19]. For thicker samples, or more highly light-scattering samples, such as the mitochondria densely packed about lipid droplets typical in adipocytes, deconvolution brightfield microscopy may be insufficient to fully resolve the structures of interest. In these cases, alternative means for acquiring optical sections may be essential.

Confocal microscopy, both laser scanning and spinning disk, provides a physical solution for excluding out of focus light from collected images to generate optical sections. The reader is directed to The Handbook of Biological Confocal Microscopy [20] and several book chapters and reviews [21–23] for in depth information regarding confocal microscopy and its application. Briefly, a high intensity light source, almost exclusively a laser, is used to excite a small volume of the sample. Fluorescence is collected while excluding out of focus photons by introducing a confocal aperture between the detector and sample. In laser scanning confocal systems this sampling volume is raster scanned across the sample while recording the fluorescence intensity with a point detector such as a photomultiplier tube, to generate an image one pixel at a time. Spinning disk systems generate a set of independent sampling volumes that are simultaneously scanned and detected on an area detector (typically a CMOS or electron multiplying CCD) to build up an image in a fraction of the time as needed for scanning. While scanning a single point at a time inherently results in longer acquisition times, it provides the ability to selectively bleach or photo-activate regions. For dynamic experiments such as fluorescence recovery after photobleaching or photoactivation of optogenetic constructs, such control is beneficial if not mandatory. Due to the physical rejection of a large portion of emitted photons, confocal methods typically use higher intensity excitation light which may result in greater phototoxicity and sample degradation.

Optical sectioning can be achieved post-acquisition as well by modulating the illumination pattern on the sample. This is referred to as optical sectioning structured illumination microscopy (OS-SIM), not to be confused with super-resolution structured illumination microscopy (SR-SIM). Recovery of optical sections by structured illumination was described as early as 1997 [24]. Such systems work by projecting a structured pattern, typically a regular grid, onto a specimen which causes only the in focus regions to be efficiently illuminated. By stepping the grid position through the full sample, each position in the image is sampled both efficiently and inefficiently allowing for the contributions of out of focus fluorescence to be effectively discarded. The Zeiss Apotome system provides this functionality in a commercially available package that has been shown to provide resolution comparable to or better than confocal methods, though with reduced performance on thicker samples [25]. These systems provide a performant alternative to confocal microscopy, especially for thin samples such as adherent mammalian cells.

The methods of image acquisition discussed above are diffraction-limited. With growing interest in mitochondrial morphology and automated morphological analysis, there has been a need for higher resolution techniques, especially for more difficult compact morphologies. Super-resolution fluorescence microscopy is the blanket term for many techniques that can generate images exceeding the theoretical maximum resolution for a microscope, approximately ½ the wavelength of light being imaged with or typically about 200–300 nm. Some systems rely primarily on hardware to overcome this limit, such as the beam shaping of stimulated emission depletion (STED) microscopy [26], or the multi-lens arrangement of a 4π microscope [27]. Other systems use a combination of hardware and software to reconstruct images at a higher resolution than would otherwise be possible. Such systems include SR-SIM [28], photoactivatable probe localization based techniques including stochastic optical reconstruction microscopy (STORM) [29] and photoactivatable localization microscopy (PALM) [30]. These methods can provide resolution ranging from roughly 100 nm down to tens of nm in the case of localizing methods. There is benefit in such high-resolution techniques, but there are also some drawbacks associated with acquisition and image characteristics. Super-resolution methods often mandate either high intensity illumination, long acquisition periods, or both. This can be a serious drawback for imaging live cells as high resolution becomes irrelevant if there is movement artifact or phototoxicity effects. This typically limits the practicality of such techniques to fixed samples, though there have been several notable live cell examples [31–35].

Motivated by the need for fast, high resolution, and gentle fluorescence microscopy techniques, light sheet microscopy, or selective plane illumination microscopy (SPIM), has been gaining more attention. There are many examples of how light sheet microscopy has already proven itself as an important tool for live imaging of larger samples [36]. Light sheet imaging is typically thought of as a performant method for imaging larger samples at cellular resolution while limiting phototoxicity by only illuminating the focal plane. However, it can be implemented to generate thinner light sheets using Bessel beams and has been demonstrated as a tool for imaging live cells with resolution similar to confocal techniques but while capturing over 100 frames/s [37]. As the resolution of light sheet microscopy is continuously improved, its ability to acquire 3D datasets at astonishing rates while remaining relatively gentle to the sample will likely stimulate the adoption of this technique for the study of mitochondrial dynamics.

### 10.2.2 Experimental Considerations

In live cell imaging, it is essential to maintain temperature and $CO_2$ levels, and minimize photodamage during acquisition. To illustrate this, we have provided results from time lapse videos of mitochondria in cells under conditions in which these variables are not controlled. Profound changes in mitochondrial networks occur rapidly (<10 min) when cell culture dishes are shifted from incubators to atmosphere when bicarbonate is a major media pH buffer (Fig. 10.1). Therefore, on-stage maintenance of $CO_2$, or use of media insensitive to $CO_2$ changes is a critical feature of a microscopy system used for mitochondrial imaging in live cells. Intense light exposure can also lead to mitochondrial damage and rapid morphological changes [33], eventually leading to cell death. Assessing phototoxicity, minimizing sample damage, and optimizing acquisition parameters are non-trivial tasks but essential for live cell investigations [38]. Ensuring samples are "happy" on the stage can be difficult, but a practical start to assessing their tolerance of

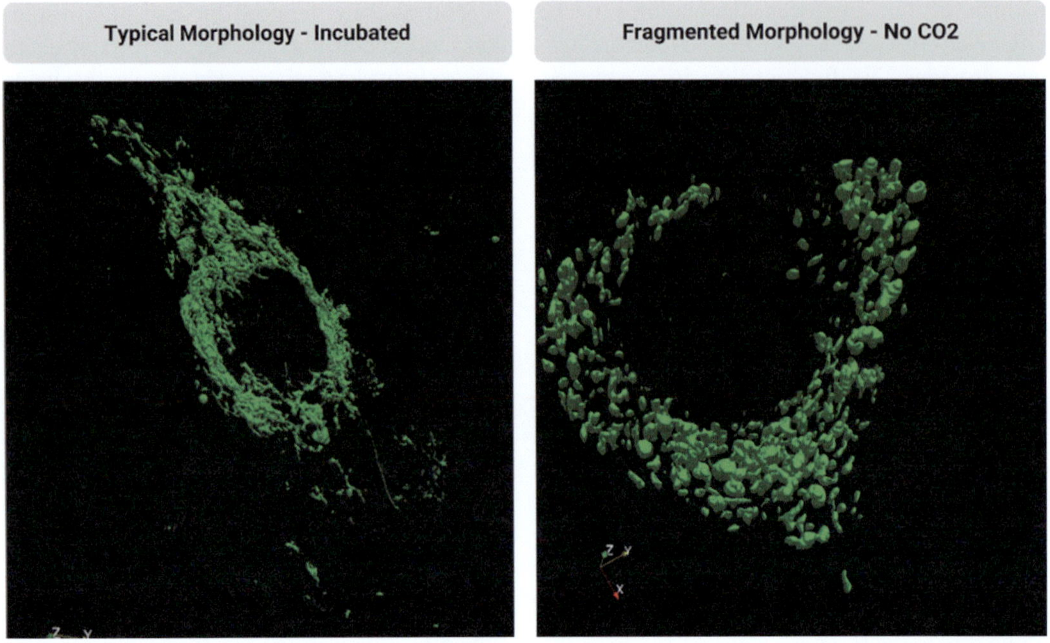


**Fig. 10.1 Morphological changes induced by sub-optimal environmental control during image acquisition**. C2C12 and PC3 cells stably transfected with mEmeral-Mito-7 (a gift from Michael Davidson, Florida State University) were grown in Dulbecco's Modified Eagle Medium (DMEM) supplemented with 10% fetal bovine serum (FBS) in a humidified 5% $CO_2$ atmosphere at 37 °C. Live cell fluorescence images were obtained with a Plan-Apochromat 63×/1.40 Oil DIC M27 microscope objective using a Carl Zeiss Axio Observer Z1 inverted light/epifluorescence microscope equipped with ApoTome.2 optical sectioning and a Hamamatsu ORCA-Flash 4.0 V2 digital camera. The microscope stage and objectives were maintained at 37 °C, with temperature control achieved through TempModule S-controlled stage heater and objective heater (PeCon, Erbach, Germany). $CO_2$ was maintained for the left image, while the right image was taken without any gas regulation on stage. Green fluorescence was detected using a fluorescence channel possessing excitation and emission wavelength filter sets of 450–490 nm and 500–550 nm, respectively. Both the intensity of fluorescence illumination and camera exposure time were held constant throughout all experiments. Z-stacks consisted of 20 slices, each 0.25 μm apart and used to produce 3D renderings using Fiji and ParaView. Cells imaged without $CO_2$ regulation demonstrate significantly altered morphology from those cells imaged under 5% $CO_2$

specific experimental conditions is to image control samples over a duration longer than that of the experiment and note if any cellular or organelle morphological changes appear outside of what one might expect.

A less appreciated aspect of experimental design in studies of mitochondrial networks in live cells is the importance of two other media constituents: oxygen and glucose. If the goal of an experiment is to model in vivo conditions, then 1–6% $O_2$ and 5 mM glucose [39] are appropriate near-physiological levels for most cell types. Glucose has been shown to affect mitochondrial morphology significantly. Under hyperglycemic conditions (20 mM glucose), mitochondrial morphology appears fragmented unlike the typically fused mitochondria in low glucose (5 mM) culture [40], a more physiologically relevant condition. Furthermore, measuring fusion by determining the extent of locally photo-activated photo-activatable GFP dispersion [41] demonstrates a reduction in fusion associated with high glucose conditions [40]. Provided that the mitochondria appear to behave vastly different under different nutrient loads, this is likely an important consideration in designing experiments investigating the mitochondrial implications of a compound or treatment.

We routinely use 5% $O_2$ as a physiologically relevant O2 setpoint and have found that cellular

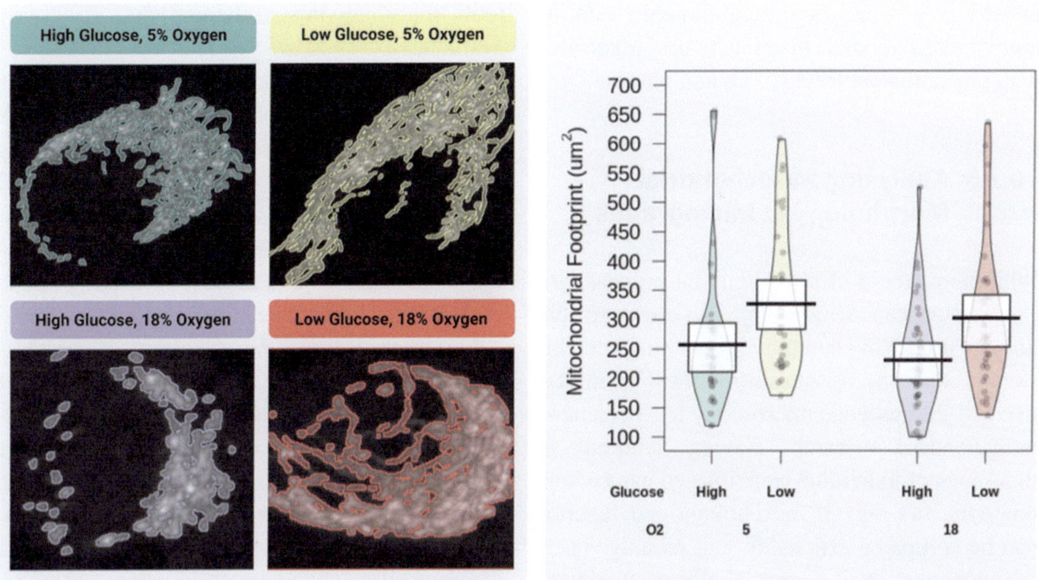


**Fig. 10.2 Media oxygen and glucose levels affect mitochondrial morphology**. C2C12 and PC3 cells stably transfected with mEmeral-Mito-7 (a gift from Michael Davidson, Florida State University) were grown in Dulbecco's Modified Eagle Medium (DMEM) supplemented with 10% fetal bovine serum (FBS) in a humidified 5% $CO_2$ atmosphere at 37 °C. $O_2$ levels in the incubators were maintained at either 5% or 18%. Glucose concentrations in DMEM were either 5 mM (low) or 25 mM (high). Cells were grown for 48 h in each condition and each condition was maintained throughout imaging. Live cell fluorescence images were obtained with a Plan-Apochromat 63×/1.40 Oil DIC M27 microscope objective using a Carl Zeiss Axio Observer Z1 inverted light/epifluorescence microscope equipped with ApoTome.2 optical sectioning and a Hamamatsu ORCA-Flash 4.0 V2 digital camera. The microscope stage and objectives were maintained at 37 °C, with temperature control achieved through TempModule S-controlled stage heater and objective heater (PeCon, Erbach, Germany). Green fluorescence was detected using a fluorescence channel possessing excitation and emission wavelength filter sets of 450–490 nm and 500–550 nm, respectively. Both the intensity of fluorescence illumination and camera exposure time were held constant throughout all experiments. Z-stacks consisted of 20 slices, each 0.25 μm apart and used to produce maximum intensity projections using Fiji. Mitochondrial footprint was determined by generating a binary mask using an automatically determined threshold via Otsu's method within Fiji following noise filtering by a median filter of radius 2 pixels and sharpened with an unsharp mask with a 2 pixel radius and 0.6 strength. The area measures are presented as pirate plots where each data point is plotted as a point, the boxes represents the 95% confidence intervals, the dark lines are the means, and the coloured bean is the density. Plots were generated using the yarrr package for R

reactive oxygen species (ROS) production, specifically hydrogen peroxide, is greater at 18% $O_2$, at which most cell culture is done. Since mitochondrial dynamics may be affected by intracellular ROS, it will in most instances be inappropriate to use 18% $O_2$ as a control for experimental comparisons. Figure 10.2 shows how continual exposure for >48 h to 5% or 18% $O_2$ affects two mitochondrial network parameters: network size (number of branches per network) and mitochondrial footprint (total area per cell in a compressed z-stack image occupied by mitochondria). Similar effects are observable on other mitochondrial morphological parameters (not shown here). Note that the effects of $O_2$ and glucose on mitochondrial morphology might be difficult to predict, as evidenced by the fact that they are not consistent between the two cell types studied here. Glucose effects are equally significant. Most eukaryotic cells in culture respond to low (physiological; 5 mM) glucose levels by increasing their reliance on oxidative phosphorylation to generate ATP, and this can increase the mitochondrial footprint (Fig. 10.2). If the goal of an experiment is to identify small molecules that can alter mitochondrial network morphology in vivo, then it may be critical to begin with the more extensive and fused network associated with physiological levels of $O_2$ and glucose.

### 10.2.3 Analyzing Mitochondrial Morphology in Micrographs

With the plethora of mitochondrial probes, variety of fluorescence microscopy systems, and continually improving computer processing and data storage solutions, it is no surprise how prominent live cell fluorescence microscopy has become in mitochondrial research. Having hundreds or thousands of individual cells imaged can be very powerful, but only if morphology and function can be compared efficiently and reliably. There are numerous open source platforms for image analysis available, such as Fiji [42], CellProfiler [43], and scikit-image [44]. Tools such as these have provided the means for generating transparent, efficient, and shareable methods.

The most basic analysis of morphology is to merely qualitatively assign descriptive keywords to images or groups - terms like fused, fragmented, and intermediate. It is easy to unintentionally introduce confirmation bias through such an analysis. This is not an issue unique to microscopy and there are ways to improve the faithfulness of such analyses. Blind analysis helps to reduce the injection of bias into qualitative observations, and its implementation may be as simple as removing identifying labels and scrambling image analysis order such that the observer is unaware of its experimental group assignation [45]. When deciding on how to classify images, keep in mind that using many categories may help capture more subtle differences, but may also hinder reproducibility. While you may find good agreement between the assignment of a couple of labels (perhaps fragmented and fused), there may be more variation between the results from different observers as more categories are introduced (such as intermediate, swollen, punctate, etc.). Blinding can be done by hand, but there are many tools available. A quick search in a browser for "ImageJ Blind Experiment" or "ImageJ Blind Analysis" will return many useful tools for doing this in ImageJ that are freely available.

Manual blinded qualitative analysis has its benefits: it is conceptually simple, trivial to explain, and is less susceptible to generating erroneous results due to sub-optimal tuning of analysis parameters or poor image quality. However, it does not allow for efficiently analyzing large datasets like those produced by plate readers and automated microscopy systems. Additionally, it limits the granularity of comparisons to a small number of qualitative bins. To get around these limitations, researchers have developed methods for automated analysis. As described by Harwig et al., approaches can largely be described as morphometric or morphological [4]. Morphometric methods are used to automatically calculate descriptive measures such as areas, perimeters, and lengths [4, 46–54], while morphological methods use this information to bin independent structures or even pixels into qualitative categories [55–60].

Automated analysis methods often follow a similar workflow. First, images are typically preprocessed to enhance contrast and reduce spurious noise. Then, images are segmented such that the signal from the mitochondrial label is separated from the background to produce a binary image. The binary representation can be used to gather various measurements for each spatially independent object or further simplified into a topological skeleton to extract topological information. A topological skeleton is a 1 pixel wide wire frame depiction of segmented structures that can be converted to a set of line segments. The skeleton provides access to measures of length and organizational parameters like the number of branching points. Morphometric approaches either hand the values acquired at this stage back over to the user or summarize them before handing them over to the user. Morphological approaches go a step further. They use the information generated to bin the structures as having one of several categorical morphologies. This is often achieved using supervised machine learning methods such as random forest classifiers or support vector

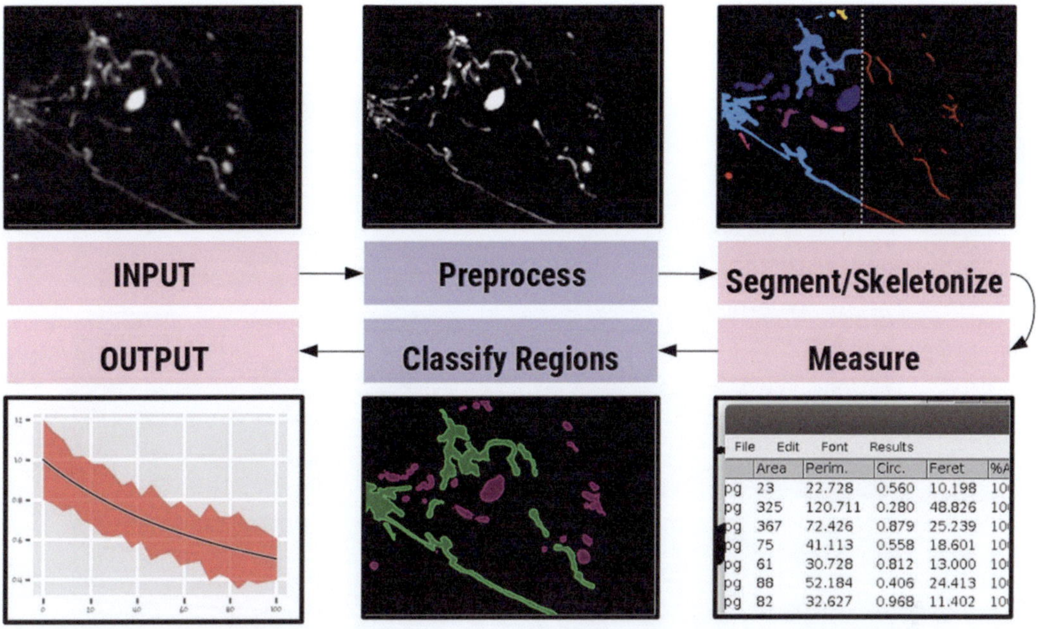


**Fig. 10.3** Morphological analysis typically follows a similar foundational recipe. First an input image is preprocessed to reduce spurious noise and enhance contrast. Then the mitochondria are segmented using a global or local threshold. This binary representation may be further simplified into a skeleton for topological analysis. Various morphometric parameters are then extracted from the independent regions of the binary and skeletal representations. These may be used for classifying the structures categorically often using supervised learning methods as is typical of morphological binning approaches. Finally, the information is summarized and provided to the user as output for further exploration and statistical analysis

machines. An overview of the generic workflow for analysis is summarized by Fig. 10.3.

Multiple processing methods have additionally sought to address the dynamic nature of mitochondria. Inspection of organizational changes has previously been addressed using colocalization and optical flow measurements. Colocalization between frames has been used as an effective comparison of which regions have or have not changed over time [47, 53]. A perfect correlation indicates no changes have occurred, while a poor correlation indicates more movement is taking place. The movement may arise from fission, fusion, or mitochondria being shuttled throughout the cell. However, it may also arise from the cell itself moving. To compensate for this, a rigid registration may be suitable to pre-align the cell from frame to frame before performing the colocalization. However, this may not fully compensate for more subtle changes in cell shape. An alternative for capturing changes to the mitochondrial ultrastructure over time is estimation of optical flow [53]. Optical flow estimation aims to measure motion in a series of images as a field of displacement and velocity vectors. The optical flow estimate is beneficial in that it also provides information regarding the directionality of the movements. Both colocalization and optical flow estimation provide a means to assess mitochondrial mobility and are demonstrated in Fig. 10.4. Such methods do not directly address whether fission or fusion is occurring, but rather changes in the organization of mitochondrial structures. Addressing such processes as fission and fusion requires additional methodologies.

To detect fission and fusion events, Westrate et al. tracked independent regions from frame to frame in time series data sets [56]. If two regions merged to generate a single region, a fusion event was recorded, while if a single region broke into two regions a fission event was recorded. This

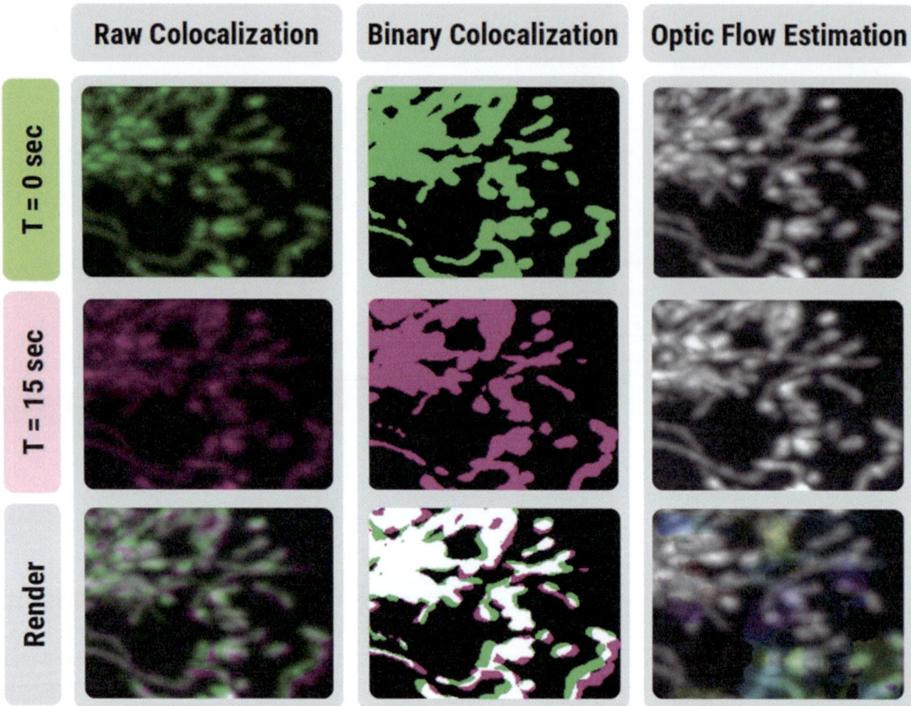


**Fig. 10.4** Besides morphology, there is interest in generating information regarding the movement of mitochondrial objects in cells. This has been investigated using colocalization and optic flow estimation. Here, an example is provided for rendering colocalization and direction maps. In a colocalization, subsequent frames can be coloured green and magenta to produce white where there is colocalization of signal. Note that colocalizations are not limited to the limiting channel merge technique, but that actual coefficients describing the frame to frame colocalization can be obtained and used as a metric for motility. In the render of optical flow, overlayed above the mitochondrial mitograph, the hue indicates the direction of motion and the brightness the velocity

was done using a scoring system that interpreted the merging or separation of regions as fission or fusion events, respectively. It is important to note that an observation of fission may be made from adjacent mitochondria moving away from each other, or fusion events detected when separated mitochondria clump together without actually fusing. Using well isolated mitochondria and validating connectivity changes by photoactivatable probes as these authors did helps to prevent possible erroneous interpretations [56].

Many of the methods implemented as scripts, extensions, macros, or plugins are publicly and freely available. A summary of these available methodologies has been summarized in Table 10.1. As some of these tools, such as MitoGraph and MiNA, are regularly and publicly updated with new features and bug fixes, it is best to check their respective websites for the most up to date functionality.

The analysis methods published to date all have strengths and weaknesses. The heterogeneity of mitochondrial morphology and organization, variation in image characteristics (such as resolution and noise) from different acquisition systems, and labeling efficacy can all affect the accuracy of an automated analysis method without further tuning. It is important that researchers validate methods for their use. Typically, representative validation images depicting segmentation, classification, or morphological skeleton results should be included in publication, at least as supplementary information. Having projects that are open source and invite collaboration, such as those hosted as GitHub repositories, are helping bring limitations to light by providing an

**Table 10.1** Publicly available methods and their currently demonstrated functionality. Note that time course analysis capability indicates functionality beyond repeated measures at multiple time points

| Publication | Availability | 2D | 3D | Time Course | High Throughput | Functional |
|---|---|---|---|---|---|---|
| Quantitative analysis of mitochondrial morphology and membrane potential in living cells using high-content imaging machine learning and morphological binning – Anthony P. Leonard *et al.*, 2015 | ***Supplementary*** | ✔ | | | ✔ | ✔ |
| Multiplexed high-content analysis of mitochondrial morphofunction using live-cell microscopy – Eligio F. Iannetti *et al.*, 2016 | ***Supplementary*** | ✔ | | | ✔ | ✔ |
| Loss of PINK1 Function Promotes Mitophagy through Effects on Oxidative Stress and Mitochondrial Fission – Ruben K. Dagda *et al.*, 2009 | http://imagejdocu.tudor.lu/doku.php?id=plugin:morphology:mitochondrial_morphology_macro_plug-in:start | ✔ | | | | ✔ |
| MitoLoc: A method for the simultaneous quantification of mitochondrial network morphology and membrane potential in single cells - Jakob Vowinckel *et al.*, 2015 | https://www.gurdon.cam.ac.uk/institute-life/downloadspublic/imaging-plugins | | ✔ | | | ✔ |
| A novel algorithm identifies stress-induced alterations in mitochondrial connectivity and inner membrane structure from confocal images - Mathieu Ouellet *et al.*, 2017 | http://www.uqtr.ca/LaboMarcGermain | ✔ | | | | |
| Methods for imaging mammalian mitochondrial morphology: A prospective on MitoGraph - Megan Cleland Harwig *et al.*, 2018 | https://github.com/vianamp/MitoGraph | ✔ | ✔ | | | |
| MyToe: automatic analysis of mitochondrial dynamics – Eero Lihavainen *et al.*, 2012 | http://www.cs.tut.fi/~sanchesr/tool_Mytoe/MyToe.html | ✔ | | ✔ | | |
| A simple ImageJ macro tool for analyzing mitochondrial network morphology in mammalian cell culture – Valente *et al.*, 2017 | https://github.com/StuartLab/MiNA | ✔ | | | | |

For example, motion analysis through inter-frame colocalization or optical flow would be considered time course functionalities. High throughput analysis indicates demonstration for use with slide scanners or high content imaging systems and functional capabilities include things such as membrane potential probe quantification

avenue for dialog and action to improve methods and the general usefulness of these tools.

## 10.3 Conclusions and Future Perspectives

The past decade has seen significant progress in our understanding of mitochondrial dynamics in the context of various disease states. Much of this relates to the identification of the proteins that mediate fusion and fission. There is still much progress to be made in refining the methods used to analyze specific aspects of mitochondrial dynamics. We have shown here the importance of maintaining temperature, $CO_2$, $O_2$, and glucose levels to avoid the introduction of artifactual effects on mitochondrial network morphology. Similarly, virtually all of the tools used to fluorescently label mitochondria in live cells can, under some circumstances, induce artifactual changes in mitochondrial form and function, so it is important to consider and appropriately control for these. Also, the very act of exposing cells to large quantities of light may induce cellular responses not relevant to in vivo physiology, so it is important to design experimental protocols with the goal of avoiding phototoxicity.

Once having obtained high quality data from well controlled and calibrated experiments, a variety of tools has been developed to support qualitative and/or quantitative analyses (Table 10.1). One advantage of using such tools is the avoidance of cognitive bias in the analysis of results. Mitochondrial network analysis tools such as those described in Table 10.1 take steps to formalize the specific parameters of interest being measured and provide summary statistics in a relatively objective way. Many of these tools have been provided as open source code to promote their continued refinement and development. These tools can be combined with high-throughput approaches for identifying small molecules that improve mitochondrial function in disease. While important developments in electron microscopic tomography have provided finer detail of mitochondrial structure, this approach requires fixing cells and generates relatively large datasets that are not conducive to high-throughput analyses. Thus, live cell fluorescence microscopy combined with a formally defined analysis that can be reproduced by others represents an effective compromise between resolution and attainability for the study of mitochondrial network morphology in disease.

One of the more exciting recent developments in mitochondrial research is the development of novel Optogenetic approaches that allow for precise spatio-temporal control of discrete mitochondrial functions. These include light-gated control of protein localization to the outer mitochondrial membrane [57–60], control of endoplasmic reticulum-mitochondria tethering [61], control of mitochondrial metabolism and $Ca^{2+}$ signaling [62], inactivation of mitochondrial electron transport chain complex II [63], and guided localization of mitochondria to discrete subcellular regions [64–67]. These approaches lend themselves to live cell fluorescence imaging of mitochondria and will dovetail naturally with computational approaches for subsequent analysis. Together, these new tools promise to push the field toward a deeper and more complete understanding of mitochondrial dynamics in health and disease.

## References

1. Lewis MR, Lewis WH (1914) Mitochondria in tissue culture. Science 39:330–333
2. Johnson LV, Walsh ML, Chen LB (1980) Localization of mitochondria in living cells with rhodamine 123. Proc Natl Acad Sci U S A 77:990–994
3. Johnson LV (1981) Monitoring of relative mitochondrial membrane potential in living cells by fluorescence microscopy. J Cell Biol 88:526–535
4. Harwig MC, Viana MP, Egner JM, Harwig JJ, Widlansky ME, Rafelski SM et al (2018) Methods for imaging mammalian mitochondrial morphology: a prospective on MitoGraph. Anal Biochem 552:81–99
5. Rafelski SM, Marshall WF (2008) Building the cell: design principles of cellular architecture. Nat Rev Mol Cell Biol 9:593–602
6. Rafelski SM (2013) Mitochondrial network morphology: building an integrative, geometrical view. BMC Biol 11:71
7. Tilokani L, Nagashima S, Paupe V, Prudent J (2018) Mitochondrial dynamics: overview of molecular mechanisms. Essays Biochem 62:341–360
8. Yu SB, Pekkurnaz G (2018) Mechanisms orchestrating mitochondrial dynamics for energy homeostasis. J Mol Biol 430(21):3922–3941. https://doi.org/10.1016/j.jmb.2018.07.027
9. Cid-Castro C, Hernández-Espinosa DR, Morán J (2018) ROS as regulators of mitochondrial dynamics in neurons. Cell Mol Neurobiol 38:995–1007
10. Schorr S, van der Laan M (2018) Integrative functions of the mitochondrial contact site and cristae organizing system. Semin Cell Dev Biol 76:191–200
11. Rambold AS, Lippincott-Schwartz J (2011) Mechanisms of mitochondria and autophagy crosstalk. Cell Cycle 10:4032–4038
12. Pareyson D, Saveri P, Sagnelli A, Piscosquito G (2015) Mitochondrial dynamics and inherited peripheral nerve diseases. Neurosci Lett 596:66–77
13. Bagli E, Zikou AK, Agnantis N, Kitsos G (2017) Mitochondrial membrane dynamics and inherited optic neuropathies. In Vivo 31:511–525
14. Filadi R, Pendin D, Pizzo P (2018) Mitofusin 2: from functions to disease. Cell Death Dis 9:330
15. Kameoka S, Adachi Y, Okamoto K, Iijima M, Sesaki H (2018) Phosphatidic acid and cardiolipin coordinate mitochondrial dynamics. Trends Cell Biol 28:67–76
16. Rehman J, Zhang HJ, Toth PT, Zhang Y, Marsboom G, Hong Z et al (2012) Inhibition of mitochondrial fission prevents cell cycle progression in lung cancer. FASEB J. Federation of American Societies for Experimental Biology Bethesda, MD, USA 26:2175–2186
17. Sarder P, Nehorai A (2006) Deconvolution methods for 3-D fluorescence microscopy images. IEEE Signal Process Mag 23:32–45
18. Goodwin PC (2014) Quantitative deconvolution microscopy. Methods Cell Biol 123:177–192
19. Biggs DSC (2010) 3D Deconvolution Microscopy. Curr Protoc Cytom 52:12.19.1–12.19.20
20. Pawley J (2010) Handbook of biological confocal microscopy. Springer Science & Business Media, New York
21. Conchello J-A, Lichtman JW (2005) Optical sectioning microscopy. Nat Methods 2:920–931
22. Gräf R, Rietdorf J, Zimmermann T (2005) Live cell spinning disk microscopy. Adv Biochem Eng Biotechnol 95:57–75
23. Paddock SW, Eliceiri KW (2013) Laser scanning confocal microscopy: history, applications, and related optical sectioning techniques. Methods Mol Biol 1075:9–47
24. Neil MA, Juskaitis R, Wilson T (1997) Method of obtaining optical sectioning by using structured light in a conventional microscope. Opt Lett 22:1905–1907
25. Weigel A, Schild D, Zeug A (2009) Resolution in the ApoTome and the confocal laser scanning microscope: comparison. J Biomed Opt 14:014022
26. Hell SW, Wichmann J (1994) Breaking the diffraction resolution limit by stimulated emission: stimulated-emission-depletion fluorescence microscopy. Opt Lett 19:780

27. Hell SW, Stelzer EHK, Lindek S, Cremer C (1994) Confocal microscopy with an increased detection aperture: type-B 4Pi confocal microscopy. Opt Lett, OL Optical Soc Am 19:222–224
28. Gustafsson MGL (2000) Surpassing the lateral resolution limit by a factor of two using structured illumination microscopy. SHORT COMMUNICATION. J Microsc 198:82–87
29. Rust MJ, Bates M, Zhuang X (2006) Sub-diffraction-limit imaging by stochastic optical reconstruction microscopy (STORM). Nat Methods 3:793–796
30. Hess ST, Girirajan TPK, Mason MD (2006) Ultra-high resolution imaging by fluorescence photoactivation localization microscopy. Biophys J 91:4258–4272
31. Fiolka R, Shao L, Rego EH, Davidson MW, Gustafsson MGL (2012) Time-lapse two-color 3D imaging of live cells with doubled resolution using structured illumination. Proc Natl Acad Sci U S A 109:5311–5315
32. Shao L, Kner P, Rego EH, Gustafsson MGL (2011) Super-resolution 3D microscopy of live whole cells using structured illumination. Nat Methods 8:1044–1046
33. Shim S-H, Xia C, Zhong G, Babcock HP, Vaughan JC et al (2012) Super-resolution fluorescence imaging of organelles in live cells with photoswitchable membrane probes. Proc Natl Acad Sci 109:13978–13983
34. Gugel H, Bewersdorf J, Jakobs S, Engelhardt J, Storz R, Hell SW (2004) Cooperative 4Pi excitation and detection yields sevenfold sharper optical sections in live-cell microscopy. Biophys J 87:4146–4152
35. Egner A, Jakobs S, Hell SW (2002) Fast 100-nm resolution three-dimensional microscope reveals structural plasticity of mitochondria in live yeast. Proc Natl Acad Sci U S A 99:3370–3375
36. Ahrens MB, Orger MB, Robson DN, Li JM, Keller PJ (2013) Whole-brain functional imaging at cellular resolution using light-sheet microscopy. Nat Methods 10:413–420
37. Planchon TA, Gao L, Milkie DE, Davidson MW, Galbraith JA, Galbraith CG et al (2011) Rapid three-dimensional isotropic imaging of living cells using Bessel beam plane illumination. Nat Methods 8:417–423
38. Laissue PP, Alghamdi RA, Tomancak P, Reynaud EG, Shroff H (2017) Assessing phototoxicity in live fluorescence imaging. Nat Methods 14:657–661
39. Güemes M, Rahman SA, Hussain K (2016) What is a normal blood glucose? Arch Dis Child 101:569–574
40. Liesa M, Shirihai OS (2013) Mitochondrial dynamics in the regulation of nutrient utilization and energy expenditure. Cell Metab 17:491–506
41. Twig G, Graf SA, Wikstrom JD, Mohamed H, Haigh SE, Elorza A et al (2006) Tagging and tracking individual networks within a complex mitochondrial web with photoactivatable GFP. Am J Physiol Cell Physiol 291:C176–C184
42. Schindelin J, Arganda-Carreras I, Frise E, Kaynig V, Longair M, Pietzsch T et al (2012) Fiji: an open-source platform for biological-image analysis. Nat Methods Nat Publ Group 9:676
43. Lamprecht MR, Sabatini DM, Carpenter AE (2007) CellProfiler: free, versatile software for automated biological image analysis. Biotechniques 42:71–75
44. van der Walt S, Schönberger JL, Nunez-Iglesias J, Boulogne F, Warner JD, Yager N et al (2014) scikit-image: image processing in python. PeerJ 2:e453
45. MacCoun R, Perlmutter S (2015) Blind analysis: Hide results to seek the truth. Nature 526:187–189
46. Iannetti EF, Smeitink JAM, Beyrath J, Willems PHGM, Koopman WJH (2016) Multiplexed high-content analysis of mitochondrial morphofunction using live-cell microscopy. Nat Protoc 11:1693–1710
47. Koopman WJH, Visch H-J, Smeitink JAM, Peter HG (2005) Simultaneous quantitative measurement and automated analysis of mitochondrial morphology, mass, potential, and motility in living human skin fibroblasts. Cytometry A 69A:1–12
48. Nikolaisen J, Nilsson LIH, Pettersen IKN, Peter HG, Lorens JB, Koopman WJH et al (2014) Automated quantification and integrative analysis of 2D and 3D mitochondrial shape and network properties. PLoS One 9:e101365
49. Dagda RK, Cherra SJ 3rd, Kulich SM, Tandon A, Park D, Chu CT (2009) Loss of PINK1 function promotes mitophagy through effects on oxidative stress and mitochondrial fission. J Biol Chem 284:13843–13855
50. Vowinckel J, Hartl J, Butler R, Ralser M (2015) MitoLoc: a method for the simultaneous quantification of mitochondrial network morphology and membrane potential in single cells. Mitochondrion 24:77–86
51. Ouellet M, Guillebaud G, Gervais V, St-Pierre DL, Germain M (2017) A novel algorithm identifies stress-induced alterations in mitochondrial connectivity and inner membrane structure from confocal images. PLoS Comput Biol 13:e1005612
52. Viana MP, Lim S, Rafelski SM (2015) Quantifying mitochondrial content in living cells. Methods Cell Biol 125:77–93
53. Lihavainen E, Mäkelä J, Spelbrink JN, Ribeiro AS (2012) Mytoe: automatic analysis of mitochondrial dynamics. Bioinformatics 28:1050–1051
54. Valente AJ, Maddalena LA, Robb EL, Moradi F, Stuart JA (2017) A simple ImageJ macro tool for analyzing mitochondrial network morphology in mammalian cell culture. Acta Histochem 119:315–326
55. Peng J-Y, Lin C-C, Chen Y-J, Kao L-S, Liu Y-C, Chou C-C et al (2011) Automatic morphological subtyping reveals new roles of caspases in mitochondrial dynamics. PLoS Comput Biol 7:e1002212
56. Westrate LM, Drocco JA, Martin KR, Hlavacek WS, MacKeigan JP (2014) Mitochondrial morphological features are associated with fission and fusion events. PLoS One 9:e95265

57. Milas A, Jagrić M, Martinčić J, Tolić IM (2018) Optogenetic reversible knocksideways, laser ablation, and photoactivation on the mitotic spindle in human cells. Methods Cell Biol 145:191–215. https://doi.org/10.1016/bs.mcb.2018.03.024. Epub 2018 Apr 26. PubMed PMID: 29957204
58. Wang H, Vilela M, Winkler A, Tarnawski M, Schlichting I, Yumerefendi H, Kuhlman B, Liu R, Danuser G, Hahn KM (2016) LOVTRAP: an optogenetic system for photoinduced protein dissociation. Nat Methods 13(9):755–758. https://doi.org/10.1038/nmeth.3926. Epub 2016 Jul 18. PubMed PMID: 27427858; PubMed Central PMCID: PMC5137947
59. Wang H, Hahn KM (2016) LOVTRAP: a versatile method to control protein function with light. Curr Protoc Cell Biol 73:21.10.1–21.10.14. https://doi.org/10.1002/cpcb.12. PubMed PMID: 27906450; PubMed Central PMCID: PMC5137945
60. Valon L, Marín-Llauradó A, Wyatt T, Charras G, Trepat X (2017) Optogenetic control of cellular forces and mechanotransduction. Nat Commun 8:14396. https://doi.org/10.1038/ncomms14396. PubMed PMID: 28186127; PubMed Central PMCID: PMC5309899
61. Shi F, Kawano F, Park SE, Komazaki S, Hirabayashi Y, Polleux F, Yazawa M (2018) Optogenetic control of endoplasmic reticulum-mitochondria tethering. ACS Synth Biol 7(1):2–9. https://doi.org/10.1021/acssynbio.7b00248. Epub 2017 Dec 4. PubMed PMID: 29172503
62. Tkatch T, Greotti E, Baranauskas G, Pendin D, Roy S, Nita LI, Wettmarshausen J, Prigge M, Yizhar O, Shirihai OS, Fishman D, Hershfinkel M, Fleidervish IA, Perocchi F, Pozzan T, Sekler I (2017) Optogenetic control of mitochondrial metabolism and Ca(2+) signaling by mitochondria-targeted opsins. Proc Natl Acad Sci U S A 114(26):E5167–E5176. https://doi.org/10.1073/pnas.1703623114. Epub 2017 Jun 13. PubMed PMID: 28611221; PubMed Central PMCID: PMC5495261
63. Wojtovich AP, Wei AY, Sherman TA, Foster TH, Nehrke K (2016) Chromophore-assisted light inactivation of mitochondrial electron transport chain complex II in *Caenorhabditis elegans*. Sci Rep 6:29695. https://doi.org/10.1038/srep29695. PubMed PMID: 27440050; PubMed Central PMCID: PMC4954975
64. Cunniff B, McKenzie AJ, Heintz NH, Howe AK (2016) AMPK activity regulates trafficking of mitochondria to the leading edge during cell migration and matrix invasion. Mol Biol Cell 27(17):2662–2674. https://doi.org/10.1091/mbc.E16-05-0286. Epub 2016 Jul 6. PubMed PMID: 27385336; PubMed Central PMCID: PMC5007087
65. Ballister ER, Ayloo S, Chenoweth DM, Lampson MA, Holzbaur ELF (2015) Optogenetic control of organelle transport using a photocaged chemical inducer of dimerization. Curr Biol 25(10):R407–R408. https://doi.org/10.1016/j.cub.2015.03.056. PubMed PMID: 25989077; PubMed Central PMCID: PMC4737597
66. Duan L, Che D, Zhang K, Ong Q, Guo S, Cui B (2015) Optogenetic control of molecular motors and organelle distributions in cells. Chem Biol 22(5):671–682. https://doi.org/10.1016/j.chembiol.2015.04.014. Epub 2015 May 9. PubMed PMID: 25963241; PubMed Central PMCID: PMC4443846
67. van Bergeijk P, Adrian M, Hoogenraad CC, Kapitein LC (2015) Optogenetic control of organelle transport and positioning. Nature 518(7537):111–114. https://doi.org/10.1038/nature14128. Epub 2015 Jan 7. PubMed PMID: 25561173; PubMed Central PMCID: PMC5063096

# Cysteine Switches and the Regulation of Mitochondrial Bioenergetics and ROS Production

11

Ryan J. Mailloux

**Abstract**

Mitochondria are dynamic organelles that perform a number of interconnected tasks that are elegantly intertwined with the regulation of cell functions. This includes the provision of ATP, reactive oxygen species (ROS), and building blocks for the biosynthesis of macromolecules while also serving as signaling platforms for the cell. Although the functions executed by mitochondria are complex, at its core these roles are, to a certain degree, fulfilled by electron transfer reactions and the establishment of a protonmotive force (PMF). Indeed, mitochondria are energy conserving organelles that extract electrons from nutrients to establish a PMF, which is then used to drive ATP and NADPH production, solute import, and many other functions including the propagation of cell signals. These same electrons extracted from nutrients are also used to produce ROS, pro-oxidants that can have potentially damaging effects at high levels, but also serve as secondary messengers at low amounts. Mitochondria are also enriched with antioxidant defenses, which are required to buffer cellular ROS. These same redox buffering networks also fulfill another important role; regulation of proteins through the reversible oxidation of cysteine switches. The modification of cysteine switches with the antioxidant glutathione, a process called protein S-glutathionylation, has been found to play an integral role in controlling various mitochondrial functions. In addition, recent findings have demonstrated that disrupting mitochondrial protein S-glutathionylation reactions can have some dire pathological consequences. Accordingly, this chapter focuses on the role of mitochondrial cysteine switches in the modulation of different physiological functions and how defects in these pathways contribute to the development of disease.

**Keywords**

Mitochondria · Redox signaling · Protein S-glutathionylation · Reactive oxygen species · Bioenergetics · Cysteine switches · Antioxidant defenses

## 11.1 Introduction

Mitochondria are considered the "power house" of the cell for good reason. This is simply because most mammalian cells rely on these organelles to meet their energy demands through the provision of ATP. However, the real source of energy in mitochondria is the electrochemical potential of

R. J. Mailloux (✉)
Department of Biochemistry, Memorial University of Newfoundland, St. John's, NL, Canada
e-mail: rjmailloux@mun.ca


A. Urbani, M. Babu (eds.), *Mitochondria in Health and in Sickness*, Advances in Experimental Medicine and Biology 1158, https://doi.org/10.1007/978-981-13-8367-0_11

protons experienced across the inner membrane (IM), more commonly referred to as the PMF [67]. All life on the planet relies on the establishment of electrochemical gradients. In mitochondria the PMF is utilized to drive ATP, ROS, and NADPH production and biosynthetic reactions as well as solute import and the conveyance of cell signals. The IM is ~5 to 7 nm thick and can generate a potential of ~150 to 180 mV [55]. This means that the potential experienced across the IM is ~300,000 V $cm^{-1}$ [55]. With this in mind, the statement "life is not much like a candle, more like a rocket launcher", which was coined in the book "The Vital Question" by Nick Lane, is certainly true when one considers the amount of energy that is being conducted by mitochondria on a regular basis [42]. When viewed from this perspective, it is the PMF, not ATP, that should be heralded as the universal energy currency.

Due to their central function in energy metabolism, mitochondria are heavily integrated into the information transmission circuitry of the cell [14, 87]. Mitochondria utilize a variety of secondary messengers to coordinate its functions with the rest of the cell. Of these molecules, one of the most important messengers happens to be ROS, specifically hydrogen peroxide ($H_2O_2$) [94]. Indeed, work conducted over the past decade has demonstrated that mitochondrial $H_2O_2$ release can modulate a wide breadth of cell functions, ranging from proliferation to stress adaptation [40]. However, mitochondria need to walk a delicate tight rope with $H_2O_2$ since higher levels can overwhelm antioxidant defenses leading to oxidative distress and cell damage [83]. Several systems, including proton leaks and supercomplex assemblies, can be used to control the strength and duration of mitochondrial $H_2O_2$ signals [82]. Another important mechanism for regulating $H_2O_2$ production is the reversible oxidation and reduction of protein cysteine switches. Redox switches are vital for controlling mitochondrial bioenergetics and functions, which includes how much ROS is formed by nutrient oxidizing enzymes and the electron transport chain (ETC) [55]. In addition, mitochondrial redox switches fulfill important physiological functions and defects in these pathways are associated with the pathogenesis of different disorders. This chapter focuses on how cysteine switches control mitochondrial bioenergetics and ROS release and the importance of these pathways in physiology and disease.

## 11.2 Mitochondrial Bioenergetics

The genesis of a PMF begins when different types of carbon (carbohydrates, lipids, amino acids) are converted into Krebs cycle intermediates. The breakdown of glucose by the glycolytic pathway, for example, produces pyruvate, which then enters the matrix of mitochondria using pyruvate carrier where it is combusted by pyruvate dehydrogenase to generate acetyl-CoA, $CO_2$, and NADH [38]. Following its production, acetyl-CoA is condensed with oxaloacetate through a Claisen-type condensation reaction catalyzed by citrate synthase, producing citrate, which then undergoes systematic oxidation and decarboxylation by Krebs cycle enzymes [86]. Similar to glycolysis, lipid oxidation and amino acid metabolism also generate Krebs cycle intermediates that can undergo further oxidation. Lipid oxidation yields acetyl-CoA [44]. Amino acid and nitrogen metabolism, on the other hand, can produce a range of intermediates, including acetyl-CoA, α-ketoglutarate, succinyl-CoA, fumarate, and oxaloacetate [44].

Oxidation of Krebs cycle intermediates is coupled to evolution of carbon dioxide ($CO_2$) and the production of the electron carriers, NADH and $FADH_2$ (Fig. 11.1) [11]. Other sources of NADH include branched-chain keto acid dehydrogenase (BCKDH), glutamate dehydrogenase (GDH), 2-oxoadipate dehydrogenase (OADH), pyruvate dehydrogenase (PDH), mitochondrial lactate dehydrogenase (LDH), β-hydroxybutyrate dehydrogenase (BDH), and the β-oxidation pathway (Fig. 11.1). Once formed, NADH and $FADH_2$ are oxidized by complexes I and II, respectively, and the free electrons are passed to ubiquinone (UQ), producing ubiquinol ($UQH_2$) [11]. Note that reduction of UQ occurs in the UQ binding pocket of either respiratory complex, which makes contact with the hydrophobic interior of the IM (Fig. 11.2). Other nutrient sources, such as pro-

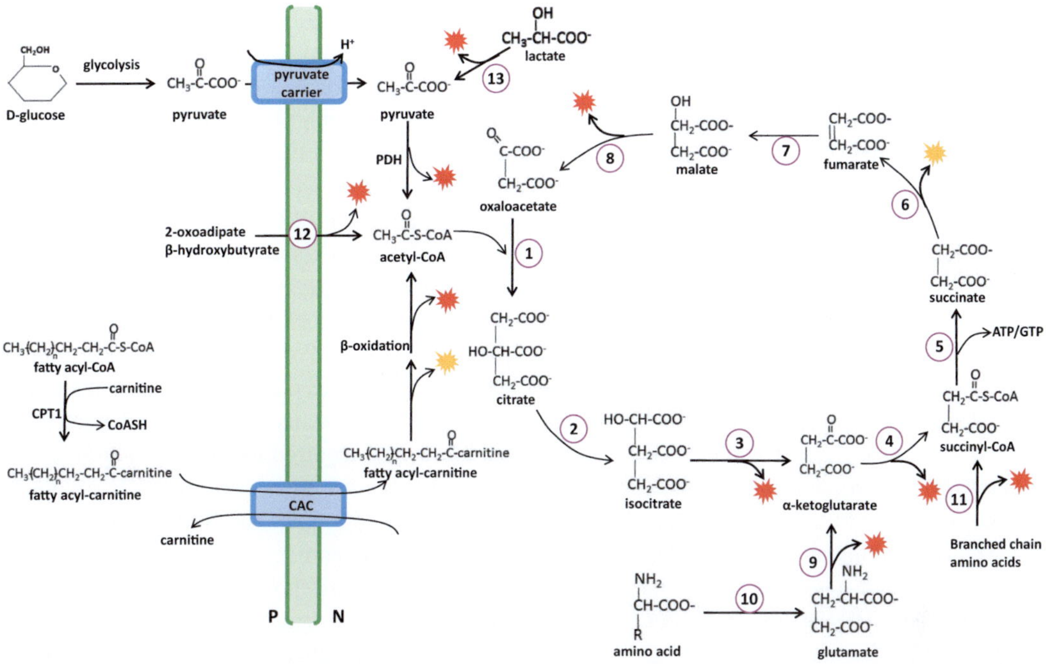


**Fig. 11.1** Mitochondrial carbon combustion and the Krebs cycle. 1: citrate synthase 2: aconitase, 3: isocitrate dehydrogenase, 4: α-ketoglutarate dehydrogenase, 5: succinyl-CoA synthetase, 6: succinate dehydrogenase/complex II, 7: fumarate, 8: malate dehydrogenase, 9: glutamate dehydrogenase, 10: amino transferase, 11: branched chain ketoacid dehydrogenase, 12: 2-oxoadipate dehydrogenase and β-hydroxybutyrate dehydrogenase, and 13: lactate dehydrogenase. P: positive side, N: negative side of the inner membrane. Red star: NADH, orange star: $FADH_2$

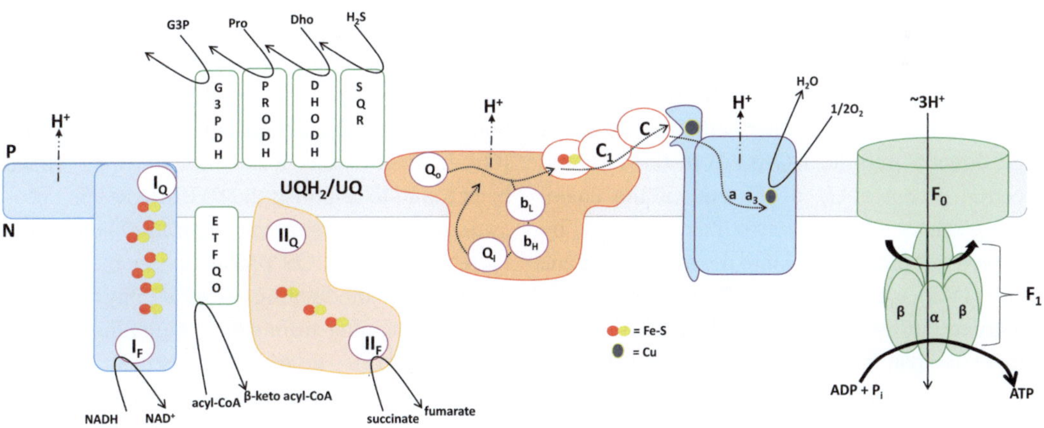


**Fig. 11.2** The electron transport chain and oxidative phosphorylation. Oxidation of NADH by complex I or different carbon sources by complex II, glycerol-3-phosphate dehydrogenase (G3PDH), proline dehydrogenase (PRODH), electron-transfer flavoprotein oxidoreductase (ETFQO), dihydroorotate dehydrogenase (DHODH), and sulfide:quinone oxidoreductase (SQR) liberates electrons that are ferried to $O_2$ at the end of the chain. This establishes a protonmotive force which is used to drive ATP production by complex V. *G3P* glycerol-3-phosphate, *Pro* proline, *Dho* dihydroorotate

line, glycerol-3-phosphate, dihydroorotate, hydrogen sulfide ($H_2S$), and acyl-CoA, can reduce the UQ pool directly, by-passing the Krebs cycle completely. This requires dedicated flavoenzymes, namely, proline dehydrogenase (PRODH), *sn*-glycerol-3-phosphate dehydrogenase (G3PDH), dihydroorate dehydrogenase (DHODH), sulfide:quinone oxidoreductase (SQO), and electron transfer flavoprotein-ubiquinone oxidoreductase (ETFQO) (Fig. 11.2) [44]. Once UQ has been reduced to $UQH_2$, it is then oxidized by complex III of the respiratory chain. Complex III contains two quinone binding pockets, termed $Q_O$ (outer leaflet of IM) and $Q_I$ (inner leaflet of IM), which are named according to their proximity to the outer or inner leaflet of the IM [8]. Ubiquinol is oxidized in the $Q_O$ pocket of complex III, resulting in the transfer of one electron through Rieske 2Fe-2S cluster protein to the electron carrier cytochrome *c* (Fig. 11.2) [8]. The lone electron in semiquinone ($UQ^{\bullet}$) is then recycled through the Q-cycle to reform $UQH_2$ (*in more detail in* [18]). Electrons from cytochrome *c* are then ferried through complex IV, reducing $O_2$ to $H_2O$ at the end of the chain.

The transfer of electrons through the chain relies on prosthetic groups, which are strategically positioned through the chain according to increasing electron affinity. This creates a redox gradient allowing for the thermodynamically favorable transfer of electrons from donors like NADH or succinate (or other carbon sources that feed directly into the UQ pool) to the terminal electron acceptor, $O_2$, at the end of the chain. Energy released from electron flux is trapped by the export of protons ($H^+$) into the intermembrane space (IMS), establishing a PMF (Fig. 11.2). This is carried out by complexes I, III, and IV, which happen to span the IM, bridging the matrix with the IMS. Complex I and IV couple the flux of electrons to the pumping of protons whereas acidification of the IMS by complex III relies on a Mitchellian redox loop. The stoichiometry for proton efflux is tightly coupled to the Gibbs free energy change associated with electron transfer reactions in the individual complexes (complexes I and IV pump 4 $H^+$ while complex III pumps 2 $H^+$). The PMF is comprised of electrical ($\Delta\Psi_M$) and chemical ($\Delta pH$). The $\Delta\Psi_M$ component makes up ~90% of the overall force experienced across the IM because the potential energy for charge separation is higher than the chemical component [7]. Once established, the PMF can be tapped to fulfill a number of tasks. The most obvious function of the PMF is to generate ATP. This is achieved by complex V (ATP synthase), which couples proton return to the matrix to the phosphorylation of ADP (Fig. 11.2) [92]. The proton gradient also drives solute import (e.g. the transfer of nascent proteins, nutrients, and other critical molecules required to support mitochondrial functions against their concentration gradient) and is required to generate NADPH through the action of nicotinamide nucleotide transhydrogenase (NNT) [70, 77]. Changes in the polarity of this gradient has been shown to modulate cell signals like autophage/mitophagy and depolarization of the membrane potential opens the mitochondrial permeability transition pore (MPTP) triggering cell death pathways [87]. Finally, the PMF regulates body temperature and ROS production, feats that are made possible by a specialized class of proteins called the uncoupling proteins (UCPs). In this particular case, UCPs catalyze proton return to the matrix, by-passing solute translocators and ATP synthase. In brown fat, the energy released from these "proton leaks" are used to make heat for maintenance of body temperature, which is catalyzed by UCP1 [66]. In other tissues, leaks through UCP2-5 have been found to diminish the PMF, decreasing protonic back pressure on the respiratory complexes thereby limiting ROS production, a mechanism crucial for modulating several physiological functions ranging from satiety signaling to insulin release from pancreatic β-cells [51]. Indeed, a non-Ohmic relationship exists between the PMF and ROS production where a small increase in membrane potential can induce a burst in ROS production. Therefore, this simple and evolutionary conserved mechanism of storing energy in an electrochemical gradient seems to serve as a driving force behind several complex physiological processes.

## 11.3 Production of Reactive Oxygen Species

### 11.3.1 ROS Are an Important Metabolite Formed by Mitochondria

Mitochondria are sources of other critical molecules, like Fe-S clusters, haem, and key sites for lipid and de novo pyrimidine biosynthesis. Another set of crucial molecules produced by mitochondria, which happen to also be generated by the same nutrient oxidizing and electron transferring pathways that make ATP, are reactive oxygen species (ROS). "ROS" is a broad term that describes all oxygen-centered radicals and non-radicals formed in nature. Therefore, it is imperative that we define "ROS" in the context of mitochondrial biology. The most important ROS formed by mitochondria are superoxide ($O_2^{\bullet-}$), $H_2O_2$, and hydroxyl radical ($OH^{\bullet}$) [34]. The first two molecules are the proximal ROS formed by mitochondria and their production is an inevitable consequence of respiration. This is attributed to fact that $O_2$ contains two lone electrons in its outer most antibonding orbital, forcing it to accept only one electron at a time [63]. Therefore, during the full reduction of $O_2$ to $H_2O$, several reactive intermediates are formed, namely, the oxyradicals $O_2^{\bullet-}$ and $OH^{\bullet}$ and the nonradical $H_2O_2$. Mitochondrial ROS production is also inevitable because its production is thermodynamically favorable. This is due to the low concentration of $O_2^{\bullet-}$and $H_2O_2$ ($[O_2^{\bullet-}] = 10^{-10} - 10^{-12}$ M and $[H_2O_2] = 10^{-7} - 10^{-9}$ M) relative to $O_2$ (~1 μM in mitochondria) in the matrix and the negative redox potential for ROS formation [63, 95]. Although both $O_2^{\bullet-}$and $H_2O_2$ can cause oxidative damage at high enough amounts, both molecules are kept in check by antioxidant defenses, which happens to be critical for harnessing the secondary signaling properties of $H_2O_2$. The ugliness of oxidative distress and tissue damage is more often associated with the genesis of $OH^{\bullet}$. Indeed, $OH^{\bullet}$ has a redox potential of ~2 V and its reactivity is limited by diffusion meaning that it can irreversibly oxidize and damage a number of molecules including DNA, lipids, and proteins [34]. Hydroxyl radical is produced by Fenton and Haber-Weiss reactions, which require a supply of free Fe and adequate $O_2^{\bullet-}$ and $H_2O_2$ levels [34]. Fortunately, cells and mitochondria are equipped with chaperones, small molecules, and antioxidants that bind Fe and quench ROS. It is only when antioxidant defenses are overwhelmed and other protective systems that limit access to Fe are disabled that a cell succumbs to oxidative distress.

Mitochondria contain several sources of ROS housed in the IMS, IM, and matrix of mitochondria. Sources include the ubiquinone binding site of complex I ($I_Q$) and the $Q_O$ pocket of complex III (Table 11.1) [10]. However, most ROS forming sites in mitochondria happen to be flavin-containing proteins, which harbor either a flavin mononucleotide (FMN) or flavin adenine dinucleotide (FAD) (Table 11.1) [10]. This capacity is based on the reactivity of flavins towards oxygen and their redox chemistry [59]. In addition, flavins have the capacity to generate either $O_2^{\bullet-}$ or $H_2O_2$ or a mixture of both, characteristics that depend on the catalytic microenvironment, the redox state of the flavin moiety, and reactivity of the flavin towards $O_2$ [10, 59]. For instance, flavin radical ($FH^{\bullet}$) can react with $O_2^{\bullet-}$ at ~$10^8$ $M^{-1}$ $s^{-1}$ to generate $H_2O_2$ [59]. In contrast, oxidation of flavin hydroperoxide (FOOH) produces $H_2O_2$ at a rate that is eight orders of magnitude lower than $FH^{\bullet}$ [59].

### 11.3.2 ROS Producing Enzymes Associated with Nutrient Oxidation and Respiration Pathways

Complex I and III of the respiratory chain are often, if not always, considered the chief ROS emitters in mammalian mitochondria (site $I_F$ of complex I and $Q_O$ of complex III are often erroneously thought to be the sole sites of production) (Table 11.1). However, over the past two decades it has become increasingly clear that several matrix and IM bound enzymes outside of complexes I and III can also generate ROS. In 2004, it was demonstrated using synaptosomes

**Table 11.1** The 12 sites for ROS production in mitochondria

| Site | Enzyme | Isopotential group | Compartment |
|---|---|---|---|
| $I_F$ | FMN – complex I | NADH/NAD$^+$ | Matrix |
| $K_F$ | FAD – α-ketoglutarate dehydrogenase (KGDH) | NADH/NAD$^+$ | Matrix |
| $P_F$ | FAD – pyruvate dehydrogenase (PDH) | NADH/NAD$^+$ | Matrix |
| $B_F$ | FAD – branched chained keto acid dehydrogenase complex (BCKDH) | NADH/NAD$^+$ | Matrix |
| $A_F$ | 2-Oxadipate dehydrogenase complex (OADH) | NADH/NAD$^+$ | Matrix |
| $I_Q$ | UQ binding site – complex I | $UQH_2$/UQ | Matrix |
| $II_F$ | FAD – complex II | $UQH_2$/UQ | Matrix |
| $III_Q$ | UQ binding site – complex III | $UQH_2$/UQ | Intermembrane space |
| $G_F$ | FAD – glycerol-3-phosphate dehydrogenase (G3PDH) | $UQH_2$/UQ | Intermembrane space |
| $P_F$ | FAD – proline dehydrogenase (PRODH) | $UQH_2$/UQ | Intermembrane space |
| $E_F$ | FAD – electron-transfer flavoprotein; ubiquinone oxidoreductase (ETFQO) | $UQH_2$/UQ | Matrix |
| $D_F$ | FAD – dihydroorotate dehydrogenase (DHODH) | $UQH_2$/UQ | Intermembrane space |

that α-ketoglutarate dehydrogenase (KGDH), an important entry point for carbon into the Krebs cycle, can display high rates of ROS production [90]. The same study also showed that PDH generates $O_2^{\bullet-}$ and $H_2O_2$ [90]. The $E_3$ subunit of both enzyme complexes, which harbors an FAD that plays an important role in their catalytic cycle, was identified as the source of ROS (Fig. 11.3) [1]. Later studies on the same enzyme complex showed that PDH and KGDH actually generate $O_2^{\bullet-}$ and $H_2O_2$, with the latter species accounting for ~75% of the ROS formed by the complex [85]. In addition, recent work has also provided evidence that the $E_1$ subunit of KGDH can generate ROS through a thiamin-derived radical [1, 65]. Around the same period, reports also started to emerge showing that complex II of the respiratory chain can also generate ROS, serving as a critical source during hypoxia [69]. Other ROS release sites associated with the respiratory chain were also identified and include ETFQO, G3PDH, PRODH, and DHODH, which have been shown to serve as crucial sites in different tissues and organisms and cancer cells [27, 29, 91].

The importance of the other sites of production (e.g. ROS sources outside of complex I and III that are required for nutrient oxidation and energy metabolism) did not crystallize until a seminal publication by Quinlan et al., where it was demonstrated for the first time that KGDH and PDH produce ~8× and ~4× more $O_2^{\bullet-}/H_2O_2$, respectively, than complex I in skeletal muscle mitochondria [74]. Moreover, in the same study, the authors also quantified the native rates of $O_2^{\bullet-}/H_2O_2$ from the other sites and found that (1) BCKDH produces ~2× more ROS than complex I when branched chain amino acids are being oxidized, (2) complex III is the highest capacity site in muscle mitochondria, (3) complex II is also a high capacity site, and (4) other sites of production, like PRODH and DHODH, make negligible contributions to the overall ROS release profile of muscle mitochondria (exceptions include cancer cells and insects) [74]. In addition, the same group demonstrated that OADH can also serve as a significant source of $O_2^{\bullet-}/H_2O_2$ in muscle mitochondria as well [25]. The importance of PDH as a critical ROS source in muscle mitochondria was then confirmed using permeabilized muscle fibers of mouse, rat, and human origin and unchecked production from this site was found to be associated with obesity and insulin resistance [21]. To date, 12 enzymes associated with nutri-

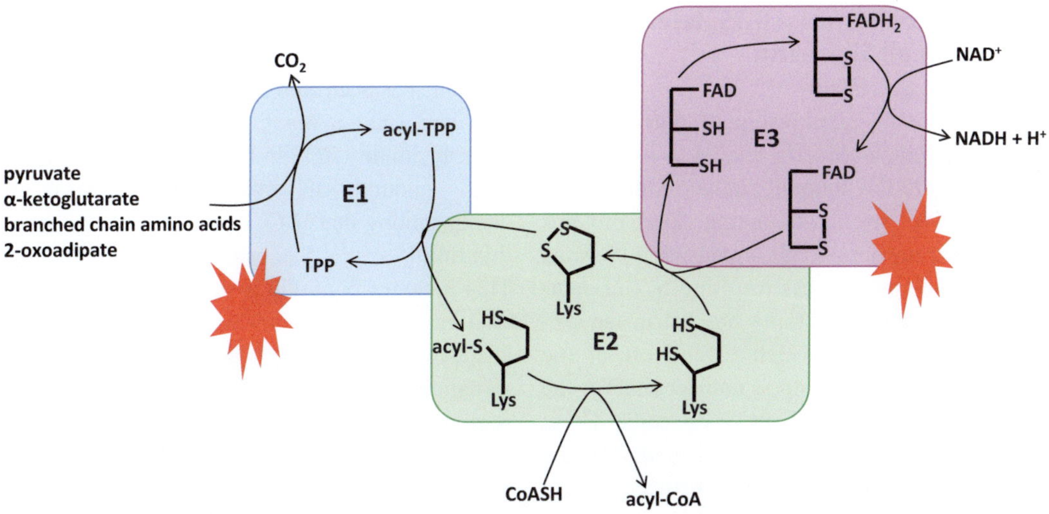


**Fig. 11.3** Basic structure and catalytic mechanism of keto acid dehydrogenases (pyruvate dehydrogenase, α-ketoglutarate dehydrogenase, branched chain keto acid dehydrogenase, and 2-oxoadipate dehydrogenase). Carbon oxidation by the E1 subunit (α-keto acid decarboxylase) transfers an acyl group to thiamine pyrophosphate (TPP), producing $CO_2$ in the process. Dihydrolipoamide acyl-transferase (E2) transfers the acyl group to a vicinal thiol in dihydrolipoamide (DLA). A thiol disulfide exchange reaction with CoASH produces acyl-CoA and a fully reduced DLA. DLA dehydrogenase (E3) oxidizes DLA and the hydrides are transferred to FAD, reducing $NAD^+$ at the end of the chain. Red stars indicate sites for ROS production (note that only KGDH has been found to produce ROS via its E1 subunits)

ent metabolism in mitochondria have been found to serve as sources of $O_2^{\bullet-}/H_2O_2$ production in different tissues (Table 11.1) [12]. Based on their findings, Quinlan et al. classified these ROS release sites based on which isopotential group facilitates $O_2^{\bullet-}/H_2O_2$ production [74]. Enzymes that generate or oxidize NADH, such as KGDH, PDH, BCKDH, OADH, or complex I are part of the NADH/$NAD^+$ isopotential group (Table 11.1). By contrast, enzymes that couple substrate oxidation to the direct reduction or oxidation of the ubiquinone pool are classified in the $UQH_2$/UQ group (Table 11.1). This taxonomy is critical since it provides insights into which mitochondrial ROS source may serve as a more significant one when different fuels are being oxidized. For instance, as denoted above, KGDH and PDH are high capacity sites when Krebs cycle metabolites are being oxidized, with other important sources, like complex I, making negligible contributions. By contrast, muscle mitochondria produce most of their ROS through complex I, II, or III when succinate, acyl-CoA, or other carbon sources that transfer electrons directly to the ubiquinone pool are being oxidized [31, 71]. Similar observations have been made with liver mitochondria. KGDH accounts for ~35% of the total ROS release from liver mitochondria oxidizing Krebs cycle metabolites, with complex III accounting for ~45% [84]. In addition, PDH made up ~12% of the total ROS release with complex I or II making minor contributions when pyruvate or α-ketoglutarate were being oxidized [84]. By contrast, complexes I and III serve as critical sources when substrates like succinate or dimethylglycine, which donate electrons directly to the ETC, served as substrates [13, 58]. Taken together, the ROS release profile of mitochondria is highly complex and dynamic, can vary significantly between different tissues and mouse strains, and the rate of release can vary significantly based on substrate supply and concentration and the redox state and concentration of the source of electrons for ROS production.

### 11.3.3 Mitochondrial ROS Release in Cell Signaling

The bifunctional relationship of mitochondrial ROS with cells is directly related to its concentration, where low amounts induce a mild form of oxidative stress, oxidizing key proteins involved in mediating cell signals, and elevated levels induce overt stress resulting in cell damage. In fact, this dichotomy has led to the creation of new terms which extend off of the definition of oxidative stress and account for the wide ranging effects of ROS on cells. These terms are *oxidative eustress* and *oxidative distress* (Fig. 11.4) [83]. Oxidative eustress refers to an oxidative challenge where $H_2O_2$ is used in adaptive signaling [82]. This is achieved through the site-specific oxidation of cysteine switches which modulates the cellular proteome, genome, and transcriptome in response to physiological cues (e.g. changes in the exposome) (Fig. 11.4) [36]. Oxidative distress, on the other hand, refers to the over production of free radicals, overwhelming antioxidant defenses resulting in the non-specific oxidation of cell macromolecules and the induction of tissue damage (Fig. 11.4) [83]. Discussing the differences between oxidative eustress and distress is critical in the context of mitochondria since these organelles are involved in both.

Mitochondrial eustress is a relatively new phenomenon considering the function of $H_2O_2$ as a mitokine has been studied for only a few decades. However, since it was first found that mitochondria can employ ROS in this capacity, a considerable amount of evidence has surfaced demonstrating that $H_2O_2$ emitted from mitochondria fulfills a variety of critical cellular functions. This includes controlling steroidogenesis, circadian rhythms, cell division and differentiation, adaptive responses such as insulin release and signaling and T-cell activation, and stress responses like hypoxic and electrophilic stress signaling and the induction of apoptosis [40, 82]. The activation of these pathways depends highly on the availability of $H_2O_2$, which is influenced by its rate of production and degradation [45, 82]. The role of mitochondrial $H_2O_2$ in cell signaling has been discussed in detail in several outstanding review articles [9, 14, 87]. Here, I will only discuss the relationship between the rate of ROS release by individual sites of production and the signaling properties of $H_2O_2$. Most, if not all, studies have focused on the role of $O_2^{\bullet-}$ from complex III in cell signaling. This is because complex III is a high capacity site for ROS production in most mammalian cells but also because it is one of the few respiratory complexes that emits $O_2^{\bullet-}$ into the cytosol [84]. The $O_2^{\bullet-}$ released into the IMS is quickly converted to $H_2O_2$ by Cu/ZnSOD

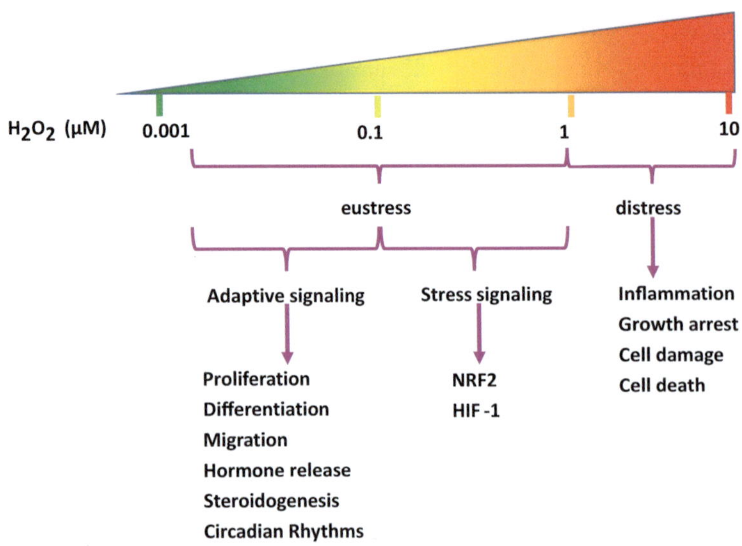


**Fig. 11.4** Oxidative eustress and distress and the diverse cellular roles fulfilled by $H_2O_2$ signaling

which is then used to transmit information from mitochondria to the rest of the cell. ROS emission from complex III has been found to modulate a plethora of cell functions ranging from T-cell activation to adipocyte differentiation, hypoxic signaling, and stress signaling through the stabilization of NF-E2p45-related factor2 (NRF2) transcription factor (Fig. 11.4) [40]. However, as noted above other nutrient oxidizing and electron transferring enzymes in mitochondria can also generate significant amounts of ROS. Therefore, it is feasible that these other sources may also supply $H_2O_2$ for cell signaling. For instance, *sn*-G3PDH serves as a high capacity site for ROS in several tissues including brown fat and muscle [10, 79]. A recent study demonstrated that UCP1 activation and induction of nonshivering thermogenesis also relies on the oxidation of protein cysteine switches by $H_2O_2$ [16]. Several studies have also shown that stimulation of brown fat thermogenesis is associated with an increase in mitochondrial ROS production [46, 79]. Although, there is no direct evidence showing that *sn*-G3PDH is the source of this ROS, it is possible that it plays a part in brown fat thermogenesis through the induction of redox signaling cascades. Complex I of the respiratory chain has also been implicated in using ROS for the regulation of mitochondrial metabolism [19]. Another study demonstrated that conditions mimicking exercise and rest conditions can dictate which mitochondrial enzymes produce the most ROS, which can play a significant role in redox signaling in response to muscle training [26]. We are still only scratching the surface in terms of how mitochondria use ROS in signaling and which enzymes serve as major sources in response to different physiological signals. Based on the accumulated evidence, mitochondrial ROS are responsible for modulating a variety of biological functions.

### 11.3.4 Reverse Electron Transfer (RET) and Mitochondrial ROS Production

The reversal of electron flow, or RET, was originally found to occur between complex II and complex I of the respiratory chain. Reverse flow of electrons requires several factors including a polarized IM, reduced ubiquinone pool, and considerable amounts of succinate [2]. It is easy to replicate these conditions in vitro by supplying isolated mitochondria with complex III and IV blockers and high succinate levels (mM range). This has also led many to question the physiological relevance of reverse electron flow regarding mitochondrial bioenergetics and ROS production. A recent study published by Chouchani et al, however, has demonstrated that succinate could potentially serve as a supplier of electrons for ROS production by complex I during ischemic-repefusion (IR) injury to the myocardium [17]. In this study, it was proposed that during ischemic conditions, succinate accumulates by the reversal of complex II through the provision of fumarate, which is generated by the purine nucleotide cycle (PNC) and malate-aspartate shuttle (MAS) [17]. The subsequent reintroduction of oxygen results in the over reduction of the UQ pool and IM repolarization, inducing a spike in $O_2^{\bullet-}$ release by RET from succinate to complex I [17]. This results in MPTP opening, oxidative distress, and myocardial tissue damage. Some considerable criticisms of this mechanism for myocardial injury have been cited, including findings showing that reperfusion conditions favor forward electron flow and that complex I is not the sole source of ROS during IR injury [2, 41]. Despite these apparent shortcomings, Chouchani et al. demonstrated for the first time that RET from complex II to I could perhaps fulfill some important physiological functions. In another study, Scialò et al. provided compelling evidence that RET from complex II may be important for extending lifespan and improving mitochondrial function in a *D. melanogaster* model for Parkinson's disease [78]. RET from succinate to complex I may also be an important driving force behind brown fat thermogenesis [62]. Overall, these studies have provided some important insights into the role of RET in physiology and disease.

RET to complex I and II from other flavoproteins that feed electrons into the UQ pool has also been documented. For example, dimethylglycine dehydrogenase produces the bulk of its ROS by RET to complex I [58]. Oxidation of acyl-CoA, proline, or dihydroorate also promotes high rates of ROS production by reverse flow to either complexes I or II [27, 31, 71]. Flavoproteins like KGDH and PDH have also been found to produce $O_2^{\bullet-}/H_2O_2$ by RET from NADH. This phenomenon was first observed in 2004, where it was demonstrated KGDH can produce $O_2^{\bullet-}/H_2O_2$ by RET from NADH [90]. This led to the development of the postulate that defects in complex I activity and diminished NADH turnover could promote oxidative distress through increased ROS production by KGDH [90]. Further studies revealed that both PDH and KGDH were both able to catalyze $O_2^{\bullet-}/H_2O_2$ by RET from NADH [49]. The Km for ROS production by RET for both KGDH and PDH is ~1.6 μM and ~21.7 μM [49]. Since the matrix ratio of $NADH/NAD^+$ is ~8 and the overall concentration of $NAD^+$ is ~400 to 500 μM, then both enzyme complexes could potentially generate ROS by RET in cells. These observations do warrant further investigation into the physiological function of RET from NADH to KGDH and PDH.

## 11.4 Controlling Mitochondrial Bioenergetics and ROS Production with Cysteine Switches

Bioenergetics refers to study of oxidation-reduction reactions that are used to establish a PMF required to drive mitochondrial functions (e.g. oxidative phosphorylation). This definition also includes the study of the redox reactions that produce ROS and how $O_2^{\bullet-}/H_2O_2$ are regulated through its rate of production and degradation [45]. Degradation of ROS is facilitated by several antioxidant systems found in the matrix and IMS or mitochondria. Superoxide is rapidly dismutated to $H_2O_2$ by SOD1 in the IMS or SOD2 in the matrix (Fig. 11.5). Once formed $H_2O_2$ is removed by the thioredoxin-2 (TRX2) or glutathione (GSH) systems, both of which use NADPH as a reducing agent to maintain their activity (Fig. 11.5). Another critical mechanism for keeping $H_2O_2$ in check is catalase, which has been shown to eliminate ROS in liver and cardiac, and potentially muscle, mitochondria [75, 84]. Cellular $H_2O_2$ levels are also controlled through the modulation of its production. This can be achieved by regulating the availability of electrons at donating centers, either through the modulation of proton return to the matrix and respiration or by controlling the entry and exit of electrons from sites of ROS formation [10]. Proton leaks are the most well-studied and controversial mechanism for regulating ROS production in mitochondria (Fig. 11.5). Returning protons to the matrix can decrease protonic back pressure, thereby increasing the rate of respiration and limiting the number of electrons available for ROS production (Fig. 11.5). Chemical uncouplers like protonophores have been found to suppress ROS release by augmenting proton return and mammalian cells harbor uncoupling proteins that can fulfill a similar duty. The role of UCP2 and UCP3 in this process is still enthusiastically debated, even though a number of studies have shown both proton leaks regulate mitochondrial ROS production and prevent oxidative distress in various cell types [50]. The second mechanism that has been documented to play a critical role in regulating mitochondrial ROS production is the oxidation of cysteine switches in proteins (Fig. 11.5). It is critical to point out that protein cysteine thiols can undergo a range of oxidative modifications, a property related to the ability of sulfur to adopt different oxidation states [54]. So far, however, only one cysteine modification has been found to fulfill all the properties of a legitimate posttranslational modification (PTM) that can regulate protein functions in response to physiological cues, and that is protein S-glutathionylation (PGLU) [54]. In this section, the importance of these reactions in regulating mitochondrial functions in response to changes in redox buffering capacity will be discussed in detail. This will include the discussion of evidence demonstrating that PGLU is required to control mitochondrial ROS release for signaling and that disruption of these pathways can have some pathological consequences.

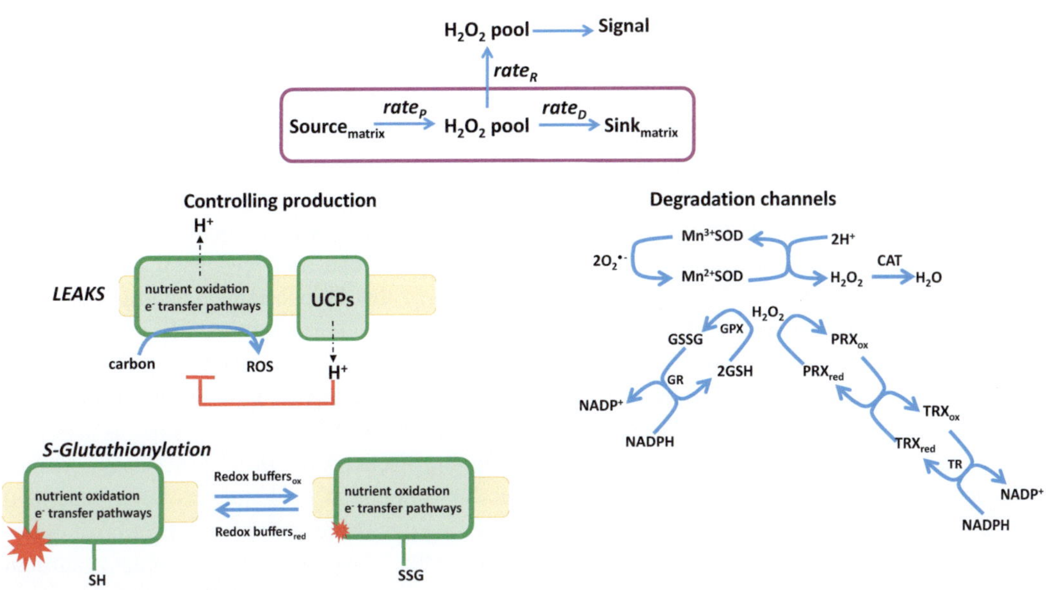


**Fig. 11.5** Controlling mitochondrial $H_2O_2$. Mitochondrial $H_2O_2$ levels and signaling is controlled by the rate of production ($rate_P$) and degradation ($rate_D$), which dictates the rate of release ($rate_R$). ROS production is controlled by proton leaks and protein S-glutathionylation. Proton leaks control ROS production by reducing protonic backpressure on electron transfer pathways, diminishing production. Protein S-glutathionylation serves as a dynamic switch, increasing and decreasing ROS production through the reversible oxidation of cysteine switches following changes in redox buffering capacity. The degradation of $H_2O_2$ signals is facilitated by antioxidant defenses

### 11.4.1 PGLU Reactions and the Regulation of Mitochondrial Proteins

Protein S-glutathionylation reactions were first found to be associated with oxidative distress [96]. This was associated with the over-oxidation of the reduced glutathione pool (GSH), resulting in the formation of sufficient amounts of glutathione disulfide (GSSG) and the nonspecific S-glutathionylation of protein cysteine thiols through spontaneous disulfide exchange reactions. A few years later, it was also observed that proteins can be S-glutathionylated by an unknown enzyme in hepatocytes, which was later identified as glutaredoxin-1(GRX1) [4]. GRX1 is a cytosolic thiol oxidoreductase that catalyzes the deglutathionylation of a target protein through a nucleophilic displacement reaction [22]. After it was identified, studies found that GRX1 plays a vital role in modulating various cell functions, ranging from energy sensing to apoptosis, in response to changes in cellular redox buffering capacity (reviewed in [28, 80]. Further studies into identifying how thiol disulfide exchange reactions can modulate cell processes in response to physiological cues led to the identification of its mitochondrial matrix homologue, GRX2 [43]. GRX2 is a small heat-stable thiol oxidoreductase that adopts a thioredoxin-fold, much like GRX1. GRX2, although only ~34% homologous to GRX1, employs a catalytic mechanism like GRX1 for the deglutathionylation of target proteins [22]. GRX2 can also catalyze the S-glutathionylation of target proteins, which is discussed in more detail below.

Protein cysteine thiols can be subjected to a range of redox modifications. Therefore, it is important to distinguish between protein S-glutathionylation and the other modifications and why the former serves as device required to control protein function in response to fluctuations in redox buffering capacity. PTMs must satisfy a series of criteria in order to serve as a regulatory device for proteins. Criteria include: (1) must modify protein function, (2) respond to physiological cues, (3) modification should be rapid, (4) modification should be site specific, and (5) the modification should be reversible [80]. When these criteria are taken into consideration, only PGLU reactions appear to be the only legitimate redox signal involved in cell adaptation. Indeed, protein S-glutathionylation reactions are reversible, which is evidenced by the observation that increases and decreases in the availability of GSH and GSSG dictate whether or not a protein will be modified [6]. These modifications are also, as indicated above, enzymatically mediated and rapid, with GRX1 and GRX2 displaying kinetics for deglutathionylation approaching ~$10^5$ $M^1$ $s^{-1}$ [35]. Proteins also have PGLU motifs, conveying specificity, and these reactions respond to physiological cues, which can range from alterations to nutrient status and exercise to stress signaling and protection from oxidative damage [39, 60, 61]. Reversible S-glutathionylation events seem to be especially prevalent in mitochondria. As demonstrated in several recent review articles, mitochondria harbor a number of PGLU targets, which includes enzymes and proteins involved in nutrient metabolism and respiration, ATP and ROS production, mitochondrial fission/fusion, solute import, permeability transition and apoptosis [52]. The overwhelming number of PGLU targets in mitochondria is related to the unique physical properties of the organelle. First, mitochondria contain high amounts of protein cysteine thiols (60–95 mM), making proteins the most concentrated sulfur-containing molecules in the matrix [64]. Second, the matrix is basic (pH ~8.2 to 8.5), which is significant since it means that thiols (PrSH) are more likely to ionize and form reactive thiolate ($PrS^-$) anions, which can be more easily modified. Finally, mitochondria contain high amounts of glutathione (~2 mM) and the matrix can be subjected massive fluctuations in redox state since it also happens to be a major source of ROS [95]. Together, this makes mitochondria a "*hot-bed*" for regulation by PGLU reactions.

### 11.4.2 Regulation of Mitochondrial Nutrient Metabolism and Oxidative Phosphorylation

Interest in understanding how cysteine residue oxidation can be used to regulate mitochondrial bioenergetics can be traced back 1999 when it was found that iodoacetamide and diethyl maleate, agents that covalently modify protein thiols, alters the activity of complex I [5]. Determining whether or not cysteine switches can modulate mitochondrial bioenergetics was accelerated after the discovery of GRX2. Since then a considerable amount of evidence has been generated showing that the reversible modification of cysteines with glutathione can modulate nutrient uptake and oxidation as well as oxidative phosphorylation. Most of the information surrounding our collective understanding for how PGLU reactions regulate mitochondrial bioenergetics has been generated with complex I. NDUFS1, which is part of the N-module of complex I and thus involved in NADH oxidation, is an S-glutathionylation target, which regulates its activity and protects it from irreversible oxidation [33]. GRX2 is required to reversibly S-glutathionylate complex I in response to changes in the availability of GSH [6]. Moreover, the reversible S-glutathionylation of complex I plays a vital role in regulating mitochondrial bioenergetics – oxidation of the glutathione pool by ROS induces GRX2-mediated S-glutathionylation of complex I lowering its activity and respiration while restoration of the reduced nature of the pool has the opposite effect (Fig. 11.6a) [57]. In total, 12 complex I subunits have been found to undergo S-glutathionylation, indicating that it is a major site for the regulation of metabolism through redox signals (Fig. 11.6a) [39]. Complexes II and V are also regulated by S-glutathionylation and defects in the redox modification of either protein has been linked to heart disease and failure [15, 93].

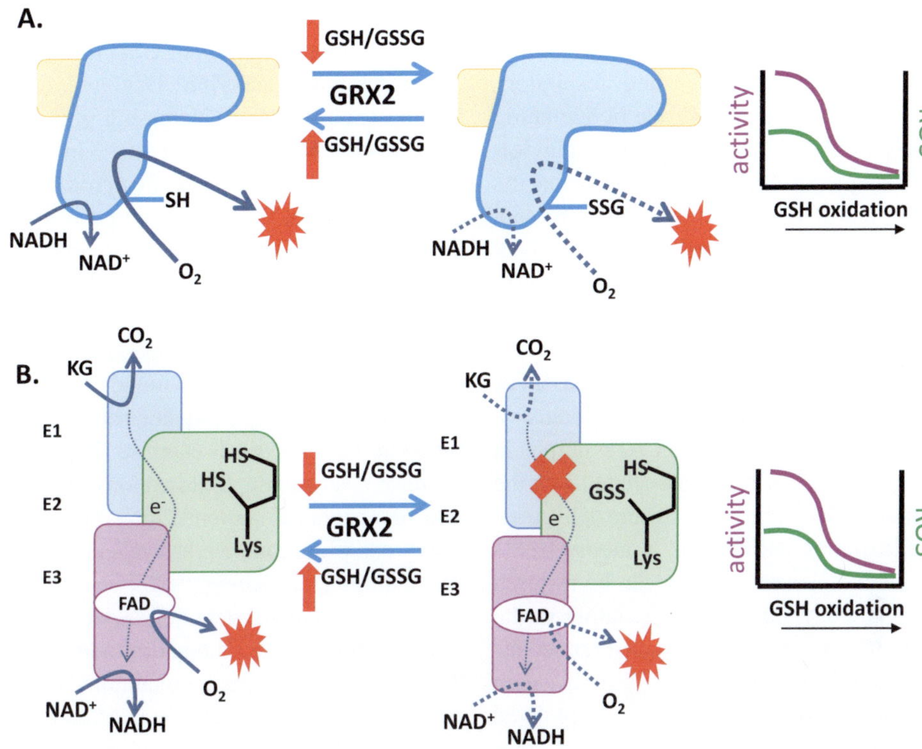


**Fig. 11.6** Protein S-glutathionylation regulates the activity and ROS release rate of complex I (**a**) and KGDH (**b**) in response to changes in redox buffering capacity

Krebs cycle enzymes PDH, aconitase (ACN), NADP+-dependent isocitrate dehydrogenase (IDH2), KGDH, succinyl-CoA synthetase (SCS), and malate dehydrogenase (MDH) also undergo protein S-glutathionylation. Overall, protein S-glutathionylation lowers carbon flux through the cycle by limiting Krebs cycle enzyme activity. This response may be required to diminish reducing equivalent production and limit mitochondrial ROS production following the oxidation of the mitochondrial glutathione pool. Indeed, S-glutathionylation of the E2 subunit on KGDH and PDH, two critical sources of NADH and entry points for carbon into the Krebs cycle, decreases the activity of both enzyme complexes (Fig. 11.6b) [3, 68]. This occurs following the artificial oxidation of the mitochondrial glutathione pool with diamide or disulfiram, an effect that can be reversed by GRX2 [68]. Similar results were collected using mice heterozygous or homozygous for the *Grx2* gene. Indeed, the partial or full deletion of the *Grx2* gene prolongs KGDH and PDH S-glutathionylation, specifically on the E2 subunit which is required for the eventual reduction of NAD+ (Fig. 11.6b) [13]. Finally, reversible S-glutathionylation of KGDH, like complex I, protects it from irreversible oxidative deactivation when ROS levels are too high [3]. Outside of KGDH and PDH, only ACN and IDH2 have been shown to be regulated by S-glutathionylation. In the case of ACN, modification of cysteine residues adjacent to the citrate binding site protects the Fe-S cluster from disassembly [30]. MDH and SCS are also documented targets for S-glutathionylation but it remains unknown if this modification can alter their activity.

Protein S-glutathionylation of Krebs cycle and respiratory complex enzymes serves as an important mechanism for limiting ROS production when the glutathione pool is oxidized while simultaneously protecting enzymes from irreversible deactivation and oxidative distress. Reversible S-glutathionylation of mitochondrial nutrient metabolism enzymes and respiratory complexes also plays a critical role in regulating oxidative phosphorylation and ATP output following changes in redox buffering capacity. Indeed, controlled and reversible protein S-glutathionylation reactions modulates ATP production by cardiac and skeletal muscle mitochondria. Moreover, disruption of these signaling pathways can compromise oxidative phosphorylation, leading to the development of disease. For example, elimination of the *Grx2* gene induces a ~50% decrease in mitochondrial ATP output in cardiac tissue [13, 57]. This is associated with the development of left ventricular hypertrophy, metabolic inflexibility, and fibrosis in mouse hearts [57]. Oxidative phosphorylation can be restored with deglutathionylating agents, pointing to the possibility that mitochondria-targeted reductants could be used to treat heart disease when S-glutathionylation reactions are deregulated [57]. It is crucial to point out as well that these effects appear to be associated with the S-glutathionylation of complex I on its NDUFS1 subunit, which can be reversed with dithiothreitol [57]. In addition, recent evidence found that variances in the gene sequence encoding human *Grx2* gene correlates with an increase in the susceptibility for developing heart disease [37]. The redox buffering capacity of mitochondria is in a constant state of flux, becoming oxidized and reduced in response to the availability of $H_2O_2$ and NADPH. Changes in this capacity could potentially serve as a critical communication mechanism that fine tunes cardiac ATP output and carbon oxidation in response to acute myocardial energy needs. However, as noted above it is critical that these reactions are short lived and controlled since prolonged S-glutathionylation of proteins can compromise metabolic efficiency leading to cardiac disease.

Intriguingly, deletion of the *Grx2* gene in mice has the opposite effect in skeletal muscle. Loss of GRX2 augments mitochondrial nutrient metabolism and oxidative phosphorylation in mouse muscle, an effect associated with the deglutathionylation and chronic activation of UCP3 [56]. UCP3 is targeted for S-glutathionylation which, following the conjugation of GSH to the protein, diminishes proton leaks [56]. Deglutathionylation has the opposite effect, increasing proton return to the matrix. Maintenance of UCP3 in a deglutathionylated and active state also correlates with an increase in whole body energy expenditure and a significant decrease in body and fat mass [56]. Our group has recently found that this effect protects mice from diet-induced obesity by increasing muscle fuel combustion by several fold. It has also been found that protein S-glutathionylation plays a key role in the adaptation of muscle towards exercise [39]. Fatiguing exercise results in the S-glutathionylation of several mitochondrial proteins, including the respiratory complexes and Krebs cycle enzymes [39]. In addition, fatiguing exercise alters the S-glutathionylation state of several proteins required for muscle contraction and relaxation. This demonstrates that S-glutathionylation reactions have opposite functions in different tissues, which may be related to the different physiological roles fulfilled by these tissues.

### 11.4.3 Regulation of Mitochondrial ROS Production

Reversible S-glutathionylation is ideal for the negative regulation of ROS production since conjugation and removal of GSH from proteins is highly response to fluctuations in the redox buffering capacity of the glutathione pool. This response is critical for inhibiting $O_2^{\bullet-}/H_2O_2$ release when ROS levels are higher than normal, protecting cells from oxidative distress. So far, $O_2^{\bullet-}/H_2O_2$ release from complex I, complex II, KGDH, and PDH has been found to be modulated by S-glutathionylation. Most studies have focused on complex I because it is often considered a major ROS source and was the first mitochondrial S-glutathionylation target that was identified. Protein S-glutathionylation of complex I subunit NDUFS1 has been found to augment $O_2^{\bullet-}/H_2O_2$ production in bovine and mouse heart samples [57, 88]. Intriguingly, other studies have yielded the opposite results, demonstrating that modification of NDUFS1 suppresses ROS production in mouse muscle and bovine heart mitochondria [6, 24]. Although this may seem counterintuitive, it can be explained by the fact that complex I harbors two ROS release sites ($I_F$ and $I_Q$) and that electrons used for $O_2^{\bullet-}/H_2O_2$ release can be provided by forward (from NADH) or reverse (from $UQH_2$) flow. For instance, modification of NDUFS1 would prevent NADH oxidation, limiting the number of electrons available for $O_2^{\bullet-}$ production. In contrast, reverse electron flow from the $UQH_2$ pool could augment ROS production due to NDUFS1 modification and accumulation of electrons in the $I_F$ or $I_Q$ sites. Complex I also contains 12 subunits in total that can be modified with glutathione [39]. Therefore, the sites that are modified could also affect the rate of ROS production. For example, modification of ND3 could prevent ROS production by complex I by blocking access to the UQ binding site. This would occlude the $I_Q$ site and also prevent reverse electron transfer to site $I_F$, curtailing ROS production. Production of ROS by the respiratory chain is also controlled at the level of complex II. S-glutathionylation of the SDHA subunit is required to suppress ROS production in cardiac mitochondria and protect complex II from irreversible oxidation [15]. Furthermore, alterations in the S-glutathionylation state of complex II augments the ROS release rate of cardiac mitochondria, contributing to ischemia-reperfusion injury to the myocardium [15].

Recent work has also identified S-glutathionylation as an important regulator of $O_2^{\bullet-}/H_2O_2$ release from KGDH and PDH. Chemical induction of S-glutathionylation lowers ROS release from KGDH and PDH by ~90% in mouse liver mitochondria [68]. Similar results were collected with KGDH and PDH purified from porcine heart [47, 68]. Modification of the E2 subunit of KGDH and PDH is responsible for almost abolishing ROS release by either enzyme complex. Indeed, blocking the dihydrolipoamide residue in the E2 subunit of either enzyme limits electron transfer to the E3 subunit, which harbors the $O_2^{\bullet-}/H_2O_2$ generating site for KGDH ($K_F$) or PDH ($P_F$) [13, 68]. Alterations in the redox buffering capacity of the glutathione pool augments ROS release from PDH in permeabilized muscle fibers [20]. Both KGDH and PDH are also targeted by GRX2. Purified GRX2 can deglutathionylate both enzyme complexes and deletion of the *Grx2* gene can prolong KGDH and PDH S-glutathionylation, lowering ROS production in liver mitochondria [13]. Evidence has also demonstrated that S-glutathionylation can also control ROS production by KGDH and PDH during RET from NADH [47, 68]. Protein S-glutathionylation of PDH augments ROS release during RET by up to ~8-fold [68]. Purified GRX2 can reverse this effect indicating that it may play a role in limiting ROS production by RET. This would be beneficial under conditions when complex I is defective and NADH turnover is slower than usual. By deglutathionylating PDH, GRX2 could protect cells from oxidative distress by limiting ROS production during RET.

Evidence collected over the past few years has indicated that protein S-glutathionylation reactions may play a critical role in regulating mitochondrial ROS signaling. Most of this evidence stems from studies conducted on UCP2 and UCP3. The S-glutathionylation of both proteins turns off proton leaks, increasing the strength of the membrane potential which in turn augments mitochondrial ROS production. As mentioned above, mitochondrial $H_2O_2$ is an important secondary messenger implicated in retrograde signaling. For instance, mitochondrial ROS is now regarded as an important factor for glucose-stimulated insulin release (GSIS). The S-glutathionylation of UCP2 in cultured MIN6 insulinoma cells or pancreatic islets was found to augment insulin release by promoting mitochondrial ROS production [48]. Deglutathionylation of UCP2 has the opposite effect, lowering GSIS by decreasing ROS production [48]. Chemical induction of UCP2 S-glutathionylation was also shown to be effective at rendering drug-resistant cancer cells more susceptible to chemotherapy [72]. This was achieved by disabling proton leaks and promoting ROS production and oxidative distress, resulting in increased cell death [72]. In regard to UCP3, deglutathionylation augments muscle fuel combustion, which may play a critical role in the adaptation of muscle towards exercise [53].

### 11.4.4 Regulation of Solute Import and Other Mitochondrial Processes

Outside of regulating the bioenergetics, protein S-glutathionylation also controls other mitochondrial functions including solute import, mitochondrial permeability transition pore (MPTP) opening and apoptosis, and mitochondrial fission and fusion. In terms of solute import, recent studies have found that protein S-glutathionylation modulates the pyruvate and fatty acyl-carnitine uptake [23, 24]. Conjugation of glutathione to pyruvate carrier almost abolishes pyruvate import by skeletal muscle mitochondria [24]. The S-glutathionylation of carnitine/acyl-carnitine carrier (CAC) also decreases fatty acyl-carnitine import, an effect that can be reversed by purified GRX1 [23]. Adenine nucleotide translocase (ANT), which is required to catalyze the antiport of ADP and ATP, has also been found to be S-glutathionylated [73]. These findings have strong implications for the regulation of nutrient metabolism. The S-glutathionylation of solute import proteins could serve as a means to limit carbon oxidation when redox buffering networks are oxidized due to higher than normal ROS. This would prevent the further production of ROS and promote recovery antioxidant defences, restoring the reduced nature of redox buffering networks and preventing oxidative stress.

Another important feature here is the modification of ANT, which is also a putative component of the MPTP. Although the actual structure of the MPTP is quite controversial, S-glutathionylation of ANT prevents pore opening [73]. This response could be utilized to prevent the induction of apoptosis in response to the acute oxidation of glutathione pools when ROS levels are higher than normal. Indeed, S-glutahtionylation of cyclophilin-D, another key component of MPTP, also curtails pore opening [32]. It is thus likely that S-glutathionylation is critical for limiting the induction of cell death programs and preventing mitochondrial depolarization and potentially mitophagy in response to redox network oxidation.

Mitochondria also take on a variety of shapes, from fused to fragmented, in response to changes in nutrient demands and alterations in bioenergetics. Acute stressors like nutrient deprivation or an increase in ROS promotes the adoption of a hyperfused state, protecting mitochondria from oxidative distress while consolidating ATP producing capacity. To this end, protein S-glutathionylation promotes the fusion of mitochondria. High GSSG levels promotes the S-glutathionylation of fusion proteins mitofusin-1 (MFN1) and mitofusin-2 (MFN2) and the oligomerization of optic atrophy protein-1 (OPA1) [81, 89]. In addition, chemical S-glutathionylation agents like diamide induce mitochondrial filamentation [76]. As indicated above, oxidation of glutathione pools promotes protein S-glutathionylation by increasing the availability of GSSG. Therefore, S-glutathionylation seems to be part of the acute response to stress, promoting mitochondrial fusion to curtail any further oxidation damage.

## 11.5 Conclusions

In the past decade, awareness of the number of mitochondrial proteins governed by oxidation of cysteine switches through protein S-glutathionylation has expanded significantly. Protein S-glutathionylation reactions have been implicated in controlling a broad range of mitochondrial functions including carbon oxidation and ATP production, ROS release, solute import, apoptosis, mitochondrial fission and fusion, and antioxidant defense. GRX2 is involved in facilitating some of these reactions, playing a key role in regulating mitochondrial metabolism, proton leaks, and ROS production. However, the importance of mitochondrial protein S-glutathionylation reactions in regulating ROS production for retrograde signaling also needs to be investigated and fully characterized. Based on the evidence so far, mitochondrial S-glutathionylation reactions fulfill vital roles in the provision of ATP for basic physiological functions and are required to modulate fuel combustion in response to changes in redox buffering capacity. Protein S-glutathionylation is also required to protect mitochondria from acute stress by (1) preventing the irreversible oxidation of proteins, (2) deactivating MPTP opening, (3) promoting mitochondrial fusion, and (4) limiting carbon combustion thereby curtailing any further ROS production.

## References

1. Ambrus A, Nemeria NS, Torocsik B, Tretter L, Nilsson M, Jordan F, Adam-Vizi V (2015) Formation of reactive oxygen species by human and bacterial pyruvate and 2-oxoglutarate dehydrogenase multienzyme complexes reconstituted from recombinant components. Free Radic Biol Med 89:642–650
2. Andrienko TN, Pasdois P, Pereira GC, Ovens MJ, Halestrap AP (2017) The role of succinate and ROS in reperfusion injury – a critical appraisal. J Mol Cell Cardiol 110:1–14
3. Applegate MA, Humphries KM, Szweda LI (2008) Reversible inhibition of alpha-ketoglutarate dehydrogenase by hydrogen peroxide: glutathionylation and protection of lipoic acid. Biochemistry 47:473–478
4. Axelsson K, Mannervik B (1983) An essential role of cytosolic thioltransferase in protection of pyruvate kinase from rabbit liver against oxidative inactivation. FEBS Lett 152:114–118
5. Balijepalli S, Annepu J, Boyd MR, Ravindranath V (1999) Effect of thiol modification on brain mitochondrial complex I activity. Neurosci Lett 272:203–206
6. Beer SM, Taylor ER, Brown SE, Dahm CC, Costa NJ, Runswick MJ, Murphy MP (2004) Glutaredoxin 2 catalyzes the reversible oxidation and glutathionylation of mitochondrial membrane thiol proteins: implications for mitochondrial redox regulation and antioxidant DEFENSE. J Biol Chem 279:47939–47951
7. Berry BJ, Trewin AJ, Amitrano AM, Kim M, Wojtovich AP (2018) Use the protonmotive force: mitochondrial uncoupling and reactive oxygen species. J Mol Biol 430:3873–3891
8. Bleier L, Drose S (2013) Superoxide generation by complex III: from mechanistic rationales to functional consequences. Biochim Biophys Acta 1827:1320–1331
9. Bohovych I, Khalimonchuk O (2016) Sending out an SOS: mitochondria as a signaling hub. Front Cell Dev Biol 4:109
10. Brand MD (2016) Mitochondrial generation of superoxide and hydrogen peroxide as the source of mitochondrial redox signaling. Free Radic Biol Med 100:14–31
11. Brand MD, Nicholls DG (2011) Assessing mitochondrial dysfunction in cells. Biochem J 435:297–312
12. Bunik VI, Brand MD (2018) Generation of superoxide and hydrogen peroxide by side reactions of mitochondrial 2-oxoacid dehydrogenase complexes in isolation and in cells. Biol Chem 399:407–420
13. Chalker J, Gardiner D, Kuksal N, Mailloux RJ (2018) Characterization of the impact of glutaredoxin-2 (GRX2) deficiency on superoxide/hydrogen peroxide release from cardiac and liver mitochondria. Redox Biol 15:216–227
14. Chandel NS (2015) Evolution of mitochondria as signaling organelles. Cell Metab 22:204–206
15. Chen YR, Chen CL, Pfeiffer DR, Zweier JL (2007) Mitochondrial complex II in the post-ischemic heart: oxidative injury and the role of protein S-glutathionylation. J Biol Chem 282:32640–32654
16. Chouchani ET, Kazak L, Jedrychowski MP, Lu GZ, Erickson BK, Szpyt J, Pierce KA, Laznik-Bogoslavski D, Vetrivelan R, Clish CB, Robinson AJ, Gygi SP, Spiegelman BM (2016) Mitochondrial ROS regulate thermogenic energy expenditure and sulfenylation of UCP1. Nature 532:112–116
17. Chouchani ET, Pell VR, Gaude E, Aksentijevic D, Sundier SY, Robb EL, Logan A, Nadtochiy SM, Ord ENJ, Smith AC, Eyassu F, Shirley R, Hu CH, Dare AJ, James AM, Rogatti S, Hartley RC, Eaton S, Costa ASH, Brookes PS, Davidson SM, Duchen MR, Saeb-Parsy K, Shattock MJ, Robinson AJ, Work LM, Frezza C, Krieg T, Murphy MP (2014) Ischaemic accumulation of succinate controls reperfusion injury through mitochondrial ROS. Nature 515:431–435
18. Crofts AR, Holland JT, Victoria D, Kolling DR, Dikanov SA, Gilbreth R, Lhee S, Kuras R, Kuras MG (2008) The Q-cycle reviewed: how well does a monomeric mechanism of the bc(1) complex account for the function of a dimeric complex? Biochim Biophys Acta 1777:1001–1019
19. Drose S, Brandt U, Wittig I (2014) Mitochondrial respiratory chain complexes as sources and targets of thiol-based redox-regulation. Biochim Biophys Acta 1844:1344–1354
20. Fisher-Wellman KH, Gilliam LA, Lin CT, Cathey BL, Lark DS, Neufer PD (2013) Mitochondrial glutathione depletion reveals a novel role for the pyruvate dehydrogenase complex as a key H2O2-emitting source under conditions of nutrient overload. Free Radic Biol Med 65:1201–1208
21. Fisher-Wellman KH, Lin CT, Ryan TE, Reese LR, Gilliam LA, Cathey BL, Lark DS, Smith CD, Muoio DM, Neufer PD (2015) Pyruvate dehydrogenase complex and nicotinamide nucleotide transhydrogenase constitute an energy-consuming redox circuit. Biochem J 467:271–280
22. Gallogly MM, Starke DW, Leonberg AK, Ospina SM, Mieyal JJ (2008) Kinetic and mechanistic characterization and versatile catalytic properties of mammalian glutaredoxin 2: implications for intracellular roles. Biochemistry 47:11144–11157
23. Giangregorio N, Palmieri F, Indiveri C (2013) Glutathione controls the redox state of the mitochondrial carnitine/acylcarnitine carrier Cys residues by glutathionylation. Biochim Biophys Acta 1830:5299–5304

24. Gill RM, O'Brien M, Young A, Gardiner D, Mailloux RJ (2018) Protein S-glutathionylation lowers superoxide/hydrogen peroxide release from skeletal muscle mitochondria through modification of complex I and inhibition of pyruvate uptake. PLoS One 13:e0192801
25. Goncalves RL, Bunik VI, Brand MD (2016) Production of superoxide/hydrogen peroxide by the mitochondrial 2-oxoadipate dehydrogenase complex. Free Radic Biol Med 91:247–255
26. Goncalves RL, Quinlan CL, Perevoshchikova IV, Hey-Mogensen M, Brand MD (2015) Sites of superoxide and hydrogen peroxide production by muscle mitochondria assessed ex vivo under conditions mimicking rest and exercise. J Biol Chem 290:209–227
27. Goncalves RL, Rothschild DE, Quinlan CL, Scott GK, Benz CC, Brand MD (2014) Sources of superoxide/H2O2 during mitochondrial proline oxidation. Redox Biol 2:901–909
28. Gorelenkova Miller O, Mieyal JJ (2018) Critical roles of glutaredoxin in brain cells-implications for Parkinson's disease. Antioxid Redox Signal 30:1352–1368
29. Hail N Jr, Chen P, Kepa JJ, Bushman LR, Shearn C (2010) Dihydroorotate dehydrogenase is required for N-(4-hydroxyphenyl)retinamide-induced reactive oxygen species production and apoptosis. Free Radic Biol Med 49:109–116
30. Han D, Canali R, Garcia J, Aguilera R, Gallaher TK, Cadenas E (2005) Sites and mechanisms of aconitase inactivation by peroxynitrite: modulation by citrate and glutathione. Biochemistry 44:11986–11996
31. Hey-Mogensen M, Goncalves RL, Orr AL, Brand MD (2014) Production of superoxide/H2O2 by dihydroorotate dehydrogenase in rat skeletal muscle mitochondria. Free Radic Biol Med 72:149–155
32. Hortelano S, Dallaporta B, Zamzami N, Hirsch T, Susin SA, Marzo I, Bosca L, Kroemer G (1997) Nitric oxide induces apoptosis via triggering mitochondrial permeability transition. FEBS Lett 410:373–377
33. Hurd TR, Requejo R, Filipovska A, Brown S, Prime TA, Robinson AJ, Fearnley IM, Murphy MP (2008) Complex I within oxidatively stressed bovine heart mitochondria is glutathionylated on Cys-531 and Cys-704 of the 75-kDa subunit: potential role of CYS residues in decreasing oxidative damage. J Biol Chem 283:24801–24815
34. Imlay JA (2013) The molecular mechanisms and physiological consequences of oxidative stress: lessons from a model bacterium. Nat Rev Microbiol 11:443–454
35. Jensen KS, Pedersen JT, Winther JR, Teilum K (2014) The pKa value and accessibility of cysteine residues are key determinants for protein substrate discrimination by glutaredoxin. Biochemistry 53:2533–2540
36. Jones DP, Sies H (2015) The redox code. Antioxid Redox Signal 23:734–746
37. Kanaan GN, Ichim B, Gharibeh L, Maharsy W, Patten DA, Xuan JY, Reunov A, Marshall P, Veinot J, Menzies K, Nemer M, HARPER ME (2018) Glutaredoxin-2 controls cardiac mitochondrial dynamics and energetics in mice, and protects against human cardiac pathologies. Redox Biol 14:509–521
38. Kornberg H (2000) Krebs and his trinity of cycles. Nat Rev Mol Cell Biol 1:225–228
39. Kramer PA, Duan J, Gaffrey MJ, Shukla AK, Wang L, Bammler TK, Qian WJ, Marcinek DJ (2018) Fatiguing contractions increase protein S-glutathionylation occupancy in mouse skeletal muscle. Redox Biol 17:367–376
40. Kuksal N, Chalker J, Mailloux RJ (2017) Progress in understanding the molecular oxygen paradox – function of mitochondrial reactive oxygen species in cell signaling. Biol Chem 398:1209–1227
41. Kuksal N, Gardiner D, Qi D, Mailloux RJ (2018) Partial loss of complex I due to NDUFS4 deficiency augments myocardial reperfusion damage by increasing mitochondrial superoxide/hydrogen peroxide production. Biochem Biophys Res Commun 498:214–220
42. Lane N (2015) The vital question. W.W. Norton and Company, New York
43. Lundberg M, Johansson C, Chandra J, Enoksson M, Jacobsson G, Ljung J, Johansson M, Holmgren A (2001) Cloning and expression of a novel human glutaredoxin (Grx2) with mitochondrial and nuclear isoforms. J Biol Chem 276:26269–26275
44. Mailloux RJ (2015) Teaching the fundamentals of electron transfer reactions in mitochondria and the production and detection of reactive oxygen species. Redox Biol 4:381–398
45. Mailloux RJ (2018) Mitochondrial antioxidants and the maintenance of cellular hydrogen peroxide levels. Oxidative Med Cell Longev 2018:7857251
46. Mailloux RJ, Adjeitey CN, Xuan JY, Harper ME (2012a) Crucial yet divergent roles of mitochondrial redox state in skeletal muscle vs. brown adipose tissue energetics. FASEB J 26:363–375
47. Mailloux RJ, Craig Ayre D, Christian SL (2016a) Induction of mitochondrial reactive oxygen species production by GSH mediated S-glutathionylation of 2-oxoglutarate dehydrogenase. Redox Biol 8:285–297
48. Mailloux RJ, Fu A, Robson-Doucette C, Allister EM, Wheeler MB, Screaton R, Harper ME (2012b) Glutathionylation state of uncoupling protein-2 and the control of glucose-stimulated insulin secretion. J Biol Chem 287:39673–39685
49. Mailloux RJ, Gardiner D, O'Brien M (2016b) 2-Oxoglutarate dehydrogenase is a more significant source of O2(.-)/H2O2 than pyruvate dehydrogenase in cardiac and liver tissue. Free Radic Biol Med 97:501–512
50. Mailloux RJ, Harper ME (2011) Uncoupling proteins and the control of mitochondrial reactive oxygen species production. Free Radic Biol Med 51:1106–1115
51. Mailloux RJ, Harper ME (2012) Mitochondrial proticity and ROS signaling: lessons from the uncoupling proteins. Trends Endocrinol Metab 23:451–458
52. Mailloux RJ, McBride SL, Harper ME (2013a) Unearthing the secrets of mitochondrial ROS and

glutathione in bioenergetics. Trends Biochem Sci 38:592–602
53. Mailloux RJ, Seifert EL, Bouillaud F, Aguer C, Collins S, Harper ME (2011) Glutathionylation acts as a control switch for uncoupling proteins UCP2 and UCP3. J Biol Chem 286:21865–21875
54. Mailloux RJ, Treberg JR (2016) Protein S-glutathionlyation links energy metabolism to redox signaling in mitochondria. Redox Biol 8:110–118
55. Mailloux RJ, Willmore WG (2014) S-glutathionylation reactions in mitochondrial function and disease. Front Cell Dev Biol 2:68
56. Mailloux RJ, Xuan JY, Beauchamp B, Jui L, Lou M, Harper ME (2013b) Glutaredoxin-2 is required to control proton leak through uncoupling protein-3. J Biol Chem 288:8365–8379
57. Mailloux RJ, Xuan JY, McBride S, Maharsy W, Thorn S, Holterman CE, Kennedy CR, Rippstein P, Dekemp R, Da Silva J, Nemer M, Lou M, Harper ME (2014) Glutaredoxin-2 is required to control oxidative phosphorylation in cardiac muscle by mediating deglutathionylation reactions. J Biol Chem 289:14812–14828
58. Mailloux RJ, Young A, Chalker J, Gardiner D, O'Brien M, Slade L, Brosnan JT (2016c) Choline and dimethylglycine produce superoxide/hydrogen peroxide from the electron transport chain in liver mitochondria. FEBS Lett 590:4318–4328
59. Massey V (1994) Activation of molecular oxygen by flavins and flavoproteins. J Biol Chem 269:22459–22462
60. McLain AL, Szweda PA, Szweda LI (2011) alpha-Ketoglutarate dehydrogenase: a mitochondrial redox sensor. Free Radic Res 45:29–36
61. Mieyal JJ, Gallogly MM, Qanungo S, Sabens EA, Shelton MD (2008) Molecular mechanisms and clinical implications of reversible protein S-glutathionylation. Antioxid Redox Signal 10:1941–1988
62. Mills EL, Pierce KA, Jedrychowski MP, Garrity R, Winther S, Vidoni S, Yoneshiro T, Spinelli JB, Lu GZ, Kazak L, Banks AS, Haigis MC, Kajimura S, Murphy MP, Gygi SP, Clish CB, Chouchani ET (2018) Accumulation of succinate controls activation of adipose tissue thermogenesis. Nature 560:102–106
63. Murphy MP (2009) How mitochondria produce reactive oxygen species. Biochem J 417:1–13
64. Murphy MP (2012) Mitochondrial thiols in antioxidant protection and redox signaling: distinct roles for glutathionylation and other thiol modifications. Antioxid Redox Signal 16:476–495
65. Nemeria NS, Ambrus A, Patel H, Gerfen G, Adam-Vizi V, Tretter L, Zhou J, Wang J, Jordan F (2014) Human 2-oxoglutarate dehydrogenase complex E1 component forms a thiamin-derived radical by aerobic oxidation of the enamine intermediate. J Biol Chem 289:29859–29873
66. Nicholls DG (2001) A history of UCP1. Biochem Soc Trans 29:751–755
67. Nicholls DG (2008) Forty years of Mitchell's proton circuit: from little grey books to little grey cells. Biochim Biophys Acta 1777:550–556
68. O'Brien M, Chalker J, Slade L, Gardiner D, Mailloux RJ (2017) Protein S-glutathionylation alters superoxide/hydrogen peroxide emission from pyruvate dehydrogenase complex. Free Radic Biol Med 106:302–314
69. Paddenberg R, Goldenberg A, Faulhammer P, Braun-Dullaeus RC, Kummer W (2003) Mitochondrial complex II is essential for hypoxia-induced ROS generation and vasoconstriction in the pulmonary vasculature. Adv Exp Med Biol 536:163–169
70. Palmieri F, Monne M (2016) Discoveries, metabolic roles and diseases of mitochondrial carriers: a review. Biochim Biophys Acta 1863:2362–2378
71. Perevoshchikova IV, Quinlan CL, Orr AL, Gerencser AA, Brand MD (2013) Sites of superoxide and hydrogen peroxide production during fatty acid oxidation in rat skeletal muscle mitochondria. Free Radic Biol Med 61:298–309
72. Pfefferle A, Mailloux RJ, Adjeitey CN, Harper ME (2013) Glutathionylation of UCP2 sensitizes drug resistant leukemia cells to chemotherapeutics. Biochim Biophys Acta 1833:80–89
73. Queiroga CS, Almeida AS, Martel C, Brenner C, Alves PM, Vieira HL (2010) Glutathionylation of adenine nucleotide translocase induced by carbon monoxide prevents mitochondrial membrane permeabilization and apoptosis. J Biol Chem 285:17077–17088
74. Quinlan CL, Goncalves RL, Hey-Mogensen M, Yadava N, Bunik VI, Brand MD (2014) The 2-oxoacid dehydrogenase complexes in mitochondria can produce superoxide/hydrogen peroxide at much higher rates than complex I. J Biol Chem 289:8312–8325
75. Radi R, Turrens JF, Chang LY, Bush KM, Crapo JD, Freeman BA (1991) Detection of catalase in rat heart mitochondria. J Biol Chem 266:22028–22034
76. Redpath CJ, Bou Khalil M, Drozdzal G, Radisic M, McBride HM (2013) Mitochondrial hyperfusion during oxidative stress is coupled to a dysregulation in calcium handling within a C2C12 cell model. PLoS One 8:e69165
77. Ronchi JA, Francisco A, Passos LA, Figueira TR, Castilho RF (2016) The contribution of nicotinamide nucleotide transhydrogenase to peroxide detoxification is dependent on the respiratory state and counterbalanced by other sources of NADPH in liver mitochondria. J Biol Chem 291:20173–20187
78. Scialo F, Sriram A, Fernandez-Ayala D, Gubina N, Lohmus M, Nelson G, Logan A, Cooper HM, Navas P, Enriquez JA, Murphy MP, Sanz A (2016) Mitochondrial ROS produced via reverse electron transport extend animal lifespan. Cell Metab 23:725–734
79. Shabalina IG, Vrbacky M, Pecinova A, Kalinovich AV, Drahota Z, Houstek J, Mracek T, Cannon B, Nedergaard J (2014) ROS production in brown adipose tissue mitochondria: the question of UCP1-dependence. Biochim Biophys Acta 1837:2017–2030
80. Shelton MD, Chock PB, Mieyal JJ (2005) Glutaredoxin: role in reversible protein s-glutathionylation and regu-

lation of redox signal transduction and protein translocation. Antioxid Redox Signal 7:348–366
81. Shutt T, Geoffrion M, Milne R, McBride HM (2012) The intracellular redox state is a core determinant of mitochondrial fusion. EMBO Rep 13:909–915
82. Sies H (2017) Hydrogen peroxide as a central redox signaling molecule in physiological oxidative stress: oxidative eustress. Redox Biol 11:613–619
83. Sies H, Berndt C, Jones DP (2017) Oxidative stress. Annu Rev Biochem 86:715–748
84. Slade L, Chalker J, Kuksal N, Young A, Gardiner D, Mailloux RJ (2017) Examination of the superoxide/hydrogen peroxide forming and quenching potential of mouse liver mitochondria. Biochim Biophys Acta 1861:1960–1969
85. Starkov AA, Fiskum G, Chinopoulos C, Lorenzo BJ, Browne SE, Patel MS, Beal MF (2004) Mitochondrial alpha-ketoglutarate dehydrogenase complex generates reactive oxygen species. J Neurosci 24:7779–7788
86. Sweetlove LJ, Beard KF, Nunes-Nesi A, Fernie AR, Ratcliffe RG (2010) Not just a circle: flux modes in the plant TCA cycle. Trends Plant Sci 15:462–470
87. Tait SW, Green DR (2012) Mitochondria and cell signalling. J Cell Sci 125:807–815
88. Taylor ER, Hurrell F, Shannon RJ, Lin TK, Hirst J, Murphy MP (2003) Reversible glutathionylation of complex I increases mitochondrial superoxide formation. J Biol Chem 278:19603–19610
89. Thaher O, Wolf C, Dey PN, Pouya A, Wullner V, Tenzer S, Methner A (2018) The thiol switch C684 in Mitofusin-2 mediates redox-induced alterations of mitochondrial shape and respiration. Neurochem Int 117:167–173
90. Tretter L, Adam-Vizi V (2004) Generation of reactive oxygen species in the reaction catalyzed by alpha-ketoglutarate dehydrogenase. J Neurosci 24:7771–7778
91. Tretter L, Takacs K, Hegedus V, Adam-Vizi V (2007) Characteristics of alpha-glycerophosphate-evoked H2O2 generation in brain mitochondria. J Neurochem 100:650–663
92. Walker JE (2013) The ATP synthase: the understood, the uncertain and the unknown. Biochem Soc Trans 41:1–16
93. Wang SB, Foster DB, Rucker J, O'Rourke B, Kass DA, van Eyk JE (2011) Redox regulation of mitochondrial ATP synthase: implications for cardiac resynchronization therapy. Circ Res 109:750–757
94. Yun J, Finkel T (2014) Mitohormesis. Cell Metab 19:757–766
95. Zhang J, Ye ZW, Singh S, Townsend DM, Tew KD (2018) An evolving understanding of the S-glutathionylation cycle in pathways of redox regulation. Free Radic Biol Med 120:204–216
96. Ziegler DM (1985) Role of reversible oxidation-reduction of enzyme thiols-disulfides in metabolic regulation. Annu Rev Biochem 54:305–329

# 12 Advances Towards Therapeutic Approaches for mtDNA Disease

Iman Al Khatib and Timothy E. Shutt

**Abstract**

Mitochondria maintain and express their own genome, referred to as mtDNA, which is required for proper mitochondrial function. While mutations in mtDNA can cause a heterogeneous array of disease phenotypes, there is currently no cure for this collection of diseases. Here, we will cover characteristics of the mitochondrial genome important for understanding the pathology associated with mtDNA mutations, and review recent approaches that are being developed to treat and prevent mtDNA disease. First, we will discuss mitochondrial replacement therapy (MRT), where mitochondria from a healthy donor replace maternal mitochondria harbouring mutant mtDNA. In addition to ethical concerns surrounding this procedure, MRT is only applicable in cases where the mother is known or suspected to carry mtDNA mutations. Thus, there remains a need for other strategies to treat patients with mtDNA disease. To this end, we will also discuss several alternative means to reduce the amount of mutant mtDNA present in cells. Such methods, referred to as heteroplasmy shifting, have proven successful in animal models. In particular, we will focus on the approach of targeting engineered endonucleases to specifically cleave mutant mtDNA. Together, these approaches offer hope to prevent the transmission of mtDNA disease and potentially reduce the impact of mtDNA mutations.

**Keywords**

Mitochondrial disease · Pathogenic threshold · Maternal inheritance · Mitochondrial replacement therapy · Heteroplasmy shifting

## 12.1 Introduction

In addition to their well-recognized role as "the power house of the cell", mitochondria have many important roles, and their proper function is integral to many cellular processes such as cell signaling, apoptosis, and various metabolic pathways. Given their importance, it should come as no surprise that mitochondrial dysfunction can lead to disease. Remarkably, the growing list of diseases and the variety of pathogenic phenotypes associated with mitochondrial dysfunction points to the importance of these often overlooked organelles. Mitochondrial diseases are a collection of heterogeneously diverse and rare

I. Al Khatib · T. E. Shutt (✉)
Deparments of Medical Genetics and Biochemistry & Molecular Biology, Cumming School of Medicine, Alberta Children's Hospital Research Institute, Hotchkiss Brain Institute, University of Calgary, Calgary, Alberta, Canada
e-mail: timothy.shutt@ucalgary.ca



A. Urbani, M. Babu (eds.), *Mitochondria in Health and in Sickness*, Advances in Experimental Medicine and Biology 1158, https://doi.org/10.1007/978-981-13-8367-0_12

disorders caused by severe mitochondrial dysfunction, typically exemplified by impairments in oxidative respiration, and which often manifest in tissues with high energy demands. While mitochondrial diseases are collectively estimated to have an incidence between 1:4000-5000 [23, 39, 72, 176], this number is likely low given the difficulties associated with proper diagnosis [176]. Meanwhile, milder mitochondrial dysfunction is also increasingly recognized to play an important role in common human pathologies such as, cancer, cardiovascular disease and neurodegenerative disease [12, 50].

### 12.1.1 The Mitochondrial Genome

Mitochondria maintain and express their own genetic material (mtDNA), which resides in the mitochondrial matrix and is evidence of their prokaryotic origin. In humans, the mtDNA genome is a circular double stranded molecule comprising 16,569 base pairs that is present in several hundred copies per cell. As such, mtDNA represents only ~0.25% of the total genetic material of the cell. The mtDNA encodes 37 genes distributed on both DNA strands [4]. These genes include 13 polypeptides that are core components of oxidative phosphorylation complexes, which explains the essential nature of the mtDNA genome. In addition, 2 rRNAs and 22 tRNAs encoded by the mtDNA are required for translation of the 13 mtDNA encoded proteins. Characterized by tight packaging of genetic information, the mtDNA comprises very little non-coding DNA. The so-called D-loop control region [205] encodes regulatory elements including the mtDNA replication initiation site, conserved sequence elements and the promoters required for transcription [4, 32, 142], making it as important as the encoded genes for proper mitochondrial function (Fig. 12.1). Despite the highly conserved nature of the mtDNA genome, naturally occurring polymorphisms arise and these have been used to define mtDNA haplogroups, reflecting their evolutionary history [26, 69, 224, 226].

### 12.1.2 Replication and Repair of the Mitochondrial Genome

Replication of the mitochondrial genome occurs throughout the cell cycle and is not linked to nuclear genome replication. However, mitochondria depend on nuclear-encoded proteins for transcription, replication and maintenance of mtDNA [189, 190]. Mechanistically, mtDNA replication proceeds via an unusual asynchronous process that utilizes two unidirectional origins of replication (one for each strand) [181]. The mitochondrial DNA polymerase $\gamma$ (Pol$\gamma$) encoded by *POLG* gene is the dedicated mitochondrial polymerase, which is required for mtDNA replication and repair [31, 95]. Notably, Pol$\gamma$ has a 3′–5′ exonuclease which is involved in proofreading and correcting replication errors in newly synthesized DNA. Replication of the mtDNA is also facilitated by key factors such as the accessory subunit Pol$\beta$ (encoded by *POLG2*) [119], the DNA helicase TWINKLE (*TWNK*) [136], topoisomerases (*i.e. TOP1MT*, *TOP2B*, *TOP3A*) that relieve DNA supercoiling and resolve mtDNA concatamers [122, 229, 238], and the mitochondrial single strand binding protein (*SSBP1*) that stabilizes single-stranded DNA during replication [42]. In addition, transcription of the mitochondrial genome is also required to prime mtDNA replication [19]. Hence, the mitochondrial RNA polymerase (*POLRMT*) and its regulators are also requisite for mtDNA replication [188].

As with the nuclear genome, DNA repair is also critical for the integrity of mtDNA. In particular, the highly oxidative environment in the mitochondrial matrix makes mtDNA especially susceptible to damage [54]. Though mitochondria were initially thought to lack DNA repair capabilities, it is now appreciated that mitochondria can perform base excision repair (BER) to fix damaged nucleotides [27, 74, 112, 158, 178]. Meanwhile, though classic homology recombination is not thought to occur in mammalian mitochondria, there are conflicting reports about whether other types of double strand break repair do occur [140]. Notably,

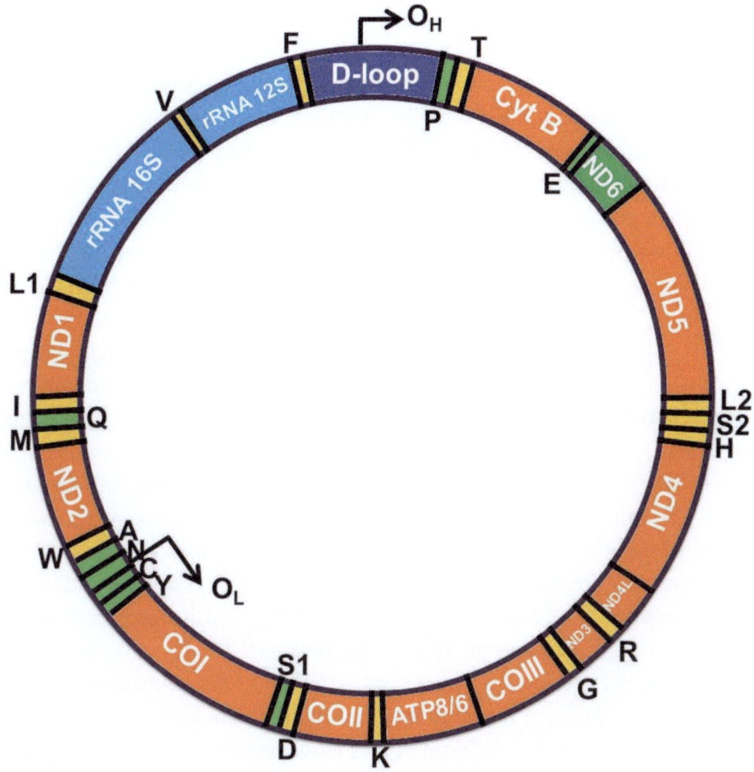


**Fig. 12.1** Map of the human mitochondrial DNA genome. The human mtDNA genome comprises 16,569 nucleotide pairs and encodes 13 mRNAs (ND1, ND2, ND3, ND4, ND4L, ND5, ND6, CytB, COI, COII, COIII, ATP6 and ATP8), two rRNAs (12S and 16S) and 22 tRNAs represented by their one letter abbreviation. In addition to the D-loop regulatory region; $O_H$ and $O_L$ represent the two origins of replication on both strands of DNA. The following are encoded by genes located on the heavy strand: rRNA 12S, rRNA 16S, ND1, ND2, ND3, ND4, ND4L, ND5, CytB, $tRNA_{Phe}$ (F), $tRNA_{Val}$ (V), $tRNA_{Leu}$-1 (L1), $tRNA_{Ile}$ (I),$tRNA_{Met}$ (M), $tRNA_{Trp}$ (W), $tRNA_{Asp}$ (D), $tRNA_{Lys}$ (K), $tRNA_{Gly}$ (G), $tRNA_{Arg}$ (R), $tRNA_{His}$ (H), $tRNA_{Ser}$-2 (S2), $tRNA_{Leu}$-2 (L2) and $tRNA_{Thr}$ (T). Whereas the light strand genes encode for: ND6, $tRNA_{Gln}$ (Q), $tRNA_{Ala}$ (A), $tRNA_{Asn}$ (N), $tRNA_{Cys}$ (C), $tRNA_{Tyr}$ (Y), $tRNA_{Ser}$-1 (S1), $tRNA_{Glu}$ (E) and $tRNA_{Pro}$ (P)

reports of non-homologous end-joining (NHEJ) [8] or microhomology-mediated end-joining (MMEJ) [107, 208], have been proposed to explain the mechanism by which mtDNA deletions occur. Nonetheless, when mtDNA double strand break repair fails, the mtDNA is thought to be degraded by mitochondrial nucleases or removed via autophagy [186]. The loss of mtDNA following double strand breaks is a key factor in our ability to develop approaches that specifically target and eliminate mutant mtDNA, which will be discussed later in this chapter.

### 12.1.3 mtDNA Copy Number

Unlike the nuclear DNA, mtDNA is present in hundreds to thousands of copies per cell [82, 191]. Notably, various factors such as cell type, age and stress can impact mtDNA copy number. For example, high energy demanding organs like the brain, muscles and liver contain high mtDNA content compared to skin fibroblast cells. However, we do not have a complete understanding of how mtDNA abundance is maintained. Nonetheless, it is well established that changes in mtDNA copy number can directly reflect the energy production in cells, and are also associated with disease. For example, point mutations in the D-loop regulatory region have been reported in many forms of cancer, and are correlated to both elevated and depleted mtDNA copy number [75, 94, 160, 231]. Meanwhile, although not the focus of this chapter, there is an entire subclass of mitochondrial disease, known as mtDNA depletion syndromes [222], which are exemplified by low levels of mtDNA.

### 12.1.4 The Mitochondrial Network and mtDNA

Mitochondria are double-membrane bound organelles that form a dynamic network throughout the entire cell, which is quite distinct from the typical textbook representation of mitochondria. Normally, the mitochondrial genome is organized into compact structures, known as nucleoids that contain one to two copies of the genome [110, 113], and which are evenly distributed throughout the mitochondrial network [82]. The balance between fusion and fission events determines the extent of the mitochondrial network, which can range between fused elongated tubules to smaller fragmented forms in response to energy demand and the health status of the cell [145, 234]. Notably, the dynamic regulation of the mitochondrial network is important for mitochondrial quality control and maintenance of mtDNA, and its impairment can lead to mtDNA depletion and accumulation of mtDNA deletions [24]. In particular, fission is important for mtDNA maintenance as it has been linked to replication of the mitochondrial genome [115]. In addition, fission is required to generate small mitochondrial fragments that can be degraded via mitochondrial autophagy (mitophagy), which is one way by which mtDNA turnover occurs [108].

### 12.1.5 Nuclear Encoded Mitochondrial Proteins

Whereas the mtDNA encodes for only 13 proteins, the remainder of the ~1000 to 1500 proteins that comprise the mitochondrial proteome are encoded in the nucleus, expressed in the cytosol and targeted to mitochondria [25, 220]. These proteins often include a mitochondrial localization sequence (MLS), an amphipathic helix located at the N-terminus, that ensures delivery and import into mitochondria [220]. Thus, both mitochondrial and nuclear genomes encode genes that are required to make fully functional mitochondria, which requires the coordinated expression of two distinct genomes [84]. Key regulators that control the coordinated biogenesis of mitochondria are the peroxisome proliferator-activated receptor γ transcription coactivators (PGC1) [52, 68, 171], which interact with various nuclear transcription factors and lead to an increase in the expression of numerous mitochondrial proteins, including those essential for mtDNA replication, transcription, and maintenance [68, 197].

In light of the coordination between mitochondria and the nucleus, it is not surprising that there is evidence for co-evolution between the nuclear and mitochondrial genomes [11], and that mtDNA haplotypes have different functional properties [154], which has important implications for human disease [224]. While mutations in either nuclear or mitochondrial encoded proteins genes can lead to disturbance in the mitochondrial function and cause mitochondrial disease, we will focus here on mitochondrial disease caused by mutations in the mitochondrial genome (mtDNA disease), which collectively represent ~70% of all mitochondrial diseases [237].

## 12.2 mtDNA Diseases

While the human mtDNA genome was first sequenced in 1981 [4], it was not until 1988 that the first pathogenic mtDNA mutations were reported. Two reports were revolutionary in how they caused researchers and clinicians to think about mitochondrial disease. The first, a missense mutation m.14459G>A converting a highly conserved arginine to a histidine at codon 340 in the ND4 subunit of complex I that is associated with Leber's hereditary optic neuropathy (LHON)

[227], and the second, an mtDNA deletion that is associated with mitochondrial myopathy and chronic progressive external ophthalmoplegia (CPEO) [79]. Since then, over 300 pathogenic mtDNA point mutations have been reported in the rRNA and tRNAs genes alone, as well as more than 300 mutations in protein coding genes and the D-loop region. In addition, over 100 large deletions ranging from 1000 to 8000 bp have been correlated to severe mitochondrial anomalies. An updated list of pathogenic mtDNA mutations is maintained at the Mitomap online database (https://www.mitomap.org) [120].

### 12.2.1 Genetics of mtDNA Diseases

With the growing list of mtDNA mutations that have been described to date, it has become evident that the pathogenicity of mtDNA mutations is very heterogeneous. While different mutations can have similar symptoms, it is also true that the same mutation can manifest with a variety of different phenotypes. For example, the most common mtDNA mutation, m.3243A>G, can cause MELAS (Mitochondrial Encephalopathy, Lactic Acidosis and Stroke-like episodes), and is also associated with diabetes, hearing loss, and retinopathy [123]. Meanwhile, the A1555G mutation is well studied with regards to an increased susceptibility to hearing loss [29]. This variability is one of the aspects that complicates the diagnosis of mtDNA disease. There are several factors that likely contribute to this phenotypic variability, which will be discussed in more detail below. For the purpose of simplicity, we will breakdown pathogenic mtDNA mutations into two classes, point mutations and large-scale deletions.

#### 12.2.1.1 MtDNA Point Mutations

The close proximity to reactive oxygen species (ROS) produced during cellular respiration, and the limited mtDNA repair mechanisms in mitochondria contribute to a higher susceptibility of mtDNA to develop mutations [54]. Point mutations are found throughout the mtDNA in rRNA, tRNA and protein-coding genes, as well as within the D-loop region and are associated with a broad array of symptoms [46, 73, 134, 147, 211, 215]. A list of the most abundant and well-studied mtDNA point mutations and their associated phenotypes can be found in Table 12.1.

**Table 12.1** Common mtDNA mutations. Genotype:phenotype correlations in the most common mtDNA point mutations and the common deletion of human mitochondrial genome

| Location | Affected gene | Syndrome | References |
|---|---|---|---|
| 1494 C>T | 12S rRNA | MID | [241] |
| 1555 A>G | 12S rRNA | MID | [218] |
| 3243 A>G | tRNA$_{Leu}$ | MELAS or MID or MHCM or MIDD or CPEO | [126, 155] |
| 3271 T>C | tRNA$_{Leu}$ | MELAS | [41] |
| 3460 G>A | ND1 (p.A52T) | LHON | [43] |
| 8344 A>G | tRNA$_{Lys}$ | MERRF | [127] |
| 8356 T>C | tRNA$_{Lys}$ | MERRF | [193] |
| 8363 G>A | tRNA$_{Lys}$ | MHCM or LS | [187] |
| 8993 T>G | ATP6 (p.L156R) | Asymptomatic or LS or NARP or complex V deficiency | [51, 125, 166] |
| 8993 T>C | ATP6 (p.L156P) | NARP or complex V deficiency or LS | [99, 172, 203] |
| 11,778 G>A | ND4 (p.R340H) | LHON | [30, 175] |
| 14,459 _>A | ND6 (frame shift) | LHON | [93] |
| 14484 T>C | ND6 (p.M64V) | LHON | [30, 80] |
| 1.9 kb deletion | From COII-COIII | KSS | [53, 177] |

The associated clinical presentations with each mutation are as follow: *CPEO* chronic progressive external ophthalmoplegia, *KSS* Kearns-Sayre syndrome, *LHON* Leber hereditary optic neuropathy, *LS* Leigh syndrome, *MELAS* mitochondrial myopathy, encephalopathy, lactic acidosis, and stroke-like episodes, *MERRF* Myoclonic epilepsy with ragged red fibers, *MID* Maternally inherited deafness, *MIDD* maternally inherited diabetes and deafness, *MILS* maternally inherited Leigh syndrome

#### 12.2.1.2 MtDNA Deletions

Deletions in the mtDNA are proposed to occur during repair of damaged mtDNA [107], or when mtDNA replication is impaired [152]. Due to its unique mode of replication, which results in long stretches of ssDNA, the mitochondrial genome is believed to be highly susceptible to large deletions. To this point, the vast majority of mtDNA deletions lie within the major replication arc [21] and are flanked by short direct repeats [107, 170]. For example the so called 'common deletion' is a 4977 bp deletion flanked by 13 bp repeats [45, 170]. A proposed mechanism leading to deletions is that following replication stalling, the replication machinery slips from one repeat to another skipping the intervening region [185]. In this regard, mutations in Polγ and Twinkle are known to cause an accumulation of mtDNA deletions [44, 230, 235]. In addition, double-strand breaks in the mtDNA have also been implicated in the generation of mtDNA deletions [107, 152, 208]. As such, most mtDNA deletions are sporadic, however cases of germline inheritance have been reported [157, 182]. As deletions affect several genes at once, patients with mtDNA deletions often have more severe pathologies. CPEO is a common clinical feature associated with mtDNA deletions, which can also be associated with additional symptoms in more complex syndromes such as Kearns-Sayre syndrome (KSS). KSS is characterized by pigmentary retinopathy, and at least one of the following: cerebellar ataxia, cardiac conduction defect, or elevated CSF protein [38, 53]. Meanwhile, Pearson Syndrome (PS) has an extremely variable phenotype affecting multiple systems and can change over time, showing either improvements and/or deterioration [53, 60].

#### 12.2.1.3 Heteroplasmy of mtDNA Mutations

The high copy number of mtDNA has important implications with regards to both the pathology and inheritance of mDNA mutations. The term heteroplasmy refers to cells that have a mixture of wild-type and mutant genomes. Meanwhile, cells that contain only one type of genome, either wild-type or mutant, are referred to as homoplasmic. Critically, the level of heteroplasmy can affect the disease phenotype and severity [148, 153]. Typically, symptoms are evident when the ratio of mutant mtDNA exceeds a certain threshold, which is usually in the range of 60–90% [3]. Below this pathogenic threshold, individuals have sufficient mitochondrial function for normal cellular metabolism and exhibit no symptoms [116]. However, as the level of heteroplasmy increases, the clinical presentation of the patients with the same mutation can vary dramatically. For example, the severity of symptoms associated with the m.8993 T>G mutation in the *ATP6* gene increases in line when the proportion of the mtDNA molecules carrying the mutation rises. At moderate levels of heteroplasmy (~80%) this mutation causes neuropathy, ataxia, and retinitis pigmentosa (NARP), while higher levels (95%) are associated with Leigh Syndrome, a severe form of mitochondrial disease caused by either nuclear or mtDNA mutations, which is characterized by progressive neurologic degeneration resulting in ataxia, bulbar palsy, with motor and intellectual developmental delay, lesions of brain-stem and basal ganglia, lactic acidosis and cerebrospinal fluid in blood [28, 29, 161, 180]. Additionally, levels of heteroplasmy can change over time, and also vary in different tissues, with high-energy demanding organs like the brain, muscles and liver more sensitive to mtDNA mutations. Finally, the issue of heteroplasmy has important relevance with regards to the inheritance of mtDNA mutations.

### 12.2.2 Inheritance of mtDNA

Unlike the nuclear genome, which receives 50% of its genetic material from each parent, mtDNA is exclusively maternally inherited [69], which is an important factor in using mtDNA haplogroups to study human evolution [26, 69]. Combined with the high copy number of mtDNA, maternal inheritance of the mitochondrial genome has important implications with regards to the pathology, transmission and inheritance of mtDNA mutations, especially in the context of heteroplasmy. Notably, it is estimated that 1 in 200 females carry pathogenic mtDNA mutations [57].

#### 12.2.2.1 The Genetic Bottleneck and mtDNA Inheritance

During the generation of primary oocytes there is a reduction in mtDNA copy number that creates a bottleneck effect where each oocyte contains a subset of mitochondria, estimated to be between 8 and 200 mtDNA genomes [116]. However, because wild-type and mutant mtDNA are randomly distributed during this process, this genetic bottleneck can result in oocytes that are either enriched or depleted of mtDNA mutations [18, 62, 105, 165, 195]. Upon oocyte maturation, the mtDNA content of these selected mitochondria is rapidly expanded to maintain mtDNA count, presenting another opportunity for mtDNA heteroplasmy to shift if there is uneven expansion of mutant or wild-type genomes. Despite the fact that mtDNA mutations can accumulate and be passed on through the germline, there is evidence for purifying selection in the female germline [200] (Fig. 12.2). However, this selection is clearly insufficient to completely prevent the transmission of pathogenic mutations.

Due to the random nature of mtDNA segregation and expansion, it is not possible to predict the level of heteroplasmy in any single oocyte, even if the heteroplasmy levels in the mother are known. Thus, there is no way to predict whether heteroplasmy levels in the offspring will be higher, lower, or the same as the mother [201, 211]. Importantly, such shifts in heteroplasmy can also lead to situations where an unaffected and undiagnosed mother with low non-pathogenic levels of mutant mtDNA can give birth to a child with a higher proportion of mutant mtDNA that leads to disease.

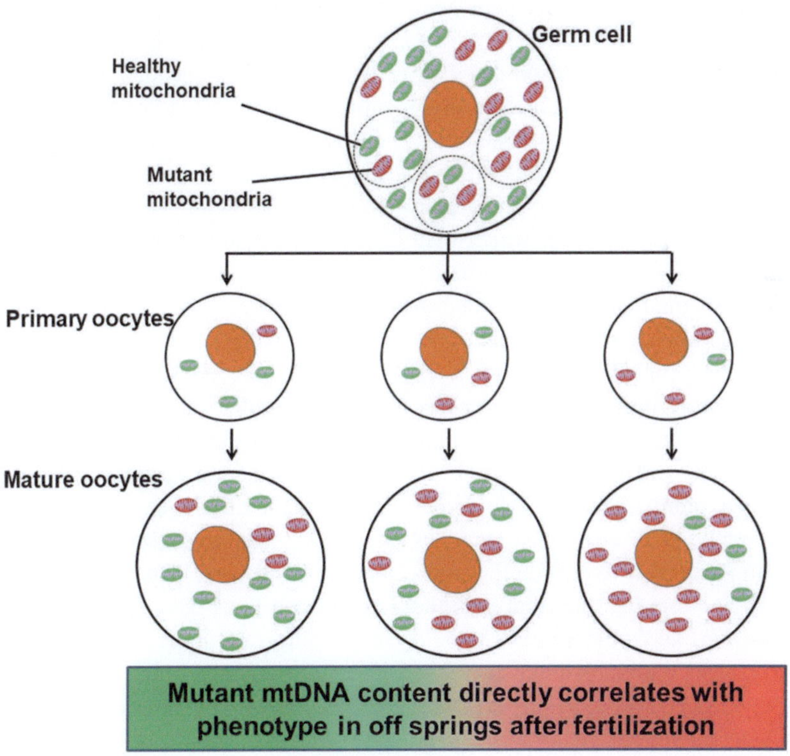


**Fig. 12.2** Generational heteroplasmy shifting. A subset of randomly selected mtDNA molecules are transferred during the production of primary oocytes, creating a genetic bottleneck where heteroplasmy levels can shift dramatically. The rapid replication and amplification of mtDNA in mature oocytes can also contribute to the random shift of mtDNA heteroplasmy. Together, these factors are responsible for the variable levels of mutant mtDNA that are maternally transmitted, which impacts the phenotypic severity of mtDNA disease

#### 12.2.2.2 Heteroplasmy Shifting

While the genetic bottleneck can explain large heteroplasmy shifts through the germline, it does not explain how changes in heteroplasmy occur over time, nor in different tissues, which has important implications for how mtDNA mutations accumulate and disease conditions. Studies performed in a heteroplasmic mouse that contains a mixture of NZB and BALB mouse mtDNA haplotypes, demonstrated how mtDNA heteroplasmy can shift in different tissues, even without the presence of pathogenic mutations [14]. Remarkably, in these mice the NZB genotype was abundant in liver and kidney, while the BALB genotype was predominant in blood and spleen [14]. The fact that the same patterns were observed repeatedly argues that these shifts were not due to random drift. Although the exact molecular mechanisms underlying these shifts remain unresolved, they are pertinent to the accumulation of mtDNA mutations and the pathogenesis of mtDNA disease.

One mechanism that has been proposed to account for shifts in heteroplasmy, is the preferential replication of one genome over the other. In this context, reports have shown that mtDNA with large deletions exhibit faster replication due to their smaller size [1, 168], which may provide a replicative advantage and explain how mtDNA deletions accumulate. Another proposal is that mtDNA deletions disrupt an as yet undefined transcriptional control mechanism such that increased transcription, which primes mtDNA replication, leads to the preferential replication of deleted mtDNA genomes [106]. However, whether point mutations also somehow offer a replicative advantage remains unknown. As an alternative to differential replication, it is also possible that differences in the turnover of wild-type and mutant mtDNA could explain heteroplasmy shifts. Nonetheless, the mechanisms regulating the selection and shifting of mtDNA heteroplasmy are not well understood at the molecular level, which imposes a challenge for diagnosis and treatment of mtDNA disease [22, 226].

### 12.2.3 Models for Studying mtDNA Mutations

Our ability to study how mtDNA mutations cause disease has been limited by a lack of techniques to specifically modify the mtDNA genome. In particular, a paucity of animal models for mtDNA disease has hampered the development of therapeutic approaches. Nonetheless, researchers have developed tools and approaches that have proven beneficial for studying mtDNA mutations.

#### 12.2.3.1 Cellular Models of mtDNA Disease

Mitochondrial cybrid cells, or cytoplasmic hybrids, are generated by fusing an enucleated cell lacking nuclear DNA (cytoplast) with a nucleated cell. The first cybrid cells were generated in 1975 as part of experiments that demonstrated chloramphenicol resistance could be transferred by enucleated cytoplasts [225]. It was later established that this resistance was mediated by the mtDNA-encoded large subunit of the mitochondrial ribosomal RNA [17]. Over a decade later, King and Attardi, modified this method by fusing cytoplasts with rho0 cells lacking mitochondrial DNA [102]. Because the resulting cybrid cell contained only mtDNA from the donor cytoplast, this powerful approach allowed researchers to study the effects of changes to the mitochondrial genome in cells with the same nuclear background. Given that nuclear modifiers are known to affect the severity of phenotypes of mtDNA disease [49, 132], the ability to generate various cybrid lines where the only genetic differences were due to the mtDNA was a key step to studying the cellular dysfunction caused by mtDNA mutations. Mitochondrial cybrid cell lines have also been instrumental in expanding our understanding of heteroplasmy, threshold effects, and compatibility between nuclear DNA and mtDNA [204]. Furthermore, the ability to transfer mitochondria set the stage for approaches to generate animal models of mtDNA, as well as mitochondrial replacement therapies, which will be discussed in detail later.

#### 12.2.3.2 Animal Models of mtDNA Disease

In vivo models of mtDNA disease are important for developing a better understand the pathophysiology that leads to the phenotypic variability associated with mtDNA mutations. In order to overcome the limitation of not being able to directly manipulate the mtDNA, researchers have introduced pre-existing mtDNA mutations found in mouse cell lines or somatic tissue into embryonic stem cells (ESC) or fertilized oocytes. However, the generation of such models is restricted to the limited number of naturally occurring pathogenic mtDNA mutations identified in mice. Nonetheless, several transmitochondrial mouse lines have been developed [59, 83, 141, 183, 184, 194]. Notably, the introduction of a chloramphenicol resistance mutation was the first demonstration that mtDNA mutations could be introduced into a mouse model [114, 131]. Additionally, the NZB/BALB heteroplasmic mouse that contains a mixture mtDNA haplotypes from NZB/BINJ and BALB/cByJ mouse strains has helped researchers study how shifts in mtDNA heteroplasmy occur [89], even in the absence of pathogenic mutations.

Alternatively, mouse models of mtDNA disease have also been developed by introducing mutations into the nuclear-encoded components of the mtDNA replication machinery, namely the mitochondrial DNA polymerase Polγ and the helicase Twinkle. The mtDNA Mutator mouse was created by introducing a mutation that reduces the 3′>5′ exonuclease activity of Polγ, resulting in diminished proofreading capabilities and subsequent accumulation of point mutations [109, 167, 214]. Meanwhile, the mtDNA Deletor mouse was created by introducing a human pathogenic mutation in Twinkle that causes CPEO and leads to the accumulation of mtDNA deletions [217]. While these models have been extremely useful in studying the selection of mtDNA mutations during embryonic development, the fact that they introduce multiple mutations makes it difficult to study genotype/phenotype correlations.

More recently, Kauppila *et al.* have developed a phenotype-driven approach that takes advantage of the Polγ Mutator mouse, where colonic crypts are screened for the clonal expansion of a single pathogenic mtDNA mutation [100]. This strategy was used to generate a mouse that harbours a heteroplasmic m.5024C>T mutation in the mitochondrial gene encoding for $tRNA_{Ala}$ gene, which is associated with a mild cardiac phenotype at older ages [6, 67]. This promising method will expand the number of pathogenic mtDNA mutations that can be studied in order to better understand mitochondrial disease.

## 12.3 Prevention and Treatment of mtDNA Diseases

Despite advances in our understanding of the genetic causes and the clinical presentations, there remains no definitive cure for mitochondrial diseases [71, 144]. Instead, most therapies are palliative and rely on supplementation with vitamins and antioxidants intended to boost the respiratory chain capacity [78] or reduce damage by reactive oxygen species [143]. Additionally, exercise has some benefit in controlling and relieving symptoms [20, 87]. However, this can be extremely challenging in individuals that already lack energy due to their mitochondrial dysfunction. These approaches only treat the symptoms, and targeted strategies that address the underlying dysfunction are required. In the context of mitochondrial diseases caused by mtDNA mutations, preventive measures that eliminate transmission of mutant mtDNA and avoid disease manifestations have been developed and will be covered herein. These range from genetic counseling to assisted reproductive therapies. In addition, we will discuss novel ways to shift the heteroplasmy levels and reduce the levels of mutant mtDNA below the pathogenic threshold as alternative approaches to treat mtDNA disease.

### 12.3.1 Preimplantation Genetic Diagnosis

For mothers diagnosed with mtDNA disease, or who are known to carry mtDNA mutations, genetic counseling can be somewhat informative [3, 169, 195, 221]. In the case of homoplasmic mutations, there will be a 100% chance of transmission. Meanwhile, when the mtDNA content is present in a heteroplasmic fashion, higher the levels of mutant mtDNA heteroplasmy are more likely to result in the offspring developing mitochondrial disease. However, as discussed previously, due to the random nature of mtDNA segregation, it is not possible to make definitive predictions as to the relative levels of heteroplasmy in the offspring.

One approach to potentially circumvent the randomness of mtDNA segregation is preimplantation genetic diagnosis (PGD), a procedure where genetic defects such as mtDNA mutations are identified in in vitro fertilized (IVF) embryos prior to implantation. In this procedure, fertilized eggs are grown until the blastocyst stage comprising 100–150 cells (~5 to 6 days), and a single cell from the developing embryo is removed for genetic analysis of heteroplasmy levels prior to transplanting the embryo into the uterus [213]. In theory, this procedure allows the selection of embryos with low levels of mutant mtDNA for implant into the mother. However, due to the inability to accurately extrapolate heteroplasmy from a single cell, and the risk of subsequent heteroplasmy shifts, there is no guarantee that PGD will eliminate the risk of disease transmission from mother to child. Furthermore, in situations where the mother carries mtDNA mutations below the pathogenic threshold and is not diagnosed with mtDNA disease, there is no reason to suspect or perform PGD, such that mutant mtDNA can be passed on and possibly manifest in disease. Finally, PGD does not help in cases of homoplasmic mtDNA mutations in affected mothers. Thus, even if PGD was completely effective in eliminating the transmission of mutant mtDNA, it would not be able to eliminate development of mtDNA disease.

### 12.3.2 Mitochondrial Replacement Therapy (MRT)

Taking advantage of the fact that the mtDNA is a distinct entity from the nuclear DNA, it is possible to physically separate these two genomes, as first demonstrated by studies with mitochondrial cybrids. As such, to prevent the maternal transmission of mutant mtDNA genomes, methods have been developed that allow replacement of the maternal mitochondria harbouring mtDNA mutations with mitochondria containing healthy mtDNA genomes. These procedures, known generally as mitochondrial replacement therapy (MRT), or mitochondrial donation, retain the nuclear genetic characteristics of both parents, but eliminate the pathogenic mtDNA mutations. There are two main approaches that are currently being used for MRT, pronuclear transfer (PNT) and mitotic spindle transfer (MST), which differ based on whether the nuclear and mitochondrial material are transferred pre or post fertilization (Table 12.2).

#### 12.3.2.1 PNT: Transfer of Nuclear Material Post Fertilization

In PNT, paternal sperm are used to fertilize eggs from both the mother carrying mtDNA mutations

**Table 12.2** Key terms related to mitochondrial replacement therapy

| | |
|---|---|
| **Oocyte**: | A female germ cell that can undergo meiosis and form a haploid ovum (egg) |
| **Zygote**: | The cell formed when a female ovum is fertilized by a male sperm, which will develop into the embryo |
| **Embryo**: | An early stage of development, after a fertilized egg (zygote) begins cell division |
| **Blastocyst**: | A structure in early embryonic development that contains a cluster of cells usually 100–150 cells |
| **Spindle**: | A microtubule structure that forms during *mitosis* or meiosis in order to segregate chromosomes into daughter cells |
| **In vitro fertilization**: | A medical procedure whereby an ovum is artificially fertilized by sperm outside the body |

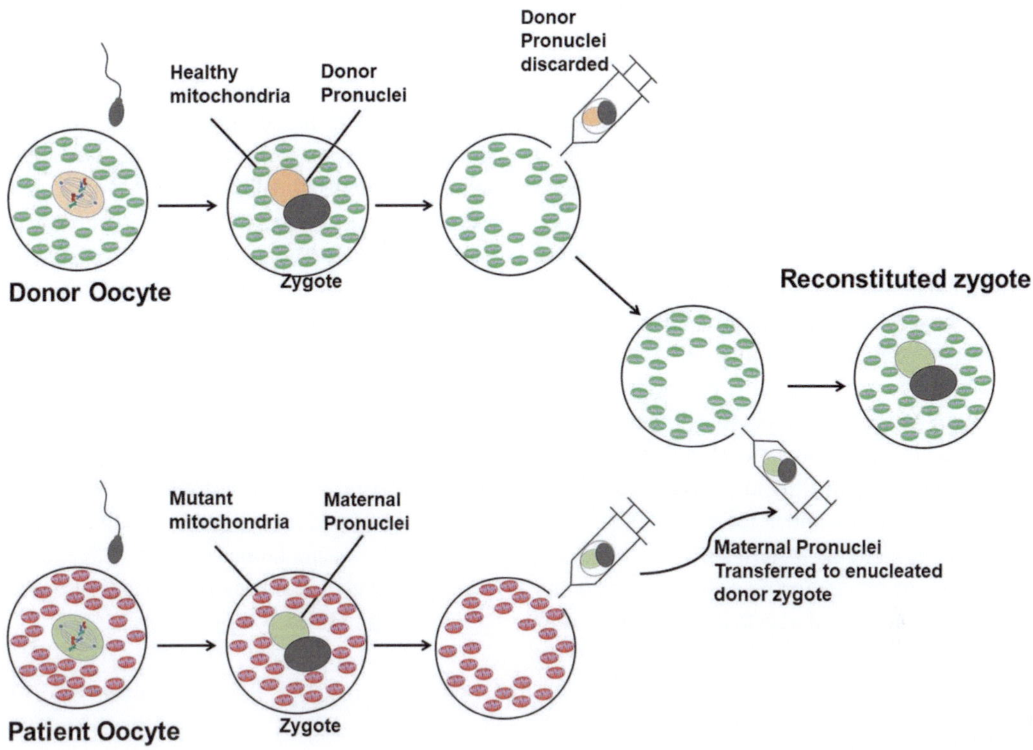


**Fig. 12.3** Pronuclear transfer (PNT). Oocytes from both the mother harbouring mtDNA mutations and the healthy donor are fertilized using paternal sperm. The nuclear genetic material is removed post-fertilization from both zygotes. The nuclear genetic material from the patient zygote is then transferred to the enucleated donor zygote, resulting in a reconstructed zygote that contains the nuclear genetic material from the intending parents and healthy mitochondria from the donor

and a donor with wild-type mtDNA [81]. Both fertilized oocytes are allowed to develop to the early zygote stage when the pronuclei become visible. During the G2 phase of the first mitotic cell cycle, the pronucleus from the donor zygote containing healthy mitochondria is removed using micromanipulation equipment, discarded, and replaced with the isolated pronucleus from the mother. The transfer of the pronucleus is performed through either electric pulses or via artificial viral envelope-lipososome delivery vectors that are derived from inactivated hemagglutinating virus of Japan as a transfer vector [3, 96, 135]. The final zygote thus contains nuclear DNA from both parents and healthy mtDNA from the donor (Fig. 12.3).

In 2005, PNT was first tested in the 'mito-mouse' mouse model that accumulates a large scale deletion in the mtDNA, and prevented transmission of the disease phenotype to the progeny [174]. PNT approaches have also been applied to fertilized human zygotes. Studies using in vitro fertilized PNT zygotes showed that embryos developed into healthy blastocysts that were similar to the controls in terms of gene expression and aneuploidy levels [81]. Notably however, a minimal amount of cytoplasm was extracted with the pronuclei, which resulted in a 2.0% of carryover of mutant mtDNA [47, 81]. Subsequently, modifications to optimize the PNT procedure have been developed. Specifically, the timing of the transfer was found to be optimal when performed directly following completion of meiosis, resulting in improved blastocyst quality and survival. Meanwhile, the use of cryopreserved patient oocytes rather than freshly extracted healthy donor oocytes was found to lower the carryover of maternal mtDNA [81].

#### 12.3.2.2 Metaphase Spindle Transfer

An alternative to PNT is metaphase spindle transfer (MST), in which transfer of the maternal nuclear DNA occurs before fertilization. In MST, oocytes from the mother and the healthy donor are closely monitored until they reach the metaphase II stage, where the chromosomes align on the metaphase plate and are attached to spindles. The nuclear material and attached spindles are then removed from both oocytes [207]. Notably, in order to avoid chromosomal DNA loss, DNA staining is required to visualize metaphase plate alignment and ensure proper attachment of chromosomes to the spindle. An enucleated healthy oocyte containing healthy mtDNA then serves as the cytoplasmic recipient for the nuclear DNA and spindles from the maternal oocyte, which are transferred by either electric pulses or viral delivery. The reconstructed oocyte is then fertilized by the paternal sperm such that the final embryo contains the nuclear material of the parents but mtDNA from the healthy donor (Fig. 12.4).

As a proof of concept, MST was initially shown to work in higher primates using monkey oocytes [207], resulting in the birth of three progeny that remained healthy and developed normally for at least 3 years [206]. Importantly, there was <1% carryover maternal mtDNA, suggesting that this approach may be superior to PNT with regards to the risk of transferring mutant mtDNA. MST was also successful using cryopreserved monkey oocytes [206], which is an important consideration as it allows the retrieval of oocytes from both the affected female and mitochondrial donor at different times and facilitates the process by avoiding issues of synchronization. In studies using human oocytes, MST fertilized eggs developed normally to blastocysts and produced normal ESC [206].

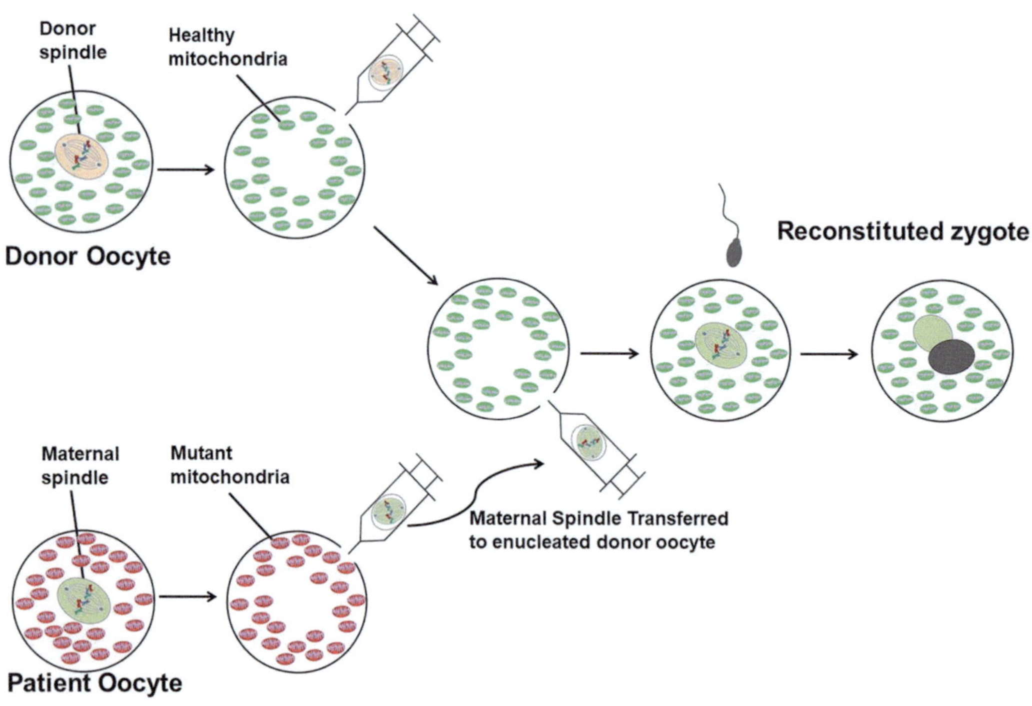


**Fig. 12.4** Maternal spindle transfer (MST). Oocytes from both the mother harbouring mtDNA mutations and the healthy donor are allowed to reach metaphase II stage where the chromosomes align on the equatorial plate. The nuclear genetic material from patient and donor oocytes is removed pre-fertilization. The nuclear genetic material from the patient oocyte is then transferred to the enucleated donor oocyte, resulting in a reconstructed oocyte that is allowed to be fertilized using paternal sperm. The resulting zygote contains the nuclear genetic material from the intending parents and normal mitochondria from the donor

#### 12.3.2.3 MRT in the Clinic

The clinical use of MRT is still in its infancy, as only a few cases have been reported to date. The first report of a child born following MST was in early 2015 [239]. In this case, the reportedly healthy mother, carrying an mtDNA mutation, had several miscarriages and gave birth to two children, both of whom died from Leigh Syndrome. Following MRT, the pregnancy yielded a baby boy that was reported healthy at 7 months of age. As the clinical application of MRT is banned in the United States, the procedure was performed under the supervision of an American team in Mexico, where at the time there were no regulations regarding MRT. Notably, the United Kingdom is leading the way with respect to developing regulations around MRT. In 2017, the United Kingdom legally approved PNT to reduce the risk of serious hereditary mtDNA diseases [85, 162], and in early 2018, the Human Fertilisation and Embryology Authority approved two cases for women diagnosed with MERRF (Myoclonic epilepsy with ragged red fibers) to proceed with MRT at Newcastle's Fertility Centre. Finally, an unpublished clinical trial in Ukraine was reported in the media, apparently producing a healthy baby using the PNT technique [162]. Although the initial reports of MRT are sparse, they are encouraging. However, it is important to note the small sample size and the relative young age of the born children, which currently precludes long term evaluations. Thus, long term monitoring with a larger cohort will need to be maintained before the full benefits of MRT can be properly evaluated.

#### 12.3.2.4 Limitations of MRT

A concern for both PNT and MST is the low level of mtDNA carryover (1–2%) that has been observed in embryos [162]. This carryover is a result of a small portion of maternal mitochondria that are transferred with the nuclear material. While this low level of mutant mtDNA is not expected to cause disease, and is certainly far lower that what would likely be inherited normally without MRT, there is still a chance that the mutant mtDNA could accumulate to pathogenic levels. This issue emerged during the initial trial of MST in human oocytes, where there was a reversion to the maternal mtDNA haplotype in ~15% of the ESC derived from embryos [97]. In these cases, the donor mitochondria appeared to be replaced by the maternal mtDNA that was carried over during the nuclear transfer. It was speculated that a haplotype-specific polymorphism led to preferential replication of the maternal haplotype. This finding raises the possibility that reversal to the mutant mtDNA could occur in some MRT children. However, matching the haplotypes of the mother and donor should prevent such reversion. Nonetheless, long term monitoring will need to be maintained, which would have been the case anyhow for a child suspected of mtDNA disease.

The issue of haplotype mismatch is also a concern with regards to potential incompatibility between the nuclear and mtDNA genomes. For example, studies have indicated that haplotype of the donated mitochondria can affect expression of nuclear genes [164]. Nonetheless, in studies using monkey oocytes, there was no effect of the different haplotypes on the development of embryos following MST, suggesting that the haplotype mismatch does not negatively impact development [206]. Regardless, concerns about haplotype compatibility can be easily addressed by matching the haplotype of the donor to that of the mother.

A concern specific to MST, was that ~50% of the embryos produced from human oocytes failed to complete meiosis and died due to triploidy [97, 206]. Since triploidy was not evident in the primate studies [207], this finding in human emphasizes differences between species and establishes the need to carefully monitor and evaluate the reconstituted embryos prior to implantation. Meanwhile, a potential concern for both approaches is the use of hemagglutinating virus of Japan to facilitate the transfer of the nuclear material. The use of this virus-derived protein poses safety concern as it has not yet been fully tested in human oocytes or embryos [3]. However, it is notable that there was no presence of any viral genomes detected in either derived ESC or placental tissue of MST offspring [207]. Finally, as MRT is often combined with assisted reproductive techniques, which can interfere with nuclear DNA methylation in oocytes and result in syndromes like Beckwith-Wiedemann syndrome [130], alterations to epigenetic modifications are also a concern for MRT. However, this phenomenon has not yet been studied in the context of MRT [3].

#### 12.3.2.5 MRT Ethical Considerations

While both PNT and MST rely on reconstituting embryos for implantation, due to the fact that MST is performed using unfertilized oocytes, it may be the preferred ethical option as it does not result in the destruction of fertilized oocytes [232]. One issue with both approaches pertains to the low levels of mutant mtDNA transferred from the affected mother. Due to the persistence of mtDNA mutations, there remains a risk of developing mtDNA disease at some point if the level of mutant genomes increases. Furthermore, in the case of female offspring, mtDNA mutations can be passed to future generations. In this regard, in the single case of MST in humans, doctors chose to implant a male embryo to avoid future transmission of the disease [239]. Such gender selection in MRT procedures raises additional ethical considerations. While no one can deny that MRT offers hope for families suffering from mtDNA diseases, it comes with ethical concerns regarding genetic manipulation and the risk of the technology being abused. As such it is essential that proper regulation be put in place.

#### 12.3.2.6 General Considerations for MRT

There is currently no consensus with regards to which procedure, PNT or MST, is preferable for MRT. However, there are several issues that should be considered. Technically, PNT is easier to perform than MST. For example, although the metaphase II spindle is smaller and theoretically easier to remove than an intact nucleus, MST requires DNA staining for visualization prior to transfer in order to guarantee that the full set of nuclear chromosomes is transferred, and that no anomalies arise from loss of nuclear material. Meanwhile, for PNT, pronuclei are larger and easier to visualize. Yet, the manipulation of pronuclei is more difficult as they are more susceptible to physical damage. Another consideration is that PNT is more compatible with the available assisted reproductive procedures currently in use, which makes it a more feasible option to operate.

Despite the encouraging outcomes of MRT to date, long term follow up is required to confirm the efficacy of these techniques, as consequences might not be evident until later in life. Additionally, it is likely that MRT approaches can be further optimized to improve their efficacy. In particular, improvements to reduce the carryover of mutant mtDNA, and avoid abnormal fertilization would be beneficial. Nonetheless, both PNT and MST techniques are potential solutions for families where the mother is known to carry mtDNA mutations. However, it is important to remember that mtDNA disease can still be inherited from an asymptomatic mother harbouring mutant mtDNA, and can also still occur due to the segregation/expansion of de novo mutations in the germline. Thus, even if MRT is adopted more generally to prevent mitochondrial disease, alternative treatment approaches will still be required.

### 12.3.3 Shifting mtDNA Heteroplasmy

An attractive strategy to treating mtDNA diseases is to reduce the number of mutant mtDNA copies below the pathogenic threshold. In order to accomplish such a shift in heteroplasmy, researchers have developed several approaches. One of the most promising of these approaches is predicated on targeting DNA nucleases to the mitochondrial matrix in order to specifically cleave mutant mtDNA. As mammalian mitochondria lack classic homologous recombination for double strand break repair [111, 140, 151], cleavage of mutant mtDNA leads to the selective depletion of these genomes, which are degraded by components of the mtDNA replication machinery upon linearization [140, 149]. Although this cleavage and degradation leads to an initial depletion of mtDNA, cells can survive

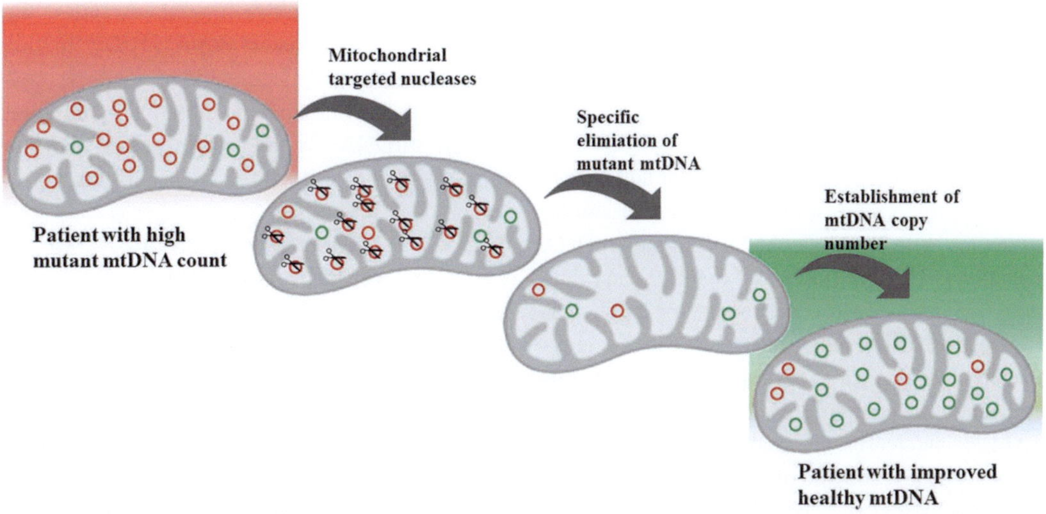


**Fig. 12.5** Heteroplasmy shifting via mitochondrial-targeted nucleases. Mutant mtDNA molecules are specifically recognized and degraded by mitochondrial-targeted nucleases, such as restriction endonucleases (*REs*), Zinc finger nucleases (*ZFNs*), and transcription activator-like effector nucleases (*TALENs*). Nucleases in the mitochondrial matrix will then specifically recognize mutant mtDNA sequences and create double-strand breaks. The linearized mtDNA molecules containing mutations are subsequently eliminated, creating a transient reduction in mtDNA copy number. Finally, mtDNA copy number is reestablished as the remaining mtDNA molecules proliferate, repopulating the mitochondria with a higher proportion of the wild-type mtDNA. The end result is a beneficial reduction heteroplasmy levels of the mutant mtDNA and an improved bioenergetic profile

with lower levels of mtDNA, and can re-establish normal copy number (Fig. 12.5). Thus, following specific depletion of mutant mtDNA and subsequent repletion, cells can be enriched for wild-type mtDNA [211]. Here we will focus primarily on several approaches to target mutant mtDNA via endonuclease-mediated cleavage and their applicability to human mtDNA disease.

#### 12.3.3.1 Targeted Restriction Endonucleases

Restriction endonucleases (RE) are naturally occurring endonucleases that recognize and cleave specific DNA sequences. Like most proteins, REs can be targeted to the mitochondria by adding a mitochondria targeting sequence, and are able to cleave mtDNA both in cultured cybrid cell lines as well as in animal models. The first example that targeting restriction endonucleases to mitochondria could cleave mtDNA and shift heteroplasmy was performed in mouse-rat cybrid cells; where only mouse mtDNA contained PstI restriction sites [199]. In these cells, targeting PstI to mitochondria led to a significant decrease in the mouse haplotype.

In addition to cell culture models, heteroplasmy shifting via restriction endonculeases has also been successful in mice. Using the NZB/BALB chimeric mouse strain, both mito-*Apa*LI [15] and mito-*Sca*I [7] effectively shifted heteroplasmy by reducing the BALB haplotype containing one *Apa*LI and three *Sca*I restriction sites that are absent in NZB. This work in an animal model was an important step forward, as it demonstrates that the initial depletion caused by removing mutant mtDNA does not cause serious side effects. To show that this approach could also effectively target mutant mtDNA where there is only a single base pair difference, a mitochondrial targeted *Sma*I restriction enzyme was effectively used to shift heteroplasmy in cybrid human cell lines containing the m.8993T>G mutation in *MT-ATP6,* which introduces a *Sma*I restriction site [209].

Although there are many known REs with defined recognition sequences, it is exceedingly rare that a pathogenic mutation in mtDNA would introduce a unique restriction site that would allow specific targeting by a RE. In addition, another concern is that REs might cause off-target cuts in the nuclear genome if the protein is mislocalized [148]. Nonetheless, these initial studies have clearly demonstrated that targeted cleavage of mutant mtDNA can successfully shift mtDNA heteroplasmy in both cellular and animal models, providing proof of principle that similar strategies will be applicable to human mtDNA mutations. Thus, in order to expand the applicability of this approach, recent research has focused on using nucleases that can be engineered to recognize a desired sequence. To this end, our understanding of how Zinc-finger and TALEN nucleases recognize specific DNA sequences has allowed us to rationally design these nucleases to target desired sequences in mtDNA.

#### 12.3.3.2 Zinc-Finger Nucleases

Zinc-finger nucleases (ZFNs) have been engineered and used for a number of molecular genetics approaches [156]. ZFNs are composed of a DNA- binding region that can be tailored to recognize a desired sequence, typically 9–18 nucleotides, which is fused to the non-specific endonuclease *Fok*I that causes a nick in the DNA strand at a specific distance from the DNA binding region [9, 63]. While a single DNA nick can be repaired, even by the limited mtDNA repair machinery in the mitochondria, when two nicks are introduced in close proximity using two distinct ZFNs, this results in a double-strand break [16, 138, 146], which is irreversible in mitochondria. Importantly, the requirement for two nucleases reduces the chance of off-target cleavage. Initial studies using a ZFN targeted to the mitochondrial matrix (mtZFN) were successful in reducing the heteroplasmy levels in cultured cells harbouring either the m.8993T>G mutation [66, 137, 138] or the 4977 bp common deletion [66]. However mtZFNs were initially reported to have toxic effects that reduced cell survival [138], as well as off-target cleavage in the nucleus [66, 137]. To overcome this limitation, Gammage *et al.*, were able to show that off-target binding and toxicity could be reduced when expression levels of the mtZFNs were carefully controlled [64]. MtZFNs have also been tested in a heteroplasmic mouse model carrying the m.5024C>T mutation in $tRNA_{Ala}$ [67]. In this work, the adenoviral delivery of mtZFNs was optimized, and shown to be effective in reducing mutant mtDNA in the heart, along with an improvement in mitochondrial respiration.

#### 12.3.3.3 Transcription Activator-Like Effector Nucleases

TALENs, or transcription activator-like effector nucleases, are comprised of a series of 34 amino acid polypeptide repeats. These repeats are capable of specifically detecting and binding a single DNA base pair, which are fused to *Fok*I restriction endonuclease [10, 77, 88]. The order of these repeats can be rationally designed to specifically bind to a particular DNA sequence. As with ZFNs, a double strand break can be produced when two distinct TALENs bind on opposing complementary DNA strands. TALENs have also been successfully targeted to the mitochondria (mitoTALENs) using a mitochondrial targeting sequence [77], and were able to shift the hetreoplasmy in cybrid cells targeting mtDNA containing the common deletion [10], as well as the m.8344A>G mutation associated with MERRF and the m.13513G>A mutation associated with Leigh Syndrome [77]. Functional evaluation showed that these mitoTALEN treated cells had improved respiratory enzyme activity and overall respiration capacity. MitoTALENs were also able to lower the mutant mtDNA load in immortalized patient cells from patients with NARP harbouring the m.9176T>C mutation [163]. Similar to both restriction endonucleases and ZFNs, mitoTALENs have also been successful in shifting mtDNA heteroplasmy in mouse models [6, 163], setting the stage for future clinical trials in humans. Most recently, Bacman *et al.* injected mitoTALENs via adenovirus directly into the muscles of mice carrying m.5024C>T mutation in $tRNA_{Ala}$ [6]. The wild type $tRNA_{Ala}$ level was restored in the injected mice indicating that this strategy has the potential to rescue mitochondrial disease phenotypes. Combined with the recent animal studies using mtZFNs [67], this work with mitoTALENs provides proof of principle for the use of targeted nucleases as a therapeutic tool to reduce levels of mutant mtDNA via simple intramuscular or intravenous injections.

#### 12.3.3.4 Germline Editing Via Targeted Nucleases

In addition to the potential application of using targeted nucleases to shift mtDNA heteroplasmy in mtDNA disease patients, another application is to target mutant mtDNA in the germline, in order to prevent transmission. In this regard, restriction endonucleases have been shown to reduce the abundance of the BALB haplotype in NZB/BALB mice when oocytes or one-cell embryos were targeted [163]. Notably implantation of these treated blastocysts resulted in the birth of normal pups with significantly reduced levels of the BALB haplotype, providing evidence that nuclease treated oocytes can go on to develop normally. Meanwhile, mitoTALENs were also effective in shifting heteroplasmy in oocytes reducing the NZB haplotype in NZB/BALB oocytes [163]. In addition, the use of mtZFNs to target mtDNA mutations in one day mouse embryos has also been proposed [133]. Despite the promise of using mtZFNs and mitoTALENs to target mtDNA mutations in mouse embryos, additional studies are required to confirm that mitoTALEN and mtZFNs treated embryos develop normally and maintain reduced heteroplasmy. Finally, with regards to human studies, mitoTALENs were effective in reducing two different pathogenic mtDNA mutations in human oocytes [163]. Although additional optimization and testing of this approach is required to further reduce the levels of mutant mtDNA, advantages over MRT include a simpler and less traumatic treatment of oocytes, as well as no requirement for a healthy donor oocyte, which bypasses some of the ethical issues of MRT. Thus, the use of targeted endonucleases to reduce the abundance of mutant mtDNA in the germline offers promise as an alternative to MRT.

#### 12.3.3.5 General Limitations for ZFN and TALENs

While TALEN and Zn-Finger nucleases have been used successfully to target, cleave and eliminate specific mtDNA sequences, they face several limitations. First, two separate proteins must be designed and delivered to mitochondria in order to provide specificity and generate a double-strand break. While this requirement reduces off-target cleavage, it also requires additional optimization. Second, the large size of these proteins presents a potential challenge when it comes to adenoviral delivery, which is the major delivery method used so far to deliver ZFN and TALENs [77]. These limitations could potentially be overcome by using other engineered nucleases such as the recently described mitoTev-TALE nuclease [150], which uses a much smaller Tev monomeric nuclease instead of *Fok*I. Using cybrid cells harbouring m.8344A>G, this smaller mitoTev-TALE nuclease, which can be easily delivered via adenovirus vectors, has also been shown to reduce heteroplasmy similar to ZFNs and TALENs.

While initial results with NZB/BALB mice suggest that combining targeted nuclease cleavage of mutant mtDNA in the germline could further reduce the maternal transfer of mutant mtDNA, there are potential complications. Notably, as mature oocytes do not replicate their mtDNA, targeting the mtDNA is expected to lower the mtDNA copy number [223], which could be a potential complication. For example, though mouse embryos with low mtDNA copy number (due to loss of the mtDNA maintenance protein TFAM) were shown to develop normally prior to implantation, they failed to implant properly [223]. However, the extent of mtDNA depletion would depend on the initial levels of heteroplasmy in the embryo and expression of the nucleases. Thus, further studies are required to understand how mtDNA depletion affects implantation, and how the use of targeted nucleases can be optimized for treating embryos.

A final limitation of targeted nucleases as a general methodology is that there are hundreds of pathogenic mtDNA mutations that have been described to date, and each one would require design, optimization and validation of a specific set of targeted endonucleases. As such, approaches such as CRISPR/Cas9, which would facilitate the design and targeting of multiple mtDNA mutations, could be more broadly applicable to mtDNA disease.

#### 12.3.3.6 CRISPR (Clustered Regularly Interspaced Short Palindromic Repeats) Technology

The CRISPR/Cas9 endonuclease system has gained notoriety for ease of design and adaptability, and would be an ideal tool to target the growing list of pathogenic mtDNA mutations as well as potentially circumvent some of the limitations associated with TALEN and ZFN nucleases. The CRISPR/Cas9 system requires the presence of two components: a nuclease Cas9 and a guide RNA. The two components form a complex where the guide RNA complements a selected DNA region and directs the nuclease Cas9 to the DNA site in order to introduce a double strand break. However, despite the potential advantages offered by using a CRISPR/Cas9 approach, to date, there have been no convincing reports of using the CRISPR-Cas9 system to target mtDNA [65, 162].

While it is possible to target the Cas-9 nuclease to the mitochondria using a mitochondrial targeting sequence [65, 91, 121], the primary limitation of applying CRISPR to mtDNA appears to be an inability to import the guide RNAs required for sequence specificity into mammalian mitochondria [65]. Although import of RNAs into mitochondria does occur, the exact mechanisms regulating this process are not well characterized. Notably, mammalian mitochondria can import a variety of small RNAs such as the 5S rRNA required for translation [58, 124, 233], the MRP RNA [33] and RNAse P RNA [159] required for RNAse activity, microRNAs [240], and even the telomerase RNA TERC [37]. As mammalian mitochondria express a full set of tRNAs required for mitochondrial translation, they do not require a tRNA import pathway [4]. However, tRNA import is required in several organisms including yeast [210, 220], and marsupials [55]. Notably, yeast tRNAs can be imported into mammalian mitochondria both in vitro and in vivo [58, 103], further demonstrating that mammalian mitochondria maintain the ability to import RNA molecules. However, while methods have been developed to import exogenous RNAs into mammalian mitochondria by attaching them to RNA import sequences from known mitochondrial RNAs [56, 228, 236], whether these methods can deliver a functional gRNA to mitochondria remains to be determined. Regardless, to date there have been no convincing reports that a CRISPR-CAS9 approach can efficiently shift mtDNA heteroplasmy.

### 12.3.4 Alternative Approaches to Shifting mtDNA Heteroplasmy

Although the use of targeted nucleases may be approaching clinical use, many additional ways to shift mtDNA heteroplasmy have been proposed, which may also eventually prove effective. However, as these approaches remain in the early stages of development and many have not yet been verified independently, they will only be briefly discussed here.

#### 12.3.4.1 Elimination of Mutant mtDNA

Although it is clear that mtDNA heteroplasmy can shift over time and in different tissues, the mechanisms regulating these shifts remain unknown. Nonetheless, several studies have shown that certain treatments are able to reduce the load of mutant mtDNA heteroplasmy in cultured cells, offering promise for patients with mtDNA disease. One of the first studies to show that heteroplasmy could be shifted in cultured cells, combined the antibiotic oligomycin with growth media containing galactose as the energy source, leading to lower levels of the m.8993T>G mutation [129]. However, given that oligomycin is a mitochondrial toxin that inhibits the ATP synthase, it is not a viable therapeutic option.

More promisingly, treatment of m.11778G>A cybrid cells with the mTOR inhibitor rapamycin resulted in a striking decline in mutant mtDNA levels and restoration of ATP levels [48]. Notably, rapamycin treatment has been shown to be beneficial in a mouse model of Leigh Syndrome caused by deficiency of the nucleus-encoded Ndufs4 protein [92] and a mouse model of mtDNA depletion [192]. However, rapamycin has not yet been tested in animal models of mtDNA mutations. The benefits of rapamycin are thought to be via an increase in mitophagy, one way by which cells can selectively degrade dysfunctional mitochondria [101]. Presumably, mitochondria harbouring mutant mtDNA are more likely to exhibit dysfunction and are thus more likely to be removed by autophagy [216]. Notably, overexpression of the mitophagy mediator Parkin also promoted selection against the mutant mtDNA in cybrid cells [202]. Meanwhile, work in *C. elegans* demonstrated a role for Parkin in regulating mtDNA heteroplasmy [219]. However, it should be noted that certain cell lines harbouring mutant mtDNA express lower levels of Parkin, which may reflect an adaptive response and may explain why normal mitophagy is unable to prevent the accumulation of mutant mtDNA genomes [70]. Nonetheless, the search for methods to modulate mitophagy and promote the removal of mutant mtDNA is a promising avenue for potential treatments of mtDNA disease.

Finally, treating cells with β-hydroxybutyrate, a metabolic product of the ketogenic diet, was shown to reduce the abundance of a 1.9 kb deletion in heteroplsmic cybrid cells derived from a KSS patient [173]. The ketogenic diet is high fat/low carbohydrate diet that is relatively easy to implement and has shown promise for mitochondrial disease [2, 98]. Nevertheless, despite the benefits of the ketogenic diet for treating the Twinkle Deletor mice, there was no change in the abundance of mtDNA deletions [2], suggesting the benefits were independent of any shift in heteroplasmy. However, it should be noted that these mice continue to generate deletions, which might explain why no shift in heteroplasmy was observed. Finally, it remains possible the ketogenic diet might also prove beneficial for other mtDNA diseases, even if it does not shift heteroplasmy levels.

#### 12.3.4.2 Mitochondrial Delivery

Another strategy to shift mtDNA heteroplasmy is simply to add exogenous healthy mitochondria. In theory, such mitochondrial delivery can be accomplished via different routes. One approach is the use of mesenchymal stem cells (MSCs), which can deliver healthy mitochondria to cells with severe mitochondrial dysfunction through formation of tunneling nanotubes [40, 90, 117, 196]. In co-culture experiments, MSCs have been shown transfer mitochondria and rescue function in rho0 cells completely lacking mitochondrial function. However, when cultured with cells carrying either m.3243A>G mutation or the 4977 bp deletion [40], MSC transfer of mitochondria was not observed, suggesting there are limitations to this approach. Recently, it was also shown that isolated mitochondria could be delivered to cultured cells using cell permeable peptides [118], and improvements in mitochondrial function in both MERRF and MELAS cell lines were reported [34–36]. However, this approach requires independent verification. Moreover, in addition to potential delivery issues of mitochondria in vivo, it is unclear whether isolated mitochondria are stable enough to make this a viable therapeutic tool.

### 12.3.5 Alternative Approaches to Treat mtDNA Disease

#### 12.3.5.1 Allotopic Expression of mtDNA Genes

As most mitochondrial proteins are encoded in the nucleus, an alternative to targeting the mtDNA directly is to express wild-type mitochondrial genes allotopically using the standard nuclear expression system. This approach has been used to express both proteins and tRNAs normally encoded in the mtDNA. For example, nuclear expression and mitochondrial targeting of NADH dehydrogenase subunit 4 (*MTND4*) improved mitochondrial function in cells carrying the m.11778G>A mutation that causes LHON [76]. The same strategy was also used to express ATP6 and improve ATP synthesis in cells homoplasmic for the m.8993T>G mutation causing NARP syndrome or Leigh Syndrome [128]. Theoretically, while allotropic expression could be applied to any of the 13 mtDNA-encoded proteins, these proteins are highly hydrophobic, which could impair proper import and assembly

[179, 212]. Meanwhile, cytoplasmic expression of tRNAs targeted to mitochondria using the *H1* RNA import sequence has been shown to rescue translation defects in MERFF and MELAS cybrid lines [228].

#### 12.3.5.2 Upregulation of Mitochondrial Biogenesis

Promoting mitochondrial biogenesis via PGC-1 is another mechanism proposed to increase mitochondrial function in cells harbouring mtDNA deletions [52, 68, 171]. In the context of mtDNA disease, the basic idea is that the average mitochondrial output is insufficient to meet the cells energy requirements. However, simply making more mitochondria, even if this includes dysfunctional mitochondria with mutant genomes, may be sufficient to meet cellular energy demands. Notably, overexpression of PGC-1α and/or PGC-1β was shown to improve mitochondrial respiration in patient fibroblasts with complex III or IV deficiency, as well as in cybrids harbouring a MELAS mutation [198]. In addition, overexpression of PGC1α in muscles of Polγ mutator mice [52] and aged mice [68] resulted in reversion of pathological phenotypes. While genetic overexpression of PGC-1 is not a viable therapeutic approach, there are several well-studied drugs that are pharmacological activators of PGC-1 activity, including pioglitazone, rosiglitazone, fenofibrate, metformin and bezafibrate among others [13, 171]. Notably, the utility of bezafibrate to improve mitochondrial function in several mouse models of mitochondrial disease has been reviewed in detail [104]. However, it should be noted that activation of PGC-1 can alter metabolism in unpredicted ways, such as producing high levels of ROS, which may cause deleterious effects [5].

#### 12.3.5.3 Hypoxic Treatment

It was recently shown that hypoxia (11% oxygen) was effective in preventing pathology in the *NDUFS4* knockout mouse model of Leigh Syndrome [61, 86, 139]. Although this finding was initially somewhat paradoxical, given that mitochondria utilize oxygen to generate ATP, it appears that the benefits of hypoxia may be due to shifting metabolism more towards glycolysis in order to meet energetic needs. However, while 11% oxygen can be tolerated by healthy individuals, it is unknown how well hypoxia would be tolerated by different mitochondrial disease patients [86]. Moreover, it is not feasible to permanently maintain hypoxic levels. Finally, there is no guarantee that such an approach will translate from mice to humans. Nonetheless, it may be possible to develop drugs that effectively mimic hypoxia, which could prove effective for mitochondrial disease in general, including mtDNA diseases.

## 12.4 Conclusion

Despite the fact that there is currently no cure for mtDNA disease, it is clear that researchers are moving ever closer to developing effective strategies to prevent and treat this heterogeneous collection of diseases. This ongoing research and development of novel therapeutic approaches targeting the mitochondrial genome offers new hope to patients and families affected by mtDNA disease.

## References


1. Aanen DK, Spelbrink JN, Beekman M (2014) What cost mitochondria? The maintenance of functional mitochondrial DNA within and across generations. Philos Trans R Soc Lond Ser B Biol Sci 369:20130438
2. Ahola-Erkkila S, Carroll CJ, Peltola-Mjosund K, Tulkki V, Mattila I, Seppanen-Laakso T, Oresic M, Tyynismaa H, Suomalainen A (2010) Ketogenic diet slows down mitochondrial myopathy progression in mice. Hum Mol Genet 19:1974–1984
3. Amato P, Tachibana M, Sparman M, Mitalipov S (2014) Three-parent in vitro fertilization: gene replacement for the prevention of inherited mitochondrial diseases. Fertil Steril 101:31–35
4. Anderson S, Bankier AT, Barrell BG, De Bruijn MH, Coulson AR, Drouin J, Eperon IC, Nierlich DP, Roe BA, Sanger F, Schreier PH, Smith AJ, Staden R, Young IG (1981) Sequence and organization of the human mitochondrial genome. Nature 290:457–465
5. Austin S, St-Pierre J (2012) PGC1alpha and mitochondrial metabolism – emerging concepts and relevance in ageing and neurodegenerative disorders. J Cell Sci 125:4963–4971

6. Bacman SR, Kauppila JHK, Pereira CV, Nissanka N, Miranda M, Pinto M, Williams SL, Larsson NG, Stewart JB, Moraes CT (2018) MitoTALEN reduces mutant mtDNA load and restores tRNA(Ala) levels in a mouse model of heteroplasmic mtDNA mutation. Nat Med 24:1696–1700
7. Bacman SR, Williams SL, Hernandez D, Moraes CT (2007) Modulating mtDNA heteroplasmy by mitochondria-targeted restriction endonucleases in a 'differential multiple cleavage-site' model. Gene Ther 14:1309–1318
8. Bacman SR, Williams SL, Moraes CT (2009) Intra- and inter-molecular recombination of mitochondrial DNA after in vivo induction of multiple double-strand breaks. Nucleic Acids Res 37:4218–4226
9. Bacman SR, Williams SL, Pinto M, Moraes CT (2014) The use of mitochondria-targeted endonucleases to manipulate mtDNA. Methods Enzymol 547:373–397
10. Bacman SR, Williams SL, Pinto M, Peralta S, Moraes CT (2013) Specific elimination of mutant mitochondrial genomes in patient-derived cells by mitoTALENs. Nat Med 19:1111–1113
11. Bar-Yaacov D, Blumberg A, Mishmar D (2012) Mitochondrial-nuclear co-evolution and its effects on OXPHOS activity and regulation. Biochim Biophys Acta 1819:1107–1111
12. Barrera G, Gentile F, Pizzimenti S, Canuto RA, Daga M, Arcaro A, Cetrangolo GP, Lepore A, Ferretti C, Dianzani C, Muzio G (2016) Mitochondrial dysfunction in cancer and neurodegenerative diseases: spotlight on fatty acid oxidation and lipoperoxidation products. Antioxidants (Basel) 5:7
13. Bastin J, Aubey F, Rotig A, Munnich A, Djouadi F (2008) Activation of peroxisome proliferator-activated receptor pathway stimulates the mitochondrial respiratory chain and can correct deficiencies in patients' cells lacking its components. J Clin Endocrinol Metab 93:1433–1441
14. Battersby BJ, Shoubridge EA (2001) Selection of a mtDNA sequence variant in hepatocytes of heteroplasmic mice is not due to differences in respiratory chain function or efficiency of replication. Hum Mol Genet 10:2469–2479
15. Bayona-Bafaluy MP, Blits B, Battersby BJ, Shoubridge EA, Moraes CT (2005) Rapid directional shift of mitochondrial DNA heteroplasmy in animal tissues by a mitochondrially targeted restriction endonuclease. Proc Natl Acad Sci USA 102:14392–14397
16. Bibikova M, Beumer K, Trautman JK, Carroll D (2003) Enhancing gene targeting with designed zinc finger nucleases. Science 300:764
17. Blanc H, Adams CW, Wallace DC (1981) Different nucleotide changes in the large rRNA gene of the mitochondrial DNA confer chloramphenicol resistance on two human cell lines. Nucleic Acids Res 9:5785–5795
18. Bolhuis PA, Bleeker-Wagemakers EM, Ponne NJ, Van Schooneveld MJ, Westerveld A, Van Den Bogert C, Tabak HF (1990) Rapid shift in genotype of human mitochondrial DNA in a family with Leber's hereditary optic neuropathy. Biochem Biophys Res Commun 170:994–997
19. Bonawitz ND, Clayton DA, Shadel GS (2006) Initiation and beyond: multiple functions of the human mitochondrial transcription machinery. Mol Cell 24:813–825
20. Broskey NT, Greggio C, Boss A, Boutant M, Dwyer A, Schlueter L, Hans D, Gremion G, Kreis R, Boesch C, Canto C, Amati F (2014) Skeletal muscle mitochondria in the elderly: effects of physical fitness and exercise training. J Clin Endocrinol Metab 99:1852–1861
21. Bua E, Johnson J, Herbst A, Delong B, McKenzie D, Salamat S, Aiken JM (2006) Mitochondrial DNA-deletion mutations accumulate intracellularly to detrimental levels in aged human skeletal muscle fibers. Am J Hum Genet 79:469–480
22. Burgstaller JP, Johnston IG, Poulton J (2015) Mitochondrial DNA disease and developmental implications for reproductive strategies. Mol Hum Reprod 21:11–22
23. Burr SP, Pezet M, Chinnery PF (2018) Mitochondrial DNA Heteroplasmy and purifying selection in the mammalian female germ line. Develop Growth Differ 60:21–32
24. Busch KB, Kowald A, Spelbrink JN (2014) Quality matters: how does mitochondrial network dynamics and quality control impact on mtDNA integrity? Philos Trans R Soc Lond Ser B Biol Sci 369:20130442
25. Calvo S, Jain M, Xie X, Sheth SA, Chang B, Goldberger OA, Spinazzola A, Zeviani M, Carr SA, Mootha VK (2006) Systematic identification of human mitochondrial disease genes through integrative genomics. Nat Genet 38:576–582
26. Cann RL, Stoneking M, Wilson AC (1987) Mitochondrial DNA and human evolution. Nature 325:31–36
27. Canugovi C, Shamanna RA, Croteau DL, Bohr VA (2014) Base excision DNA repair levels in mitochondrial lysates of Alzheimer's disease. Neurobiol Aging 35:1293–1300
28. Carelli V, Baracca A, Barogi S, Pallotti F, Valentino ML, Montagna P, Zeviani M, Pini A, Lenaz G, Baruzzi A, Solaini G (2002) Biochemical-clinical correlation in patients with different loads of the mitochondrial DNA T8993G mutation. Arch Neurol 59:264–270
29. Casano RA, Bykhovskaya Y, Johnson DF, Hamon M, Torricelli F, Bigozzi M, Fischel-Ghodsian N (1998) Hearing loss due to the mitochondrial A1555G mutation in Italian families. Am J Med Genet 79:388–391
30. Catarino CB, Ahting U, Gusic M, Iuso A, Repp B, Peters K, Biskup S, Von Livonius B, Prokisch H, Klopstock T (2017) Characterization of a Leber's hereditary optic neuropathy (LHON) family harboring two primary LHON mutations m.11778G>A and m.14484T>C of the mitochondrial DNA. Mitochondrion 36:15–20

31. Chan SS, Copeland WC (2009) DNA polymerase gamma and mitochondrial disease: understanding the consequence of POLG mutations. Biochim Biophys Acta 1787:312–319
32. Chang DD, Clayton DA (1984) Precise identification of individual promoters for transcription of each strand of human mitochondrial DNA. Cell 36:635–643
33. Chang DD, Clayton DA (1989) Mouse RNAase MRP RNA is encoded by a nuclear gene and contains a decamer sequence complementary to a conserved region of mitochondrial RNA substrate. Cell 56:131–139
34. Chang JC, Hoel F, Liu KH, Wei YH, Cheng FC, Kuo SJ, Tronstad KJ, Liu CS (2017) Peptide-mediated delivery of donor mitochondria improves mitochondrial function and cell viability in human cybrid cells with the MELAS A3243G mutation. Sci Rep 7:10710
35. Chang JC, Liu KH, Chuang CS, Su HL, Wei YH, Kuo SJ, Liu CS (2013a) Treatment of human cells derived from MERRF syndrome by peptide-mediated mitochondrial delivery. Cytotherapy 15:1580–1596
36. Chang JC, Liu KH, Li YC, Kou SJ, Wei YH, Chuang CS, Hsieh M, Liu CS (2013b) Functional recovery of human cells harbouring the mitochondrial DNA mutation MERRF A8344G via peptide-mediated mitochondrial delivery. Neurosignals 21:160–173
37. Cheng Y, Liu P, Zheng Q, Gao G, Yuan J, Wang P, Huang J, Xie L, Lu X, Tong T, Chen J, Lu Z, Guan J, Wang G (2018) Mitochondrial trafficking and processing of telomerase RNA TERC. Cell Rep 24:2589–2595
38. Chinnery PF (1993) Mitochondrial disorders overview. In: Adam MP, Ardinger HH, Pagon RA, Wallace SE, Bean LJH, Stephens K, Amemiya A (eds) GeneReviews((R)). University of Washington, Seattle. University of Washington, Seattle. GeneReviews is a registered trademark of the University of Washington, Seattle. All rights reserved
39. Chinnery PF (2015) Mitochondrial disease in adults: what's old and what's new? EMBO Mol Med 7:1503–1512
40. Cho YM, Kim JH, Kim M, Park SJ, Koh SH, Ahn HS, Kang GH, Lee JB, Park KS, Lee HK (2012) Mesenchymal stem cells transfer mitochondria to the cells with virtually no mitochondrial function but not with pathogenic mtDNA mutations. PLoS One 7:e32778
41. Chou HF, Liang WC, Zhang Q, Goto Y, Jong YJ (2008) Clinical and genetic features in a MELAS child with a 3271T>C mutation. Pediatr Neurol 38:143–146
42. Ciesielski GL, Bermek O, Rosado-Ruiz FA, Hovde SL, Neitzke OJ, Griffith JD, Kaguni LS (2015) Mitochondrial single-stranded DNA-binding proteins stimulate the activity of DNA polymerase gamma by organization of the Template DNA. J Biol Chem 290:28697–28707
43. Cock HR, Cooper JM, Schapira AH (1999) Functional consequences of the 3460-bp mitochondrial DNA mutation associated with Leber's hereditary optic neuropathy. J Neurol Sci 165:10–17
44. Copeland WC (2012) Defects in mitochondrial DNA replication and human disease. Crit Rev Biochem Mol Biol 47:64–74
45. Cortopassi GA, Shibata D, Soong NW, Arnheim N (1992) A pattern of accumulation of a somatic deletion of mitochondrial DNA in aging human tissues. Proc Natl Acad Sci USA 89:7370–7374
46. Craigen WJ (2012) Mitochondrial DNA mutations: an overview of clinical and molecular aspects. Methods Mol Biol 837:3–15
47. Craven L, Tuppen HA, Greggains GD, Harbottle SJ, Murphy JL, Cree LM, Murdoch AP, Chinnery PF, Taylor RW, Lightowlers RN, Herbert M, Turnbull DM (2010) Pronuclear transfer in human embryos to prevent transmission of mitochondrial DNA disease. Nature 465:82–85
48. Dai Y, Zheng K, Clark J, Swerdlow RH, Pulst SM, Sutton JP, Shinobu LA, Simon DK (2014) Rapamycin drives selection against a pathogenic heteroplasmic mitochondrial DNA mutation. Hum Mol Genet 23:637–647
49. Davidson MM, Walker WF, Hernandez-Rosa E, Nesti C (2009) Evidence for nuclear modifier gene in mitochondrial cardiomyopathy. J Mol Cell Cardiol 46:936–942
50. De Moura MB, Dos Santos LS, Van Houten B (2010) Mitochondrial dysfunction in neurodegenerative diseases and cancer. Environ Mol Mutagen 51:391–405
51. De Praeter C, Vanlander A, Vanhaesebrouck P, Smet J, Seneca S, De Sutter P, Van Coster R (2015) Extremely high mutation load of the mitochondrial 8993 T>G mutation in a newborn: implications for prognosis and family planning decisions. Eur J Pediatr 174:267–270
52. Dillon LM, Williams SL, Hida A, Peacock JD, Prolla TA, Lincoln J, Moraes CT (2012) Increased mitochondrial biogenesis in muscle improves aging phenotypes in the mtDNA mutator mouse. Hum Mol Genet 21:2288–2297
53. Dimauro S, Hirano M (1993) Mitochondrial DNA deletion syndromes. In: Adam MP, Ardinger HH, Pagon RA, Wallace SE, Bean LJH, Stephens K, Amemiya A (eds) GeneReviews((R)). University of Washington, Seattle, Seattle. University of Washington, Seattle. GeneReviews is a registered trademark of the University of Washington, Seattle. All rights reserved
54. Dimauro S, Schon EA (2001) Mitochondrial DNA mutations in human disease. Am J Med Genet 106:18–26
55. Dorner M, Altmann M, Paabo S, Morl M (2001) Evidence for import of a lysyl-tRNA into marsupial mitochondria. Mol Biol Cell 12:2688–2698
56. Dovydenko I, Heckel AM, Tonin Y, Gowher A, Venyaminova A, Tarassov I, Entelis N (2015) Mitochondrial targeting of recombinant RNA. Methods Mol Biol 1265:209–225

57. Elliott HR, Samuels DC, Eden JA, Relton CL, Chinnery PF (2008) Pathogenic mitochondrial DNA mutations are common in the general population. Am J Hum Genet 83:254–260
58. Entelis NS, Kolesnikova OA, Dogan S, Martin RP, Tarassov IA (2001) 5 S rRNA and tRNA import into human mitochondria. Comparison of in vitro requirements. J Biol Chem 276:45642–45653
59. Fan W, Waymire KG, Narula N, Li P, Rocher C, Coskun PE, Vannan MA, Narula J, Macgregor GR, Wallace DC (2008) A mouse model of mitochondrial disease reveals germline selection against severe mtDNA mutations. Science 319:958–962
60. Farruggia P, Di Marco F, Dufour C (2018) Pearson syndrome. Expert Rev Hematol 11:239–246
61. Ferrari M, Jain IH, Goldberger O, Rezoagli E, Thoonen R, Cheng KH, Sosnovik DE, Scherrer-Crosbie M, Mootha VK, Zapol WM (2017) Hypoxia treatment reverses neurodegenerative disease in a mouse model of Leigh syndrome. Proc Natl Acad Sci USA 114:E4241–e4250
62. Floros VI, Pyle A, Dietmann S, Wei W, Tang WCW, Irie N, Payne B, Capalbo A, Noli L, Coxhead J, Hudson G, Crosier M, Strahl H, Khalaf Y, Saitou M, Ilic D, Surani MA, Chinnery PF (2018) Segregation of mitochondrial DNA heteroplasmy through a developmental genetic bottleneck in human embryos. Nat Cell Biol 20:144–151
63. Gaj T, Gersbach CA, Barbas CF, 3RD (2013) ZFN, TALEN, and CRISPR/Cas-based methods for genome engineering. Trends Biotechnol 31:397–405
64. Gammage PA, Gaude E, Van Haute L, Rebelo-Guiomar P, Jackson CB, Rorbach J, Pekalski ML, Robinson AJ, Charpentier M, Concordet JP, Frezza C, Minczuk M (2016) Near-complete elimination of mutant mtDNA by iterative or dynamic dose-controlled treatment with mtZFNs. Nucleic Acids Res 44:7804–7816
65. Gammage PA, Moraes CT Minczuk M (2017) Mitochondrial genome engineering: the revolution may not be CRISPR-Ized. Trends Genet
66. Gammage PA, Rorbach J, Vincent AI, Rebar EJ, Minczuk M (2014) Mitochondrially targeted ZFNs for selective degradation of pathogenic mitochondrial genomes bearing large-scale deletions or point mutations. EMBO Mol Med 6:458–466
67. Gammage PA, Viscomi C, Simard ML, Costa ASH, Gaude E, Powell CA, Van Haute L, McCann BJ, Rebelo-Guiomar P, Cerutti R, Zhang L, Rebar EJ, Zeviani M, Frezza C, Stewart JB, Minczuk M (2018) Genome editing in mitochondria corrects a pathogenic mtDNA mutation in vivo. Nat Med 24:1691–1695
68. Garcia S, Nissanka N, Mareco EA, Rossi S, Peralta S, Diaz F, Rotundo RL, Carvalho RF Moraes CT (2018) Overexpression of PGC-1alpha in aging muscle enhances a subset of young-like molecular patterns. Aging Cell 17
69. Giles RE, Blanc H, Cann HM, Wallace DC (1980) Maternal inheritance of human mitochondrial DNA. Proc Natl Acad Sci USA 77:6715–6719
70. Gilkerson RW, De Vries RL, Lebot P, Wikstrom JD, Torgyekes E, Shirihai OS, Przedborski S, Schon EA (2012) Mitochondrial autophagy in cells with mtDNA mutations results from synergistic loss of transmembrane potential and mTORC1 inhibition. Hum Mol Genet 21:978–990
71. Gorman GS, Chinnery PF, Dimauro S, Hirano M, Koga Y, McFarland R, Suomalainen A, Thorburn DR, Zeviani M, Turnbull DM (2016) Mitochondrial diseases. Nat Rev Dis Primers 2:16080
72. Gorman GS, Schaefer AM, NG Y, Gomez N, Blakely EL, Alston CL, Feeney C, Horvath R, Yu-Wai-Man P, Chinnery PF, Taylor RW, Turnbull DM, McFarland R (2015) Prevalence of nuclear and mitochondrial DNA mutations related to adult mitochondrial disease. Ann Neurol 77:753–759
73. Greaves LC, Reeve AK, Taylor RW, Turnbull DM (2012) Mitochondrial DNA and disease. J Pathol 226:274–286
74. Gredilla R, Stevnsner T (2012) Mitochondrial base excision repair assays. Methods Mol Biol 920:289–304
75. Guo W, Yang D, Xu H, Zhang Y, Huang J, Yang Z, Chen X, Huang Z (2013) Mutations in the D-loop region and increased copy number of mitochondrial DNA in human laryngeal squamous cell carcinoma. Mol Biol Rep 40:13–20
76. Guy J, Qi X, Pallotti F, Schon EA, Manfredi G, Carelli V, Martinuzzi A, Hauswirth WW, Lewin AS (2002) Rescue of a mitochondrial deficiency causing Leber hereditary optic neuropathy. Ann Neurol 52:534–542
77. Hashimoto M, Bacman SR, Peralta S, Falk MJ, Chomyn A, Chan DC, Williams SL, Moraes CT (2015) MitoTALEN: a general approach to reduce mutant mtDNA loads and restore oxidative phosphorylation function in mitochondrial diseases. Mol Ther 23:1592–1599
78. Hirano M, Emmanuele V, Quinzii CM (2018) Emerging therapies for mitochondrial diseases. Essays Biochem 62:467–481
79. Holt IJ, Harding AE, Morgan-Hughes JA (1988) Deletions of muscle mitochondrial DNA in patients with mitochondrial myopathies. Nature 331:717–719
80. Hudson G, Carelli V, Spruijt L, Gerards M, Mowbray C, Achilli A, Pyle A, Elson J, Howell N, La Morgia C, Valentino ML, Huoponen K, Savontaus ML, Nikoskelainen E, Sadun AA, Salomao SR, Belfort R Jr, Griffiths P, Yu-Wai-Man P, De Coo RF, Horvath R, Zeviani M, Smeets HJ, Torroni A, Chinnery PF (2007) Clinical expression of Leber hereditary optic neuropathy is affected by the mitochondrial DNA-haplogroup background. Am J Hum Genet 81:228–233

81. Hyslop LA, Blakeley P, Craven L, Richardson J, Fogarty NM, Fragouli E, Lamb M, Wamaitha SE, Prathalingam N, Zhang Q, O'Keefe H, Takeda Y, Arizzi L, Alfarawati S, Tuppen HA, Irving L, Kalleas D, Choudhary M, Wells D, Murdoch AP, Turnbull DM, Niakan KK, Herbert M (2016) Towards clinical application of pronuclear transfer to prevent mitochondrial DNA disease. Nature 534:383–386
82. Iborra FJ, Kimura H, Cook PR (2004) The functional organization of mitochondrial genomes in human cells. BMC Biol 2:9
83. Inoue K, Nakada K, Ogura A, Isobe K, Goto Y, Nonaka I, Hayashi JI (2000) Generation of mice with mitochondrial dysfunction by introducing mouse mtDNA carrying a deletion into zygotes. Nat Genet 26:176–181
84. Isaac RS, McShane E, Churchman LS (2018) The multiple levels of Mitonuclear Coregulation. Annu Rev Genet 52:511–533
85. Ishii T, Hibino Y (2018) Mitochondrial manipulation in fertility clinics: regulation and responsibility. Reprod Biomed Soc Online 5:93–109
86. Jain IH, Zazzeron L, Goli R, Alexa K, Schatzman-Bone S, Dhillon H, Goldberger O, Peng J, Shalem O, Sanjana NE, Zhang F, Goessling W, Zapol WM, Mootha VK (2016) Hypoxia as a therapy for mitochondrial disease. Science 352:54–61
87. Jang Y, Kwon I, Song W, Cosio-Lima LM, Taylor S, Lee Y (2018) Modulation of mitochondrial phenotypes by endurance exercise contributes to neuroprotection against a MPTP-induced animal model of PD. Life Sci 209:455–465
88. Jankele R, Svoboda P (2014) TAL effectors: tools for DNA targeting. Brief Funct Genomics 13:409–419
89. Jenuth JP, Peterson AC, Shoubridge EA (1997) Tissue-specific selection for different mtDNA genotypes in heteroplasmic mice. Nat Genet 16:93–95
90. Jiang D, Gao F, Zhang Y, Wong DS, Li Q, Tse HF, Xu G, Yu Z, Lian Q (2016) Mitochondrial transfer of mesenchymal stem cells effectively protects corneal epithelial cells from mitochondrial damage. Cell Death Dis 7:e2467
91. Jo A, Ham S, Lee GH, Lee YI, Kim S, Lee YS, Shin JH, Lee Y (2015) Efficient mitochondrial genome editing by CRISPR/Cas9. Biomed Res Int 2015:305716
92. Johnson SC, Yanos ME, Kayser EB, Quintana A, Sangesland M, Castanza A, Uhde L, Hui J, Wall VZ, Gagnidze A, Oh K, Wasko BM, Ramos FJ, Palmiter RD, Rabinovitch PS, Morgan PG, Sedensky MM, Kaeberlein M (2013) mTOR inhibition alleviates mitochondrial disease in a mouse model of Leigh syndrome. Science 342:1524–1528
93. Jun AS, Trounce IA, Brown MD, Shoffner JM, Wallace DC (1996) Use of transmitochondrial cybrids to assign a complex I defect to the mitochondrial DNA-encoded NADH dehydrogenase subunit 6 gene mutation at nucleotide pair 14459 that causes Leber hereditary optic neuropathy and dystonia. Mol Cell Biol 16:771–777
94. Kabekkodu SP, Bhat S, Mascarenhas R, Mallya S, Bhat M, Pandey D, Kushtagi P, Thangaraj K, Gopinath PM, Satyamoorthy K (2014) Mitochondrial DNA variation analysis in cervical cancer. Mitochondrion 16:73–82
95. Kaguni LS (2004) DNA polymerase gamma, the mitochondrial replicase. Annu Rev Biochem 73:293–320
96. Kaneda Y, Nakajima T, Nishikawa T, Yamamoto S, Ikegami H, Suzuki N, Nakamura H, Morishita R, Kotani H (2002) Hemagglutinating virus of Japan (HVJ) envelope vector as a versatile gene delivery system. Mol Ther 6:219–226
97. Kang E, Wu J, Gutierrez NM, Koski A, Tippner-Hedges R, Agaronyan K, Platero-Luengo A, Martinez-Redondo P, Ma H, Lee Y, Hayama T, Van Dyken C, Wang X, Luo S, Ahmed R, Li Y, Ji D, Kayali R, Cinnioglu C, Olson S, Jensen J, Battaglia D, Lee D, WU D, Huang T, Wolf DP, Temiakov D, Belmonte JC, Amato P, Mitalipov S (2016) Mitochondrial replacement in human oocytes carrying pathogenic mitochondrial DNA mutations. Nature 540:270–275
98. Kang HC, Lee YM, Kim HD, Lee JS, Slama A (2007) Safe and effective use of the ketogenic diet in children with epilepsy and mitochondrial respiratory chain complex defects. Epilepsia 48:82–88
99. Kara B, Arikan M, Maras H, Abaci N, Cakiris A, Ustek D (2012) Whole mitochondrial genome analysis of a family with NARP/MILS caused by m.8993T>C mutation in the MT-ATP6 gene. Mol Genet Metab 107:389–393
100. Kauppila JHK, Baines HL, Bratic A, Simard ML, Freyer C, Mourier A, Stamp C, Filograna R, Larsson NG, Greaves LC, Stewart JB (2016) A phenotype-driven approach to generate mouse models with pathogenic mtDNA mutations causing mitochondrial disease. Cell Rep 16:2980–2990
101. Kim I, Rodriguez-Enriquez S, Lemasters JJ (2007) Selective degradation of mitochondria by mitophagy. Arch Biochem Biophys 462:245–253
102. King MP, Attardi G (1989) Human cells lacking mtDNA: repopulation with exogenous mitochondria by complementation. Science 246:500–503
103. Kolesnikova OA, Entelis NS, Mireau H, Fox TD, Martin RP, Tarassov IA (2000) Suppression of mutations in mitochondrial DNA by tRNAs imported from the cytoplasm. Science 289:1931–1933
104. Komen JC, Thorburn DR (2014) Turn up the power – pharmacological activation of mitochondrial biogenesis in mouse models. Br J Pharmacol 171:1818–1836
105. Koopman WJ, Beyrath J, Fung CW, Koene S, Rodenburg RJ, Willems PH, Smeitink JA (2016) Mitochondrial disorders in children: toward development of small-molecule treatment strategies. EMBO Mol Med 8:311–327
106. Kowald A, Kirkwood TB (2014) Transcription could be the key to the selection advantage of mitochondrial deletion mutants in aging. Proc Natl Acad Sci USA 111:2972–2977

107. Krishnan KJ, Reeve AK, Samuels DC, Chinnery PF, Blackwood JK, Taylor RW, Wanrooij S, Spelbrink JN, Lightowlers RN, Turnbull DM (2008) What causes mitochondrial DNA deletions in human cells? Nat Genet 40:275–279
108. Kubli DA, Gustafsson AB (2012) Mitochondria and mitophagy: the yin and yang of cell death control. Circ Res 111:1208–1221
109. Kujoth GC, Hiona A, Pugh TD, Someya S, Panzer K, Wohlgemuth SE, Hofer T, Seo AY, Sullivan R, Jobling WA, Morrow JD, Van Remmen H, Sedivy JM, Yamasoba T, Tanokura M, Weindruch R, Leeuwenburgh C, Prolla TA (2005) Mitochondrial DNA mutations, oxidative stress, and apoptosis in mammalian aging. Science 309:481–484
110. Kukat C, Wurm CA, Spahr H, Falkenberg M, Larsson NG, Jakobs S (2011) Super-resolution microscopy reveals that mammalian mitochondrial nucleoids have a uniform size and frequently contain a single copy of mtDNA. Proc Natl Acad Sci USA 108:13534–13539
111. Kyriakouli DS, Boesch P, Taylor RW, Lightowlers RN (2008) Progress and prospects: gene therapy for mitochondrial DNA disease. Gene Ther 15:1017–1023
112. Leandro GS, Sykora P, Bohr VA (2015) The impact of base excision DNA repair in age-related neurodegenerative diseases. Mutat Res 776:31–39
113. Lee SR, Han J (2017) Mitochondrial nucleoid: shield and switch of the mitochondrial genome. Oxidative Med Cell Longev 2017:8060949
114. Levy SE, Waymire KG, Kim YL, MacGregor GR, Wallace DC (1999) Transfer of chloramphenicol-resistant mitochondrial DNA into the chimeric mouse. Transgenic Res 8:137–145
115. Lewis SC, Uchiyama LF, Nunnari J (2016) ER-mitochondria contacts couple mtDNA synthesis with mitochondrial division in human cells. Science 353:aaf5549
116. Li M, Rothwell R, Vermaat M, Wachsmuth M, Schroder R, Laros JF, Van Oven M, De Bakker PI, Bovenberg JA, Van Duijn CM, Van Ommen GJ, Slagboom PE, Swertz MA, Wijmenga C, Kayser M, Boomsma DI, Zollner S, De Knijff P, Stoneking M (2016) Transmission of human mtDNA heteroplasmy in the genome of the Netherlands families: support for a variable-size bottleneck. Genome Res 26:417–426
117. Li X, Michaeloudes C, Zhang Y, Wiegman CH, Adcock IM, Lian Q, Mak JCW, Bhavsar PK, Chung KF (2018) Mesenchymal stem cells alleviate oxidative stress-induced mitochondrial dysfunction in the airways. J Allergy Clin Immunol 141:1634–1645.e5
118. Liu CS, Chang JC, Kuo SJ, Liu KH, Lin TT, Cheng WL, Chuang SF (2014) Delivering healthy mitochondria for the therapy of mitochondrial diseases and beyond. Int J Biochem Cell Biol 53:141–146
119. Longley MJ, Clark S, Yu Wai Man C, Hudson G, Durham SE, Taylor RW, Nightingale S, Turnbull DM, Copeland WC, Chinnery PF (2006) Mutant POLG2 disrupts DNA polymerase gamma subunits and causes progressive external ophthalmoplegia. Am J Hum Genet 78:1026–1034
120. Lott MT, Leipzig JN, Derbeneva O, Xie HM, Chalkia D, Sarmady M, Procaccio V, Wallace DC (2013) mtDNA variation and analysis using Mitomap and Mitomaster. Curr Protoc Bioinformatics 44:1.23.1–1.2326
121. Loutre R, Heckel AM, Smirnova A, Entelis N, Tarassov I (2018) Can mitochondrial DNA be CRISPRized: pro and contra. IUBMB Life 70:1233–1239
122. Low RL, Orton S, Friedman DB (2003) A truncated form of DNA topoisomerase IIbeta associates with the mtDNA genome in mammalian mitochondria. Eur J Biochem 270:4173–4186
123. Ma YY, Wu TF, Liu YP, Wang Q, Li XY, Song JQ, Shi XY, Zhang WN, Zhao M, Hu LY, Yang YL, Zou LP (2013) Heterogeneity of six children and their mothers with mitochondrial DNA 3243 A>G mutation. Mitochondrial DNA 24:297–302
124. Magalhaes PJ, Andreu AL, Schon EA (1998) Evidence for the presence of 5S rRNA in mammalian mitochondria. Mol Biol Cell 9:2375–2382
125. Mak SC, Chi CS, Liu CY, Pang CY, Wei YH (1996) Leigh syndrome associated with mitochondrial DNA 8993 T–>G mutation and ragged-red fibers. Pediatr Neurol 15:72–75
126. Mancuso M, Orsucci D, Angelini C, Bertini E, Carelli V, Comi GP, Donati A, Minetti C, Moggio M, Mongini T, Servidei S, Tonin P, Toscano A, Uziel G, Bruno C, Ienco EC, Filosto M, Lamperti C, Catteruccia M, Moroni I, Musumeci O, Pegoraro E, Ronchi D, Santorelli FM, Sauchelli D, Scarpelli M, Sciacco M, Valentino ML, Vercelli L, Zeviani M, Siciliano G (2014) The m.3243A>G mitochondrial DNA mutation and related phenotypes. A matter of gender? J Neurol 261:504–510
127. Mancuso M, Orsucci D, Angelini C, Bertini E, Carelli V, Comi GP, Minetti C, Moggio M, Mongini T, Servidei S, Tonin P, Toscano A, Uziel G, Bruno C, Caldarazzo Ienco E, Filosto M, Lamperti C, Martinelli D, Moroni I, Musumeci O, Pegoraro E, Ronchi D, Santorelli FM, Sauchelli D, Scarpelli M, Sciacco M, Spinazzi M, Valentino ML, Vercelli L, Zeviani M, Siciliano G (2013) Phenotypic heterogeneity of the 8344A>G mtDNA "MERRF" mutation. Neurology 80:2049–2054
128. Manfredi G, Fu J, Ojaimi J, Sadlock JE, Kwong JQ, Guy J, Schon EA (2002) Rescue of a deficiency in ATP synthesis by transfer of MTATP6, a mitochondrial DNA-encoded gene, to the nucleus. Nat Genet 30:394–399
129. Manfredi G, Gupta N, Vazquez-Memije ME, Sadlock JE, Spinazzola A, De Vivo DC, Schon EA (1999) Oligomycin induces a decrease in the cellular content of a pathogenic mutation in the human mitochondrial ATPase 6 gene. J Biol Chem 274:9386–9391

130. Manipalviratn S, Decherney A, Segars J (2009) Imprinting disorders and assisted reproductive technology. Fertil Steril 91:305–315
131. Marchington DR, Barlow D, Poulton J (1999) Transmitochondrial mice carrying resistance to chloramphenicol on mitochondrial DNA: developing the first mouse model of mitochondrial DNA disease. Nat Med 5:957–960
132. Marian AJ (2002) Modifier genes for hypertrophic cardiomyopathy. Curr Opin Cardiol 17:242–252
133. McCann BJ, Cox A, Gammage PA, Stewart JB, Zernicka-Goetz M, Minczuk M (2018) Delivery of mtZFNs into early mouse embryos. Methods Mol Biol 1867:215–228
134. McFarland R, Taylor RW, Turnbull DM (2002) The neurology of mitochondrial DNA disease. Lancet Neurol 1:343–351
135. McGrath J, Solter D (1983) Nuclear transplantation in the mouse embryo by microsurgery and cell fusion. Science 220:1300–1302
136. Milenkovic D, Matic S, Kuhl I, Ruzzenente B, Freyer C, Jemt E, Park CB, Falkenberg M, Larsson NG (2013) TWINKLE is an essential mitochondrial helicase required for synthesis of nascent D-loop strands and complete mtDNA replication. Hum Mol Genet 22:1983–1993
137. Minczuk M, Papworth MA, Kolasinska P, Murphy MP, Klug A (2006) Sequence-specific modification of mitochondrial DNA using a chimeric zinc finger methylase. Proc Natl Acad Sci USA 103:19689–19694
138. Minczuk M, Papworth MA, Miller JC, Murphy MP, Klug A (2008) Development of a single-chain, quasi-dimeric zinc-finger nuclease for the selective degradation of mutated human mitochondrial DNA. Nucleic Acids Res 36:3926–3938
139. Mootha VK, Chinnery PF (2018) Oxygen in mitochondrial disease: can there be too much of a good thing? J Inherit Metab Dis 41:761–763
140. Moretton A, Morel F, Macao B, Lachaume P, Ishak L, Lefebvre M, Garreau-Balandier I, Vernet P, Falkenberg M, Farge G (2017) Selective mitochondrial DNA degradation following double-strand breaks. PLoS One 12:e0176795
141. Nakada K, Sato A, Sone H, Kasahara A, Ikeda K, Kagawa Y, Yonekawa H, Hayashi J (2004) Accumulation of pathogenic DeltamtDNA induced deafness but not diabetic phenotypes in mito-mice. Biochem Biophys Res Commun 323:175–184
142. Nicholls TJ, Minczuk M (2014) In D-loop: 40 years of mitochondrial 7S DNA. Exp Gerontol 56:175–181
143. Nicolson GL (2007) Metabolic syndrome and mitochondrial function: molecular replacement and antioxidant supplements to prevent membrane peroxidation and restore mitochondrial function. J Cell Biochem 100:1352–1369
144. Nightingale H, Pfeffer G, Bargiela D, Horvath R, Chinnery PF (2016) Emerging therapies for mitochondrial disorders. Brain 139:1633–1648
145. Nunnari J, Suomalainen A (2012) Mitochondria: in sickness and in health. Cell 148:1145–1159
146. Pabo CO, Peisach E, Grant RA (2001) Design and selection of novel Cys2His2 zinc finger proteins. Annu Rev Biochem 70:313–340
147. Park CB, Larsson NG (2011) Mitochondrial DNA mutations in disease and aging. J Cell Biol 193:809–818
148. Patananan AN, Wu TH, Chiou PY, Teitell MA (2016) Modifying the mitochondrial genome. Cell Metab 23:785–796
149. Peeva V, Blei D, Trombly G, Corsi S, Szukszto MJ, Rebelo-Guiomar P, Gammage PA, Kudin AP, Becker C, Altmuller J, Minczuk M, Zsurka G, Kunz WS (2018) Linear mitochondrial DNA is rapidly degraded by components of the replication machinery. Nat Commun 9:1727
150. Pereira CV, Bacman SR, Arguello T, Zekonyte U, Williams SL, Edgell DR Moraes CT (2018) mitoTev-TALE: a monomeric DNA editing enzyme to reduce mutant mitochondrial DNA levels. EMBO Mol Med 10
151. Pereira CV, Moraes CT (2017) Current strategies towards therapeutic manipulation of mtDNA heteroplasmy. Front Biosci (Landmark Ed) 22:991–1010
152. Phillips AF, Millet AR, Tigano M, Dubois SM, Crimmins H, Babin L, Charpentier M, Piganeau M, Brunet E, Sfeir A (2017) Single-molecule analysis of mtDNA replication uncovers the basis of the common deletion. Mol Cell 65:527–538.e6
153. Picard M, Zhang J, Hancock S, Derbeneva O, Golhar R, Golik P, O'Hearn S, Levy S, Potluri P, Lvova M, Davila A, Lin CS, Perin JC, Rappaport EF, Hakonarson H, Trounce IA, Procaccio V, Wallace DC (2014) Progressive increase in mtDNA 3243A>G heteroplasmy causes abrupt transcriptional reprogramming. Proc Natl Acad Sci USA 111:E4033–E4042
154. Pichaud N, Ballard JW, Tanguay RM, Blier PU (2012) Naturally occurring mitochondrial DNA haplotypes exhibit metabolic differences: insight into functional properties of mitochondria. Evolution 66:3189–3197
155. Pickett SJ, Grady JP, Ng YS, Gorman GS, Schaefer AM, Wilson IJ, Cordell HJ, Turnbull DM, Taylor RW, McFarland R (2018) Phenotypic heterogeneity in m.3243A>G mitochondrial disease: the role of nuclear factors. Ann Clin Transl Neurol 5:333–345
156. Porteus MH, Carroll D (2005) Gene targeting using zinc finger nucleases. Nat Biotechnol 23:967–973
157. Poulton J, Deadman ME, Ramacharan S, Gardiner RM (1991) Germ-line deletions of mtDNA in mitochondrial myopathy. Am J Hum Genet 48:649–653
158. Prakash A, Doublie S (2015) Base excision repair in the mitochondria. J Cell Biochem 116:1490–1499
159. Puranam RS, Attardi G (2001) The RNase P associated with HeLa cell mitochondria contains an essential RNA component identical in sequence to that of the nuclear RNase P. Mol Cell Biol 21:548–561

160. Qiao L, Ru G, Mao Z, Wang C, Nie Z, Li Q, Huang-Yang Y, Zhu L, Liang X, Yu J, Jiang P (2017) Mitochondrial DNA depletion, mitochondrial mutations and high TFAM expression in hepatocellular carcinoma. Oncotarget 8:84373–84383
161. Rahman S, Blok RB, Dahl HH, Danks DM, Kirby DM, Chow CW, Christodoulou J, Thorburn DR (1996) Leigh syndrome: clinical features and biochemical and DNA abnormalities. Ann Neurol 39:343–351
162. Rai PK, Craven L, Hoogewijs K, Russell OM, Lightowlers RN (2018) Advances in methods for reducing mitochondrial DNA disease by replacing or manipulating the mitochondrial genome. Essays Biochem 62:455–465
163. Reddy P, Ocampo A, Suzuki K, Luo J, Bacman SR, Williams SL, Sugawara A, Okamura D, Tsunekawa Y, Wu J, Lam D, Xiong X, Montserrat N, Esteban CR, Liu GH, Sancho-Martinez I, Manau D, Civico S, Cardellach F, Del Mar O'Callaghan M, Campistol J, Zhao H, Campistol JM, Moraes CT, Izpisua Belmonte JC (2015) Selective elimination of mitochondrial mutations in the germline by genome editing. Cell 161:459–469
164. Reinhardt K, Dowling DK, Morrow EH (2013) Medicine. Mitochondrial replacement, evolution, and the clinic. Science 341:1345–1346
165. Richardson J, Irving L, Hyslop LA, Choudhary M, Murdoch A, Turnbull DM, Herbert M (2015) Concise reviews: assisted reproductive technologies to prevent transmission of mitochondrial DNA disease. Stem Cells 33:639–645
166. Rojo A, Campos Y, Sanchez JM, Bonaventura I, Aguilar M, Garcia A, Gonzalez L, Rey MJ, Arenas J, Olive M, Ferrer I (2006) NARP-MILS syndrome caused by 8993 T>G mitochondrial DNA mutation: a clinical, genetic and neuropathological study. Acta Neuropathol 111:610–616
167. Ross JM, Stewart JB, Hagstrom E, Brene S, Mourier A, Coppotelli G, Freyer C, Lagouge M, Hoffer BJ, Olson L, Larsson NG (2013) Germline mitochondrial DNA mutations aggravate ageing and can impair brain development. Nature 501:412–415
168. Russell OM, Fruh I, Rai PK, Marcellin D, Doll T, Reeve A, Germain M, Bastien J, Rygiel KA, Cerino R, Sailer AW, Lako M, Taylor RW, Mueller M, Lightowlers RN, Turnbull DM, Helliwell SB (2018) Preferential amplification of a human mitochondrial DNA deletion in vitro and in vivo. Sci Rep 8:1799
169. Sallevelt SC, Dreesen JC, Drusedau M, Hellebrekers DM, Paulussen AD, Coonen E, Van Golde RJ, Geraedts JP, Gianaroli L, Magli MC, Zeviani M, Smeets HJ, De Die-Smulders CE (2017) PGD for the m.14487 T>C mitochondrial DNA mutation resulted in the birth of a healthy boy. Hum Reprod 32:698–703
170. Samuels DC, Schon EA, Chinnery PF (2004) Two direct repeats cause most human mtDNA deletions. Trends Genet 20:393–398
171. Sanchis-Gomar F, Garcia-Gimenez JL, Gomez-Cabrera MC, Pallardo FV (2014) Mitochondrial biogenesis in health and disease. Molecular and therapeutic approaches. Curr Pharm Des 20:5619–5633
172. Santorelli FM, Mak SC, Vazquez-Memije ME, Shanske S, Kranz-Eble P, Jain KD, Bluestone DL, De Vivo DC, Dimauro S (1996) Clinical heterogeneity associated with the mitochondrial DNA T8993C point mutation. Pediatr Res 39:914–917
173. Santra S, Gilkerson RW, Davidson M, Schon EA (2004) Ketogenic treatment reduces deleted mitochondrial DNAs in cultured human cells. Ann Neurol 56:662–669
174. Sato A, Kono T, Nakada K, Ishikawa K, Inoue S, Yonekawa H, Hayashi J (2005) Gene therapy for progeny of Mito-mice carrying pathogenic mtDNA by nuclear transplantation. Proc Natl Acad Sci USA 102:16765–16770
175. Sawano T, Tanaka M, Ohno K, Yoneda M, Ota Y, Terasaki H, Awaya S, Ozawa T (1996) Mitochondrial DNA mutations associated with the 11778 mutation in Leber's disease. Biochem Mol Biol Int 38:693–700
176. Schaefer AM, Taylor RW, Turnbull DM, Chinnery PF (2004) The epidemiology of mitochondrial disorders – past, present and future. Biochim Biophys Acta 1659:115–120
177. Schon EA, Rizzuto R, Moraes CT, Nakase H, Zeviani M, Dimauro S (1989) A direct repeat is a hotspot for large-scale deletion of human mitochondrial DNA. Science 244:346–349
178. Scott TL, Rangaswamy S, Wicker CA, Izumi T (2014) Repair of oxidative DNA damage and cancer: recent progress in DNA base excision repair. Antioxid Redox Signal 20:708–726
179. Seibel P, Trappe J, Villani G, Klopstock T, Papa S, Reichmann H (1995) Transfection of mitochondria: strategy towards a gene therapy of mitochondrial DNA diseases. Nucleic Acids Res 23:10–17
180. Sgarbi G, Baracca A, Lenaz G, Valentino LM, Carelli V, Solaini G (2006) Inefficient coupling between proton transport and ATP synthesis may be the pathogenic mechanism for NARP and Leigh syndrome resulting from the T8993G mutation in mtDNA. Biochem J 395:493–500
181. Shadel GS, Clayton DA (1997) Mitochondrial DNA maintenance in vertebrates. Annu Rev Biochem 66:409–435
182. Shanske S, Tang Y, Hirano M, Nishigaki Y, Tanji K, Bonilla E, Sue C, Krishna S, Carlo JR, Willner J, Schon EA, Dimauro S (2002) Identical mitochondrial DNA deletion in a woman with ocular myopathy and in her son with Pearson syndrome. Am J Hum Genet 71:679–683
183. Shimizu A, Mito T, Hashizume O, Yonekawa H, Ishikawa K, Nakada K, Hayashi J (2015) G7731A mutation in mouse mitochondrial tRNALys regulates late-onset disorders in transmitochondrial mice. Biochem Biophys Res Commun 459:66–70

184. Shimizu A, Mito T, Hayashi C, Ogasawara E, Koba R, Negishi I, Takenaga K, Nakada K, Hayashi J (2014) Transmitochondrial mice as models for primary prevention of diseases caused by mutation in the tRNA(Lys) gene. Proc Natl Acad Sci USA 111:3104–3109
185. Shoffner JM, Lott MT, Voljavec AS, Soueidan SA, Costigan DA, Wallace DC (1989) Spontaneous Kearns-Sayre/chronic external ophthalmoplegia plus syndrome associated with a mitochondrial DNA deletion: a slip-replication model and metabolic therapy. Proc Natl Acad Sci USA 86:7952–7956
186. Shokolenko IN, Wilson GL, Alexeyev MF (2013) Persistent damage induces mitochondrial DNA degradation. DNA Repair (Amst) 12:488–499
187. Shtilbans A, Shanske S, Goodman S, Sue CM, Bruno C, Johnson TL, Lava NS, Waheed N, Dimauro S (2000) G8363A mutation in the mitochondrial DNA transfer ribonucleic acidLys gene: another cause of Leigh syndrome. J Child Neurol 15:759–761
188. Shutt TE, Bestwick M, Shadel GS (2011) The core human mitochondrial transcription initiation complex: it only takes two to tango. Transcription 2:55–59
189. Shutt TE, Gray MW (2006) Twinkle, the mitochondrial replicative DNA helicase, is widespread in the eukaryotic radiation and may also be the mitochondrial DNA primase in most eukaryotes. J Mol Evol 62:588–599
190. Shutt TE, Lodeiro MF, Cotney J, Cameron CE, Shadel GS (2010) Core human mitochondrial transcription apparatus is a regulated two-component system in vitro. Proc Natl Acad Sci USA 107:12133–12138
191. Shutt TE, Shadel GS (2010) A compendium of human mitochondrial gene expression machinery with links to disease. Environ Mol Mutagen 51:360–379
192. Siegmund SE, Yang H, Sharma R, Javors M, Skinner O, Mootha V, Hirano M, Schon EA (2017) Low-dose rapamycin extends lifespan in a mouse model of mtDNA depletion syndrome. Hum Mol Genet 26:4588–4605
193. Silvestri G, Moraes CT, Shanske S, Oh SJ, Dimauro S (1992) A new mtDNA mutation in the tRNA(Lys) gene associated with myoclonic epilepsy and ragged-red fibers (MERRF). Am J Hum Genet 51:1213–1217
194. Sligh JE, Levy SE, Waymire KG, Allard P, Dillehay DL, Nusinowitz S, Heckenlively JR, MacGregor GR, Wallace DC (2000) Maternal germ-line transmission of mutant mtDNAs from embryonic stem cell-derived chimeric mice. Proc Natl Acad Sci USA 97:14461–14466
195. Smeets HJ, Sallevelt SC, Dreesen JC, De Die-Smulders CE, De Coo IF (2015) Preventing the transmission of mitochondrial DNA disorders using prenatal or preimplantation genetic diagnosis. Ann N Y Acad Sci 1350:29–36
196. Spees JL, Olson SD, Whitney MJ, Prockop DJ (2006) Mitochondrial transfer between cells can rescue aerobic respiration. Proc Natl Acad Sci USA 103:1283–1288
197. Spiegelman BM (2007) Transcriptional control of mitochondrial energy metabolism through the PGC1 coactivators. Novartis Found Symp 287:60–63. discussion 63-9
198. Srivastava S, Diaz F, Iommarini L, Aure K, Lombes A, Moraes CT (2009) PGC-1alpha/beta induced expression partially compensates for respiratory chain defects in cells from patients with mitochondrial disorders. Hum Mol Genet 18:1805–1812
199. Srivastava S, Moraes CT (2001) Manipulating mitochondrial DNA heteroplasmy by a mitochondrially targeted restriction endonuclease. Hum Mol Genet 10:3093–3099
200. Stewart JB, Freyer C, Elson JL, Larsson NG (2008) Purifying selection of mtDNA and its implications for understanding evolution and mitochondrial disease. Nat Rev Genet 9:657–662
201. Stewart JB, Larsson NG (2014) Keeping mtDNA in shape between generations. PLoS Genet 10:e1004670
202. Suen DF, Narendra DP, Tanaka A, Manfredi G, Youle RJ (2010) Parkin overexpression selects against a deleterious mtDNA mutation in heteroplasmic cybrid cells. Proc Natl Acad Sci USA 107:11835–11840
203. Suzuki Y, Wada T, Sakai T, Ishikawa Y, Minami R, Tachi N, Saitoh S (1998) Phenotypic variability in a family with a mitochondrial DNA T8993C mutation. Pediatr Neurol 19:283–286
204. Swerdlow RH (2007) Mitochondria in cybrids containing mtDNA from persons with mitochondriopathies. J Neurosci Res 85:3416–3428
205. Taanman JW (1999) The mitochondrial genome: structure, transcription, translation and replication. Biochim Biophys Acta 1410:103–123
206. Tachibana M, Amato P, Sparman M, Woodward J, Sanchis DM, Ma H, Gutierrez NM, Tippner-Hedges R, Kang E, Lee HS, Ramsey C, Masterson K, Battaglia D, Lee D, Wu D, Jensen J, Patton P, Gokhale S, Stouffer R, Mitalipov S (2013) Towards germline gene therapy of inherited mitochondrial diseases. Nature 493:627–631
207. Tachibana M, Sparman M, Sritanaudomchai H, Ma H, Clepper L, Woodward J, Li Y, Ramsey C, Kolotushkina O, Mitalipov S (2009) Mitochondrial gene replacement in primate offspring and embryonic stem cells. Nature 461:367–372
208. Tadi SK, Sebastian R, Dahal S, Babu RK, Choudhary B, Raghavan SC (2016) Microhomology-mediated end joining is the principal mediator of double-strand break repair during mitochondrial DNA lesions. Mol Biol Cell 27:223–235
209. Tanaka M, Borgeld HJ, Zhang J, Muramatsu S, Gong JS, Yoneda M, Maruyama W, Naoi M, Ibi T, Sahashi K, Shamoto M, Fuku N, Kurata M, Yamada Y, Nishizawa K, Akao Y, Ohishi N, Miyabayashi S, Umemoto H, Muramatsu T, Furukawa K, Kikuchi A, Nakano I, Ozawa K, Yagi K (2002) Gene therapy

for mitochondrial disease by delivering restriction endonuclease SmaI into mitochondria. J Biomed Sci 9:534–541
210. Tarassov IA, Martin RP (1996) Mechanisms of tRNA import into yeast mitochondria: an overview. Biochimie 78:502–510
211. Taylor RW, Turnbull DM (2005) Mitochondrial DNA mutations in human disease. Nat Rev Genet 6:389–402
212. Torchilin VP, Khaw BA, Weissig V (2002) Intracellular targets for DNA delivery: nuclei and mitochondria. Somat Cell Mol Genet 27:49–64
213. Treff NR, Campos J, Tao X, Levy B, Ferry KM, Scott RT Jr (2012) Blastocyst preimplantation genetic diagnosis (PGD) of a mitochondrial DNA disorder. Fertil Steril 98:1236–1240
214. Trifunovic A, Wredenberg A, Falkenberg M, Spelbrink JN, Rovio AT, Bruder CE, Bohlooly YM, Gidlof S, Oldfors A, Wibom R, Tornell J, Jacobs HT, Larsson NG (2004) Premature ageing in mice expressing defective mitochondrial DNA polymerase. Nature 429:417–423
215. Tuppen HA, Blakely EL, Turnbull DM, Taylor RW (2010) Mitochondrial DNA mutations and human disease. Biochim Biophys Acta 1797:113–128
216. Twig G, Elorza A, Molina AJ, Mohamed H, Wikstrom JD, Walzer G, Stiles L, Haigh SE, Katz S, Las G, Alroy J, WU M, Py BF, Yuan J, Deeney JT, Corkey BE, Shirihai OS (2008) Fission and selective fusion govern mitochondrial segregation and elimination by autophagy. EMBO J 27:433–446
217. Tyynismaa H, Mjosund KP, Wanrooij S, Lappalainen I, Ylikallio E, Jalanko A, Spelbrink JN, Paetau A, Suomalainen A (2005) Mutant mitochondrial helicase Twinkle causes multiple mtDNA deletions and a late-onset mitochondrial disease in mice. Proc Natl Acad Sci USA 102:17687–17692
218. Usami S, Abe S, Kasai M, Shinkawa H, Moeller B, Kenyon JB, Kimberling WJ (1997) Genetic and clinical features of sensorineural hearing loss associated with the 1555 mitochondrial mutation. Laryngoscope 107:483–490
219. Valenci I, Yonai L, Bar-Yaacov D, Mishmar D, Ben-Zvi A (2015) Parkin modulates heteroplasmy of truncated mtDNA in *Caenorhabditis elegans*. Mitochondrion 20:64–70
220. Verechshagina NA, Konstantinov YM, Kamenski PA, Mazunin IO (2018) Import of proteins and nucleic acids into mitochondria. Biochemistry (Mosc) 83:643–661
221. Verlinsky Y, Cieslak J, Ivakhnenko V, Evsikov S, Wolf G, White M, Lifchez A, Kaplan B, Moise J, Valle J, Ginsberg N, Strom C, Kuliev A (1997) Prepregnancy genetic testing for age-related aneuploidies by polar body analysis. Genet Test 1:231–235
222. Viscomi C, Zeviani M (2017) MtDNA-maintenance defects: syndromes and genes. J Inherit Metab Dis 40:587–599
223. Wai T, Ao A, Zhang X, Cyr D, Dufort D, Shoubridge EA (2010) The role of mitochondrial DNA copy number in mammalian fertility. Biol Reprod 83:52–62
224. Wallace DC (2015) Mitochondrial DNA variation in human radiation and disease. Cell 163:33–38
225. Wallace DC, Bunn CL, Eisenstadt JM (1975) Cytoplasmic transfer of chloramphenicol resistance in human tissue culture cells. J Cell Biol 67:174–188
226. Wallace DC, Chalkia D (2013) Mitochondrial DNA genetics and the heteroplasmy conundrum in evolution and disease. Cold Spring Harb Perspect Biol 5:a021220
227. Wallace DC, Singh G, Lott MT, Hodge JA, Schurr TG, Lezza AM, Elsas LJ 2nd, Nikoskelainen EK (1988) Mitochondrial DNA mutation associated with Leber's hereditary optic neuropathy. Science 242:1427–1430
228. Wang G, Shimada E, Zhang J, Hong JS, Smith GM, Teitell MA, Koehler CM (2012) Correcting human mitochondrial mutations with targeted RNA import. Proc Natl Acad Sci USA 109:4840–4845
229. Wang Y, Lyu YL, Wang JC (2002) Dual localization of human DNA topoisomerase IIIalpha to mitochondria and nucleus. Proc Natl Acad Sci USA 99:12114–12119
230. Wanrooij S, Luoma P, Van Goethem G, Van Broeckhoven C, Suomalainen A, Spelbrink JN (2004) Twinkle and POLG defects enhance age-dependent accumulation of mutations in the control region of mtDNA. Nucleic Acids Res 32:3053–3064
231. Warowicka A, Kwasniewska A, Gozdzicka-Jozefiak A (2013) Alterations in mtDNA: a qualitative and quantitative study associated with cervical cancer development. Gynecol Oncol 129:193–198
232. Wrigley A, Wilkinson S, Appleby JB (2015) Mitochondrial replacement: ethics and identity. Bioethics 29:631–638
233. Yoshionari S, Koike T, Yokogawa T, Nishikawa K, Ueda T, Miura K, Watanabe K (1994) Existence of nuclear-encoded 5S-rRNA in bovine mitochondria. FEBS Lett 338:137–142
234. Youle RJ, Van Der Bliek AM (2012) Mitochondrial fission, fusion, and stress. Science 337:1062–1065
235. Young MJ, Copeland WC (2016) Human mitochondrial DNA replication machinery and disease. Curr Opin Genet Dev 38:52–62
236. Zelenka J, Jezek P (2016) Import of fluorescent RNA into mitochondria of living cells. Methods Mol Biol 1351:175–181
237. Zghaib T, Nazarian S (2018) 100 – arrhythmia in neurological disease. In: Zipes DP, Jalife J, Stevenson WG (eds) Cardiac electrophysiology: from cell to bedside, 7th edn. Elsevier, Amsterdam
238. Zhang H, Barcelo JM, Lee B, Kohlhagen G, Zimonjic DB, Popescu NC, Pommier Y (2001) Human mitochondrial topoisomerase I. Proc Natl Acad Sci USA 98:10608–10613

239. Zhang J, Liu H, Luo S, Lu Z, Chavez-Badiola A, Liu Z, Yang M, Merhi Z, Silber SJ, Munne S, Konstantinidis M, Wells D, Tang JJ, Huang T (2017) Live birth derived from oocyte spindle transfer to prevent mitochondrial disease. Reprod Biomed Online 34:361–368
240. Zhang X, Zuo X, Yang B, Li Z, Xue Y, Zhou Y, Huang J, Zhao X, Zhou J, Yan Y, Zhang H, Guo P, Sun H, Guo L, Zhang Y, Fu XD (2014) MicroRNA directly enhances mitochondrial translation during muscle differentiation. Cell 158:607–619
241. Zhao H, Li R, Wang Q, Yan Q, Deng JH, Han D, Bai Y, Young WY, Guan MX (2004) Maternally inherited aminoglycoside-induced and nonsyndromic deafness is associated with the novel C1494T mutation in the mitochondrial 12S rRNA gene in a large Chinese family. Am J Hum Genet 74:139–152

# Mitochondrial Genetics

# 13

Vanessa F. Gonçalves

**Abstract**

The maternally inherited mitochondrial DNA (mtDNA) is located inside every mitochondrion, in variable number of copies, and it contains 37 crucial genes for cellular bioenergetics. This chapter will discuss the unique features of this circular genome including heteroplasmy, haplogroups, among others, along with the corresponding clinical relevance for each. The discussion also covers the nuclear-encoded mitochondrial genes (N > 1000) and the epistatic interactions between mtDNA and the nuclear genome. Examples of mitochondrial diseases related to specific mtDNA mutation sites of relevance for humans are provided. This chapter aims to provide an overview of mitochondrial genetics as an emerging hot topic for the future of medicine.

**Keywords**

Mitochondrial DNA · Mitonuclear epistatis · Single Nucleotide Polymorphism · Mitochondrial haplogroups · Epigenetics

V. F. Gonçalves (✉)
Molecular Brain Sciences Department, Centre for Addiction and Mental Health, Toronto, Canada
e-mail: Vanessa.Goncalves@camh.ca

## 13.1 Introduction

Mitochondria are intracellular organelles primarily responsible for the conversion of energy-carrying molecules into ATP through the process of oxidative phosphorylation (OXPHOS). In addition to their energy production function, mitochondria regulate other cellular processes such as reactive oxygen species (ROS) generation, calcium flux, and apoptosis [45]. Unlike other organelles in animal cells, mitochondria carry their own genetic information, termed mitochondrial DNA (mtDNA). In humans, 37 genes are encoded by the mtDNA, while the vast majority of genes involved in the mitochondrial function are located in the nucleus. Indeed, proteins encoded by over 1100 genes operate within mitochondria [3].

Genetic variation in mitochondrial system consists of single nucleotide polymorphisms (SNPs), deletions, mtDNA copy number variants, heteroplasmy (a mix of mutant and normal mtDNA within a cell), mitonuclear communication (interaction between mtDNA and chromosomal DNA), and epigenetics. The identification, as well as the functional characterization of mitochondrial variants is relevant for basic and clinical research, since they hold the potential to influence mitochondrial gene expression and function [41, 51]. As proof of its importance, mtDNA mutations cause many inherited diseases, such as MELAS (mitochondrial encepha-



A. Urbani, M. Babu (eds.), *Mitochondria in Health and in Sickness*, Advances in Experimental Medicine and Biology 1158, https://doi.org/10.1007/978-981-13-8367-0_13

lomyopathy, lactic acidosis, and stroke-like episodes) [7] and Leigh syndrome [30]. Acquired (*de novo*) mtDNA mutations contribute to multifactorial conditions such as cancer and aging [51]. Furthermore, loss of mtDNA integrity leads to dysfunctional mitochondria and is associated with neurodegenerative diseases such as Parkinson's and Alzheimer's disease [8, 33].

This chapter will provide a review of the current knowledge on the genetic architecture of the mitochondrial system. It will also provide an overview of mitochondrial genetic variants while citing analytical tools currently used in the field of mitochondrial genetics.

## 13.2 The Mitochondrial DNA

The human mtDNA is a circular double-stranded molecule measuring 16,569 kb in length. The mtDNA strands differ from each other based on their nucleotide composition: the heavy strand (H-strand) is guanine rich, while the light strand (L-strand) is cytosine rich. Furthermore, this genome has unique features that differentiate it from the nuclear genome. Unlike nuclear genome, mtDNA follows an exclusively maternal pattern for genetic inheritance. Each mitochondrion possesses many copies of the mtDNA, and a non-universal genetic code is used for protein translation; for more details see [41].

The mtDNA coding region (between the nucleotide positions 577-16023) encodes 37 genes in humans: 13 genes code for peptides involved in the OXPHOS protein complexes, 22 genes code for transfer RNAs (tRNAs), and 2 genes code for ribosomal RNAs (rRNAs) [40, 43], as shown in (Fig. 13.1). Specifically, the mtDNA genes include seven genes of complex I (ND1–6), one gene of complex III (CYTB), three genes of complex IV (COI–III), and two genes of complex V (ATP synthase; ATP6 and ATP8). There is also the control region (between the nucleotides positions 16024-576) that contains the origin of replication of the heavy strand, as well as the origin of transcription for both strands.

The mtDNA is organized in protein-DNA structures called mitochondrial nucleoids, which are the mitochondrial units of inheritance. The mtDNA replication machinery is located within this structure and consists of five (known) main proteins: (1) the TWINKLE helicase has a 5′ to 3′ DNA helicase activity, facilitating mtDNA synthesis, (2) the mitochondrial single-stranded binding protein that stabilizes the single stranded conformation, (3) the DNA polymerase gamma (POLG) is responsible for DNA synthesis and repair, (4) the mitochondria RNA polymerase that synthesizes the RNA primer for light strand DNA synthesis, and (5) the mitochondrial transcription factor A (TFAM) that binds to the mtDNA and acts as a chaperone (among other known functions) [6, 53]. mtDNA replication is a continuous process, which occurs independently from the cell cycle, see Yasukawa and Kang [56] for an overview of proposed mtDNA replication mechanisms.

The transcription of human mtDNA starts from promoters in both heavy and light strands. Briefly, each strand is transcribed as a polycistronic precursor mRNA molecule. The transcript elongation is performed by the mitochondrial RNA polymerase, and two transcription factors (TFAM and mitochondrial transcription factor B2, TFB2M). The termination of mature transcripts is performed by mitochondrial termination factor 1 (see D'Souza and Minczuk [12] for more detailed description).

The final products are two long polycistronic transcripts, which are processed into individual tRNAs, rRNAs, and mRNAs, through endonucleolytic cleavage, although this step is not applicable to all mitochondrial mRNAs [12]. Post-transcriptional editions (maturation) include chemical modifications (such as a "wobble" base at position 34) and the addition of the CCA sequence in the tRNAs, as well as polyadenylation (polyA tail) in rRNAs and mRNAs (except MT-ND6). Aminoacyl tRNA-synthetases (many of them are unique to mitochondria) attach the appropriate amino acid to its mt-tRNAs [12].

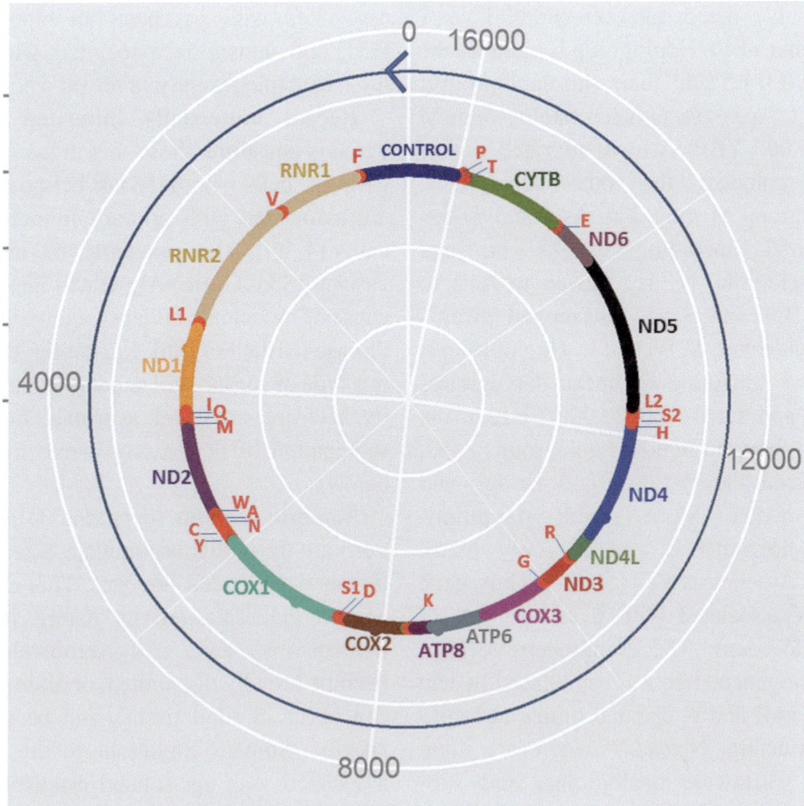


**Fig. 13.1** Map of human mtDNA with genes marked. The mtDNA genes names are shown. The tRNAs are marked in red and control region in blue. Numbers correspond to nucleotide positions in the mtDNA molecule

### 13.2.1 The Mitochondrial DNA Variants

The mtDNA is present in multiple copies inside each mitochondrion, and there are hundreds to thousands of mitochondria in each of our cells (although numbers vary across cell types). mtDNA has a high mutation rate (100–1000 fold higher than nuclear genome) with an accumulation of neutral/mildly deleterious variants, most likely due to its effective small population size and haploid inheritance [26]. Notwithstanding, highly deleterious variants are eliminated by a (yet poorly understood) germline selection, which prevents their fixation in the mtDNA of an organism's progeny [14, 38, 46].Thus, the mitochondrion is a mutation-prone system, and mutations are constantly arising (including pathogenic mutations) [51]. The prevalence in adults of primary mtDNA disease is 1:5000 [34].

There are two types of mtDNA variants: **homoplasmic** when the mutation affects all mtDNA copies and **heteroplasmic** when the mutation is present in some copies. Homoplasmic variants are further classified into three types: ancient adaptive, recent maternally inherited deleterious and somatic variants.

**Ancient adaptive** variants are those accumulated during human migration across the globe and were important for the adaptation to different climatic environments. Thus, these variants are geographically specific, and they are ancestry markers for mtDNA [19]. Mitochondrial **haplogroups** are defined as an array of sequences sharing specific ancient adaptive variants due to common matrilineal ancestor and by convention are annotated by combinations of letters and numbers.

Briefly, African mtDNA sequences are the most ancient ones, and they form the macrohaplogroup "L", dated 130,000–200,000 years before present (YBP). Haplogroup L includes the haplogroups L0-L5 and their sub-haplogroups. Haplogroup L3 gave rise to macrohaplogroups M and N (~65,000 YBP), which migrated out of Africa to colonize the other continents. Macrohaplogroup M moved to Asia and generated M1-M80, including C, D, G, and Z. Macrohaplogroup N also moved to Asia to generate A, B, F, and O, and also moved towards Europe to generate I, X, W, and R. Haplogroup R generated the remaining European haplogroups H, V, J, U, and T. Around 20,000 YBP, in the northeastern Siberian region, haplogroups A, C, and D crossed the Bering Land Bridge and reached America. More recent migrations brought haplogroups B and X (only North America) to America as well [50, 51]. Thus, each continent is associated with a specific set of related haplogroups. The comprehensive haplogroup phylogenetic tree was established by van Oven et al. [44] and is updated online at http://www.phylotree.org. Nowadays, there are more than 5400 worldwide haplogroups and sub-haplogroups currently described at Phylotree (build 17).

It is also important to mention that the impact of a mtDNA variant is influenced by the mitochondrial background (haplogroup) where the variant arose [51]. A classical example, the mutant 3394C (Y30H) is found in both haplogroups N and M. When it arises in haplogroup N, it is associated with decreased complex I activity and increased penetrance of milder complex I variants associated with the Leber hereditary optic neuropathy (LHON). When it arises on haplogroup M in populations living in high altitude, it is associated with increased complex I activity and appears to be crucial for the adaptation to high altitude (revised by Wallace and Chalkia [51]).

Haplogroups are functionally distinct from each other [17, 20], and several studies have found haplogroup-specific susceptibility to a variety of phenotypes ranging from neuropsychiatric disorders to aging profiles [31, 47]. For example, the Asian haplogroup N9a has been associated with type 2 diabetes [16]. The haplogroup U is associated with psychosis in bipolar disorder [15], and autism is associated with European mtDNA haplogroups in a recent study [4].

**Recent maternally inherited deleterious** variants constitute those that arose in the female germline cells within the last ten generations and are associated with primary mitochondrial diseases [4, 49]. As an example, the heteroplasmic variant A8344G (tRNALys) was identified as the cause of myoclonic epilepsy and ragged red fiber disease (MERRF) [36]. In general, the impact of this type of variant will depend on its pathogenicity, heteroplasmy level, as well as the replicative segregation of the heteroplasmic mutation (see below).

**Somatic variants** (deletions or base substitution) are those that accumulate over time within tissues or stem cell lineages. This class of variants is heteroplasmic by nature [4]. Somatic mutants may arise early in development and become broadly distributed, or alternatively, they can occur in adult tissues and be tissue/organ-specific. Somatic mutations in the mtDNA are associated with age-related diseases [48]. Both genetic and environmental factors appear to influence the accumulation of somatic variants in the mtDNA [49].

A **heteroplasmic** variant is present in only a subset of the mtDNA molecules and can arise in both somatic and germline cells. Heteroplasmic variants segregate to daughter cells through a mechanism termed **replicative segregation**. This cellular process (although not completely understood) consists of a random shift in heteroplasmic allele frequency during meiotic or mitotic cell division [51]. As result, the proportion of mutant-to-normal mtDNA (heteroplasmy level) can vary among tissues and organs [39, 51].

Briefly, in *meiotic replicative segregation* (or *vegetative segregation*), when a cell divides, the daughter cells can receive different proportions of mutant mtDNA copies, and, as a consequence, maternal relatives will show a spectrum of clinical phenotypes. In *mitotic segregation* (or *relaxed segregation*), the percentage of mutant copies will vary across tissues in the same individual, derived originally from a heteroplasmic oocyte. Interestingly, the level of heteroplasmy within a given cell can increase, decrease or remain stable over time, and this effect is often tissue-specific. The underlying mechanisms are not yet known, but it is possible that nuclear genes influence this regulation. See [39] for an overview of heteroplasmy segregation mechanisms.

There is evidence for quality control mechanisms restricting the propagation of damaged or heteroplasmic mtDNA molecules. For example, according to the *bottleneck* hypothesis, there is a reduction in the number of mtDNA molecules in the primordial germline cells in order to restrict inherited heteroplasmy [11], although the exact mechanism remains unclear. Thus, heteroplasmy levels shift between generations, influencing their clinical manifestations among family members.

Pathogenic heteroplasmic variants can be found to be associated with mitochondrial genetic disorders [35], and the clinical symptomatology usually correlates with the degree of heteroplasmy [39]. Indeed, there is also a threshold (tissue-specific) that is crucial for the clinical manifestation of the diseases or biochemical defects associated with the mutation [41], namely **threshold effect** (in general, this threshold level is 80% [39]).

Overall, hundreds of pathogenic variants (homoplasmic and heteroplasmic) have been reported for mtDNA. These variants are discovered mainly (and more accurate) through next generation sequencing approach, and nowadays there are many specific tools for variants call and data quality control, such as mtDNA-server [54]. After identification, mitochondrial variants can be loaded on specific databases such as the MITOMAP-MITOMASTER website maintains a comprehensive catalogue of these variants and is publicly available at https://www.mitomap.org/MITOMAP.

## 13.3 Nuclear-Encoded Mitochondrial Gene Variants

There are more than 1000 mitochondrial genes that are encoded by the nuclear genome. A list of these genes can be found through the MitoCarta dataset, which consists of an inventory of human genes with strong evidence of mitochondrial localization [3]. Variants inside nuclear-encoded mitochondrial genes have been identified (see MITOMAP-MITOMASTER website for a catalogue of these variants).

Nuclear-encoded mitochondrial gene variants have been associated with primary and secondary mitochondrial diseases. For example, Leigh Syndrome is a pediatric mitochondrial disease with more than 75 known mutations in both nuclear-encoded mitochondrial genes and mtDNA [21]. Examples of diseases with evidence for a mitochondrial component (i.e, secondary mitochondrial disease) include psychiatric and brain diseases such as schizophrenia [18], bipolar disorder [1], Alzheimer's and Parkinson's diseases [27], and Autism spectrum disorders [4]. This class of mitochondrial genes is analyzed like any other nuclear gene, and they are included in many large and global genomics consortiums for a range of complexes diseases.

## 13.4 Mitochondrial Pseudogenes

Mitochondrial pseudogenes (NUMTs) are sequences of mtDNA that have been incorporated into the nuclear genome. NUMTs are not expressed and they are also nonfunctional, however, they pose a challenge to the amplification and further analysis of mtDNA. The polymerase chain reaction (PCR) primers designed to amplify mtDNA may inadvertently hybridize to NUMTs as well [58], sometimes preferentially [9]. Currently, NUMTs are a challenging technical nuisance that have led to false positive results in mtDNA analysis [52]. Many of the tools available for analysis of the mtDNA, such as mtDNA-Server [54] do not filter out NUMTs yet, and researchers need to be aware about this potential confounder.

It is also worth mention that several studies have correlated non-mitochondrial pseudogenes with risk for several diseases such as cancer [28]. It is likely that NUMTs will play a role in the future of mitochondrial genetic analysis.

## 13.5 Mitonuclear Communication

As discussed above, the mitochondrial system is bi-genomic, requiring a tight co-evolution of the interactions between mtDNA and the nuclear genome [37]. Thus, variants in epistatic interactions between mtDNA and nuclear DNA can lead to disease. It is suggested that even mild nuclear variants may become clinically relevant when interacting with an incompatible mtDNA [51].

Indeed, there are reports for variants in mitonuclear interactions causing disease. For example, in a case report study, the G32R mutation in *NDUFA1* gene was associated with a complex I deficiency in two subjects (siblings), and its effect appears to be influenced by epistasis interaction with additional mtDNA variants from mtDNA complex I genes [29]. Another study reported an interaction between the *APOE* nuclear gene and the mitochondrial haplogroup K with respect to influencing severity for traumatic brain injury [2]. Furthermore, a study of late onset Alzheimer's disease found significant interaction between nuclear variant *APOE4+* status and mitochondrial sub-haplogroup H5 [23]. Genetic variation in the mitonuclear interaction is proposed to explain variation in the penetrance, spectrum of symptoms, as well as the severity of mitochondrial diseases or diseases with a mitochondrial component [55].

There are also incompatibilities leading to loss of fitness due to differences in ancestry between nuclear DNA and mtDNA. This scenario is named the mitochondrial-nuclear DNA mismatch (or misalignment) hypothesis and it is based on the fact that ancient adaptive variants in mtDNA co-evolved with the nuclear genome in their native population. These variants might be harmful when interacting with a nuclear DNA context with a distinct ancestry, potentially altering the cellular physiology and leading to diseases. The importance of this phenomenon is well demonstrated in crossbreeding studies in animal models [22, 25, 32, 57]. For example, conplastic mice generated probands with altered performances in cognition functioning (learning, exploration) when compared with the wild strains [32]. In human, Crawford et al. [10] showed that spontaneous preterm birth was more prevalent in infants with divergent mitochondrial-nuclear DNA ancestries; however, more studies are warranted to fully understand the consequences of ancestry mismatch between nuclear DNA and mtDNA.

## 13.6 Mitochondria and Epigenetics

The role of mitochondria in the regulation of nuclear epigenetic mechanisms is an emerging aspect of mitonuclear communication. It is known that transcription and translation in both mtDNA and nuclear DNA are co-regulated, in a way that mitochondria function is controlled by the nucleus to meet the energetic cellular demand. On the other hand, mitochondria control nuclear gene expression for metabolic reprogramming [24].

There is evidence for mitochondria regulating nuclear epigenetics. Briefly, some of the mitochondrial intermediate metabolites, such as acetyl CoA, S-adenosyl methionine (SAM), and NAD+, play a role in the histones and nuclear DNA epigenetic modifications. For example, histone acetyltransferases (HATs) are dependent of the levels of acetyl-CoA, which is an intermediate metabolite of the mitochondrial tricarboxylic acid cycle. Acetyl-CoA is important for histone acetylation, and then, transcriptional activation. Another mitochondria-mediated epigenetic modification is histone methylation. Histone methyltransferases use SAM as a precursor in methyl group transfer [42]. SAM is produced through methionine-homocysteine cycle, which depends on mitochondrial folate cycle and ATP. Also, nuclear DNA methylation (particularly 5-methylcytosine – 5mC) is another epigenetic mechanism to alter chromatin structure and control gene expression. Mitochondria are also involved in this mechanism since three DNA-methyltransferases (DNMT1, 3A and 3B) uses SAM as a methyl donor. On the other hand, there is also evidence for mitochondria in the demethylation of DNA and histones, as well as deacetylation of histones (revised at Metilainen et al. [24]).

There is also evidence for epigenetic modifications (methylation) for the mtDNA. The set of evidence include the presence of 435 CpG sites in the human mtDNA. Also, the DNMT1 was showed to translocate to the mitochondria as well as the TET1 and TET2 (two DNA demethylation) were also found in mitochondria [5, 24]. Furthermore, the mtDNA methylation status is proposed to be a future biomarker for the detection and diagnosis of diseases involving altered epigenetics [24]. For example, an analysis of the human mtDNA methylome using postmortem brain schizophrenia samples reported tissue-specific patterns between blood, cerebellum and cortex of patients, and also reported regions differentially methylated between cases and matched controls [13].

The understanding of the mitonuclear communication and epigenetics holds the potential to shed light on clinical interventions in diseases with a mitochondrial component, which might be managed by therapeutical targeting of the epigenome.

## 13.7 Conclusions

This chapter discussed the main features and complexity of the mitochondrial genetics, including variants, haplogroups, mitonuclear communication and epigenetics. Currently, the field is in fast expansion, and novel methodological approaches targeting mitochondria have been developed. The future will sure integrate mitochondrial system as part of etiology of a range of complex disorders, and mitochondrial system holds the potential to be used as a molecular tool for diagnosis and personalized medicine.

**Acknowledgments** I would like to thank Mr. Ari Cuperfain , Dr. Emiko Koyama and Ayeshah Mohiuddin. I am also thankful for Judy and Larry Tanenbaum Family Foundation for support.

## References

1. Andreazza AC, Duong A, Young LT (2018) Bipolar disorder as a mitochondrial disease. Biol Psychiatry 83:720–721
2. Bulstrode H, Nicoll JA, Hudson G, Chinnery PF, Di Pietro V, Belli A (2014) Mitochondrial DNA and traumatic brain injury. Ann Neurol 75:186–195
3. Calvo SE, Clauser KR, Mootha VK (2016) MitoCarta2.0: an updated inventory of mammalian mitochondrial proteins. Nucleic Acids Res 44(D1):D1251–D1257
4. Chalkia D, Singh LN, Leipzig J, Lvova M, Derbeneva O, Lakatos A, Hadley D, Hakonarson H, Wallace DC (2017) Association between mitochondrial DNA Haplogroup variation and autism spectrum disorders. JAMA Psychiat 74:1161–1168
5. Chinnery PF, Elliott HR, Hudson G, Samuels DC, Relton CL (2012) Epigenetics, epidemiology and mitochondrial DNA diseases. Int J Epidemiol 41:177–187
6. Chinnery PF, Hudson G (2013) Mitochondrial genetics. Br Med Bull 106:135–159
7. Ciafaloni E, Ricci E, Shanske S, Moraes CT, Silvestri G, Hirano M, Simonetti S, Angelini C, Donati MA, Garcia C, ET AL (1992) MELAS: clinical features, biochemistry, and molecular genetics. Ann Neurol 31:391–398
8. Cieri D, Brini M, Cali T (2017) Emerging (and converging) pathways in Parkinson's disease: keeping mitochondrial wellness. Biochem Biophys Res Commun 483:1020–1030
9. Collura RV, Stewart C-B (1995) Insertions and duplications of mtDNA in the nuclear genomes of Old World monkeys and hominoids. Nature 378:485
10. Crawford N, Prendergast D, Oehlert JW, Shaw GM, Stevenson DK, Rappaport N, Sirota M, Tishkoff SA, Sondheimer N (2018) Divergent patterns of mitochondrial and nuclear ancestry are associated with the risk for preterm birth. J Pediatr 194:40–46 e4
11. Cree LM, Samuels DC, De Sousa Lopes SC, Rajasimha HK, Wonnapinij P, Mann JR, Dahl H-HM, Chinnery PF (2008) A reduction of mitochondrial DNA molecules during embryogenesis explains the rapid segregation of genotypes. Nat Genet 40:249–254
12. D'Souza AR, Minczuk M (2018) Mitochondrial transcription and translation: overview. Essays Biochem 62:309–320
13. Devall M, Smith RG, Jeffries A, Hannon E, Davies MN, Schalkwyk L, Mill J, Weedon M, Lunnon K (2017) Regional differences in mitochondrial DNA methylation in human post-mortem brain tissue. Clin Epigenetics 9:47
14. Fan W (2008) A mouse model of mitochondrial disease reveals germline selection against severe mtDNA mutations. Science 319:958–962

15. Frye MA, Ryu E, Nassan M, Jenkins GD, Andreazza AC, Evans JM, Mcelroy SL, Oglesbee D Jr, Highsmith WE, Biernacka JM (2017) Mitochondrial DNA sequence data reveals association of haplogroup U with psychosis in bipolar disorder. J Psychiatr Res 84:221–226
16. Fuku N, Park KS, Yamada Y, Nishigaki Y, Cho YM, Matsuo H, Segawa T, Watanabe S, Kato K, Yokoi K, Nozawa Y, Lee HK, Tanaka M (2007) Mitochondrial haplogroup N9a confers resistance against type 2 diabetes in Asians. Am J Hum Genet 80:407–415
17. Gomez-Duran A, Pacheu-Grau D, Lopez-Gallardo E, Diez-Sanchez C, Montoya J, Lopez-Perez MJ, Ruiz-Pesini E (2010) Unmasking the causes of multifactorial disorders: OXPHOS differences between mitochondrial haplogroups. Hum Mol Genet 19:3343–3353
18. Gonçalves VF, Cappi C, Hagen CM, Sequeira A, Vawter MP, Derkach A, Zai CC, Hedley PL, Bybjerg-Grauholm J, Pouget JG, Cuperfain AB, Sullivan PF, Christiansen M, Kennedy JL, Sun L (2018) A comprehensive analysis of nuclear-encoded mitochondrial genes in schizophrenia. Biol Psychiatry 83:780–789
19. Gonçalves VF, Carvalho CM, Bortolini MC, Bydlowski SP, Pena SD (2008) The phylogeography of African Brazilians. Hum Hered 65:23–32
20. Ji F, Sharpley MS, Derbeneva O, Alves LS, Qian P, Wang Y, Chalkia D, Lvova M, Xu J, Yao W, Simon M, Platt J, Xu S, Angelin A, Davila A, Huang T, Wang PH, Chuang LM, Moore LG, Qian G, Wallace DC (2012) Mitochondrial DNA variant associated with Leber hereditary optic neuropathy and high-altitude Tibetans. Proc Natl Acad Sci U S A 109:7391–7396
21. Lake NJ, Compton AG, Rahman S, Thorburn DR (2016) Leigh syndrome: one disorder, more than 75 monogenic causes. Ann Neurol 79:190–203
22. Latorre-Pellicer A, Moreno-Loshuertos R, Lechuga-Vieco AV, Sanchez-Cabo F, Torroja C, Acin-Perez R, Calvo E, Aix E, Gonzalez-Guerra A, Logan A, Bernad-Miana ML, Romanos E, Cruz R, Cogliati S, Sobrino B, Carracedo A, Perez-Martos A, Fernandez-Silva P, Ruiz-Cabello J, Murphy MP, Flores I, Vazquez J, Enriquez JA (2016) Mitochondrial and nuclear DNA matching shapes metabolism and healthy ageing. Nature 535:561–565
23. Maruszak A, Safranow K, Branicki W, Gaweda-Walerych K, Pospiech E, Gabryelewicz T, canter j a, Barcikowska m, Zekanowski C (2011) The impact of mitochondrial and nuclear DNA variants on late-onset Alzheimer's disease risk. J Alzheimers Dis 27:197–210
24. Matilainen O, Quiros PM, Auwerx J (2017) Mitochondria and epigenetics – crosstalk in homeostasis and stress. Trends Cell Biol 27:453–463
25. Nagao Y, Totsuka Y, Atomi Y, Kaneda H, Lindahl KF, Imai H, Yonekawa H (1998) Decreased physical performance of congenic mice with mismatch between the nuclear and the mitochondrial genome. Genes Genet Syst 73:21–27
26. Neiman M, Taylor DR (2009) The causes of mutation accumulation in mitochondrial genomes. Proc Biol Sci 276:1201–1209
27. Onyango IG, Khan SM, Bennett JP Jr (2017) Mitochondria in the pathophysiology of Alzheimer's and Parkinson's diseases. Front Biosci (Landmark Ed) 22:854–872
28. Pink RC, Wicks K, Caley DP, Punch EK, Jacobs L, Carter DRF (2011) Pseudogenes: pseudo-functional or key regulators in health and disease? RNA 17:792–798
29. Potluri P, Davila A, Ruiz-Pesini E, Mishmar D, O'Hearn S, Hancock S, Simon M, Scheffler IE, Wallace DC, Procaccio V (2009) A novel NDUFA1 mutation leads to a progressive mitochondrial complex I-specific neurodegenerative disease. Mol Genet Metab 96:189–195
30. Rahman S, Blok RB, Dahl HH, Danks DM, Kirby DM, Chow CW, Christodoulou J, Thorburn DR (1996) Leigh syndrome: clinical features and biochemical and DNA abnormalities. Ann Neurol 39:343–351
31. Rishishwar L, Jordan IK (2017) Implications of human evolution and admixture for mitochondrial replacement therapy. BMC Genomics 18:140
32. Roubertoux PL, Sluyter F, Carlier M, Marcet B, Maarouf-Veray F, Cherif C, Marican C, Arrechi P, Godin F, Jamon M, Verrier B, Cohen-Salmon C (2003) Mitochondrial DNA modifies cognition in interaction with the nuclear genome and age in mice. Nat Genet 35:65–69
33. Santos RX, Correia SC, Zhu X, Smith MA, Moreira PI, Castellani RJ, Nunomura A, Perry G (2013) Mitochondrial DNA oxidative damage and repair in aging and Alzheimer's disease. Antioxid Redox Signal 18:2444–2457
34. Schaefer AM, Mcfarland R, Blakely EL, HE L, Whittaker RG, Taylor RW, Chinnery PF, Turnbull DM (2008) Prevalence of mitochondrial DNA disease in adults. Ann Neurol 63:35–39
35. Sequeira A, Martin MV, Rollins B, Moon EA, Bunney WE, Macciardi F, Lupoli S, Smith EN, Kelsoe J, Magnan CN (2012) Mitochondrial mutations and polymorphisms in psychiatric disorders. Front Genet 3:103
36. Shoffner JM, Lott MT, Lezza AM, Seibel P, Ballinger SW, Wallace DC (1990) Myoclonic epilepsy and ragged-red fiber disease (MERRF) is associated with a mitochondrial DNA tRNA(Lys) mutation. Cell 61:931–937
37. Sloan DB, Fields PD, Havird JC (2015) Mitonuclear linkage disequilibrium in human populations. Proc Biol Sci 282
38. Stewart JB (2008) Strong purifying selection in transmission of mammalian mitochondrial DNA. PLoS Biol 6:e10
39. Stewart JB, Chinnery PF (2015) The dynamics of mitochondrial DNA heteroplasmy: implications for human health and disease. Nat Rev Genet 16:530–542

40. Taanman JW (1999) The mitochondrial genome: structure, transcription, translation and replication. Biochim Biophys Acta 1410:103–123
41. Taylor RW, Turnbull DM (2005) Mitochondrial DNA mutations in human disease. Nat Rev Genet 6:389–402
42. Teperino R, Schoonjans K, Auwerx J (2010) Histone methyl transferases and demethylases; can they link metabolism and transcription? Cell Metab 12:321–327
43. Vafai SB, Mootha VK (2012) Mitochondrial disorders as windows into an ancient organelle. Nature 491:374–383
44. Van Oven M, Kayser M (2009) Updated comprehensive phylogenetic tree of global human mitochondrial DNA variation. Hum Mutat 30:E386–E394
45. Verge B, Alonso Y, Valero J, Miralles C, Vilella E, Martorell L (2011) Mitochondrial DNA (mtDNA) and schizophrenia. Eur Psychiatry 26:45–56
46. Wai T, Teoli D, Shoubridge EA (2008) The mitochondrial DNA genetic bottleneck results from replication of a subpopulation of genomes. Nat Genet 40:1484–1488
47. Wallace DC (2005) A mitochondrial paradigm of metabolic and degenerative diseases, aging, and cancer: a dawn for evolutionary medicine. Annu Rev Genet 39:359–407
48. Wallace DC (2010) Mitochondrial DNA mutations in disease and aging. Environ Mol Mutagen 51:440–450
49. Wallace DC (2013) A mitochondrial bioenergetic etiology of disease. J Clin Invest 123:1405–1412
50. Wallace DC (2015) Mitochondrial DNA variation in human radiation and disease. Cell 163:33–38
51. Wallace DC, Chalkia D (2013) Mitochondrial DNA genetics and the heteroplasmy conundrum in evolution and disease. Cold Spring Harb Perspect Med 3:a021220
52. Wallace DC, Stugard C, Murdock D, Schurr T, Brown MD (1997) Ancient mtDNA sequences in the human nuclear genome: a potential source of errors in identifying pathogenic mutations. Proc Natl Acad Sci 94:14900–14905
53. Wanrooij S, Falkenberg M (2010) The human mitochondrial replication fork in health and disease. Biochim Biophys Acta 1797:1378–1388
54. Weissensteiner H, Forer L, Fuchsberger C, Schopf B, Kloss-Brandstatter A, Specht G, Kronenberg F, Schonherr S (2016) mtDNA-server: next-generation sequencing data analysis of human mitochondrial DNA in the cloud. Nucleic Acids Res 44:W64–W69
55. Wolff JN, Ladoukakis ED, Enriquez JA, Dowling DK (2014) Mitonuclear interactions: evolutionary consequences over multiple biological scales. Philos Trans R Soc Lond Ser B Biol Sci 369:20130443
56. Yasukawa T, Kang D (2018) An overview of mammalian mitochondrial DNA replication mechanisms. J Biochem 164:183–193
57. Yu X, Gimsa U, Wester-Rosenlof L, Kanitz E, Otten W, Kunz M, Ibrahim SM (2009) Dissecting the effects of mtDNA variations on complex traits using mouse conplastic strains. Genome Res 19:159–165
58. Zhang D-X, Hewitt GM (1996) Nuclear integrations: challenges for mitochondrial DNA markers. Trends Ecol Evol 11:247–251

# Heteroplasmy Shifting as Therapy for Mitochondrial Disorders

14

Mansur M. Naeem and Neal Sondheimer

**Abstract**

Mitochondrial disease can arise due to pathogenic sequence variants in the mitochondrial DNA (mtDNA) that prevent cells from meeting their energy demands. Mitochondrial diseases are often fatal and currently there are no treatments directed towards the underlying cause of disease. Pathogenic variants in mtDNA often exist in a state of heteroplasmy, with coexistence of pathogenic and wild type mtDNA. The load of heteroplasmy, defined as the relative amount of pathogenic mtDNA to wild type mtDNA, corresponds to timing and symptom severity. Thus, changing the heteroplasmy load may lead to a shift in disease onset and symptom severity. Here we review techniques aimed at preventing inheritance of pathogenic mtDNA via mitochondrial replacement therapy (MRT) and strategies geared toward shifting of heteroplasmy in individuals with active mitochondrial disease. MRT strategies seek to create embryos with the nuclear genetic makeup of the intended parents and wild type mtDNA from a donor in order to avoid known maternal pathogenic variants. Heteroplasmy shift approaches in patients are of two categories: nuclease dependent and nuclease independent strategies. Despite initial success in mouse models and patient cells, these techniques have not reached clinical use. Translational attempts in this area are urgently needed to improve therapies for a currently untreatable set of disorders.

**Keywords**

Mitochondrial disease · Mitochondrial DNA · Heteroplasmy · Mitochondrial replacement therapy

## 14.1 Introduction

Mitochondria are semi-autonomous organelles found in all nucleated eukaryotic cells. They are responsible for a wide range of cellular functions including heme synthesis, calcium sequestration and fatty acid oxidation. However, the informational content of the mitochondria is directed exclusively toward electron transport and oxidative phosphorylation (collectively termed OXPHOS). Human mtDNA has numerous features that set it apart from the nuclear chromosomes. It is circular, densely coding, maternally inherited, lacks introns and has differences in underlying genetic code compared to the nucleus. Beyond the 13 encoded polypeptides, it also houses the ribosomal and transfer RNAs needed for mitochondrial translation.

M. M. Naeem · N. Sondheimer (✉)
Institute of Medical Science, The University of Toronto, Toronto, ON, Canada
e-mail: neal.sondheimer@sickkids.ca





A. Urbani, M. Babu (eds.), *Mitochondria in Health and in Sickness*, Advances in Experimental Medicine and Biology 1158, https://doi.org/10.1007/978-981-13-8367-0_14

There is a required synergy between the nuclear DNA (nDNA) and mitochondrial DNA (mtDNA), because the mtDNA only encodes a subset of the polypeptides that are required for OXPHOS. Mitochondrial diseases can arise from pathogenic variants in either mtDNA or nDNA-encoded proteins with mitochondrial function. Because mitochondria play key roles in virtually all cells and tissues, mitochondrial diseases have heterogeneous and complex presentations. Tissues with high energy demand such as the brain, skeletal muscle, kidneys and heart are often the most highly affected [5]. In total mitochondrial diseases have a prevalence of 1 in 5000 and of these approximately half are due to mtDNA variants [43]. Unfortunately, there are no targeted treatments for virtually any of the forms of mitochondrial disease and care for patients is largely limited to symptomatic management [41].

## 14.2 MtDNA Pathogenic Variants and Heteroplasmy

Unlike nDNA, there are multiple copies of mtDNA within a cell and approximately 2–10 mtDNA per mitochondrion [41]. Due to its multicopy nature, pathogenic variants in mtDNA typically present with the co-existence of both wild type and variant bearing genomes, known as heteroplasmy. In classical Mendelian genetics individuals may possess two wild type alleles, one wild type allele and one allele with a pathogenic variant (conceptually a 50% load) or two pathogenic alleles (100% load). By contrast, heteroplasmic mtDNA variants can affect any fraction of the total mtDNA population, allowing heteroplasmy to lie in a continuum between 0% and 100%. Complete presence of a pathogenic variant (known as homoplasmy) is only seen for variants that cause disorders that are either incompletely penetrant or are not reproductively lethal. The classic example of a homoplasmic disease is Leber's Hereditary Optic Neuropathy, a painless, sequential blindness that is environmentally triggered [25].

For all other disorders associated with pathogenic mtDNA variants, higher loads of heteroplasmy are associated with more severe disease. For instance, patients who harbor 70–90% m.8993T>G variant present with neuropathy, ataxia and retinitis pigmentosa (NARP), a late onset, debilitating, but non-fatal condition. However, if the pathogenic load is above 90% then patients may present with maternally inherited Leigh syndrome (MILS), an early onset and typically rapidly fatal encephalopathy [42]. Since disease severity, timing, and symptoms all depend on the heteroplasmy level, techniques that would directionally shift heteroplasmy have been sought for many years ([50]).

Heteroplasmy levels are not uniform across tissues or over time, and heteroplasmy can also undergo drastic changes between generations. Mitochondrial replication is independent of cell cycle and is guided by cellular energy needs. As such, the percentage of pathogenic mtDNA passed to daughter cell during division may not reflect the percent heteroplasmy of the parent cell. Moreover, cells with very high variant load may be selected against in highly mitotic cells such as leukocytes. For example, in patients with the m.3243A>G variant, the blood level of pathogenic mtDNA decreases over time [36].

In contrast, pathogenic heteroplasmies may be more likely to accumulate in post-mitotic high energy tissues such as muscle [47]. The mechanism for redistribution of heteroplasmy in non-dividing cells is mitophagy (mitochondrial autophagy). Mitophagy is process of selective mitochondrial degradation, promoting cellular health by blocking the accumulation of dysfunctional mitochondria [58]. In the absence of mitophagy, damaged mitochondria release cytochrome C which activates caspases and leads to apoptosis [9].

mtDNA is more susceptible to somatic mutation than nDNA. Although it was initially thought that this was due to a loss of most repair mechanisms, recent studies indicate that these repair mechanisms may be more widely intact [23, 46]. Instead, the increased mutation rate of mtDNA may be due to the oxidative environment of the mitochondrial matrix and high frequency of DNA replication [5]. Indeed the integrity of mitochondrial DNA is under constant threat by reactive oxidative species (ROS) that are generated during OXPHOS [17]. ROS are known to be cytotoxic and mutagenic. mtDNA replication occurs frequently and every replication cycle presents opportunities for the introduction of mutation. The highly oxidative environment and increased replication leads to accumulation of somatic mutations with age [45].

The accumulation of mtDNA variants over time would be catastrophic for reproduction if unchecked. The impact of mutation in asexual, non-recombining genomes is known as Muller's ratchet, and leads to the loss of essential genetic information. Hence there are mechanisms in place to reduce accumulation of pathogenic mtDNA variants. First, mothers with high level of mtDNA variants have decreased fertility [8, 39]. Second, profoundly pathogenic mtDNA variants can be counter-selected in the developing embryo [13]. Nonetheless, pathogenic variants can shift heteroplasmy rapidly within the germline because of a bottleneck for mtDNA that separates generations. This bottleneck is caused by the amplification of a small number of initial mtDNA copies in the oocyte lineage which leads to distributions of resulting heteroplasmy in the next generation (Fig. 14.1) [56].

Since there is little non-coding mtDNA, mutations often fall in coding regions and may have detrimental effects on mitochondrial activity. We tolerate most low-level mtDNA variants well. Indeed, a substantial level of pathogenic-variant bearing mtDNA needs to accumulate before clinical manifestations appear, a concept known as

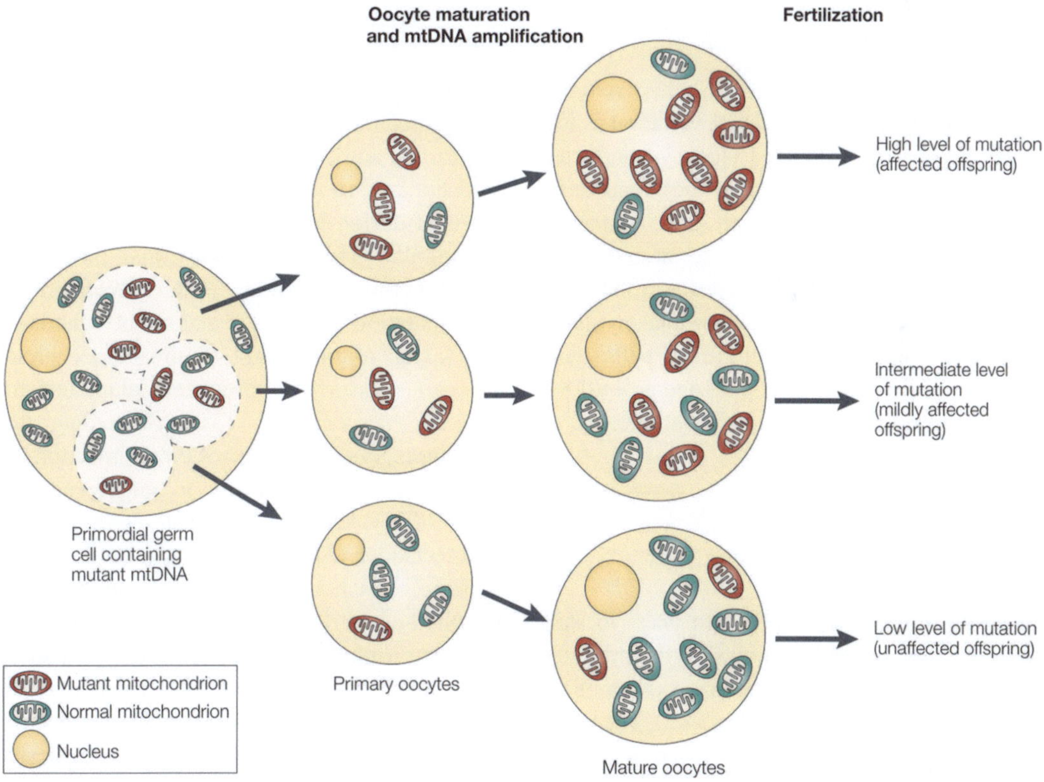


**Fig. 14.1** Adapted: Taylor and Turnbull [52]. The mitochondrial genetic bottleneck. During the production of primary oocytes, only a subset of maternal mitochondrial DNA (mtDNA) molecules are transferred into each developing oocyte, which is extensively replicated to generate mature oocyte. Since the total population of mtDNA in mature oocyte comes only from a subset of mtDNA selected from primordial germ cell, this bottleneck can lead to random shift of mtDNA mutational load between generations. Mitochondria with pathogenic mtDNA are shown in red, those with wild type mtDNA are shown in green

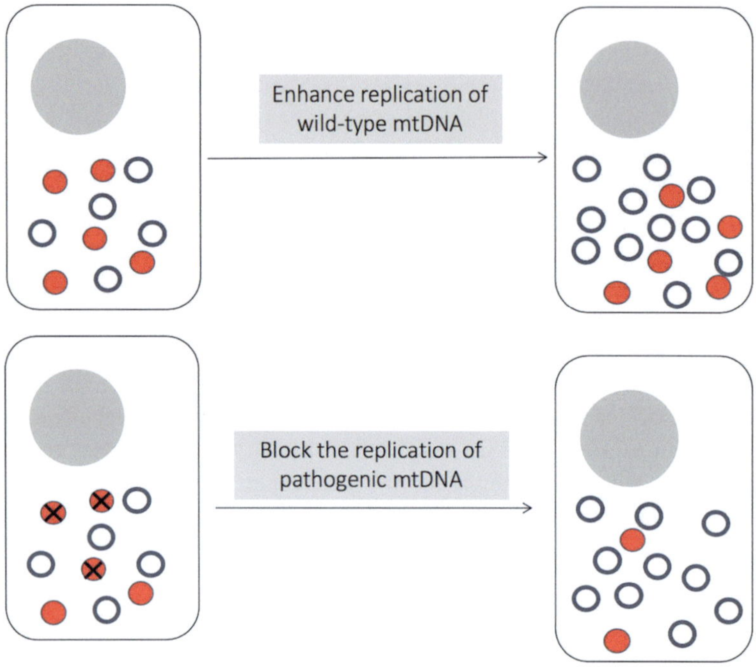


**Fig. 14.2** Conceptual diagram for heteroplasmy shift. Filled (red) circle indicate pathogenic mtDNA and hollow (blue) circles represent wild type mtDNA. Shift can be achieved by enhancing replication of wild type mtDNA or selective inhibition of replication of pathogenic mtDNA

the critical threshold. Low-level variation is tolerated because it is compensated by unaffected mtDNA. Indeed, the presence of occasional damaged mtDNA is likely a small price to pay to avoid the expense of maintaining an arsenal of mtDNA repair proteins [22]. Hence in therapies aimed at directional shift in heteroplasmy, there is no need to completely eliminate pathogenic mtDNA, and a decrease of pathogenic mtDNA below critical threshold would be sufficient.

## 14.3 Mitochondrial Transplantation

There are only conceptually two ways to shift heteroplasmy, either by increasing the presence of wild type mtDNA or impairing the replication of pathogenic mtDNA (Fig. 14.2). Therapies under development attempt to insert wild type mtDNA or healthy mitochondria into cells with known or potential mitochondrial dysfunction. A technique described as mitochondrial transplantation was developed for patients suffering from mitochondrial injury during myocardial ischemia and reperfusion injury [33]. Mitochondrial function and structure is severely damaged during myocardial ischemia and this compromises cellular viability and post-ischemic functional recovery [27]. Mitochondrial transplantation involves isolation of functionally intact mitochondria from non-ischemic autologous tissue and direct transplantation to ischemic myocardium [31]. According to the authors, one advantage of direct injection of mitochondria is the ability to target specific region of myocardium [10]. The technique has been tested on pediatric patients suffering from myocardial ischemia reperfusion injury [11]. Although an intriguing approach, it is unlikely that this technique would be feasible for patients with mitochondrial disease. Such patients are globally affected, so isolating mitochondria for auto-transplantation would present significant challenges. Furthermore, mitochondrial diseases typically affect multiple systems simultaneously and direct injection to all target organs may not be feasible.

## 14.4 Mitochondrial Replacement Therapy (MRT)

Although individuals with mitochondrial disease may not directly benefit from mitochondrial transplantation, an oocyte directed form of MRT can prevent germ line transmission of pathogenic mtDNA variants (Fig. 14.3) [19, 59]. The purpose of MRT is to eliminate the risk of transmission of mtDNA-associated disease from mothers who are known to carry pathogenic variants. These are typically women who have had affected children or other maternally related family members with disease. All techniques involve the transfer of parental nuclear DNA (either maternal or both parents) to enucleated oocyte of a donor with wild type mtDNA. This allows the creation of embryos that are nuclear offspring of their intended biological parents who are at limited risk for inheritance of mitochondrial disease due to pathogenic mtDNA.

The most likely pre-fertilization MRT technique is maternal spindle transfer (MST) [49]. Post fertilization, MRT could also be accomplished by pronuclear transfer (PNT) [6]. Additional approaches considered involve the use of polar bodies either prior to, or following fertilization [29, 55]. These techniques were developed using mouse models and human zygotes that were developed to blastocyst stage. MST can take advantage of meiosis spindles that are still present since oocytes are arrested in metaphase II, allowing transplant of spindles with associated chromosomes [19], while PBT may offer an opportunity to minimize carry-over of pathogenic mtDNA.

A single case of MRT through oocyte spindle transfer has been reported [59]. The mother was an asymptomatic 36 year old woman with two previous children affected by Leigh syndrome due to heteroplasmy for m.8993T>G. She had 24.5% heteroplasmy in her blood. Following MRT she delivered a boy, and testing of this infant showed low loads of m.8993T>G in his tissues. At 7 months of age the child was reported to be healthy. It is important to note that there were serious concerns voiced about the manner in which this work was conducted [37]. The family has not agreed to ongoing evaluations of the infant, which creates challenges in observing for long-term effects of the procedure. In addition, the US Food and Drug Administration notified the authors of the study that they were in violation of their agreement not to market this procedure to patients [30]. A more measured approach to the introduction of MRT is being undertaken in the United Kingdom, where the technique will be used as part of formal studies conducted in partnership with government regulatory agencies [54].

Many potential problems and challenges remain for MRT before it can become the standard of care. These include the carry-over of pathogenic mtDNA, reversal, and mito-nuclear divergence. Carry-over refers to the sustained presence of maternally inherited pathogenic mtDNA in spite of the transfer of the maternal spindle or pronucleus to a donor egg. This is largely a technical challenge due to the inability to isolate nuclear material without mitochondrial contamination. Despite the attempts to minimize pathogenic mtDNA carryover, some (<5%) of pathogenic mtDNA is carried over to the donor oocyte [21].

Although small amount of pathogenic mtDNA carry-over does not directly threaten mitochondrial function, it has been shown in some studies that in a minority of embryonic stem cells (15–20%) the proportion of pathogenic mtDNA increases to nearly 100% [16, 57]. It is unclear whether the observed phenomenon, reversal, is a cell culture anomaly or has in vivo relevance. The molecular basis of reversal remains unknown and its clinical consequences are yet to be determined.

Since mitochondrial function emerges from the synergy of nuclear and mitochondrial components, one other potential problem is the creation of novel combinations of mitochondrial and nuclear alleles through the MRT procedure. The apparently neutral polymorphisms that separate populations are known to impact mitochondrial activity, rather than being simple markers of

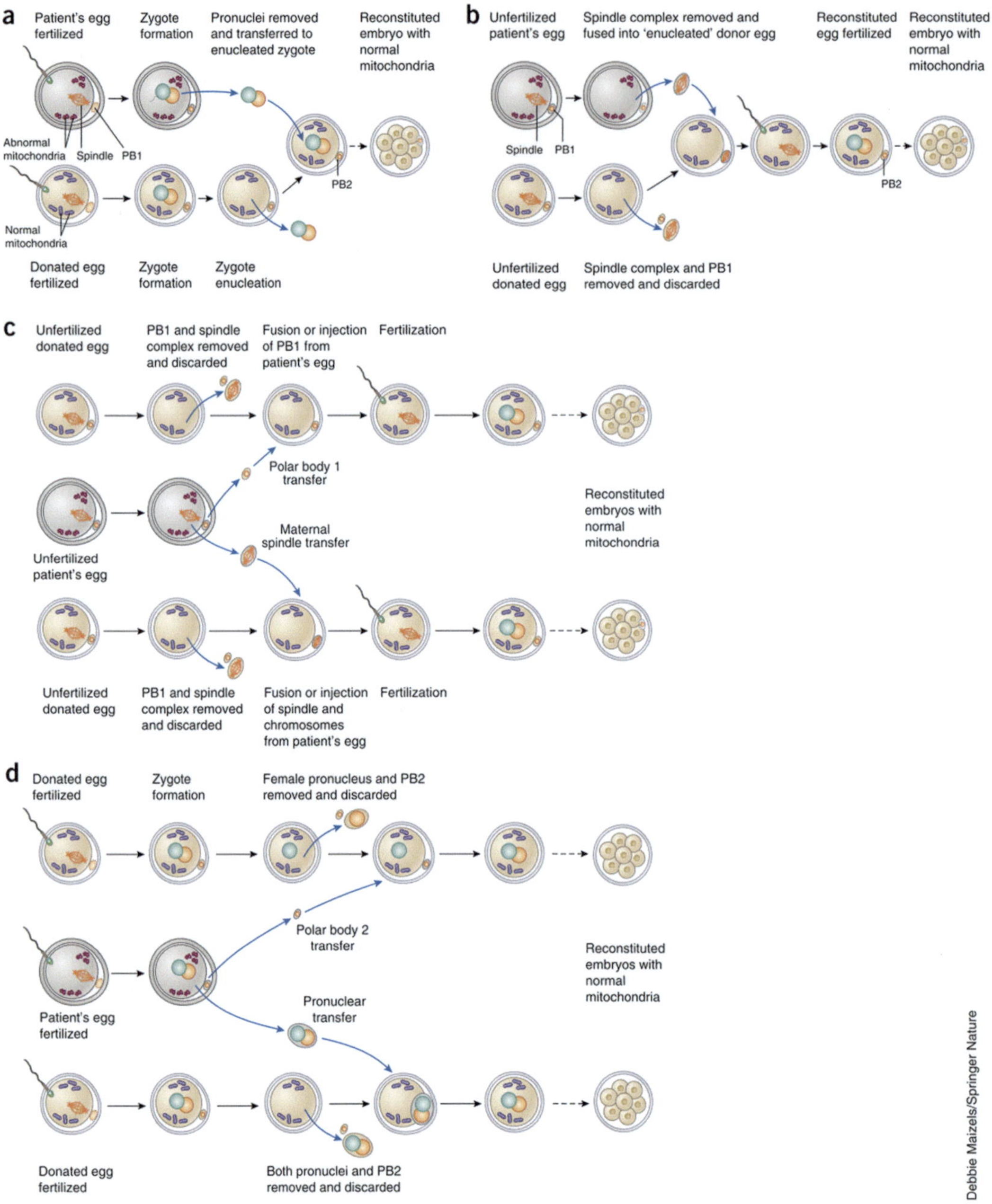


**Fig. 14.3** Adopted: Greenfield et al. [16]. Mitochondrial replacement therapy protocols. (**a**) The basic scheme of PNT. (**b**) MST with fertilization occurring after transfer of the spindle. (**c**) Spindle transfer using the first polar body. (**d**) Pronuclear transfer using the second polar body

matrilineal inheritance [48]. It is well recognized that the creation of unexpected combinations of mitochondrial and nuclear polymorphisms can lead to the emergence of phenotypes in mice [26]. This incompatibility is also observed in humans. One study found that difference in infants ancestral inheritance of nuclear and mitochondrial genome (divergence) is associated with preterm birth [7]. Another observed direct impacts on Complex I function caused by poor interactions between specific mitochondrial and nuclear polymorphisms [15]. To address this, it has been proposed that haplogroup matching be used, wherever possible, in MRT [40].

In addition to technical challenges, there remain many ethical questions about MRT that may slow its acceptance. These include concerns about germ line transmission of induced genetic changes, parental consent and parental identity. A common concern is that allowing MRT creates a "slippery slope" that may lead to the acceptance of nDNA germ line editing for potentially nefarious purposes. Since the donor contributes genetically to the offspring, role of the donor needs to be clarified [32]. Furthermore, there are concerns regarding trans-generational introduction of divergence. To address this concern it has been suggested that only male embryos should be selected for pregnancy (http://nationalacademies.org/HMD/reports/2016/Mitochondrial-Replacement-Techniques).

## 14.5 Current State of Mitochondrial Medicine

Mitochondrial replacement therapy shows promise in preventing intergenerational transmission of pathogenic mtDNA, but strategies are also needed to treat patients with active disease. The most common current treatment option available to patients with mitochondrial disease is a "mitochondrial cocktail." This consists of antioxidants and other co-factors of the OXPHOS complexes and may include alpha-lipoic acid, vitamins C and E, and coenzyme-$Q_{10}$. The therapy attempts to alleviate harmful effects of free radicals generated due to dysfunctional mitochondria, and support mitochondrial health. However, with few exceptions [38], these therapies have little evidence basis from well controlled trials.

Ketogenic diets are posited to stimulate mitochondrial beta-oxidation and produce ketones which serve as alternative source of energy for the brain, heart and skeletal muscles [53]. An increased concentration of ketone bodies induces mitochondrial biogenesis [35]. This therapy would be most obviously helpful when individuals harbor variants affecting Complex I of the ETC. However, these therapies have not been widely studied or used in randomized control trials. In general, current treatments are focused on symptomatic management rather than the underlying genetic cause.

Numerous groups are working on strategies to address the underlying genetic cause by inducing directional shift in heteroplasmy. These strategies can be broadly categorized into nuclease dependent strategies and nuclease independent strategies. At the heart of nuclease dependent strategy is the selective elimination of pathogenic mtDNA. This includes cleavage by zinc finger nucleases (ZFNs), transcription activator-like effector nucleases (TALENs), or CRISPR-Cas9. Nuclease independent strategies include endurance training and mitochondrial unfolded protein response ($UPR^{mt}$).

Studies using restriction enzymes, ZFNs and TALENs have all demonstrated successful shift in mitochondrial heteroplasmy in animal models and human cells (reviewed in [53]). In one of the earliest attempts, mitochondrial targeted restriction enzymes were used to decrease heteroplasmy. Treatment of heteroplasmic mice that harbored two different mtDNA populations, one containing a restriction site, led to shift in heteroplasmy level *in vivo* [3]. Shortly thereafter, it was shown that patient cells harboring m.8993T>C variant could also be shifted using a mitochondrially targeted *SmaI* [1]. Thus, these endonucleases are able to degrade pathogenic mtDNA and leave wild type mtDNA intact leading to its preferential replication.

Although restriction enzymes were effective, they are limited to variants that introduce single-cutting restriction endonuclease sites. The development of ZFNs and TALENs for this purpose has greatly expanded its potential. ZFNs are a group of chimeric proteins in which DNA binding domain is conjugated to the C- terminal catalytic subunit of the type II restriction enzyme FokI [24]. Each zinc finger module recognizes three nucleotides, and a designed arrangement of appropriate zinc finger modules can recognize complex, non-palindromic DNA sequences. Similarly, TALENs have a DNA binding domain fused to FokI nuclease which can be used to cleave specific DNA sequences [2]. ZFNs and TALENs can be targeted to the mitochondria by adding mitochondrial targeting signal (MTS) near the N-terminal domain of the protein [2]. Finally, one group reported successful mtDNA genome editing using CRISPR-Cas9, but this study failed to show convincing genetic evidence of heteroplasmy shift and has yet to be replicated [20]. The concept of using CRISPR has been disputed by other groups who note that mitochondria are not obviously competent for guide RNAs [14].

Although nuclease-directed approaches showed promising preliminary results in animal models, they have not yet reached clinical use. As with any genetic therapy which involves delivery or expression of non-endogenous proteins, there are limitations and difficulties to overcome for use in patients. The multi-copy nature of mtDNA requires high expression to shift heteroplasmy and the widespread introduction of nucleases raise concerns due to non-specific interactions. Additionally, the delivery to all affected tissues in patients maybe prove challenging.

Alternative means for heteroplasmy shift are being considered that circumvent these issues. It has been suggested that high heteroplasmic load in muscle cells may not uniformly be reflected in myogenic progenitor satellite cells [44]. Hence, activation of these progenitor satellite cells may result shift in heteroplasmy, as the satellite cells differentiate and repopulate the muscle tissue. However, the most thorough study of this technique showed that exercise therapy and endurance training improved OXPHOS function, but did not lead to a shift in heteroplasmy [51]. Moreover, given that myopathy, hypotonia, and muscle weakness are common symptoms of mitochondrial disease this therapy may not be possible for all patients.

A recent study in *Caenorhabditis elegans* identified modification of the $UPR^{mt}$ as another possible approach to heteroplasmy correction [28]. $UPR^{mt}$ is activated in response to stress signals within the mitochondrial matrix, driving a transcriptional response program within the nucleus to support mitochondrial protein folding and protein import [34]. Surprisingly, the inhibition of $UPR^{mt}$ led to lower heteroplasmy for a large mitochondrial deletion, suggesting that the suppression of mitochondrial defense mechanisms could be used to reduce pathogenic heteroplasmy. Translational studies in this area could suggest new pathways to heteroplasmy reduction.

Another potential mechanism for directional heteroplasmy shift is by selective inhibition of pathogenic mtDNA replication. Due to asymmetric base composition and its long single-stranded time during replication, the mtDNA H-strand is prone to G-quadruplex (GQ) formation. G-quadruplexes are noncanonical DNA structure whereby guanine on the same DNA strand fold back on itself and forms Hoogsteen hydrogen bonds and act as physical barrier to DNA processive enzymes [4]. Direct evidence of GQ in has been demonstrated by confocal microscopy and immunoprecipitation with GQ specific antibodies [12, 18]. Therefore, it is conceivable to reduce heteroplasmy level in GQ promoting mtDNA variants by stabilizing GQs and inhibiting replication of pathogenic mtDNA.

Despite the lack of treatment, mitochondrial disease due to mtDNA variants provides a unique opportunity for genetic therapy. Since most mtDNA point variants are heteroplasmic, there is residual wild type mtDNA present in almost all cases. Unlike genetic therapies for nDNA, there is no need for introduction of new genetic material or correction of existing alleles. Because heteroplasmy levels correlate to symptom severity and onset of disease, a shift in heteroplasmy can drive improvement. Furthermore, because of high critical threshold even small shift in heteroplasmy can lead to satisfactory clinical results. Despite promising preliminary results, treatments for mitochondrial disease have not made it to clinical use. One bottleneck for translation of basic research to clinical trials is lack of appropriate animal models for mitochondrial disease. Future work in developing these models may hold the key to the introduction of new and successful therapies.

## References


1. Bacman SR, Williams SL, Hernandez D, Moraes CT (2007) Modulating mtDNA heteroplasmy by mitochondria-targeted restriction endonucleases in a "differential multiple cleavage-site" model. Gene Ther 14(18):1309–1318. https://doi.org/10.1038/sj.gt.3302981
2. Bacman SR, Williams SL, Pinto M, Peralta S, Moraes CT (2013) Specific elimination of mutant mitochondrial genomes in patient-derived cells by mito-TALENs. Nat Med 19(9):1111–1113. https://doi.org/10.1038/nm.3261
3. Bayona-bafaluy MP, Blits B, Battersby BJ, Shoubridge EA, Moraes CT (2005) Rapid directional shift of mitochondrial DNA heteroplasmy in animal tissues by a mitochondrially targeted restriction endonuclease. PNAS 102(40):14392–14397
4. Burge S, Parkinson GN, Hazel P, Todd AK, Neidle S (2006) Quadruplex DNA: sequence, topology and structure. Nucleic Acids Res 34(19):5402–5415. https://doi.org/10.1093/nar/gkl655
5. Chinnery PF, Hudson G (2013) Mitochondrial genetics. Br Med Bull 106(1):135–159. https://doi.org/10.1093/bmb/ldt017
6. Craven L, Tuppen HA, Greggains GD, Harbottle SJ, Murphy JL, Cree LM et al (2010) Pronuclear transfer in human embryos to prevent transmission of mitochondrial DNA disease. Nature 465(7294):82–85. https://doi.org/10.1038/nature08958
7. Crawford N, Prendergast D, Oehlert JW, Shaw GM, Stevenson DK, Rappaport N et al (2017) Divergent patterns of mitochondrial and nuclear ancestry are associated with the risk for preterm birth. J Pediatr 194:40–46.e4. https://doi.org/10.1016/j.jpeds.2017.10.052
8. de Laat P, Fleuren LHJ, Bekker MN, Smeitink JAM, Janssen MCH (2015) Obstetric complications in carriers of the m.3243A>G mutation, a retrospective cohort study on maternal and fetal outcome. Mitochondrion 25:98–103. https://doi.org/10.1016/j.mito.2015.10.005
9. Dejean LM, Martinez-Caballero S, Kinnally KW (2006) Is MAC the knife that cuts cytochrome c from mitochondria during apoptosis? Cell Death Differ 13(8):1387–1395. https://doi.org/10.1038/sj.cdd.4401949
10. Emani SM, McCully JD (2018) Mitochondrial transplantation: applications for pediatric patients with congenital heart disease. Transl Pediatr 7(2):169–175. https://doi.org/10.21037/tp.2018.02.02
11. Emani SM, Piekarski BL, Harrild D, del Nido PJ, McCully JD (2017) Autologous mitochondrial transplantation for dysfunction after ischemia-reperfusion injury. J Thorac Cardiovasc Surg 154(1):286–289. https://doi.org/10.1016/j.jtcvs.2017.02.018
12. Falabella M, Kolesar JE, Xiang IM, Wang T, Horne W, Wallace C et al (2018) G-quadruplex dynamics contribute to epigenetic regulation of mitochondrial function. BioRXiv. https://doi.org/10.1101/359703
13. Fan W, Waymire KG, Narula N, Li P, Rocher C, Coskun PE et al (2008) A mouse model of mitochondrial disease reveals germline selection against severe mtDNA mutations. Science 319(5865):958–962. https://doi.org/10.1126/science.1147786
14. Gammage PA, Moraes CT, Minczuk M (2017) Mitochondrial genome engineering: the revolution may not be CRISPR-Ized. Trends Genet 34(2):101–110. https://doi.org/10.1016/j.tig.2017.11.001
15. Gershoni M, Levin L, Ovadia O, Toiw Y, Shani N, Dadon S et al (2014) Disrupting mitochondrial-nuclear coevolution affects OXPHOS complex i integrity and impacts human health. Genome Biol Evol 6(10):2665–2680. https://doi.org/10.1093/gbe/evu208
16. Greenfield A, Braude P, Flinter F, Lovell-Badge R, Ogilvie C, Perry ACF (2017) Assisted reproductive technologies to prevent human mitochondrial disease transmission. Nat Biotechnol 35(11):1059–1068. https://doi.org/10.1038/nbt.3997
17. Hamilton ML, Guo Z, Fuller CD, Van Remmen H, Ward WF, Austad SN et al (2001) A reliable assessment of 8-oxo-2-deoxyguanosine levels in nuclear and mitochondrial DNA using the sodium iodide method to isolate DNA. Nucleic Acids Res 29(10):2117–2126. https://doi.org/10.1093/nar/29.10.2117
18. Hänsel-Hertsch R, Beraldi D, Lensing SV, Marsico G, Zyner K, Parry A et al (2016) G-quadruplex structures mark human regulatory chromatin. Nat Genet 48(10):1267–1272. https://doi.org/10.1038/ng.3662
19. Herbert M, Turnbull D (2018) Progress in mitochondrial replacement therapies. Nat Rev Mol Cell Biol 19(2):71–72. https://doi.org/10.1038/nrm.2018.3
20. Jo A, Ham S, Lee GH, Lee Y, Kim S, Lee Y et al (2015) Efficient mitochondrial genome editing by CRISPR/Cas9. Biomed Res Int 201.(Article ID 305716:1–10. https://doi.org/10.1155/2015/305716
21. Kang E, Wu J, Gutierrez NM, Koski A, Tippner-Hedges R, Agaronyan K et al (2016) Mitochondrial replacement in human oocytes carrying pathogenic mitochondrial DNA mutations. Nature 540(7632):270–275. https://doi.org/10.1038/nature20592
22. Kaufman BA, Van Houten B (2017) POLB: a new role of DNA polymerase beta in mitochondrial base excision repair. DNA Repair 60.(November:A1–A5. https://doi.org/10.1016/j.dnarep.2017.11.002
23. Kazak L, Reyes A, Holt IJ (2012) Minimizing the damage: repair pathways keep mitochondrial DNA intact. Nat Rev Mol Cell Biol 13(10):659–671. https://doi.org/10.1038/nrm3439
24. Kim Y-G, Cha J, Chandrasegaran S (1996) Hybrid restriction enzymes: zinc finger fusions to Fok I cleavage domain. PNAS 93.(February:1156–1160
25. Kirkman MA, Yu-Wai-Man P, Korsten A, Leonhardt M, Dimitriadis K, De Coo IF et al (2009) Gene-environment interactions in Leber hereditary optic neuropathy. Brain 132(9):2317–2326. https://doi.org/10.1093/brain/awp158

26. Latorre-Pellicer A, Moreno-Loshuertos R, Lechuga-Vieco AV, Sánchez-Cabo F, Torroja C, Acín-Pérez R et al (2016) Mitochondrial and nuclear DNA matching shapes metabolism and healthy ageing. Nature 535(7613):561–565. https://doi.org/10.1038/nature18618
27. Lesnefsky EJ, Chen Q, Slabe TJ, Stoll MSK, Minkler PE, Hassan MO et al (2004) Ischemia, rather than reperfusion, inhibits respiration through cytochrome oxidase in the isolated, perfused rabbit heart: role of cardiolipin. Am J Physiol Heart Circ Physiol 287(1):H258–H267. https://doi.org/10.1152/ajpheart.00348.2003
28. Lin Y-F, Schulz AM, Pellegrino MW, Lu Y, Shaham S, Haynes CM (2016) Maintenance and propagation of a deleterious mitochondrial genome by the mitochondrial unfolded protein response. Nature 533(7603):1–8. https://doi.org/10.1038/nature17989
29. Ma H, O'Neil RC, Marti Gutierrez N, Hariharan M, Zhang ZZ, He Y et al (2017) Functional human oocytes generated by transfer of polar body genomes. Cell Stem Cell 20(1):112–119. https://doi.org/10.1016/j.stem.2016.10.001
30. Malarky MA (2017, August 4). Retrieved October 4, 2018, from https://www.fda.gov/downloads/BiologicsBloodVaccines/GuidanceComplianceRegulatoryInformation/ComplianceActivities/Enforcement/UntitledLetters/UCM570225.pdf
31. Masuzawa A, Black KM, Pacak CA, Ericsson M, Barnett RJ, Drumm C et al (2013) Transplantation of autologously derived mitochondria protects the heart from ischemia-reperfusion injury. AJP: Heart Circ Physiol 304(7):H966–H982. https://doi.org/10.1152/ajpheart.00883.2012
32. Matthew Liao S (2017) Do mitochondrial replacement techniques affect qualitative or numerical identity? Bioethics 31(1):20–26. https://doi.org/10.1111/bioe.12308
33. McCully JD, Cowan DB, Pacak CA, Toumpoulis IK, Dayalan H, Levitsky S (2009) Injection of isolated mitochondria during early reperfusion for cardioprotection. Am J Phys Heart Circ Phys 296(1):H94–H105. https://doi.org/10.1152/ajpheart.00567.2008
34. Melber A, Haynes CM (2018) UPR mt regulation and output: a stress response mediated by mitochondrial-nuclear communication. Cell Res 28(3):281–295. https://doi.org/10.1038/cr.2018.16
35. Nunnari J, Suomalainen A (2012) Mitochondria: in sickness and in health. Cell 148(6):1145–1159. https://doi.org/10.1016/j.cell.2012.02.035
36. Rahman S, Poulton J, Marchington D, Suomalainen A (2001) Decrease of 3243 A→G mtDNA mutation from blood in MELAS syndrome: a longitudinal study. Am J Hum Genet 68(1):238–240. https://doi.org/10.1086/316930
37. Reardon S (2017) Genetic details of controversial "three-parent baby" revealed. Nature 544:17–18. https://doi.org/10.1038/nature.2017.21761
38. Rodriguez MC, MacDonald JR, Mahoney DJ, Parise G, Beal MF, Tarnopolsky MA (2007) Beneficial effects of creatine, CoQ10, and lipoic acid in mitochondrial disorders. Muscle Nerve 35(2):235–242. https://doi.org/10.1002/mus.20688
39. Ross JM, Stewart JB, Hagström E, Brené S, Mourier A, Coppotelli G et al (2013) Germline mitochondrial DNA mutations aggravate ageing and can impair brain development. Nature 501(7467):412–415. https://doi.org/10.1038/nature12474
40. Røyrvik EC, Burgstaller JP, Johnston IG (2016) mtDNA diversity in human populations highlights the merit of haplotype matching in gene therapies. Mol Hum Reprod 22(11):809–817. https://doi.org/10.1093/molehr/gaw062
41. Saneto RP, Sedensky MM (2013) Mitochondrial disease in childhood: MtDNA encoded. Neurotherapeutics 10(2):199–211. https://doi.org/10.1007/s13311-012-0167-0
42. Santorelli FM, Shanske S, Macaya A, DeVivo DC, DiMauro S (1993) The mutation at nt 8993 of mitochondrial DNA is a common cause of Leigh's syndrome. Ann Neurol 34(6):827–834. https://doi.org/10.1002/ana.410340612
43. Schaefer AM, McFarland R, Blakely EL, He L, Whittaker RG, Taylor RW et al (2008) Prevalence of mitochondrial DNA disease in adults. Ann Neurol 63(1):35–39. https://doi.org/10.1002/ana.21217
44. Smith PM, Lightowlers RN (2011) Altering the balance between healthy and mutated mitochondrial DNA. J Inherit Metab Dis 34(2):309–313. https://doi.org/10.1007/s10545-010-9122-6
45. Sondheimer N, Glatz CE, Tirone JE, Deardorff MA, Krieger AM, Hakonarson H (2011) Neutral mitochondrial heteroplasmy and the influence of aging. Hum Mol Genet 20(8):1653–1659. https://doi.org/10.1093/hmg/ddr043
46. Stein A, Sia EA (2017) Mitochondrial DNA repair and damage tolerance. Front Biosci (Landmark Edition) 22:920–943. Retrieved fromhttp://www.ncbi.nlm.nih.gov/pubmed/27814655
47. Sue CM, Quigley A, Katsabanis S, Kapsa R, Crimmins DS, Byrne E, Morris JGL (1998) Detection of MELAS A3243G point mutation in muscle, blood and hair follicles. J Neurol Sci 161(1):36–39. https://doi.org/10.1016/S0022-510X(98)00179-8
48. Suissa S, Wang Z, Poole J, Wittkopp S, Feder J, Shutt TE et al (2009) Ancient mtDNA genetic variants modulate mtDNA transcription and replication. PLoS Genetics 5(5). https://doi.org/10.1371/journal.pgen.1000474
49. Tachibana M, Amato P, Sparman M, Woodward J, Sanchis DM, Ma H et al (2013) Towards germline gene therapy of inherited mitochondrial diseases. Nature 493(7434):627–631. https://doi.org/10.1038/nature11647
50. Taivassalo T, Fu K, Johns T, Arnold D, Karpati G, Shoubridge EA (1999) Gene shifting: a novel therapy for mitochondrial myopathy. Hum Mol Genet 8(6):1047–1052

51. Taivassalo T, Gardner JL, Taylor RW, Schaefer AM, Newman J, Barron MJ et al (2006) Endurance training and detraining in mitochondrial myopathies due to single large-scale mtDNA deletions. Brain 129(12):3391–3401. https://doi.org/10.1093/brain/awl282
52. Taylor RW, Turnbull DM (2005) Mitochondrial DNA mutations in human disease. Nat Rev Genet 6(5):389–402. https://doi.org/10.1038/nrg1606
53. Viscomi C, Bottani E, Zeviani M (2015) Emerging concepts in the therapy of mitochondrial disease. Biochim Biophys Acta Bioenerg 1847(6–7):544–557. https://doi.org/10.1016/j.bbabio.2015.03.001
54. Watts G, Braude P, Flinter F, Harding S, Lewens T, Parker M (2012) Novel techniques for the prevention of mitochondrial DNA disorders: an ethical review. Nuffield Council on Bioethics, London
55. Wang T, Sha H, Ji D, Zhang HL, Chen D, Cao Y, Zhu J (2014) Polar body genome transfer for preventing the transmission of inherited mitochondrial diseases. Cell 157(7):1591–1604. https://doi.org/10.1016/j.cell.2014.04.042
56. Wonnapinij P, Chinnery PF, Samuels DC (2008) The distribution of mitochondrial DNA heteroplasmy due to random genetic drift. Am J Hum Genet 83(5):582–593. https://doi.org/10.1016/j.ajhg.2008.10.007
57. Yamada M, Emmanuele V, Sanchez-Quintero MJ, Sun B, Lallos G, Paull D et al (2016) Genetic drift can compromise mitochondrial replacement by nuclear transfer in human oocytes. Cell Stem Cell 18(6):749–754. https://doi.org/10.1016/j.stem.2016.04.001
58. Youle RJ, Narendra DP (2011) Mechanisms of mitophagy. Nat Rev Mol Cell Biol 12(1):9–14. https://doi.org/10.1038/nrm3028
59. Zhang J, Liu H, Luo S, Lu Z, Chávez-Badiola A, Liu Z et al (2017) Live birth derived from oocyte spindle transfer to prevent mitochondrial disease. Reprod Biomed Online 35(6):750. https://doi.org/10.1016/j.rbmo.2017.07.008

# Application of CRISPR-Cas9 Screening Technologies to Study Mitochondrial Biology in Healthy and Disease States

15

David Sharon and Steven M. Chan

**Abstract**

Mitochondria play a central role in maintaining normal cellular homeostasis as well as contributing to the pathogenesis of numerous disease states. The advent of CRISPR-Cas9 screening technologies has greatly accelerated the study of mitochondrial biology. In this chapter, we review the various CRISPR-Cas9 screening platforms that are currently available and prior studies that leveraged this technology to identify genes involved in mitochondrial biology in both healthy and disease states. In addition, we discuss the challenges associated with current CRISPR-Cas9 platforms and potential solutions to further enhance this promising technology.

**Keywords**

CRISPR · Cas9 · Functional genomics · Mitochondria · Synthetic lethality · Oxidative phosphorylation · Knockout screens

## 15.1 Introduction

Mitochondria are intracellular organelles that play a central role in multiple metabolic pathways including oxidative phosphorylation (OXPHOS), β-oxidation of fatty acids, steroidogenesis, ammonia detoxification, and amino acid metabolism. In addition, they are involved in various non-metabolic processes including reactive oxygen species (ROS) production, calcium homeostasis, and apoptosis. Given the central role of mitochondria in many metabolic and non-metabolic processes, it is perhaps not surprising that cancer cells often manifest dysregulation of one or several mitochondrial functions. A classic example of this dysregulation is the Warburg effect which refers to the tendency of cancer cells to undergo aerobic glycolysis rather than OXPHOS [1]. Although this reprogramming of metabolism is not the underlying cause of cancer as originally hypothesized by Otto Warburg, it is considered one of the hallmarks of cancer as proposed by Douglas Hanahan and Robert Weinberg [2, 3]. These changes in mitochondrial function and properties promote the neoplastic phenotype in cancer cells and therefore, represent dependencies that can potentially be therapeutically targeted. While enforced reversal of these changes could work, this approach is often not feasible due to the complexities of dysregulated pathways involved in driving these changes. As a result, there is broad interest in alternative strategies that exploit synthetic lethality to indirectly target these metabolic changes. Synthetic lethality is based on the concept of non-oncogene addiction, wherein cells expressing a particular oncogenic mutation exhibit heightened dependence on a subset of non-oncogenes for survival. A classic

D. Sharon · S. M. Chan (✉)
Princess Margaret Cancer Centre,
Toronto, ON, Canada
e-mail: Steven.Chan@uhnresearch.ca


A. Urbani, M. Babu (eds.), *Mitochondria in Health and in Sickness*, Advances in Experimental Medicine and Biology 1158, https://doi.org/10.1007/978-981-13-8367-0_15

example is the dependency on poly ADP ribose polymerase (PARP) activity for survival in cells with *BRCA1* or *BRCA2* mutations [4]. Although synthetic lethality is traditionally thought of as the interaction between two genes, the concept can be extended to the interaction between a drug that disrupts a specific pathway or process and the genes that are required to maintain cell survival in the presence of the drug.

One of the challenges to identifying synthetic lethal genes associated with mitochondrial dysfunction is the sheer number of potential targets. It has been estimated that the mitochondrial proteome alone consists of over 1000 proteins of which 13 are encoded within the mitochondrial genome [5]. The remainder are nuclear encoded and imported into the organelle. The function, submitochondrial localization, and protein interaction partners of the majority of mitochondrial proteins have yet to be elucidated. To add to this complexity, proteins localized outside of the organelle also regulate mitochondrial function and properties. Thus, identification of dependencies associated with a specific mitochondrial defect could be a daunting task. Traditional hypothesis-driven experimental approaches are time-consuming and generally limited in scope. Given these challenges, many groups have leveraged the power of genome-wide loss-of-function screening technologies to tackle this problem.

The first generation of loss-of-function genomics screens were based on RNA interference (RNAi) technology. Although RNAi-based screening platforms have proved to be valuable tools for identifying essential genes and synthetic lethal partners and characterizing the function of mitochondrial proteins [6, 7], this technology is limited by a relatively high probability of off-target effects and low knockdown efficiency. The use of optimized target sequences and multiple RNAi reagents per gene has minimized the constraints of these limitations. However, since the development of Clustered Regularly-Interspaced Short Palindromic Repeats (CRISPR) – Cas9 technology for genome editing in the early 2010s, it has rapidly become the tool of choice for studying gene function with important advantages over RNAi technology [8]. The CRISPR-Cas9 system can generate complete and permanent gene knockout as opposed to RNAi which suppresses gene expression with varying degrees of efficiency. Although both techniques are susceptible to off-target effects, recent evidence suggest that CRISPR-Cas9 technology is superior in this respect [9–12]. To apply this technique for functional genomics screens, several groups independently generated genome-wide CRISPR-Cas9 gene knockout libraries [13–15]. These libraries have since been used to study multiple biological processes including those related to mitochondrial metabolism. In this review, we will summarize the literature on CRISPR-Cas9 screens that have previously been performed to study mitochondrial biology in healthy and disease states with a focus on cancer.

## 15.2 CRISPR-Cas9 Technology

The development of CRISPR-Cas9 into a reliable and user-friendly technique for genome editing has enabled the functional study of a large number of previously uncharacterized genes. The majority of the studies thus far have utilized the type II CRISPR system which consists of the Cas9 endonuclease derived from *Streptococcus pyogenes* (SpCas9) and a single guide RNA (sgRNA) molecule. The latter is composed of a scaffold sequence necessary for Cas9 binding and a 20 nucleotide targeting-sequence which directs Cas9 to the desired cut position in the genome. The Cas9-sgRNA complex searches for DNA sequences that are complementary to the targeting sequence and once found, cleaves the DNA substrate creating a double-stranded DNA break (DSB) [16]. DSBs are most commonly repaired through non-homologous end joining (NHEJ) which often introduces short insertions or deletions (indels) resulting in frameshifts and consequently, early truncation of the protein product at the carboxyl terminus.

Two types of CRISPR libraries are currently available through commercial sources or the non-profit plasmid sharing repository – Addgene. The first type is an arrayed format in which each well in a microtiter plate contains synthetic sgRNAs or lentiviral sgRNA vectors targeting a single gene. The second type is a pooled library composed of a mixture of hundreds to thousands of lentiviral sgRNA vectors. Infections are performed at a low multiplicity of infection (MOI) ratio to maximize the proportion of cells with single sgRNA integration. In the pooled format, transduced cells are either positively or negatively selected over time depending of impact of the specific gene knockout on cell growth. The relative abundance of each sgRNA construct is determined through next generation sequencing (NSG) of the targeting sequence which serves as a barcode. Pooled screens are most useful for the identification of genes that affect cell proliferation, survival, and/or death. In contrast, the arrayed format is better suited for screening more complex phenotypic readouts such as changes in cell or organelle morphology, subcellular localization of molecular components, or cell motility because it allows for the use of high-content imaging systems. However, such microplate-based screens are not regularly conducted at the genomic scale due to the expense, labor and automation expertise required. Since most reported genome-wide screens have focused on the identification of genes that impact cell growth, researchers have preferred the use of pooled libraries over arrayed libraries. In the next two sections, we will review the available literature on studies focusing on genome-wide CRISPR-Cas9 knockout screens in which mitochondrial function was perturbed either genetically or pharmacologically.

## 15.3 CRISPR-Cas9 Screens for Interactions with Mitochondrial Inhibitors

One of the main applications of functional genomics screens is the identification of genes that regulate sensitivity to a specific drug. The results of such screens can help define the mechanism of drug action and identify synergistic or synthetic lethal combination partners. The Sabatini lab was one of the first groups to perform genome-wide pooled CRISPR-Cas9 screens. In one of their studies, they used a library consisting of ~30,000 sgRNAs targeting ~3000 genes involved in metabolic pathways to investigate why inhibition of mitochondrial electron transport chain (ETC) activity suppresses cell proliferation [17]. They performed a negative selection screen for genes that upon inactivation sensitized Jurkat cells, a human T cell leukemia cell line, to phenformin, an inhibitor of complex I of the ETC. The authors found that multiple genes encoding components of the ETC including complex I subunits were selectively required for cell growth in the presence of phenformin. Interestingly, the top negatively selected gene was *GOT1* which encodes an enzyme involved in the malate-aspartate shuttle. Aspartate is normally synthesized in the mitochondrial matrix and transported to the cytosol for protein and nucleotide synthesis. A fraction of this aspartate is consumed by GOT1 to transfer reducing equivalents to the mitochondrial matrix. The authors found that upon ETC inhibition, mitochondrial synthesis of aspartate is impaired, and GOT1 operates in the reverse direction to generate aspartate and sustain cell growth. This dependency explains the synthetic lethal interaction between loss of GOT1 and ETC dysfunction. This study highlights the power of genome-wide CRISPR-Cas9 screens to unveil novel biological mechanisms or pathways that would otherwise be difficult to predict based on prior literature alone. Although not explicitly stated by the authors, their findings indicate that GOT1 is a potential therapeutic target against cancer cells with mitochondrial dysfunction (e.g., those that harbor pathogenic mitochondrial DNA mutations). Furthermore, pharmacologic inhibitors of GOT1 may synergize with known ETC inhibitors such as metformin and arsenic trioxide to suppress cancer cell proliferation.

CRISPR-Cas9 screens can also be used to search for genes that upon inactivation, are protective of the negative effects of mitochondrial dysfunction. The results of such screens may inform strategies to treat human diseases associated with mitochondrial dysfunction (e.g. inherited mitochondrial diseases). One such study was performed by the Mootha laboratory in 2016 [18]. They performed a CRISPR-Cas9 enrichment screen to identify genes that when deleted, promoted the survival of K562 cells (a chronic myeloid leukemia blast crisis cell line) in the presence of antimycin A, an inhibitor of complex III of the ETC. They used a pooled lentiviral library consisting of ~65,000 sgRNAs targeting ~18,000 genes. Following 21 days of treatment with either pyruvate alone, antimycin A alone, or a combination of antimycin A and pyruvate, cells were harvested and sequenced to identify sgRNAs that were enriched during the selection period. The highest ranked gene in both antimycin A treatment groups was the Von-Hippel-Lindau Tumor Suppressor (*VHL*) gene. The *VHL* gene encodes an E3 ubiquitin ligase that catalyzes the oxygen-dependent ubiquitination and proteasomal degradation of HIF-1α and HIF-2α, subunits of transcription factors involved in the hypoxia response [19]. Thus, inactivation of *VHL* would activate the hypoxia transcriptional program, and the authors hypothesized that this response might be protective against mitochondrial toxicity. To confirm this, they showed that pharmacological stabilization of HIF-1α using a compound known as, FG-4592, promoted survival of cells and zebrafish treated with antimycin A and other mitochondrial ETC inhibitors. Consistent with these findings, they demonstrated that exposure to chronic hypoxia improved survival and neuropathology in a genetic mouse model of Leigh syndrome, a common manifestation of mitochondrial disease. This study represents another example of the power of functional genomics to uncover previously recognized inter-pathway connections that have translational and clinical relevance.

## 15.4 CRISPR-Cas9 Screens for Interactions with Mutations

In the previous section, we discussed two illustrative examples of how CRISPR-Cas9 screens can be used to discover genes that modify cellular fitness upon treatment with inhibitors of mitochondrial respiration. Similar approaches can be used to identify genes that positively or negatively impact cell growth in the presence of mutations that alter mitochondrial respiration and related metabolic pathways. Of particular interest is the identification of genes that are synthetic lethal against oncogenic mutations that are difficult to target using conventional approaches (e.g., loss-of-function mutations). In a study by the Elledge group in 2017, they performed a genome-wide CRISPR-Cas9 screen to identify genes that were required for survival of cells with an activating K-Ras mutation but not isogenic cells with wild-type K-Ras [20]. They used a pooled lentiviral CRISPR library with ~90,000 sgRNAs targeting ~18,000 genes. They found that many of the top synthetic lethal genes in the screen encoded proteins involved in mitochondrial translation, indicative of a heightened dependency on OXPHOS in K-Ras mutant cells. To validate these findings, they treated their paired isogenic lines with tigecycline, a tetracycline analog that inhibits mitochondrial translation, and found that K-Ras mutant cells were more sensitive to tigecycline compared to their wildtype counterparts. They further showed that combination treatment with tigecycline and VLX600, a potent and selective OXPHOS inhibitor [21], decreased the in vivo growth of a murine colon carcinoma cell line (CT26) with an activating G12D K-Ras mutation. This study illustrates the utility of CRISPR-Cas9 screens to find synthetic lethal interactions that enable indirect targeting of oncogenic driver mutations that would otherwise not be easily druggable.

As discussed in the previous section, CRISPR-Cas9 screens can also be used to find genes that upon inactivation rescue cells from the lethal consequences of mitochondrial diseases. The screen performed by the Mootha group used antimycin A to mimic the deleterious effects of pathogenic mitochondrial DNA mutations on ETC function. The Puigserver group employed perhaps a more faithful model of mitochondrial diseases by taking advantage of cybrid cells that carried a pathogenic mutation (A3796G) in the mitochondrial-encoded protein MT-ND1 [22]. This specific mutation is found in a subset of patients with adult-onset dystonia [23]. They performed a genome-wide CRISPR/Cas9 genome-wide positive selection screen to identify genes that upon inactivation promoted the survival of cybrid cells cultured in media with galactose instead of glucose. The MT-ND1 mutant cells would normally die within 72 h in galactose media, presumably due to the inability to generate sufficient ATP from OXPHOS. The screen showed that deletion of bromodomain-containing protein 4 (*BRD4*), which encodes a member of the BET (bromodomain and extra terminal domain) family, conferred the highest survival advantage to mutant cybrid cells. Treatment with I-BET 525762A, a pan BET inhibitor, similarly rescued galactose-induced cell death. The mechanism of action was found to be mediated through an increase in the expression of OXPHOS genes regulated by the transcriptional activator PGC-1α and consequent improvement in ETC activity in mutant cybrid cells. Furthermore, they showed that survival of I-BET-treated mutant cybrids was dependent on glutamine oxidation since glutamine removal or pharmacological inhibition of glutaminase by BPTES completely prevented their survival. These results indicate that in addition to exposure to hypoxia and stabilization of HIF-1α as discussed above, inhibition of BET proteins may be another strategy to treat inherited mitochondrial diseases.

It is of interest to note that treatment with BET inhibitors have demonstrated activity against a range of cancer cell types [24]. The main mechanism of action in cancer is thought to be suppression of MYC-regulated gene transcription [25]. However, as illustrated in the above study, BET inhibition can also enhance activity of PGC-1α which is a master transcriptional regulator of mitochondrial biogenesis [26] and increase mitochondrial respiratory chain function. This rewiring of metabolic pathways may subvert one of the hallmarks of cancer, namely aerobic glycolysis, and reduce cancer growth. Indeed, enforced enhancement of ETC activity has been shown to be deleterious in some cancer models. For instance, expression of the single-subunit yeast NADH dehydrogenase, *Ndi1*, which enhanced complex I activity in human breast cancer cells, inhibited tumor growth and metastasis in mice [27]. Thus, it is tempting to speculate that the anti-tumor activity of BET inhibitors may, at least in part, be due to enhancement of mitochondrial ETC function and suppression of aerobic glycolysis.

## 15.5 CRISPR-Cas9 Screens to Identify Novel Genes Involved in OXPHOS

CRISPR-Cas9 screens are most commonly performed by comparing the composition of sgRNAs after selection in the presence and absence of a drug/inhibitor or mutation. The comparison can also be made between two different culture conditions to identify genes that are essential for a specific metabolic process. For the identification of genes involved in OXPHOS, several groups have taken advantage of the finding that cells growing in glucose can utilize both glycolysis and OXPHOS for ATP generation, whereas cells growing in galactose are almost entirely dependent on OXPHOS. In fact, cells deficient in OXPHOS are viable in glucose-rich medium but die when glucose is replaced with galactose. Thus, cells harboring sgRNAs that target genes essential for OXPHOS would be preferentially depleted when grown in galactose compared with glucose. The Mootha group used this approach to systematically catalog genes that are essential for OXPHOS function [28]. However, rather than doing a negative selection screen which would be the conventional approach for doing this type of screen, they performed a positive selection CRISPR-Cas9 screen by isolating dying cells based on positive Annexin V staining. They used a genome-wide lentiviral library consisting ~75,000 sgRNAs targeting ~18,000 human genes to transduce K562 cells. The infected cells were initially grown in glucose-containing medium for 11 days to allow for gene knockout to take place and then separated the cells into two cultures with one containing glucose and the other galactose as the sole carbon source. After 24 h of selection, the dead or dying cells from each culture were enriched using Annexin V-conjugated beads and subjected to next-generation deep sequencing to determine the abundance of each sgRNA. Using this approach, the authors identified 191 high-confidence genes required for OXPHOS. Many of the genes were known to be involved in OXPHOS and included genes that encode components of the ETC and mitochondrial ribosomes as well as mitochondrial tRNA synthetases. Importantly, they also identified a large number of previously uncharacterized genes. One such gene was *TMEM261* (also known as *DMAC1*) which encodes a protein that is required for the assembly of complex I and OXPHOS [29]. In addition to mitochondrial proteins, the death screen also found several non-mitochondrial hits that were essential for cell survival in galactose but not in glucose. Some of these genes are involved in the AMP kinase (AMPK) pathway, indicating that AMPK activation is required for survival when grown in galactose. To validate this finding, they treated cells with a small molecule AMPK inhibitor (Compound C) and showed that it triggered cell death selectively in galactose but not in glucose. Indeed, AMPK has previously been shown to promote mitochondrial biogenesis and function through induction of mitochondrial gene expression [30]. This study exemplifies the power of CRISPR-Cas9 screens to comprehensively identify genes that are involved in a complex cellular process.

The Sabatini group described a different approach to identifying genes involved in a common cellular process or pathway. In their study describing this approach, they performed genome-wide CRISPR-Cas9 screens on a panel of 14 acute myeloid leukemia (AML) cells lines [31]. They identified the genes that were essential for each cell line by comparing the sgRNA representation at baseline and after 14 population doublings. Most of the essential genes highly overlapped between lines, but some genes showed substantial variability in the degree of essentiality between cell lines. The authors found that genes acting in the same cellular pathway showed similar patterns of essentiality across the cell lines. For example, the genes encoding hexokinase 2 (HK2) and glucose-6-phosphate isomerase (GPI) which are enzymes in the glycolysis pathway shared a highly correlated pattern of essentiality. Based on the analysis of gene sets of known functional relationships, they attempted to use this approach to predict the function of previously uncharacterized genes. One such gene was *C17orf89* which shared the same pattern of essentiality with a large number of mitochondrial genes. Based on this correlation, they predicted that C17orf89 was a mitochondrial protein. In their validation experiments, they demonstrated that C17orf89 was indeed localized in the mitochondria and directly interacted with NDUFAF5, a complex I assembly factor. The study shows that analysis of correlated gene essentiality may be useful in predicting the function of previously uncharacterized genes.

## 15.6 Challenges with Current CRISPR-Cas9 Screening Platforms

In the previous sections, we reviewed studies that leveraged the power of CRISPR-Cas9 screens to study genes that were involved in or interacted with mitochondrial function. Despite the burgeoning interest in this technology, there are several important challenges associated with the use of CRISPR-Cas9 screens, and they should be taken into consideration when designing screening experiments.

Negative selection screens are dependent on the depletion of sgRNAs that reduce cell growth relative to the remaining unaffected population. This reduction in growth can be due cell death and/or slowing of cell proliferation. Reliance on sgRNA abundance alone as the primary readout is unable to definitively differentiate between these two outcomes. Nevertheless, dropouts that occur at earlier time points are more likely to be due to cell death and at late time points, reduction in cell proliferation. However, the interpretation of results at later time points can be complicated by skewing of the sgRNA distribution with underrepresentation of a large proportion of sgRNAs. This phenomenon may obscure the effects of sgRNAs with mild or moderate negative impact on cell growth. The skewing is likely due to the positive selection of sgRNAs that substantially improve cell growth leading to their vast overrepresentation at the expense of the remaining population at later time points. Thus, the time point at which the determination of sgRNA representation is performed is a critical parameter to consider when designing screening experiments in order to maximize the number of true positive hits.

Another challenge associated with the CRISPR-Cas9 system is activation of the DNA repair response following Cas9-mediated DSBs. Several studies have shown that CRISPR-Cas9 genome editing can cause gene-independent effects due to p53 activation in response to DSBs [32, 33]. Gene-independent effects have also been observed in cancer cells as a consequence of Cas9-mediated cleavage of amplified DNA regions [34–36]. To overcome this problem, several groups have generated variants of Cas9 that lack intrinsic endonuclease activity but can still regulate the expression of target genes through fusion to transcriptional activators [37], repressors [38, 39], and epigenetic regulators [40]. For instance, a nuclease-dead version of Cas9 fused to the KRAB domain, which functions as a transcription repressor, has been used to silence gene expression without introducing DSBs [39, 41]. This system termed CRISPR interference (CRISPRi) was found to have less off-target effects than RNAi-mediated knockdown since CRISPRi retains the target specificity of the CRISPR system [9, 42]. Another approach to inactivate gene expression without introducing DSBs is the use of CRISPR-mediated base editing techniques (e.g., fusion of nuclease-dead Cas9 to cytosine deaminase) to introduce premature stop codons in the target gene [43–46]. The Cas9 variants described above should further expand the versatility of the CRISPR-Cas9 system for studying gene function.

Genome-wide CRISPR-Cas9 pooled screens generally rely on changes in the relative abundance of sgRNAs as the primary readout. While this approach is useful for monitoring the net effect of each sgRNA on cell survival and proliferation, it does not readily permit the assessment of other cellular parameters (e.g., metabolite level) or processes (e.g. autophagy). A potential strategy to address this issue is the use of dye-based or genetically-encoded fluorescent biosensors that report cellular properties at the single cell level. Subpopulations of cells with different fluorescent properties can then be sorted by flow cytometry and analyzed for differences in their sgRNA representation. This approach is exemplified by Morita et al. in which the authors successfully identified novel genes involved in the regulation of autophagy by performing a genome-wide CRISPR-Cas9 screen using the autophagic flux reporter GFP-LC3-RFP [47]. A number of genetically-encoded fluorescent biosensors have recently been developed to monitor mitochondrial properties and processes including mtAT1.03 (a reporter of mitochondrial ATP levels) [48], MitoTimer (a reporter of mitochondrial protein turnover rate) [49], and mtKeima (a reporter of mitophagy) [50]. The use of these biosensors in combination with genome-wide CRISPR-Cas9 screens should enable the systematic and comprehensive discovery of genes involved in the regulation of mitochondrial properties and function.

## 15.7 Conclusions

Since the initial development of CRISPR-Cas9 technology for genome editing, it has become the technique of choice for studying gene function and has been adapted for numerous applications including genome-wide functional genomics screens. It is apparent that the use of CRISPR-Cas9 screening approaches will greatly expand our understanding of the genes involved in the regulation of mitochondrial function and their interactions with other cellular pathways in healthy and disease states. We envision that advancements in the development of novel Cas9 variants, sgRNA libraries with improved specificities, and the use of novel fluorescent biosensors as functional readout will further enhance the versatility and reliability of genome-wide CRISPR-Cas9 screens for the study of mitochondrial biology.

## References


1. Liberti MV, Locasale JW (2016) The Warburg effect: how does it benefit cancer cells? Trends Biochem Sci 41(3):211–218
2. Hanahan D, Weinberg RA (2011) Hallmarks of cancer: the next generation. Cell 144(5):646–674
3. Hanahan D, Weinberg RA (2000) The hallmarks of cancer. Cell 100(1):57–70
4. Pilie PG, Tang C, Mills GB, Yap TA (2018) State-of-the-art strategies for targeting the DNA damage response in cancer. Nat Rev Clin Oncol
5. Calvo SE, Clauser KR, Mootha VK (2016) MitoCarta2.0: an updated inventory of mammalian mitochondrial proteins. Nucleic Acids Res 44(D1):D1251–D1257
6. Jiang D, Zhao L, Clapham DE (2009) Genome-wide RNAi screen identifies Letm1 as a mitochondrial Ca2+/H+ antiporter. Science 326(5949):144–147
7. Lanning NJ, Looyenga BD, Kauffman AL, Niemi NM, Sudderth J, DeBerardinis RJ et al (2014) A mitochondrial RNAi screen defines cellular bioenergetic determinants and identifies an adenylate kinase as a key regulator of ATP levels. Cell Rep 7(3):907–917
8. Doudna JA, Charpentier E (2014) Genome editing. The new frontier of genome engineering with CRISPR-Cas9. Science 346(6213):1258096
9. Smith I, Greenside PG, Natoli T, Lahr DL, Wadden D, Tirosh I et al (2017) Evaluation of RNAi and CRISPR technologies by large-scale gene expression profiling in the Connectivity Map. PLoS Biol 15(11):e2003213
10. Doench JG, Fusi N, Sullender M, Hegde M, Vaimberg EW, Donovan KF et al (2016) Optimized sgRNA design to maximize activity and minimize off-target effects of CRISPR-Cas9. Nat Biotechnol 34(2):184–191
11. Doench JG, Hartenian E, Graham DB, Tothova Z, Hegde M, Smith I et al (2014) Rational design of highly active sgRNAs for CRISPR-Cas9-mediated gene inactivation. Nat Biotechnol 32(12):1262–1267
12. Hart T, Tong AHY, Chan K, Van Leeuwen J, Seetharaman A, Aregger M et al (2017) Evaluation and design of genome-wide CRISPR/SpCas9 knockout screens. G3 (Bethesda) 7(8):2719–2727
13. Wang T, Birsoy K, Hughes NW, Krupczak KM, Post Y, Wei JJ et al (2015) Identification and characterization of essential genes in the human genome. Science 350(6264):1096–1101
14. Sanjana NE, Shalem O, Zhang F (2014) Improved vectors and genome-wide libraries for CRISPR screening. Nat Methods 11(8):783–784
15. Hart T, Chandrashekhar M, Aregger M, Steinhart Z, Brown KR, MacLeod G et al (2015) High-resolution CRISPR screens reveal fitness genes and genotype-specific cancer liabilities. Cell 163(6):1515–1526
16. Jinek M, Chylinski K, Fonfara I, Hauer M, Doudna JA, Charpentier E (2012) A programmable dual-RNA-guided DNA endonuclease in adaptive bacterial immunity. Science 337(6096):816–821
17. Birsoy K, Wang T, Chen WW, Freinkman E, Abu-Remaileh M, Sabatini DM (2015) An essential role of the mitochondrial electron transport chain in cell proliferation is to enable aspartate synthesis. Cell 162(3):540–551
18. Jain IH, Zazzeron L, Goli R, Alexa K, Schatzman-Bone S, Dhillon H et al (2016) Hypoxia as a therapy for mitochondrial disease. Science 352(6281):54–61
19. Haase VH (2009) The VHL tumor suppressor: master regulator of HIF. Curr Pharm Des 15(33):3895–3903
20. Martin TD, Cook DR, Choi MY, Li MZ, Haigis KM, Elledge SJ (2017) A role for mitochondrial translation in promotion of viability in K-Ras mutant cells. Cell Rep 20(2):427–438
21. Ashton TM, McKenna WG, Kunz-Schughart LA, Higgins GS (2018) Oxidative phosphorylation as an emerging target in cancer therapy. Clin Cancer Res 24(11):2482–2490
22. Barrow JJ, Balsa E, Verdeguer F, Tavares CD, Soustek MS, Hollingsworth LR et al (2016) Bromodomain inhibitors correct bioenergetic deficiency caused by mitochondrial disease complex I mutations. Mol Cell 64(1):163–175
23. Simon DK, Friedman J, Breakefield XO, Jankovic J, Brin MF, Provias J et al (2003) A heteroplasmic mitochondrial complex I gene mutation in adult-onset dystonia. Neurogenetics 4(4):199–205
24. Doroshow DB, Eder JP, LoRusso PM (2017) BET inhibitors: a novel epigenetic approach. Ann Oncol Off J Eur Soc Med Oncol 28(8):1776–1787
25. Abedin SM, Boddy CS, Munshi HG (2016) BET inhibitors in the treatment of hematologic malignancies: current insights and future prospects. Oncotarget Ther 9:5943–5953
26. Scarpulla RC (2011) Metabolic control of mitochondrial biogenesis through the PGC-1 family regulatory network. Biochim Biophys Acta 1813(7):1269–1278
27. Santidrian AF, Matsuno-Yagi A, Ritland M, Seo BB, LeBoeuf SE, Gay LJ et al (2013) Mitochondrial complex I activity and NAD+/NADH balance regulate breast cancer progression. J Clin Invest 123(3):1068–1081
28. Arroyo JD, Jourdain AA, Calvo SE, Ballarano CA, Doench JG, Root DE et al (2016) A genome-wide CRISPR death screen identifies genes essential for oxidative phosphorylation. Cell Metab 24(6):875–885
29. Horlbeck MA, Xu A, Wang M, Bennett NK, Park CY, Bogdanoff D et al (2018) Mapping the genetic landscape of human cells. Cell 174(4):953–67 e22
30. Marin TL, Gongol B, Zhang F, Martin M, Johnson DA, Xiao H et al (2017) AMPK promotes mitochondrial biogenesis and function by phosphorylating the epigenetic factors DNMT1, RBBP7, and HAT1. Sci Signal 10(464)
31. Wang T, Yu H, Hughes NW, Liu B, Kendirli A, Klein K et al (2017) Gene essentiality profiling reveals gene networks and synthetic lethal interactions with oncogenic Ras. Cell 168(5):890–903 e15
32. Haapaniemi E, Botla S, Persson J, Schmierer B, Taipale J (2018) CRISPR-Cas9 genome editing

induces a p53-mediated DNA damage response. Nat Med 24(7):927–930
33. Ihry RJ, Worringer KA, Salick MR, Frias E, Ho D, Theriault K et al (2018) p53 inhibits CRISPR-Cas9 engineering in human pluripotent stem cells. Nat Med 24(7):939–946
34. Aguirre AJ, Meyers RM, Weir BA, Vazquez F, Zhang CZ, Ben-David U et al (2016) Genomic copy number dictates a gene-independent cell response to CRISPR/Cas9 targeting. Cancer Discov 6(8):914–929
35. Munoz DM, Cassiani PJ, Li L, Billy E, Korn JM, Jones MD et al (2016) CRISPR screens provide a comprehensive assessment of cancer vulnerabilities but generate false-positive hits for highly amplified genomic regions. Cancer Discov 6(8):900–913
36. Iorio F, Behan FM, Goncalves E, Bhosle SG, Chen E, Shepherd R et al (2018) Unsupervised correction of gene-independent cell responses to CRISPR-Cas9 targeting. BMC Genomics 19(1):604
37. Maeder ML, Linder SJ, Cascio VM, Fu Y, Ho QH, Joung JK (2013) CRISPR RNA-guided activation of endogenous human genes. Nat Methods 10(10):977–979
38. Yeo NC, Chavez A, Lance-Byrne A, Chan Y, Menn D, Milanova D et al (2018) An enhanced CRISPR repressor for targeted mammalian gene regulation. Nat Methods 15(8):611–616
39. Gilbert LA, Larson MH, Morsut L, Liu Z, Brar GA, Torres SE et al (2013) CRISPR-mediated modular RNA-guided regulation of transcription in eukaryotes. Cell 154(2):442–451
40. Xu X, Tao Y, Gao X, Zhang L, Li X, Zou W et al (2016) A CRISPR-based approach for targeted DNA demethylation. Cell Discov 2:16009
41. Gilbert LA, Horlbeck MA, Adamson B, Villalta JE, Chen Y, Whitehead EH et al (2014) Genome-scale CRISPR-mediated control of gene repression and activation. Cell 159(3):647–661
42. Stojic L, Lun ATL, Mangei J, Mascalchi P, Quarantotti V, Barr AR et al (2018) Specificity of RNAi, LNA and CRISPRi as loss-of-function methods in transcriptional analysis. Nucleic Acids Res 46(12):5950–5966
43. Gaudelli NM, Komor AC, Rees HA, Packer MS, Badran AH, Bryson DI et al (2017) Programmable base editing of A∗T to G∗C in genomic DNA without DNA cleavage. Nature 551(7681):464–471
44. Komor AC, Kim YB, Packer MS, Zuris JA, Liu DR (2016) Programmable editing of a target base in genomic DNA without double-stranded DNA cleavage. Nature 533(7603):420–424
45. Billon P, Bryant EE, Joseph SA, Nambiar TS, Hayward SB, Rothstein R et al (2017) CRISPR-mediated base editing enables efficient disruption of eukaryotic genes through induction of STOP codons. Mol Cell 67(6):1068–79 e4
46. Kuscu C, Parlak M, Tufan T, Yang J, Szlachta K, Wei X et al (2017) CRISPR-STOP: gene silencing through base-editing-induced nonsense mutations. Nat Methods 14(7):710–712
47. Morita K, Hama Y, Izume T, Tamura N, Ueno T, Yamashita Y et al (2018) Genome-wide CRISPR screen identifies TMEM41B as a gene required for autophagosome formation. J Cell Biol 217:3817–3828
48. Imamura H, Nhat KP, Togawa H, Saito K, Iino R, Kato-Yamada Y et al (2009) Visualization of ATP levels inside single living cells with fluorescence resonance energy transfer-based genetically encoded indicators. Proc Natl Acad Sci U S A 106(37):15651–15656
49. Hernandez G, Thornton C, Stotland A, Lui D, Sin J, Ramil J et al (2013) MitoTimer: a novel tool for monitoring mitochondrial turnover. Autophagy 9(11):1852–1861
50. Katayama H, Kogure T, Mizushima N, Yoshimori T, Miyawaki A (2011) A sensitive and quantitative technique for detecting autophagic events based on lysosomal delivery. Chem Biol 18(8):1042–1052

MIX
Papier aus verantwortungsvollen Quellen
Paper from responsible sources
FSC® C105338

If you have any concerns about our products,
you can contact us on
**ProductSafety@springernature.com**

In case Publisher is established outside the EU,
the EU authorized representative is:
**Springer Nature Customer Service Center GmbH**
**Europaplatz 3, 69115 Heidelberg, Germany**

Printed by Libri Plureos GmbH
in Hamburg, Germany